Probiotics and Prebiotics in Foods

Probiotics and Prebiotics in Foods

Challenges, Innovations, and Advances

ADRIANO GOMES DA CRUZ
Professor, Department of Food, Federal Institute of Rio de Janeiro (IFRJ), Rio de Janeiro, Brazil

C. SENAKA RANADHEERA
Lecturer, University of Melbourne, Parkville, VIC, Australia

FILOMENA NAZZARO
Senior Scientist, The Institute of Food Science of CNR (CNR-ISA), Avellino, Italy

AMIR MORTAZAVIAN
Professor of Dairy Science and Technology, Shahid Beheshti University of Medical Sciences, Tehran, Iran

ELSEVIER

ACADEMIC PRESS
An imprint of Elsevier

Academic Press is an imprint of Elsevier
125 London Wall, London EC2Y 5AS, United Kingdom
525 B Street, Suite 1650, San Diego, CA 92101, United States
50 Hampshire Street, 5th Floor, Cambridge, MA 02139, United States
The Boulevard, Langford Lane, Kidlington, Oxford OX5 1GB, United Kingdom

Library of Congress Cataloging-in-Publication Data
A catalog record for this book is available from the Library of Congress

British Library Cataloguing-in-Publication Data
A catalogue record for this book is available from the British Library

ISBN 978-0-12-819662-5

For information on all Academic Press publications
visit our website at https://www.elsevier.com/books-and-journals

Publisher: Charlotte Cockle
Acquisitions Editor: Megan R. Ball
Editorial Project Manager: Andrea Dulberger
Production Project Manager: Niranjan Bhaskaran
Cover Designer: Alan Studholme

Typeset by SPi Global, India

Contents

Contributors

The editor would like to acknowledge and offer grateful thanks for the input of all contributors, without whom this first edition would not have been possible.

Marzia Albenzio
Department of Agriculture, Food, Natural Science and Engineering, University of Foggia, Foggia, Italy

Cleide Oliveira de Almeida Møller
Department of Food Science, University of Copenhagen, Frederiksberg, Denmark

Rahul Kumar Anurag
Agricultural Structures and Environmental Control Division, ICAR-Central Institute of Post-Harvest Engineering and Technology, Ludhiana, Punjab, India

S. Mithul Aravind
Department of Basic and Applied Science, National Institute of Food Technology, Entrepreneurship and Management, Sonipat, Haryana, India

Patricia Burns
Instituto de Lactología Industrial (INLAIN, UNL-CONICET), Faculty of Chemical Engineering; General Microbiology, Faculty of Biochemistry and Biological Sciences, Universidad Nacional del Litoral, Santa Fe, Argentina

Flávia Aline Andrade Calixto
Departamento de Tecnologia de Alimentos, Universidade Federal Fluminense/UFF; Mestrado Profissional em Ciência e Tecnologia de Alimentos (PCTA)/IFRJ; Fundação Instituto de Pesca do Estado do Rio de Janeiro/Fiperj, Rio de Janeiro, Brazil

Débora Andreia Campos
Portuguese Catholic University, CBQF—Centre for Biotechnology and Fine Chemistry—Associated Laboratory, School of Biotechnology, Porto, Portugal

Mariangela Caroprese
Department of Agriculture, Food, Natural Science and Engineering, University of Foggia, Foggia, Italy

S. Chakkaravarthi
Department of Basic and Applied Science, National Institute of Food Technology, Entrepreneurship and Management, Sonipat, Haryana, India

Mei Zhi Alcine Chan
Department of Food Science & Technology, National University of Singapore, Singapore, Singapore

Maria Giovanna Ciliberti
Department of Agriculture, Food, Natural Science and Engineering, University of Foggia, Foggia, Italy

Carlos Humberto Corassin
Department of Food Engineering, School of Animal Science and Food Engineering, University of São Paulo, Pirassununga, Brazil

Adriano G. Cruz
Department of Food, Federal Institute of Science and Technology of Rio de Janeiro (IFRJ), Rio de Janeiro, Brazil

Juliana Saraiva dos Anjos
Nutrition Faculty, Department of Nutrition and Dietary; Medical School, Graduate Program in Cardiovascular Sciences, Fluminense Federal University (UFF), Niterói, RJ, Brazil

Maria Carmela Kasnowski Holanda Duarte
Faculty of Veterinary, Fluminense Federal University (UFF), Niterói, Brazil

Erick A. Esmerino
Faculty of Veterinary Medicine, Federal Fluminense University (UFF), Niterói; Department of Food, Federal Institute of Science and Technology of Rio de Janeiro (IFRJ), Rio de Janeiro, Brazil

Eliana de Fátima Marques de Mesquita
Departamento de Tecnologia de Alimentos, Universidade Federal Fluminense/UFF, Rio de Janeiro, Brazil

Mounir Ferchichi
Unit of Functional Proteomics and Nutraceutical Potential of Biodiversity in Tunisia, Higher Institute of Applied Biological Sciences of Tunis, University of Tunis El Manar, Tunis, Tunisia

Lucia Figliola
Department of Agriculture, Food, Natural Science and Engineering, University of Foggia, Foggia, Italy

Mônica Q. Freitas
Faculty of Veterinary, Fluminense Federal University (UFF), Niterói, Brazil

Flávia Furlaneto
Department of Oral and Maxillofacial Surgery and Periodontology, School of Dentistry of Ribeirao Preto, University of Sao Paulo—USP, Ribeirao Preto, SP, Brazil

Jonas T. Guimarães
Faculty of Veterinary, Fluminense Federal University (UFF), Niterói, Brazil

Murni Halim
Department of Bioprocess Technology; Bioprocessing and Biomanufacturing Research Center, Faculty of Biotechnology and Biomolecular Sciences, Universiti Putra Malaysia, Serdang, Selangor, Malaysia

Elena S. Inguglia
Department of Food Chemistry and Technology, Teagasc Food Research Centre, Dublin, Ireland

Marcos Invernici
Department of Oral and Maxillofacial Surgery and Periodontology, School of Dentistry of Ribeirao Preto, University of Sao Paulo—USP, Ribeirao Preto, SP, Brazil

Erhan Keyvan
Faculty of Veterinary Medicine, Department of Food Hygiene and Technology, Burdur Mehmet Akif Ersoy University, Burdur, Turkey

Débora Lemos Lino
Department of Food, Federal Institute of Science and Technology of Rio de Janeiro (IFRJ), Rio de Janeiro, Brazil

Shao-Quan Liu
Department of Food Science & Technology, National University of Singapore, Singapore, Singapore; National University of Singapore (Suzhou) Research Institute, Suzhou, China

Luciana Prado Maia
Graduate Program in Dentistry (GPD-Master's Degree), University of Western Sao Paulo-UNOESTE, Presidente Prudente, SP, Brazil

Mario Roberto Marostica Junior
School of Food Engineering, University of Campinas (UNICAMP), Campinas, São Paulo, Brazil

Michel Reis Messora
Department of Oral and Maxillofacial Surgery and Periodontology, School of Dentistry of Ribeirao Preto, University of Sao Paulo—USP, Ribeirao Preto, SP, Brazil

Khurram Muaz
Institute of Food Science & Nutrition, Bahauddin Zakariya University, Multan, Pakistan

Carlos Augusto Fernandes de Oliveira
Department of Food Engineering, School of Animal Science and Food Engineering, University of São Paulo, Pirassununga, Brazil

Arthur C. Ouwehand
International Flavors & Fragrances Inc., Kantvik, Finland

Roberto de Paula do Nascimento
School of Food Engineering, University of Campinas (UNICAMP), Campinas, São Paulo, Brazil

Maria Manuela Pintado
Portuguese Catholic University, CBQF—Centre for Biotechnology and Fine Chemistry—Associated Laboratory, School of Biotechnology, Porto, Portugal

Fergal P. Rattray
Department of Food Science, University of Copenhagen, Frederiksberg, Denmark

Leonardo Rios-Solis
School of Engineering, Institute for Bioengineering, The University of Edinburgh, Edinburgh, United Kingdom

Ramon S. Rocha
Department of Food, Federal Institute of Science and Technology of Rio de Janeiro (IFRJ), Rio de Janeiro, Brazil

Sérgio Salvador
Department of Clinical Analyses, School of Pharmaceutical Sciences of Ribeirao Preto, University of Sao Paulo—USP, Ribeirao Preto, SP, Brazil

Antonella Santillo
Department of Agriculture, Food, Natural Science and Engineering, University of Foggia, Foggia, Italy

Hugo Scudino
Faculty of Veterinary, Fluminense Federal University (UFF), Niterói, Brazil

Swati Sethi
Food Grains & Oilseeds Processing Division, ICAR-Central Institute of Post-Harvest Engineering and Technology, Ludhiana, Punjab, India

Pedro Henrique Felix Silva
Department of Oral and Maxillofacial Surgery and Periodontology, School of Dentistry of Ribeirao Preto, University of Sao Paulo—USP, Ribeirao Preto, SP, Brazil

Arturo B. Soro
Department of Food Chemistry and Technology, Teagasc Food Research Centre, Dublin, Ireland

Milena Barcza Stockler-Pinto
Nutrition Faculty, Department of Nutrition and Dietary; Medical School, Graduate Program in Cardiovascular Sciences, Fluminense Federal University (UFF), Niterói, RJ, Brazil

Rodrigo Takata
Fundação Instituto de Pesca do Estado do Rio de Janeiro/Fiperj, Rio de Janeiro; Programa de Pós-graduação em Aquicultura e Recursos Aquáticos Tropicais/UFRA, Pará, Brazil

Nur Imanina Abdullah Thaidi
Department of Bioprocess Technology; Bioprocessing and Biomanufacturing Research Center, Faculty of Biotechnology and Biomolecular Sciences, Universiti Putra Malaysia, Serdang, Selangor, Malaysia

Brijesh K. Tiwari
Department of Food Chemistry and Technology, Teagasc Food Research Centre, Dublin, Ireland

Mingzhan Toh
Department of Food Science & Technology, National University of Singapore, Singapore, Singapore

Gabriel Vinderola
Instituto de Lactología Industrial (INLAIN, UNL-CONICET), Faculty of Chemical Engineering, Universidad Nacional del Litoral, Santa Fe, Argentina

Glenise Bierhalz Voss
Portuguese Catholic University, CBQF—Centre for Biotechnology and Fine Chemistry—Associated Laboratory, School of Biotechnology, Porto, Portugal

Mohamed Zommiti
Unit of Functional Proteomics and Nutraceutical Potential of Biodiversity in Tunisia, Higher Institute of Applied Biological Sciences of Tunis, University of Tunis El Manar, Tunis, Tunisia; Laboratory of Microbiology, Signals and Microenvironment (LMSM), University of Rouen, Evreux, France

CHAPTER 1

The Biotics Family

GABRIEL VINDEROLA[a] • PATRICIA BURNS[a,b]
[a]Instituto de Lactología Industrial (INLAIN, UNL-CONICET), Faculty of Chemical Engineering, Universidad Nacional del Litoral, Santa Fe, Argentina, [b]General Microbiology, Faculty of Biochemistry and Biological Sciences, Universidad Nacional del Litoral, Santa Fe, Argentina

PROBIOTIC, PREBIOTIC, AND SYNBIOTIC DEFINITIONS

The terms probiotic, prebiotic, and synbiotic are particularly well known in the field of functional foods by researchers, regulatory experts, and health professionals and, maybe, by a growing group of consumers interested in a healthier lifestyle. Some of these terms have evolved in the last century and matured in the last decade.

The link between the consumption of fermented foods and health is ancestral and, in the 19th century, some microbiologists, biologists, and pediatricians associated longevity with the gut microbiota, or the presence of some particular Y-shaped bacteria with the prevention of diarrhea in breast-fed infants (WGO, 2017). Moreover, dietary oligosaccharides had long been used to impart health benefits, principally in Asia (Gibson et al., 2017).

The probiotic concept has evolved along time and, in 2001, an Expert Consultation of international scientists working on behalf of the Food and Agriculture Organization of the United Nations (FAO) and the World Health Organization (WHO) defined probiotics as "*live microorganisms which when administered in adequate amounts confer a health benefit on the host*" (FAO/WHO, 2001). Later on, in 2014, this definition was revised and endorsed by a consensus panel of 12 experts called out by the International Scientific Organization of Probiotics and Prebiotics (ISAPP). The panel noted, however, that a more grammatically correct definition would be worded as, "*live microorganisms that, when administered in adequate amounts, confer a health benefit on the host*" (Hill et al., 2014).

In 1995, Gibson and Roberfroid defined a prebiotic as "*a nondigestible food ingredient that beneficially affects the host by selectively stimulating the growth and/or activity of one or a limited number of bacteria in the colon and thus improves host health.*" In order for a food ingredient to be classified as a prebiotic, it must (1) be neither hydrolyzed nor absorbed in the upper part of the gastrointestinal tract; (2) be a selective substrate for one or a limited number of beneficial bacteria commensal to the colon, which are stimulated to grow and/or are metabolically activated; (3) consequently, be able to alter the colonic flora in favor of a healthier composition; and (4) induce luminal or systemic effects that are beneficial to the host health. Moreover, in the same work, Gibson and Roberfroid (1995) first mentioned the combination of both probiotics and prebiotics as synbiotics that were defined as "*a mixture of probiotics and prebiotics that beneficially affects the host by improving the survival and implantation of live microbial dietary supplements in the gastrointestinal tract, by selectively stimulating the growth and/or by activating the metabolism of one or a limited number of health-promoting bacteria, and thus improving host welfare.*" According to the World Health Organization (https://www.gutmicrobiotaforhealth.com/wp-content/uploads/2019/11/GMFH-BESTOF-SYNBIOTICS-2019.pdf), synbiotics are appropriate combinations of prebiotics and probiotics that are considered gut microbiota-management tools for improving host health.

So far, the term synbiotic lacks a concise, modern definition. However, in May 2019, ISAPP convened a panel of experts to develop a consensus around this concept. https://isappscience.org/synbiotics_panel/Currently, the updated definition of a synbiotic is "a mixture comprising live microorganisms and substrate(s) selectively utilized by host microorganisms that confers a health benefit on the host" (Swanson et al., 2020).

In the same way, as previously mentioned for probiotics, the original definition of prebiotics (Gibson and Roberfroid, 1995) was revised and expanded in 2017 by another 12-member consensus panel organized by the ISAPP (Gibson et al., 2017) and led by Prof. Gibson, one of the authors of the original definition. At this time, the definition of a prebiotic was modified to read as "*a substrate that is selectively utilized by host microorganisms conferring a health benefit.*" These health effects include defense against pathogens, immune modulation, mineral absorption, bowel function, metabolic effects, and satiety (Sanders et al., 2019).

Probiotics and Prebiotics in Foods. https://doi.org/10.1016/B978-0-12-819662-5.00014-8

The value of conveying a panel of more than 10 people with different backgrounds and experts in different fields is to provide diverse perspectives, allowing a comprehensive discussion on the relevant aspects of the topic, presenting different points of view, debating the proposed questions, and reaching a consensus on a given definition and its implications.

GENERAL ASPECTS OF PROBIOTICS, PREBIOTICS, AND SYNBIOTICS

Probiotics

The definition of probiotics comprises a broad range of microorganisms and applications and implies certain characteristics: microbial cells, identity, viability, and a health benefit demonstrated by a clinical trial. The source of isolation of the strains (from the same species of the target host, or not) is not an issue (Hill et al., 2014). This means that the old criterion that, for example, probiotics for human use should be isolated from human sources is not relevant anymore.

The strains most widely used as probiotics belong mainly to the genera *Lactobacillus* and, to a lesser extent, *Bifidobacterium*. Other genera such as *Bacillus*, *Propionibacterium*, *Streptococcus*, and *Escherichia* are also used. Recently, Sanders et al. (2019) reported that *Roseburia* spp., *Akkermansia* spp., and *Faecalibacterium* spp. are genera with promising properties that could be used in the future. Notwithstanding most of the studies refer to bacteria as probiotic microorganisms, some yeast strains from the genus *Saccharomyces* has also a long history of safe and effective use (Cani and van Hul, 2015). A meta-analysis reported that the administration of *Saccharomyces boulardii* may reduce the overall risk of antibiotic-associated diarrhea in patients treated with antibiotics compared to placebo or no treatment (Szajewska and Kolodziej, 2015).

Zheng et al. (2020) have recently proposed a reclassification of the genus *Lactobacillus* into 25 genera (23 of which are completely new genera). Some commercial well-known probiotic strains such as *Lactobacillus rhamnosus* GG, *Lactobacillus casei* Shirota, or *Lactobacillus plantarum* 299v will have their names changed. The new names are *Lacticaseibacillus rhamnosus*, *Lacticaseibacillus casei*, and *Lactiplantibacillus plantarum*, respectively. Besides, *Lactobacillus paracasei*, *Lactobacillus brevis*, *Lactobacillus salivarius*, *Lactobacillus fermentum*, and *Lactobacillus reuteri* were also renamed as *Lacticaseibacillus paracasei*, *Levilactobacillus brevis*, *Ligilactobacillus salivarius*, *Limosilactobacillus fermentum*, and *Limosilactobacillus reuteri*, respectively. This reclassification reflects the phylogenetic position of the microorganisms and groups lactobacilli into robust clades with shared ecological and metabolic characteristics (Zheng et al., 2020). All new genera proposed for probiotic lactobacilli begin with "L," so the abbreviated form of genus/species—such as *L. casei*—remains unchanged. Moreover, species names and strain designations have not changed (https://isappscience.org/new-names-for-important-probiotic-lactobacillus-species/). Other probiotic lactobacilli names, however, have not changed (e.g., *Lactobacillus acidophilus*, *Lactobacillus gasseri*, *Lactobacillus crispatus*, *Lactobacillus delbrueckii* subsp. *bulgaricus*, *Lactobacillus helveticus*, and *Lactobacillus johnsonii*) (Zheng et al., 2020).

In 2018, a panel of 30 experts from 13 countries convened by the ISAPP proposed a set of minimum criteria that should be met by a microorganism to be properly regarded as "probiotic." Following these criteria would contribute to the development of trustable products. The minimum criteria are: (1) characterization sufficient to identify the probiotic to the genus, species, and strain level; (2) probiotic named according to scientifically valid nomenclature; (3) probiotic name includes a strain designation; (4) safety of probiotic demonstrated for the intended use; (5) probiotic strain deposited in an international culture collection; (6) probiotic has health benefit demonstrated from at least one human study; and (7) products using the probiotic provide until the end of shelf life sufficient level of live microbes to deliver a health benefit (https://isappscience.org/minimum-criteria-probiotics/).

Health benefits must be demonstrated in at least one clinical human trial if the strain is for human use, but there are no requirements for probiotics to exhibit properties in preclinical assessments (such as adherence to the gut epithelium, colonization capacity, ability to survive intestinal transit in in vitro models) (Sanders et al., 2018b). These authors argue that although many researchers utilize a series of such in vitro or ex vivo tests to screen for potential probiotics, these tests have never been validated, and the results of these assays do not always correlate, in a straightforward manner, with the in vivo probiotic efficacy (Vinderola et al., 2017). Thus, they concluded that requiring such tests by regulators is not scientifically justified (Sanders et al., 2018b).

The field of probiotics has grown enormously, and it is known that there is vast information available on the Web. This information is sometimes confusing, not only for consumers but also for researchers, health professionals, and regulatory staff. For example, sometimes, probiotics and microbes present in fermented foods are treated as equivalents. It is important to mention that not all fermented foods are probiotic foods and probiotics are not only available in fermented foods. Microbiological undefined products do not meet the probiotic definition. Thus, establishing differences is important to avoid misunderstandings (Reid et al., 2019).

Yogurt is an exception since the European Food Safety Authority (EFSA Panel on Dietetic Products 2010) has approved a health claim, having determined that yogurt bacteria (*Streptococcus thermophilus* and *Lactobacillus delbrueckii* subsp. *bulgaricus*) at the species level, that is irrespective of the strain considered, can help to improve digestion of lactose among people with lactose intolerance. This was confirmed in several controlled studies with individuals consuming yogurt (Sanders et al., 2018b; WGO, 2017).

Hill et al. (2014) pointed out some recommendations for all probiotic stakeholders to have a common criterion when referring to probiotics. One of the recommendations was to keep outside of the probiotic framework both undefined fecal microbiota transplants and live cultures (traditionally associated with fermented products) without strain identity and evidence of a health benefit. On the contrary, it is considered that new commensals and consortia comprising defined strains from human samples, with proven safety and efficacy, and microbial species that have been shown in properly controlled studies to confer benefits to health are "probiotics." Finally, regarding health claims, the recommendation of any specific claim beyond "contains probiotics" must be further substantiated (Hill et al., 2014).

Probiotic microorganisms include a wide range of uses (modulation of the intestinal microbiota, improvement of the gut barrier, prevention of antibiotic-associated diarrhea, and immunomodulation, among others), different routes of administration (oral, topical, and intravaginal), and different target hosts (human, production, and companion animals) (Gómez Gallego and Salminen, 2016; Sánchez et al., 2017). The concept of strain specificity for efficacy has been accepted for decades. In a meta-analysis including 228 trials (from 1970 to 2017), McFarland et al. (2018) found evidence for strain specificity but also for disease specificity in the efficacy of specific probiotic strains. Beyond strain specificities, currently, the idea of "core" benefits of certain species is gaining attention among scientists (Hill et al., 2014). Scientific advances report that certain taxonomic groups share some common mechanisms of action, suggesting that species-specific or genus-specific probiotic effects may exist. For example, the ability to produce short-chain fatty acids (SCFAS) is a characteristic shared by many different microbial taxa and probably plays a key role in probiotic-mediated health benefits (Sanders et al., 2018a). In this sense, Hill et al. (2014) suggested a possible distribution of mechanisms among probiotics. They reported that some mechanisms, such as regulation of intestinal transit, normalization of altered microbiota, competition against pathogens, might be widespread among commonly studied probiotic genera; others might be frequently observed among most strains of a probiotic species (vitamin synthesis, bile salt metabolism, and enzymatic activity); and, finally, some mechanisms may be rare and are likely narrowly distributed among a few strains of a given species (endocrinological, immunological, and neurological effects or production of bioactive compounds). Health Canada considers the general benefit of "*contributing to a healthy gut microbiota*" to be a core benefit for *Bifidobacterium adolescentis, B. animalis, B. bifidum, B. breve,* and *B. longum* and *Lactobacillus acidophilus, L. casei, L. fermentum, L. gasseri, L. johnsonii, L. paracasei, L. plantarum, rhamnosus,* and L. *salivarius*). In this sense, the regulatory body of Canada has accepted that when some of these species are delivered at 1×10^9 colony forming units (CFU) per serving (food or food supplement), they could be labeled as probiotics, for which nonstrain-specific claims can be made (Hill et al., 2014).

In the last years, the expressions "next-generation probiotics" (NGP) and "next-generation beneficial microbes" emerged to refer to novel bacterial genera with promising probiotic properties. *Akkermansia muciniphila,* one of the most abundant species in the human intestinal microbiota (0.5%–5% of the total bacteria), has been studied as a novel probiotic candidate targeting metabolic syndrome. Its abundance in the gut has been inversely associated with obesity, diabetes, cardiometabolic diseases, and low-grade inflammation (Cani and Van Hul, 2015; Cani and de Vos, 2017). *Clostridium butyricum, Bacteroides xylanisolvens, B. dorei,* and *Faecalibacterium prausnitzii* have also been considered as NGP (Martín et al., 2018; O'Toole et al., 2017; Saarela, 2019; Tan et al., 2019). The studies with *C. butyricum* MIYAIRI 588 (CBM 588) are the most advanced among the NGP and this strain is commercialized as a supplement. Contrary to traditional probiotic genera *Lactobacillus* and *Bifidobacterium,* NGP seem to be more ambivalent regarding their potential to cause disease. Thus, safety-focused animal and human trials are needed (Saarela, 2019).

Beyond the broad range of commercially available probiotic foods and supplements, two successful social projects of fermented foods added with probiotics are being applied in Argentina and several countries in Africa. The Argentinian social program (implemented in 2008) is called "Scholar Yogurito, the social probiotic" and the probiotic product is "Yogurito," yogurt that contains *Lactobacillus rhamnosus* CRL1505, a strain with well-documented functional and technological properties (Salva et al., 2010, 2011; Villena et al., 2012).

"Yogurito Social Program" benefits some 300,000 schoolchildren in the province of Tucumán and some 50,000 in other provinces and municipalities of Argentina (https://isappscience.org/yogurito/). This social transfer project that improves the quality of life of highly vulnerable populations is a paradigm of interaction between the state and the scientific and manufacturing sectors, a sociotechnical complex alliance (Bortz and Thomas, 2017).

In Africa, Yoba for Life (Y4L) and Western Heads East (WHE) collaborated with local partners Heifer, Uganda; Heifer, Tanzania; and JKUAT, Kenya, on the production and distribution of the Yoba-Fiti starter cultures, training of local communities, and the impact measurement on how many people were reached. The project aims to facilitate daily access to affordable probiotic dairy products to 250,000 people in East Africa (http://www.yoba4life.org/about-yoba-for-life/collaboration-with-western-heads-east/). The fermented milk is produced using *Streptococcus thermophilus* C106 and *Lactobacillus rhamnosus* Yoba 2012 (the world's first generic probiotic strain a clone of L. *rhamnosus* GG) (Kort and Sybesma, 2012; Reid et al., 2018). Health benefits associated with the consumption of Yoba fermented milk are reduction in fatigue, diarrhea, and skin rashes, and, in some cases, improvement in CD4 counts in HIV patients (Reid et al., 2018).

Prebiotics

In 2004, the concept of prebiotics (first defined in 1995) was expanded to include beneficial effects not just to the colon, but also to other aspects of the gastrointestinal tract (Gibson et al., 2004). By that time, the criteria for a substance to be considered a prebiotic were: (1) to resist gastric acidity, hydrolysis by mammalian enzymes, and gastrointestinal absorption; (2) to be fermented by the intestinal microbiota; and (3) to selectively stimulate the growth and/or activity of intestinal bacteria associated with health and well-being (Scott et al., 2019).

Advances in scientific knowledge during more than 20 years in this field expanded the scope of the concept of prebiotic substances. In this sense, in the update (2017), the main points were that the microorganisms responding to prebiotics should be health-promoting bacteria, without specifying which ones. Moreover, the anatomic target site is no longer limited to the gastrointestinal tract. However, it is maintained that effects need to be demonstrated in vivo on the target host (Scott et al., 2019).

During the last 20 years, fructans and galactans (fructooligosaccharides, inulin, and galactooligosaccharides) fitted the definition of prebiotics, promoting *Lactobacillus* and/or *Bifidobacterium* populations in the gut. It was first thought that the growth of bifidobacteria was selectively stimulated by these nondigestible food ingredients, and "bifidogenesis" was considered a prebiotic effect (Gibson et al., 2017). Furthermore, most studies focused upon inulin-type fructans reported that after a short feeding period, at doses ranging from 4 to 30 g/day in adults, these prebiotics exerted selective stimulation of gut bacterial genera, mainly *Bifidobacterium* (Scott et al., 2019).

Nowadays, based on the number of studies about their efficacy, galactooligosaccharides (GOS) dominate the prebiotic category (Gibson et al., 2017). The development and application of sequencing techniques allowed scientists to expand the knowledge about the microbiota and so evidence is building on the prebiotic effects of other substances (e.g., oligomers of mannose, glucose, xylose, pectin, starches, human milk oligosaccharides, and polyphenols) over a broader spectrum of microorganisms, including *Roseburia*, *Eubacterium* or *Faecalibacterium* spp. (Gibson et al., 2017).

It is important to mention that, for a prebiotic, the target microbiota of the host should also be considered. For example, a substance such as cellulose can demonstrate a prebiotic effect in ruminants but not in humans (David et al., 2015), or xylitol can act as a prebiotic in the oral cavity but does not affect the microbiota of the gut or the skin. Thus, reproducible randomized controlled studies demonstrating the prebiotic effect on health are needed in the specific target host (Scott et al., 2019).

Human milk oligosaccharides (HMOs) are among the first substances recognized for their ability to influence gastrointestinal health (Gibson et al., 2017). HMOs are a complex combination of bioactive glycans (which structure comprises lactose) in breast milk with demonstrated benefits, influencing both the development of the intestinal microbiota and the immune system of the breast-fed neonate. It has been shown that HMOs have prebiotic effects on beneficial bacteria and antiadhesive/antimicrobial functions against pathogens. Furthermore, in vitro and in vivo studies showed HMOs can modulate immune and metabolic functions and inflammation (Hirschmugl et al., 2019). Xylans and pectins, which are polysaccharides present in plant cell walls, are gaining popularity as candidate prebiotics since they are no-digestible in the upper gastrointestinal tract but fermented by colonic microorganisms (Scott et al., 2019).

Another emerging type of substances that could be considered prebiotics is plant polyphenols, even though more studies in the target host are needed. Thilakarathna et al. (2018) reported that nondigestible polyphenols acting as prebiotics may shift the microbiome to favor the colonization of beneficial gut microbes. Most of the studies referring to polyphenol consumption indicate that health benefits depend on microbial utilization and the metabolites released, instead of the original compounds. These findings allow the concept of prebiotic to be extended to other substances beyond nondigestible oligosaccharides FOS and GOS (Gibson et al., 2017). Due to the dose required for a prebiotic to exert a health benefit (commonly more than 6 g/day for adults), it is not always possible for an oligosaccharide to exert a significant prebiotic effect through the ingestion of significantly high amounts of whole foods that contain them. One strategy is the fortification of frequently consumed foods—which is the premise of functional foods (Kolida and Gibson, 2011; Scott et al., 2019). Thus, most prebiotics are used as food ingredients in biscuits, cereals, chocolate, spreads, and dairy products.

Synbiotics

As mentioned before, during the 2019 ISAPP meeting, an expert panel revised the definition of synbiotics, and one concept emerging from this discussion is allowing flexibility of the synbiotic term to be used with combinations of live microbes and selectively utilized substrates, which, separately, may not meet the definition of probiotic and prebiotic, but may work together to deliver a health benefit through a synbiotic effect (Gut Microbiota for Health, 2019). Moreover, within the ISAPP updated definition, 'host' microorganisms involve both autochthonous (resident) and allochthonous (externally applied) microorganisms, either of which can be targets for the substrate contained in the synbiotic (Swanson et al., 2020).

Kolida and Gibson (2011) proposed two approaches in which a synbiotic might act to enhance the effects of its components: (1) complementary action: the probiotic is chosen based on an expected health effect on the host, and the prebiotic is independently chosen to selectively promote beneficial host-microbiota components. Thus, the prebiotic may promote the effect of the probiotic, but only indirectly as part of its target range; (2) synergistic action: in this case, the prebiotic is selected to have a higher affinity for the probiotic and is chosen to improve its survival and growth in the host. It may also increase the levels of beneficial host gastrointestinal microbiota, but the primary target is the ingested probiotic. Both approaches may, directly or indirectly, fit with the currently synbiotic definition (Kolida and Gibson, 2011).

An alternative concept termed "second-generation synbiotic agents" was proposed by Sharma and Padwad (2020), which argues that prebiotic agents could be defined based on their physiological effects or functional capacities in the host rather than their specific microbial targets. Plant polyphenols are emerging as potential prebiotics and synbiotic candidates that may fulfill these criteria. Time will tell if this concept succeeds or will only be part of the wide variety of emerging terms.

EXPANDING THE BIOTICS FAMILY: TRUE INNOVATION OR POTENTIAL RISK OF MUDDYING THE WATER?

In 2013, the term psychobiotic was defined as a *live organism that, when ingested in adequate amounts, produces a health benefit in patients suffering from psychiatric illness* (Dinan et al., 2013), though it was first mentioned a year before (Shanahan et al., 2012). Being a class of probiotic, these bacteria are capable of producing and delivering neuroactive substances such as gamma-aminobutyric acid and serotonin, which act on the brain-gut axis (Dinan et al., 2013). Later on, the same authors proposed the definition of psychobiotics be expanded beyond probiotics and prebiotics to include other means of influencing the microbiome (Sarkar et al., 2016); then, the term psychobiotics would no longer be a class, or subgroup, of probiotics, what could be perhaps misleading or confusing for readers due to the proximity of the expressions and in the absence of a consensus definition. A definition, and a proper description of its limits, is important to know what should be considered a psychobiotic and what should be not, as established by probiotics, for example (Hill et al., 2014; Reid et al., 2019). By April 2020, 116 papers could be tracked in PubMed that mentioned the word psychobiotics.

Coining new terms that may cover unattended areas or new niches may be useful for the development of new fields for the application of microbes with technological, agroindustrial, ecological, or health purposes. Splitting existing and defined terms, such as probiotics, into classes or subgroups depending on the benefit targeted sounds interesting for marketing purposes, but may be misleading or confusing for consumers, health professionals, the industry or may add additional work for regulatory bodies, being the continuous incorporation of new terms a challenge for the local regulation. As it was mentioned before, psychobiotics are probiotics targeting mental health. In this line, immunobiotics would be probiotics targeting the immune system; enterobiotics may be microbes promoting enteric health by preventing diarrhea; oralbiotics could be probiotics able to prevent dental caries; and oncobiotics, for example, would be useful as adjuvants in cancer therapies. At the same time, one known probiotic such as *Lactobacillus rhamnosus* GG (Capurso, 2019) may fall under different categories, as it would be an oralbiotic (Ahola et al., 2002), an enterobiotic (Szajewska et al., 2001), or maybe an immunobiotic too (Davidson et al., 2011), among many other possible specific applications of this strain, that would coin a new term again and again.

In 2019, another list of definitions was proposed (Martín and Langella, 2019):

Live Biotherapeutic Product (LBP): a biological product that contains live organisms; is applicable to the prevention, treatment, or cure of a disease or condition of human beings; and is not a vaccine.

Next-Generation Probiotic (NGP): live microorganisms identified on the basis of comparative microbiota analyses that, when administered in adequate amounts, confer a health benefit on the host.

Pharmabiotics: bacterial cells of human origin, or their products, with a proven pharmacological role in health or disease.

Postbiotics: nonviable bacterial products or metabolic products from microorganisms that have biologic activity in the host.

Paraprobiotic/ghost probiotics/inactivated probiotics: nonviable microbial cells (intact or broken) or crude cell extracts that, when administered (orally or topically) in adequate amounts, confer a benefit to the human or animal consumer.

Probioceuticals/probiotaceuticals: probiotic derived factors.

Biogenic: products made by or of life forms including secretions and metabolites.

A quick view of these definitions makes it evident that some definitions overlap between them; for instance, postbiotics and paraprobiotics both refer to nonviable cells, or biogenics overlaps with pharmabiotics, as both refer to metabolites. The mentioned work falls short of providing clear direction for the field and, indeed, may well add up to confusion by introducing unnecessary, new, and poorly defined terms. One of the difficulties with sorting through these terms is the lack of any consistent basis for defining some of them. Some terms, such as pharmabiotics and LBPs, are linked to specific regulatory categories. Others are defined by the nature of how they are comprised: live cells, cell components, or fermentation-end products. Others are defined by their physiological benefit: psychobiotic, oncobiotic, and immunobiotic. Even still, others are defined by their state of innovation: traditional vs. next-generation probiotics. This state of affairs makes it difficult to develop a logical framework for categorizing them. Instead, we are left with a long list of substances that might be related but have little real value. Where does it all stop? Next, we will have to go through thera-/metabo-/gen-/retro-/plas-/func-biotics or any other pointless terms that can be arbitrarily slapped in front of "biotic" (https://isappscience.org/defining-emerging-biotics). Certainly, there is nothing to prevent anyone from coining new terms for a niche development. It is maybe time to make an appeal to scientists to refrain from muddying the water by proposing new terms, and instead, use existing terms with appropriate modifiers. For example, use "immune-active probiotic" instead of "immunobiotic," or "psycho probiotics" instead of "psychobiotics." Keeping in everybody's mind, the term probiotic will certainly contribute to developing the science on probiotics.

Early 2020 and given the fact that "*we haven't had one new definition in at least a month,*" according to the author's own words in the blogspot, and with a touch of humor, another new term was coined: globobiotics, to refer to "*live microorganisms, microbial products or substrates selectively utilized by microorganisms, that are used in a manner that contributes to the sustainability of our planet*" (https://isappscience.org/probiotics-prebiotics-and-globobiotics). Is there room for more and more definitions without biting our own tail?

A FAMILY OF TERMS TO COVER THE EFFECTS OF NONVIABLE MICROBES AND THEIR METABOLITES

Postbiotics, paraprobiotics, ghost probiotics, inactivated probiotics, nonviable probiotics, heat-killed probiotics, or tyndallized probiotics have been all used here and there in the last two decades to refer to the health effects of inanimate microbes and/or their metabolites and cell fragments (Fig. 1). One of the issues of having different terms to refer to the same biotic is that it may impair or makes it more difficult to gather all the proper papers when a meta-analysis is needed, while risking to leave out some important contributions if they are not captured in the search, overlooked if inappropriate keywords are used in the search.

In addition, some terms were more frequently used in some periods than others. For example, two waves in the use of the term "heat-killed probiotics" peaked in

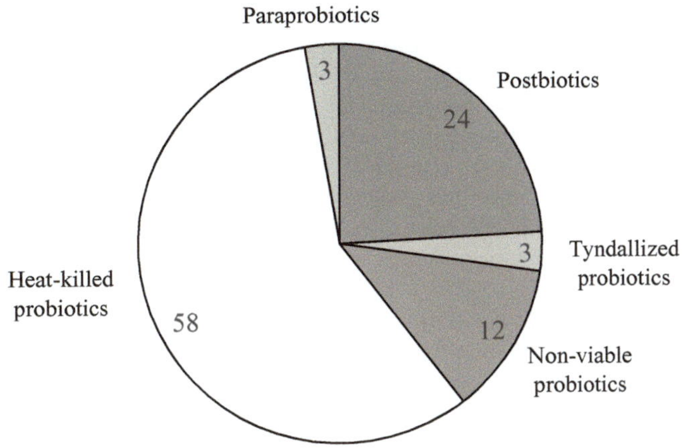

FIG. 1 Percentage of time that the different terms referring to nonviable microorganisms were used in the last two decades, according to a search in PubMed.

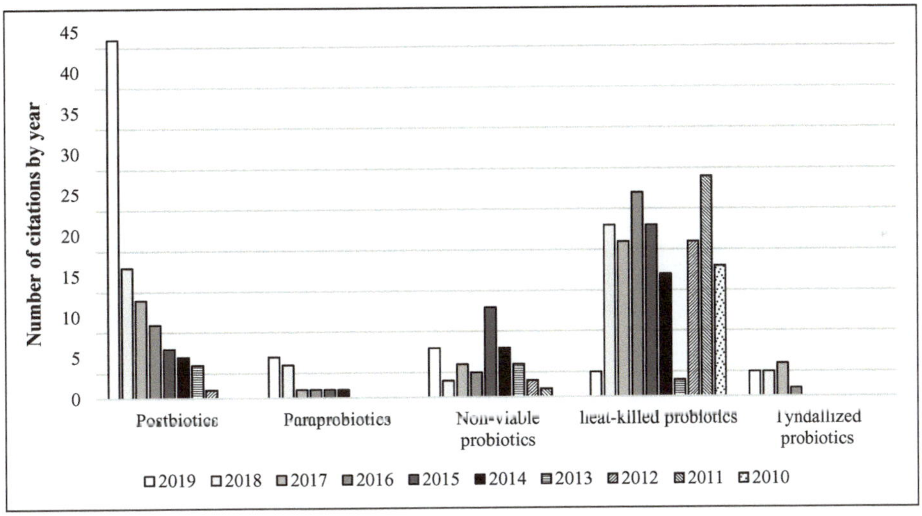

FIG. 2 Number of citations by year of the different terms referring to nonviable microorganisms in the last two decades, according to a search in PubMed.

2011 and 2016, leaving then the way out to a sustained rise in the use of the term postbiotic, the most used one in 2019 and employed more than 8 times compared to the rest of the terms last year (Fig. 2).

The term "ghost probiotics" was first described as empty cell envelopes of Gram-negative bacteria, devoid of cytoplasmic content (Lubitz et al., 2009). One possible application of this concept is to empty cells of the Gram-negative probiotic *Escherichia coli* Nissle 1917 (Schultz, 2008), being, in fact, already explored as a delivery system to target ocular surface diseases (Stein et al., 2013). However, one major limitation of this definition is that being restricted to Gram-negative bacteria, it leaves out empty cells of *Lactobacillus* and *Bifidobacterium*, the two genera the most used ones to develop probiotic cultures, as they are Gram-positive microbes. In 2011, the concept of "ghost cells" was modified to include complete nonviable cells, including their cytoplasm, and the term "paraprobiotic," as a synonym of "ghost probiotics" was defined as "nonviable microbial cells" (intact or broken) or crude cell extracts (i.e., with complex chemical composition), which, when administered (orally or topically) in adequate amounts, confer a benefit on the human or animal consumer (Taverniti and Guglielmetti, 2011). In this case, the term paraprobiotics includes nonviable cells and their metabolites.

Table 1 displays most of the definitions of the term postbiotics that were proposed during the last decade by different groups of authors. Most of the definitions seem to mean that postbiotics are components present in the supernatant of a microbial culture able to confer a health benefit. One of them may sound even strange as it defines postbiotics as nonviable metabolites produced by probiotics. Being metabolites cell components without intrinsic viability, this definition may be misleading to readers. Other definitions in Table 1 include nonviable cells, as well as cell components or metabolites. Cell-free supernatants are generally obtained by centrifugation, which may leave diverse quantities of live and nonviable cells in the supernatants, depending on the operation conditions (Doran, 2013), being not completely cell-free fractions. Then, the term postbiotics referring to cell metabolites irretrievably and intrinsically will include some proportion of viable and nonviable cells that are left in the supernatant after biomass separation, blurring the specificity of the concept if it is targeted to cell metabolites only. Another limitation of some of these definitions is that postbiotics are linked to probiotics. In other words, they must derive from probiotics according to those definitions. In many cases, a postbiotic may derive from a microbe that is not necessarily a probiotic. Therefore, being a probiotic should not be necessarily a prerequisite for being a postbiotic.

In case the discussion is on whole microbial cells, the case of *Akkermansia muciniphila* can be discussed. *A. muciniphila* is not allowed yet by the EFSA, in the European Union, to be administered as a viable microbe (Brodmann et al., 2017), and then, it could not be considered a probiotic presently. However, when this microbe is administered as a pasteurized culture, it is able to improve several metabolic parameters in obese people (Depommier et al., 2019). In this case, being a probiotic was not necessary first, to be considered a postbiotic then. Moreover, the fact of being exclusively produced, or being a cell component from, a probiotic should neither be a condition for being a postbiotic component. For example, a purified membrane protein from *A. muciniphila* improved metabolism in obese and diabetic mice (Plovier et al., 2017). In brief, many postbiotic microbes or components may derive from microbial strains that were never studied for their probiotic properties or that are not presently regarded as probiotics. As if new definitions were never enough, another term was used, twice already, to address to metabolites produced by probiotics:

TABLE 1
Definitions of the Term Postbiotics in the Last Decade.

Year	Definition	Authors Involved	Reference
2013	Any factor resulting from the metabolic activity of a **probiotic** or any released molecule capable of conferring beneficial effects to the host in a direct or indirect way	2	Tsilingiri and Rescigno (2013)
2018	Soluble factors (products or metabolic by-products), secreted by live bacteria, or released after bacterial lysis, such as enzymes, peptides, teichoic acids, peptidoglycan-derived muropeptides, polysaccharides, cell surface proteins, and organic acids	7	Aguilar-Toalá et al. (2018)
2019	Functional bioactive compounds, generated in a matrix during fermentation, which may be used to promote health	5	Wegh et al. (2019)
2019	Nonviable metabolites produced by **probiotics** that exert biological effects on the hosts	4	Foo et al. (2019)
2019	Nonviable bacterial products or metabolic by-products from **probiotic** microorganisms that have positive effects on the host or microbiota	6	Johnson et al. (2019)
2019	Compounds produced by microorganisms, released from food components or microbial constituents, including nonviable cells that, when administered in adequate amounts promote health and well-being	3	Collado et al. (2019)
2020	*A consensus panel was announced in June 2019, conveyed by the International Scientific Association for Probiotics and Prebiotics, to produce a definition for 2020*	12	https://isappscience.org/defining-emerging-biotics/

The limiting prerequisite of deriving from a probiotic is highlighted in bolds.

proteobiotics (Nordeste et al., 2017; Tarsillo and Priefer, 2020). Similarly, as was previously mentioned for pro- pre-, and synbiotics, again an 11-expert panel was conveyed in December 2019 in London by ISAPP, and a consensus definition on postbiotics is expected to be published at any moment in 2021.

The term probiotic, as we know it, has been in use for almost 20 years, gaining little by little attention, respect and acceptance by consumers, health professionals, stakeholders, and regulatory bodies, due to the accumulated evidence, in many cases supported by meta-analysis. To start splitting the term depending on the specific health benefit targeted is interesting for marketing purposes in the short term but may weaken the topic and be misleading or confusing in the long term, especially when overlapping, poorly defined or even contradictory new definitions are introduced.

For the growing concept of postbiotics, the term most used in 2019 to refer to the health benefits achieved by the administration of nonviable microbes and their metabolites, a consensus definition is still needed. In a period of 6 years, at least 6 different definitions were proposed. Again, having a consensus definition, and attaching to it in the future, will help to develop and strengthen this area, as well as to gain acceptance and use by consumers, health practitioners, the industry, and regulators, strengthening the biotics family composed by probiotics, prebiotics, synbiotics, and postbiotics.

CONCLUSION

This chapter describes many emerging terms used in the "biotics" field and highlights the importance of having clear, consensus, and scientifically supported definitions in order to avoid misleading, confusing, and ambiguous interpretations of new terms by researchers, health professionals, regulatory bodies, and consumers. For the moment, we believe the terms probiotics, prebiotics, synbiotics, and postbiotics, all have, or will soon have, a consensus definition for each of them, and by properly using them, all cases of the biotics family can be adequately covered.

REFERENCES

Aguilar-Toalá, J.E., Garcia-Varela, R., Garcia, H.S., et al., 2018. Postbiotics: an evolving term within the functional foods field. Trends Food Sci. Technol. 75, 105–114.

Ahola, A.J., Yli-Knuuttila, H., Suomalainen, T., et al., 2002. Short-term consumption of probiotic-containing cheese and its effect on dental caries risk factors. Arch. Oral Biol. 47 (11), 799–804.

Bortz, G., Thomas, H., 2017. Biotechnologies for inclusive development: scaling up, knowledge intensity and empowerment (the case of the probiotic yoghurt 'Yogurito' in Argentina). Innov. Develop. 7 (1), 37–61.

Brodmann, T., Endo, A., Gueimonde, M., et al., 2017. Safety of novel microbes for human consumption: practical examples of assessment in the European Union. Front. Microbiol. 8, 1725. https://doi.org/10.3389/fmicb.2017.01725.

Cani, P.D., de Vos, W.M., 2017. Next-generation beneficial microbes: the case of *Akkermansia muciniphila*. Front. Microbiol. 8, 1765. https://doi.org/10.3389/fmicb.2017.01765.

Cani, P.D., Van Hul, M., 2015. Novel opportunities for next-generation probiotics targeting metabolic syndrome. Curr. Opin. Biotechnol. 32, 21–27.

Capurso, L., 2019. Thirty years of *Lactobacillus rhamnosus* GG – a review. J. Clin. Gastroenterol. 53, S1–S41.

Collado, M.C., Vinderola, G., Salminen, S., 2019. Postbiotics: facts and open questions. A position paper on the need for a consensus definition. Benef. Microbes 10 (7), 711–719.

David, B.Y., Dassa, B., Borovok, I., et al., 2015. Ruminococcal cellulosome systems from rumen to human. Environ. Microbiol. 17 (9), 3407–3426.

Davidson, L.E., Fiorino, A.M., Snydman, D.R., et al., 2011. *Lactobacillus* GG as an immune adjuvant for live-attenuated influenza vaccine in healthy adults: a randomized double-blind placebo-controlled trial. Eur. J. Clin. Nutr. 65, 501–507.

Depommier, C., Everard, A., Druart, C., et al., 2019. Supplementation with *Akkermansia muciniphila* in overweight and obese human volunteers: a proof-of-concept exploratory study. Nat. Med. 25, 1096–1103.

Dinan, T.G., Stanton, C., Cryan, J.F., 2013. Psychobiotics: a novel class of psychotropic. Biol. Psychiatry 74 (10), 720–726.

Doran, P.M., 2013. Unit operations. In: Bioprocess Engineering Principles, second ed. Academic Press, Elsevier, pp. 445–595.

Foo, H.L., Loh, T.C., Abdul Mutalib, N.E., et al., 2019. The myth and therapeutic potentials of postbiotics. In: Faintuch, J., Faintuch, S. (Eds.), Microbiome and Metabolome in Diagnosis, Therapy, and Other Strategic Applications. Academic Press, Elsevier, pp. 201–211.

Food and Agricultural Organization of the United Nations and World Health Organization, 2001. Health and nutritional properties of probiotics in food including powder milk with live lactic acid bacteria. World Health Organization. http://www.who.int/foodsafety/publications/fs_management/en/probiotics.pdf.

Gibson, G.R., Roberfroid, M.B., 1995. Dietary modulation of the human colonic microbiota: introducing the concept of prebiotics. J. Nutr. 125 (6), 1401–1412.

Gibson, G.R., Probert, H.M., van Loo, J.A.E., et al., 2004. Dietary modulation of the human colonic microbiota: updating the concept of prebiotics. Nutr. Res. Rev. 17, 259–275.

Gibson, G.R., Hutkins, R., Sanders, M.E., et al., 2017. Expert consensus document: the international scientific Association for Probiotics and Prebiotics (ISAPP) consensus statement on the definition and scope of prebiotics. Nat. Rev. Gastroenterol. Hepatol. 14, 491–502.

Gómez Gallego, C., Salminen, S., 2016. Novel probiotics and prebiotics: how can they help in human gut microbiota dysbiosis? Appl. Food Biotechnol. 3 (2), 72–81.

Gut Microbiota for Health, 2019. Synbiotics. A selection of content from the Gut Microbiota for Health 2017–2019. https://www.gutmicrobiotaforhealth.com/wp-content/uploads/2019/11/GMFH-bestof-synbiotics-2019.pdf.

Hill, C., Guarner, F., Reid, G., et al., 2014. Expert consensus document: the international scientific association for probiotics and prebiotics consensus statement on the scope and appropriate use of the term probiotic. Nat. Rev. Gastroenterol. Hepatol. 11, 506–514.

Hirschmugl, B., Brandl, W., Csapo, B., 2019. Evidence of human milk oligosaccharides in cord blood and maternal-to-fetal transport across the placenta. Nutrients 11, 2640. https://doi.org/10.3390/nu11112640.

Johnson, C.N., Kogut, M.H., Genovese, K., et al., 2019. Administration of a postbiotic causes immunomodulatory responses in broiler gut and reduces disease pathogenesis following challenge. Microorganisms 7 (8), 268. https://doi.org/10.3390/microorganisms7080268.

Kolida, S., Gibson, G.R., 2011. Synbiotics in health and disease. Annu. Rev. Food Sci. Technol. 2, 373–393. https://doi.org/10.1146/annurev-food-022510-133739.

Kort, R., Sybesma, W., 2012. Probiotics for every body. Trends Biotechnol. 30 (12), 613–615.

Lubitz, P., Mayr, U.B., Lubitz, W., 2009. Applications of bacterial ghosts in biomedicine. In: Guzmán, C.A., Feuerstein, G.Z. (Eds.), Pharmaceutical Biotechnology. Advances in Experimental Medicine and Biology. Vol. 655. Springer, New York, NY, pp. 159–170.

Martín, R., Langella, P., 2019. Emerging health concepts in the probiotics field: streamlining the definitions. Front. Microbiol. 10, 1047. https://doi.org/10.3389/fmicb.2019.01047.

Martín, R., Bermúdez-Humarán, L.G., Langella, P., 2018. Searching for the bacterial effector: the example of the multi-skilled commensal bacterium *Faecalibacterium prausnitzii*. Front. Microbiol. 9, 346. https://doi.org/10.3389/fmicb.2018.00346.

McFarland, L.V., Evans, C.T., Goldstein, E.J.C., 2018. Strain-specificity and disease-specificity of probiotic efficacy: a systematic review and meta-analysis. Front. Med. 5, 124. https://doi.org/10.3389/fmed.2018.00124.

Nordeste, R., Tessema, A., Sharma, S., et al., 2017. Molecules produced by probiotics prevent enteric colibacillosis in pigs. BMC Vet. Res. 13, 335. https://doi.org/10.1186/s12917-017-1246-6.

O'Toole, P.W., Marchesi, J.R., Hill, C., 2017. Next-generation probiotics: the spectrum from probiotics to live biotherapeutics. Nat. Microbiol. 2 (5), 17057. https://doi.org/10.1038/nmicrobiol.2017.57.

Plovier, H., Everard, A., Druart, C., et al., 2017. A purified membrane protein from *Akkermansia muciniphila* or the pasteurized bacterium improves metabolism in obese and diabetic mice. Nat. Med. 23, 107–113.

Reid, R., Kort, R., Alvarez, S., et al., 2018. Expanding the reach of probiotics through social enterprises. Benef. Microbes 9 (5), 707–715.

Reid, G., Gadir, A.A., Dhir, R., 2019. Probiotics: reiterating what they are and what they are not. Front. Microbiol. 10, 424. https://doi.org/10.3389/fmicb.2019.00424.

Saarela, M.H., 2019. Safety aspects of next generation probiotics. Curr. Opin. Food Sci. 30, 8–13.

Salva, S., Villena, J., Alvarez, S., 2010. Immunomodulatory activity of *Lactobacillus rhamnosus* strains isolated from goat milk: impact on intestinal and respiratory infections. Int. J. Food Microbiol. 141, 82–89.

Salva, S., Nuñez, M., Villena, J., et al., 2011. Development of a fermented goats' milk containing *Lactobacillus rhamnosus: in vivo* study of health benefits. J. Sci. Food Agric. 91, 2355–2362.

Sánchez, B., Delgado, S., Blanco-Míguez, A., et al., 2017. Probiotics, gut microbiota, and their influence on host health and disease. Mol. Nutr. Food Res. 61 (1), 1600240. https://doi.org/10.1002/mnfr.201600240.

Sanders, M.E., Benson, A., Lebeer, S., et al., 2018a. Shared mechanisms among probiotic taxa: implications for general probiotic claims. Curr. Opin. Biotechnol. 49, 207–216.

Sanders, M.E., Merenstein, D., Merrifield, C.A., et al., 2018b. Probiotics for human use. Nutr. Bull. 43, 212–225.

Sanders, M.E., Merenstein, D.J., Reid, G., et al., 2019. Probiotics and prebiotics in intestinal health and disease: from biology to the clinic. Nat. Rev. Gastroenterol. Hepatol. 16, 605–616.

Sarkar, A., Lehto, S.M., Harty, S., et al., 2016. Psychobiotics and the manipulation of bacteria-gut-brain signals. Trends Neurosci. 39 (11), 763–781.

Schultz, M., 2008. Clinical use of *E. coli* Nissle 1917 in inflammatory bowel disease. Inflamm. Bowel Dis. 14 (7), 1012–1018.

Scott, K.P., Grimaldi, R., Cunningham, M., et al., 2019. Developments in understanding and applying prebiotics in research and practice—an ISAPP conference paper. J. Appl. Microbiol. 128, 934–949.

Shanahan, F., Dinan, T.G., Ross, P., et al., 2012. Probiotics in transition. Clin. Gastroenterol. Hepatol. 10 (11), 1220–1224.

Sharma, R., Padwad, Y., 2020. Plant polyphenol-based second-generation synbiotic agents: emerging concepts, challenges, and opportunities. Nutrition 77. https://doi.org/10.1016/j.nut.2020.110785, 110785.

Stein, E., Inic-Kanada, A., Belij, S., et al., 2013. In vitro and in vivo uptake study of *Escherichia coli* Nissle 1917 bacterial ghosts: cell-based delivery system to target ocular surface diseases. Invest. Ophth. Vis. Sci. 54 (9), 6326–6333.

Swanson, K.S., Gibson, G.R., Hutkins, R., et al., 2020. The International Scientific Association for Probiotics and Prebiotics (ISAPP) consensus statement on the definition and scope of synbiotics. Nat. Rev. Gastroenterol. Hepatol. 17, 687–701.

Szajewska, H., Kolodziej, M., 2015. Systematic review with meta-analysis: *Saccharomyces boulardii* in the prevention of antibiotic-associated diarrhoea. Aliment Pharm. Ther. 42, 793–801.

Szajewska, H., Kotowska, M., Mrukowicz, J.Z., et al., 2001. Efficacy of *Lactobacillus* GG in prevention of nosocomial diarrhea in infants. J. Pediatr. 138 (3), 361–365.

Tan, H., Zhai, Q., Chen, W., 2019. Investigations of *Bacteroides* spp. towards next-generation probiotics. Food Res. Int. 116, 637–644.

Tarsillo, B., Priefer, R., 2020. Proteobiotics as a new antimicrobial therapy. Microb. Pathog. 142. https://doi.org/10.1016/j.micpath.2020.104093, 104093.

Taverniti, V., Guglielmetti, S., 2011. The immunomodulatory properties of probiotic microorganisms beyond their viability (ghost probiotics: proposal of paraprobiotic concept). Genes Nutr. 6, 261–274.

Thilakarathna, W.P.D.W., Langille, M.G.I., Rupasinghe, H.P.V., 2018. Polyphenol-based prebiotics and synbiotics: potential for cancer chemoprevention. Curr. Opin. Food Sci. 20, 51–57.

Tsilingiri, K., Rescigno, M., 2013. Postbiotics: what else? Benef. Microbes 4 (1), 101–107.

Villena, J., Salva, S., Núñez, M., et al., 2012. Probiotics for everyone! The novel immunobiotic *Lactobacillus rhamnosus* CRL1505 and the beginning of social probiotic programs in Argentina. Int. J. Biotechnol. Wellness Ind. 1, 189–198.

Vinderola, G., Gueimonde, M., Gomez-Gallego, C., et al., 2017. Correlation between *in vitro* and *in vivo* assays in selection of probiotics from traditional species of bacteria. Trends Food Sci. Technol. 68, 83–90.

Wegh, C.A.M., Geerlings, S.Y., Knol, J., et al., 2019. Postbiotics and their potential applications in early life nutrition and beyond. Int. J. Mol. Sci. 20, 4673. https://doi.org/10.3390/ijms20194673.

World Gastroenterology Organisation (WGO), 2017. Global Guidelines Probiotics and prebiotics. https://www.worldgastroenterology.org/UserFiles/file/guidelines/probiotics-and-prebiotics-english-2017.pdf.

Zheng, J., Wittouck, S., Salvetti, E., et al., 2020. A taxonomic note on the genus *Lactobacillus*: Description of 23 novel genera, emended description of the genus *Lactobacillus* Beijerinck 1901, and union of *Lactobacillaceae* and *Leuconostocaceae*. Int. J. Syst. Evol. Microbiol. 70 (4), 2782–2858. https://doi.org/10.1099/ijsem.0.004107.

CHAPTER 2

Emerging Prebiotics: Nutritional and Technological Considerations

ROBERTO DE PAULA DO NASCIMENTO • MARIO ROBERTO MAROSTICA JUNIOR
School of Food Engineering, University of Campinas (UNICAMP), Campinas, São Paulo, Brazil

INTRODUCTION

Fructooligosaccharides (FOSs), galactooligosaccharides (GOSs), and inulin are probably the most researched (Davani-Davari et al., 2019) and commercialized (Hexa Research, 2018) prebiotics worldwide. Their proven capacity to modulate the intestinal microbiota and provide beneficial health effects against chronic diseases has brought the attention of international markets for their commercialization. Additionally, these prebiotics had also been utilized by the industry to improve processes and the physical-chemical and nutritional characteristics of foods in order to make them more sensorially interesting and healthier for general consumers (Farias et al., 2019; Mohammadi and Mortazavian, 2011).

Other commonly researched and/or commercialized prebiotics include mannan-oligosaccharides (MOSs), a largely produced product for animal feeding purposes (Grand View Research, 2018), and lactulose and polydextrose, which may be considered as established prebiotics, due to numerous researches attesting its prohealth effects in clinical trials (Patel and Goyal, 2012).

In 2015, the global market of prebiotics, majority represented by FOSs, GOSs, inulin, and MOSs, generated approximately 3 billion dollars, and by 2026, may reach an estimated eight to 10 billion dollars (Grand View Research, 2018; Hexa Research, 2018; Kunal Ahuja, 2019). This expected increase in the prebiotic business is possibly related to a growing interesting in their exploitation as a profit product, and to the inclusion of emerging prebiotics with technological and nutritional health-related properties.

Although poorly described, few researchers suggest a classification of prebiotics based on a criterion of ordinariness. According to that, it appears that prebiotics may be (1) established: extensively studied, commercially available, and/or safe for consumption; or (2) emerging: poorly studied and commercialized and/or not yet tested for safety (Corzo et al., 2015; Gallego and Salminen, 2016; Kumar et al., 2015; Mano et al., 2018; Míguez et al., 2016; Patel and Goyal, 2012). However, it seems that this segregation is not official and differs between authors, but will be assumed for this book chapter.

The prebiotics FOSs, GOSs, inulin, lactulose, MOSs, and polydextrose will not be considered as emerging (Fig. 1), since they already possess well-known commercial products in the market and/or have been tested in several clinical studies. Regarding the classification of polydextrose, despite being considered in the past as a new prebiotic, for this chapter, this product will not be discussed, as authors are already considering it as established (Lugani and Sooch, 2017; Patel and Goyal, 2012) and there is a growing global market associated with it (Grand View Research, 2017). Therefore, for this section, the following prebiotics will be considered as "emerging": algae carbohydrates, bacterial exopolysaccharides (EPSs), isomaltooligosaccharides (IMOSs), pectin and pectic oligosaccharides (POSs), polyphenols, resistant starch (RS), sugar alcohols (SAs), xylooligosaccharides (XOSs), and others (glucomannan, soybean oligosaccharides, etc.) (Fig. 1). Although there are plenty of studies and even commercial products with RS, this chapter will still consider this product as an emerging prebiotic, since, usually, RS is mainly discussed as a dietary fiber. Additionally, authors from recent studies still consider RS as emerging (Li et al., 2018), potential (Zaman and Sarbini, 2016), or a new prebiotic (Guarino et al., 2020). Finally, although the term prebiotic defines a compound that can be selectively utilized or confer changes to any host microbial ecosystem (Gibson et al., 2017), this chapter will not present the effects of prebiotics other than in the gut microbiome and related-health implications.

Probiotics and Prebiotics in Foods. https://doi.org/10.1016/B978-0-12-819662-5.00016-1

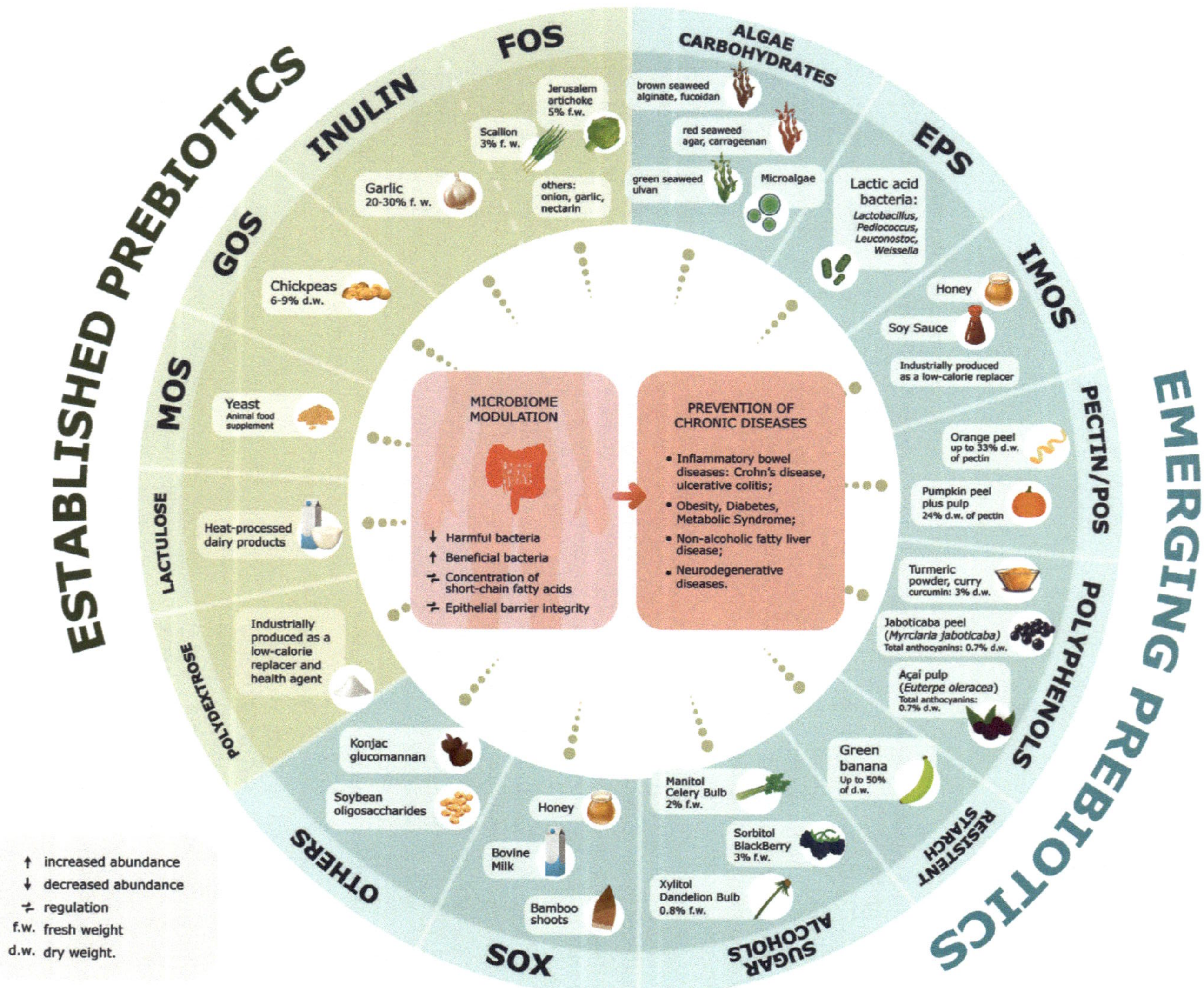

FIG. 1 Prebiotics, main food sources and/or industrial application, and potential health-related effects. Abbreviations: *EPSs*, bacterial exopolysaccharides; *FOSs*, fructooligosaccharides; *GOSs*, galactooligosaccharides; *IMOSs*, isomaltooligosaccharides; *MOSs*, mannan-oligosaccharides; *POSs*, pectic oligosaccharides; *XOSs*, xylooligosaccharides. Complementary references: (Jovanovic-Malinovska et al., 2014; Judprasong et al., 2011; Maru et al., 2015; Olano and Corzo, 2009; Cheong et al., 2018; Leite-Legatti et al., 2012; Tribess et al., 2009). Soy sauce, blackberry, celery, turmeric powder, honey, Jerusalem artichoke, and milk images: designed by macrovector/Freepik; green seaweed image: designed by photographeeasia/Freepik; garlic and white powder images: designed by vectorpocket/Freepik; and açaí image: designed by Freepik.

At the end of the chapter, Table 1 shows the latest clinical trials with emerging prebiotics in healthy or chronic ill individuals, regarding their effects on bowel function and on the microbiome.

ALGAE CARBOHYDRATES

Marine algae are mainly commercialized and consumed in Asian countries, especially China and Indonesia. In Asia, production increased by almost 50% between 2011 and 2016, reaching 30 million tones, approximately (Ferdouse et al., 2018). The inclusion of algae in food products, such as noodles, saltine crackers, cookies, drinks, and sweets, appears to be increasing and may represent a future trend not limited to Asian countries (Ferdouse et al., 2018).

Seaweeds (or macroalgae) and microalgae are the main representative groups of marine algae. They are rich sources of polysaccharides (agar, alginate, arabinoxylans, β-glucans, carrageenans, exopolysaccharides, fucoidans, inulin, laminarin, porphyran, ulvan, and xylan), oligosaccharides (agarose-derived, alginate-derived, GOSs, FOSs, neoagaro-oligosaccharides, and XOSs), lignin, and/or phenolic compounds with potential prebiotic properties (Camacho et al., 2019; De Jesus Raposo et al., 2016; O'Sullivan et al., 2010) (Fig. 1). Most of these can be considered as dietary insoluble or soluble fibers since they are not digested or absorbed in the human small intestine (DeVries et al., 2001; De Jesus Raposo et al., 2016).

General industrial application of marine algae is currently mostly limited to the polysaccharides alginate, agar, and carrageenan, which are utilized to stabilize commercial products and to produce viscous aqueous solutions, water-soluble films, and versatile gels (Gomez-Zavaglia et al., 2019). Other useful properties from algae carbohydrates still in development include antibiofilm activity, cosmeceutical, food ingredient, and pharmacological (Gomez-Zavaglia et al., 2019; Jutur et al., 2016). Additionally, initial studies in vitro (Ligen et al., 2018; Strain et al., 2020) and with animals (mice, rat, pig, and chicken) (Bouwhuis et al., 2017; Kulshreshtha et al., 2014; du Preez et al., 2020; Zhang et al., 2018b) suggest that the consumption of algae prebiotics may positively modulate the microbiota and/or augment the formation of short-chain fatty acids (SCFAs), therefore contributing to intestinal health (Cherry et al., 2019). Studies with humans are still scarce, but the ones available demonstrated the potential of algae to improve microbiota parameters (Fields et al., 2020) (Table 1), to ameliorate obesity (Kim et al., 2019), diabetes (Lee and Jeon, 2015), and nonalcoholic fatty liver disease (Ebrahimi-Mameghani et al., 2017). It remains uncertain, however, from most studies, which components certainly contributed to the beneficial effects found, since algae are rich sources of several nutrients and bioactive compounds, including poly- and oligosaccharides with prebiotic potential.

Starting from seaweeds, these are usually divided into brown, red, and green types (Fig. 1). Specially, brown seaweeds are highly produced in Asia, including species *Laminaria japonica* (or *Saccharina japonica*), *Undaria pinnatifida*, and *Sargassum fusiforme* (FAO, 2018). Potential prebiotics alginate, fucoidan, and laminarin are said to be found in abundance in some species of brown algae, depending on the species, season, habitat, and extraction method (García-Ríos et al., 2012; Múzquiz De La Garza et al., 2019; Rodrigues et al., 2015). Characterization studies reporting the percentage content of polysaccharides from brown algae are not so common, with just a few reliable claims attesting that alginate and fucoidan quantities may reach up to 50% and 32%, respectively, in a dry weight basis (Chee et al., 2011; García-Ríos et al., 2012).

Alginate is a linear polysaccharide composed of β-D-mannuronic and α-L-guluronic acids linked via 1 → 4 glycosidic linkages (Fig. 2) (Alba and Kontogiorgos, 2018). Such carbohydrate has been widely utilized in the industry to improve the technical properties of foods, for example by providing (1) thickness on sauces, syrups, and ice cream toppings; (2) stable emulsions of water and oil in mayonnaise and salad dressings; and (3) better texture, body, and sheen of yogurt (Pangestuti and Kim, 2015). Also, sodium alginate, a commercially extracted form of alginate, has been successfully tested by preclinical studies to microencapsulate oils and probiotics in order to increase its bioaccessibility (Bastos et al., 2020; Luo et al., 2020).

Alginate alone represents a potent prebiotic, as a few studies suggest that it can selectively modulate the

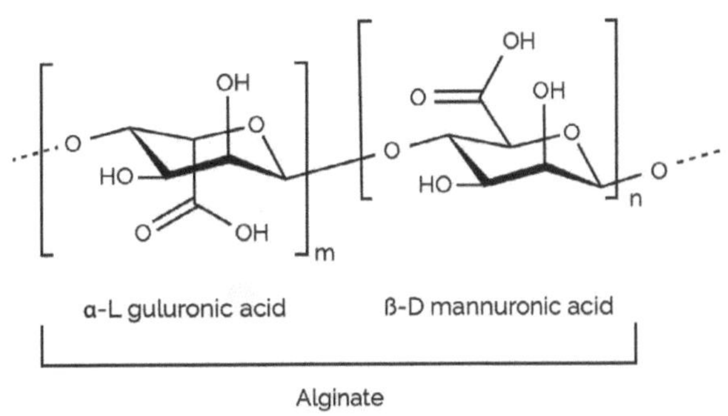

FIG. 2 Chemical structure of alginate.

microbiota, leading to positive health effects (Li et al., 2020; Wang et al., 2020b). According to results of a research lead by Wang et al. (2020b), 5% alginate-derived oligosaccharides added to a high-fat diet are able to augment the fecal concentration of six SCFAs, including butyric, acetic, and propionic, and to reverse the dysbiotic implications of obesity by increasing the abundance of well-known beneficial bacteria *Akkermansia muciniphila* and *Lactobacillus reuteri*. Additionally, alginate can reduce the concentration of proinflammatory cytokines and the accumulation of lipids in the liver and serum of animals (Wang et al., 2020b).

Another microbiota-related study shown that a probiotic mixture in association with *Laminaria japonica*, a brown seaweed rich in alginate and fucoidan, provided an augmentation in the abundance of *Bifidobacterium* and *Lactobacillus* in the feces of healthy individuals. Most importantly, this study suggests that the combination of both pro- and prebiotics, and not only the last one, may provide a better modulatory effect in the microbiota, protecting against overall digestive symptoms and later complications (Ko et al., 2014) (Table 1). Positive modulatory effects in the microbiota had also been found recently by treatments with fucoidan (100 mg/kg of body weight) in dyslipidemic rats (Chen et al., 2019) and laminarin (1% in water) in mice with obesity (Nguyen et al., 2016).

In addition to seaweeds, microalgae have also shown technological and health-related potential, some attributed to their carbohydrate contents. Most common produced microalgae worldwide include *Spirulina*, *Chlorella*, and *Nannochloropsis* (FAO, 2018). The biomass of these and other microalgae has been tested in studies that related its content in carbohydrates to interesting food applications and potential prebiotic effects, such as (1) improvement of yogurt color, stability, and texture; (2) improvement of cheese texture; (3) reduction of harmful bacteria, such as *Proteus vulgaris*, *Bacillus subtilis*, and *Bacillus pumilus*; and (4) prevention of gastric ulcers and constipation (Camacho et al., 2019; Gomez-Zavaglia et al., 2019; De Jesus Raposo et al., 2016; Wells et al., 2017). More studies are needed, including uncommon randomized double-blind trials, in order to confirm the beneficial effects of carbohydrates from marine algae as potential prebiotics.

BACTERIAL EXOPOLYSACCHARIDES

Many strains of bacteria, fungi, and algae are able to produce polysaccharides as part of their metabolism, including EPSs (Liu et al., 2017). These can be synthetized by extracellular or intracellular pathways, being loosely bound to the cell surface or launched into the extracellular medium (Kumar et al., 2007; Zhou et al., 2019). Production of such polysaccharides may depend on several factors, including medium composition, temperature, pH, and growth phase of the organism (Rahbar Saadat et al., 2019).

Specially, EPSs produced by lactic acid bacteria, such as *Weissella*, *Leuconostoc*, *Lactobacillus*, and *Pediococcus* genera (Fig. 1), possess interesting food applications, the most known ones related to their capacity to improve texture, retain flavor, and increase the shelf life of fermentable foods and bakery products. Additionally, EPSs have been associated with microbiota modulation by selectively enhancing the growth of beneficial bacteria (Hamet et al., 2016; Hongpattarakere et al., 2012; Lynch et al., 2018a,b; Zhou et al., 2019). Types of EPSs related to health implications include the ones that are (1) homopolysaccharides (e.g., dextran, inulin, and levan), meaning that they are formed by repetitive units of monosaccharides, and (2) heteropolysaccharides (e.g., kefiran), known to possess between three to eight different units of monosaccharides (Fig. 3) (Korcz et al., 2018; Rahbar Saadat et al., 2019; Salazar et al., 2016).

Bacteria-producing EPSs have been recognized as potential tools by the food industry to improve the sensorial properties of dairy products and ameliorate the technological characteristics of modified healthier food alternatives. For example, *Lactobacillus* spp. that produce EPSs are capable to improve the quality of fat-reduced cheddar cheese by (1) increasing moisture content, a common problem related to products low in fat that consequently possess a compact protein network, and (2) improving textural properties like adhesiveness and softness, without worsening the sensorial acceptance (Costa et al., 2010). Beneficial rheological effects of adding bacteria-producing EPSs can also be observed in fermented milk. High molecular weight EPSs from specific species of *Streptococcus* and *Lactobacillus* have been shown to increase some desirable characteristics of milk during fermentation, including viscosity, firmness, and whey protein retention (Gentès et al., 2011).

EPS functionalities are not limited to the dairy industry and can also be utilized in cereal-based foods to increase loaf volume and crumb softness of sourdough bread (Katina et al., 2009). These features are especially important when there is an interest in improving the quality of gluten-free products that suffer from bad structure, gas holding abilities, and staling rate. Specially, dextran EPSs from *Weissella cibaria* may represent a future possibility as a natural additive in bakery products that aim to meet the needs of celiac populations (Galle et al., 2012; Schwab et al., 2008).

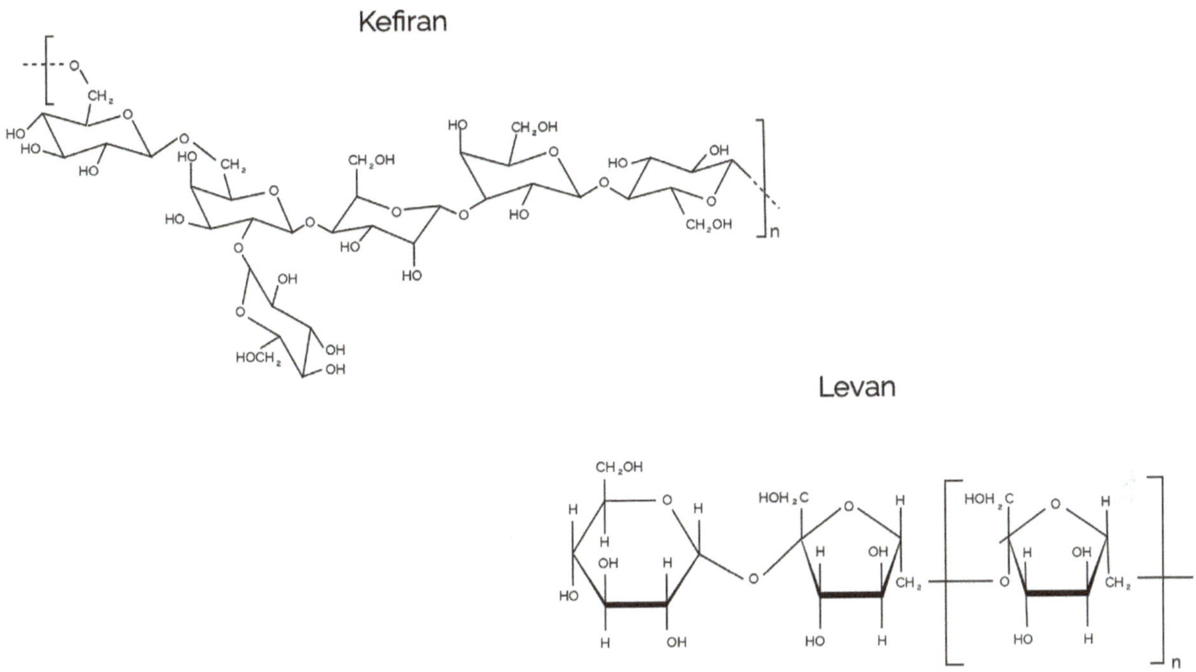

FIG. 3 Chemical structures of bacterial exopolysaccharides.

Besides previously presented technological properties, numerous experimental studies in vitro and in vivo show that EPSs can potentially help prevent or treat against chronic diseases due its antioxidant, antiinflammatory, cholesterol-lowering, antihypertensive, immunomodulatory, and antiproliferative effects (Korcz et al., 2018; Lynch et al., 2018a; Rahbar Saadat et al., 2019). Also, despite very few shreds of evidence, researches associate EPSs to gut health and microbiota modulation, to be discussed next.

Studies in vitro utilizing simulations of the colonic environment indicate that EPSs from *Bifidobacterium* spp. are not degraded in gastric and small intestine, but only in colonic conditions, which may facilitate their fermentation by the microbiota. During and after fermentation, these EPSs can act by (1) increasing the abundance of favorable bacteria genus, like *Coprococcus* and *Lactobacillus*, usually found reduced in patients with ulcerative colitis and Parkinsońs disease (Liu et al., 2019; Zhong et al., 2019), and (2) augmenting the production of propionic acid (Liu et al., 2019; Salazar et al., 2009), a SCFA produced by microbiota, and associated to the prevention of impaired metabolic mechanisms related to obesity and type 2 diabetes in humans (Al-Lahham et al., 2010). Experiments with in vitro colonic models using feces of infants have also been recently performed. In a study by Moroeanu et al. (2015), for example, a higher synthesis of EPSs (15–20 mg/L) in descending colon, produced by *Lactobacillus rhamnosus* and *Weissella paramesenteroides* species, was associated with an equilibrium in microbiota composition, suggesting both pro- and prebiotic roles of bacteria producing EPSs (Moroeanu et al., 2015).

Regarding animal studies, a recent investigation found that freeze-dried kefiran, a branched hydrosoluble glucogalactan from kefir milk, consumed at a dosage of 1 mg per mice, significantly increased the abundance of *Bifidobacterium* population. According to the authors, this may be related to an increment in the number of goblet cells, since species of *Bifidobacterium*, like *Bifidobacterium bifidum*, use proteins (mucins) produced by these cells as nutrients (Hamet et al., 2016). Goblet cell depletion and mucin alterations are commonly related to intestinal infections and chronic intestinal diseases, which indicates that EPSs from lactic acid bacteria may be considered in the future studies as possible mucosal healing compounds. Furthermore, as EPS modulates goblet cell-related mechanisms, it may also protect against epithelial barrier dysfunction, a mechanism associated with immunological and inflammatory ignition, and theorized to cause and/or exacerbate ulcerative colitis, an inflammatory bowel disease (Zhou et al., 2018).

Finally, it appears that only one study to date attested successfully the microbiome-modulating effects of EPS in a randomized double-blind placebo-controlled study. In that case, EPS produced by one billion viable cells of *Streptococcus thermophilus* protected against epithelial barrier dysfunction in healthy humans due to improved mechanisms of colonic permeability. Such species of bacteria probably colonized intestinal mucosa, contributing positively to the host's microbiota and producing abundant amounts of EPSs, which were found augmented in the feces from the treated group. In conclusion, continuous production and secretion of EPSs in the lumen appear to be related to colonic health and mucosal homeostasis (Del Piano et al., 2014) (Table 1); however, more studies with humans are needed to confirm that.

ISOMALTOOLIGOSACCHARIDES

IMOSs consist of up to eight glucose monomers formed by α-(1 → 6) glucosidic linkages, such as isomaltose, isomaltotriose, and panose, which are common short-chain carbohydrates identified in IMOS mixtures (Fig. 4) (Carlson et al., 2018). Natural sources of IMOSs include honey and fermentable foods, such as miso (Hondo and Mochizuki, 1979), soy sauce (Nishino et al., 1974), sake (Kohmoto et al., 1988), and beer (Vinogradov and Bock, 1998) (Fig. 1). However, IMOSs are most known for their utilization by the food industry, as they can be added to food or replace dietary carbohydrates, improving the quality of commercial products (Goffin et al., 2011). Also, IMOSs may increase the healthy potential of foods, as scientific data suggest their prebiotic potential (Yen et al., 2011) (Table 1). Currently, IMOSs are *Generally Recognized as Safe* by the US Food and Drug Administration (Gaynor, 2016) and commercial products from Canada, China, Japan, and the United States are reaching the market with the allegation of sweetener and prebiotic (Madsen et al., 2017).

IMOSs can be industrially obtained by enzymatic reaction from cornstarch, specially utilizing α-amylase derived from *Bacillus licheniformis* or *Bacillus stearothermophilus*. Its most known technological application is related to the capacity of replacing high-calorie sugars and regulating sweetness without negatively changing the acceptability of products, such as honey and syrups (Goffin et al., 2011). Additionally, IMOSs can also (1) increase proof volume, moisture, and softness of bread from frozen dough (Park et al., 2016); (2) improve the sensorial acceptance of sponge cake regarding crust color, tenderness, and overall liking (Lee et al., 2008); and (3) help compose a fermentable orange juice with reduced sugar content, potential sensorial acceptance, and with interest as a functional food (Rolim et al., 2019).

Although still poorly studied in comparison with GOSs and FOSs, first observations on the possible prebiotic roles of IMOSs are dated from the beginning of the 21st century. For example, in an in vitro study by Rycroft et al. (2001), IMOSs and GOSs shown to possess similar beneficial effects in the environment of common gut bacteria by increasing lactate production, lowering gas release, and effectively augmenting *Bifidobacterium* abundance, a genus usually diminished in the elderly population (Rycroft et al., 2001). On other studies from the same year, the capacity of IMOSs to reduce constipation was attested in elderly man (Chen et al., 2001; Wang et al., 2001) (Table 1), indicating a possible targeting role for this carbohydrate, which were later confirmed by a reliable controlled study by Yen et al. (2011). From this study, by increasing the content of dietary fibers or compounds with similar biological functions, like IMOSs, given 10 g/day for elderly, feces became bulkier and wetter, facilitating spontaneous defecation (Yen et al., 2011) (Table 1).

Additionally to the most frequent studies regarding constipation, IMOSs can also increase the abundance of beneficial bacteria *Akkermansia muciniphila* and *Roseburia* spp., as well as the concentration of SCFAs in

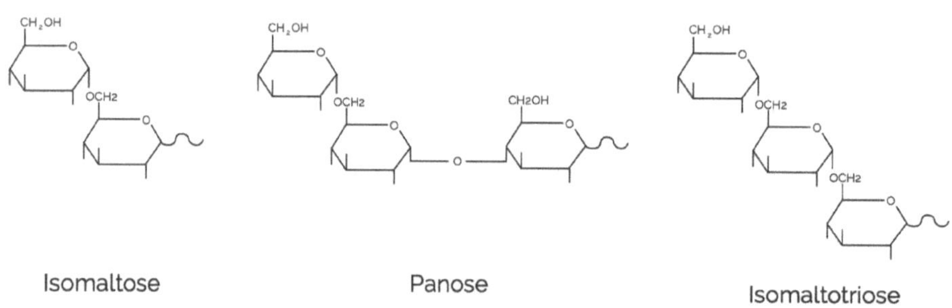

FIG. 4 Chemical structures of isomaltooligosaccharides.

feces (Wang et al., 2017b). Other health-related effects of IMOSs, administrated together with polyphenols, are associated with the prevention of metabolic impairment and inflammation coming from consumption of a high-fat diet (Singh et al., 2017a,b, 2018). Considering that there are only a few studies confirming the health effects of IMOSs, it seems inappropriate that several companies are already claiming their IMOS-rich products as prebiotic without strong scientific background.

It is important to remember that interventions with animals and humans utilizing IMOSs should probably consider high dosages, since this carbohydrate can be partially digested by the gastrointestinal tract and therefore may have its fermentation and functionally reduced in the colon (Oku and Nakamura, 2003). In fact, by being partially degraded before reaching the colon, IMOSs may not be considered a fully realized prebiotic. Also, proof of its selective utilization by the microbiota is lacking. Further studies are needed with IMOSs.

PECTIN AND PECTIC OLIGOSACCHARIDES

Pectin is a structurally highly complex polysaccharide (Fig. 5) present in the cell walls of fruits and vegetables, specially industrial by-products such as apple pomace, citrus waste (mainly orange and lemon), sugar beet pulp, olive pomace, potato pulp, and others (tomato pomace, purple berries, passion fruit waste, and soybean) (Babbar et al., 2016; Gullón et al., 2013). A recent study attesting the amount of pectin in plant wastes found a high percentage of this polysaccharide (dry weight basis) in orange peel (25%), pumpkin peel plus pulp (24%), and leek leaves (15%) (Müller-Maatsch et al., 2016). Quantities found, however, may depend on the different extraction methods utilized in literature, as other studies already found up to 33% of pectin in dry orange peel, which appears to be the maximum found so far in a food product (El-Nawawi and Shehata, 1987; Hosseini et al., 2019; Su et al., 2019) (Fig. 1).

The depolymerization of pectin can be obtained by either degradation followed by fermentation from colonic bacteria (Dongowski et al., 2000) or industrially mostly by enzymatic processes. These polysaccharides usually resist digestion (Holloway et al., 1983) and their components, including POSs, are not reported to be found naturally free in foods. The breakdown of pectin causes the release of POSs, which are selectively fermented by the microbiota (Onumpai et al., 2011; Tingirikari, 2018). They are represented predominantly by oligogalacturonides, arabinooligosaccharides, rhamnogalacturonoligosaccharides, xylooligogalacturonides, and arabinogalactooligosaccharides (Gullón et al., 2013).

Pectin extraction from the industry's by-products has important technological applications, serving, for example, (1) as a food additive to increase the gelling (jams, jellies, and marmalades) and emulsifying (cheese, mayonnaise, and yogurt low in fat) characteristics of commercial products; (2) to create innovative food packages capable of retaining the quality of foods (Vanitha and

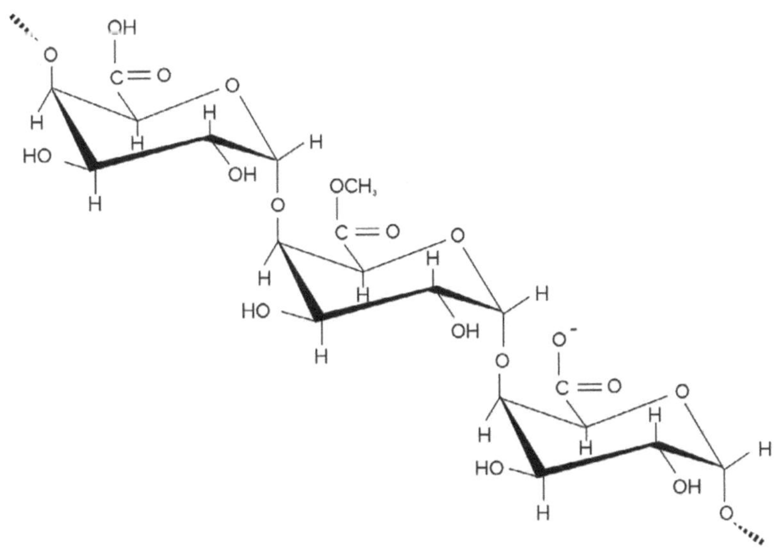

FIG. 5 Chemical structure of pectin.

Khan, 2020); (3) as a matrix intending the delivery of drugs (Sriamornsak, 2007); and (4) for the manufacturing of vaccines (Rosales-Mendoza et al., 2016).

Specifically, POSs can also be extracted for utilization in the food, feed, and pharmaceutical industries, but studies attesting that are still scarce. In the feed industry, POSs may be useful to prevent against pathogenic bacterial infestation in poultry flocks, therefore reducing the risk for contaminated chicken products (Babbar et al., 2016). Also, POSs may be considered as elicitors to improve the postharvest performance of grapefruit (Vera-Guzmán et al., 2019) and strawberry (Virgen-Ortiz et al., 2019) and the color development and antioxidant potential of grapes (Ochoa-Villarreal et al., 2011), contributing to extended shelf life and avoiding the use of synthetic products.

Claims of the literature associate POSs with their health and prebiotic effects. Differences between constituents, degree of polymerization and molecular weight, and degree of methylesterification may influence the fermentation process by intestinal bacteria, as in vitro studies suggest (Gullón et al., 2013). Specially, it appears that modified pectins and/or POSs with lower molecular weights, like the ones found in sugar beet (Al-Tamimi et al., 2006), artichoke, and citrus sources (Ferreira-Lazarte et al., 2018; Onumpai et al., 2011), are preferable by *Bifidobacterium* spp., therefore increasing and maintaining the fermentation process for a longer time. In addition, they also promote an augmentation in the concentration of SCFAs, like propionate (Al-Tamimi et al., 2006) and, if deesterified, these structures may protect against adhesiveness from pathogenic bacteria, like *Escherichia coli* (Di et al., 2017).

The prebiotic effects of long-chain POSs had also been shown in a few studies. According to Di et al. (2017), a POS rich in arabinose with a high molecular weight of 800 kDa has bifidogenic effects and is able to increase the concentration of butyrate (Di et al., 2017). Butyrate is an essential SCFA that maintains colonic health and may prevent against alterations that led to colorectal cancer, like cellular oxidative damage and inflammation (Hamer et al., 2008). In another study, by Wilkowska et al. (2019), the positive effects of a higher order POSs from apple pomace with a degree of polymerization of 7–10 were also confirmed from observations of high *Bifidobacterium* growth and SCFA modulation (Wilkowska et al., 2019). Clinical trials are mostly available for pectin, which has shown some significant modulatory effects in the microbiota of individuals with ulcerative colitis (Wei et al., 2016) and functional constipation (Zhang et al., 2018a) that went through fecal microbiota transplantation (Table 1).

Other health-related effects of the consumption of POSs are still scarce, despite its potential to prevent or help treat against chronic diseases due to previously mentioned microbiome mechanisms. Some studies in vitro indicate that POSs may possess beneficial effects against cancer by inducing the apoptosis of colonic adenocarcinoma cells (Olano-Martin et al., 2003) and by blocking molecules related to increased survival of melanoma cells, like galectin-3 and survivin (Mallikarjuna and Dharmesh, 2018). Further studies with animals and humans need to be performed in order to confirm and deepen the beneficial effects of POSs.

POLYPHENOLS

Phenolic compounds or simply polyphenols, such as phenolic acids, flavonoids, and phenolic amides, are widely distributed in the plant kingdom, being most commonly found in fruits, vegetables, grains, and derived commercial products (Tsao, 2010). They are secondary metabolites produced by plants aiming their protection against biotic (fungi, bacteria, insects, and weeds) and abiotic (virus, dehydration, excessive watering, extreme temperatures, salinity, and mineral toxicity) stresses (Zhang and Tsao, 2016).

Chemically, polyphenols possess one or more aromatic rings attached to hydroxyl groups (Fig. 6). Their chemical structure facilitates the donation of electron or hydrogen atoms to other molecules, neutralizing, for example, reactive oxygen species and acting foremost as antioxidants. Additionally, polyphenols may also interfere positively in signaling pathways related to (1) oxidative stress, by regulating the function of antioxidant enzymes, such as glutathione peroxidase and superoxide dismutase; and (2) inflammation, by correcting the activation of nuclear factor-kappa B and gene expression of proinflammatory cytokines (Tsao, 2010; Zhang and Tsao, 2016).

The antioxidant and antiinflammatory potentials of polyphenols have been observed by several in vitro and in vivo studies, highlighting their role as preventive or healing tools against chronic diseases, like obesity, diabetes, inflammatory bowel diseases, and cancer. Also, such compounds have been shown to interfere in intestinal health, including microbiota (Dueñas et al., 2015; Singh et al., 2019). In fact, according to the 2017 consensus of The International Scientific Association for Probiotics and Prebiotics, polyphenols can be considered as prebiotics, since most part (90%–95%) is not absorbed by the small intestine, therefore reaching the colon and potentially being utilized by the microbiota, causing alterations in the microbiome (Gibson

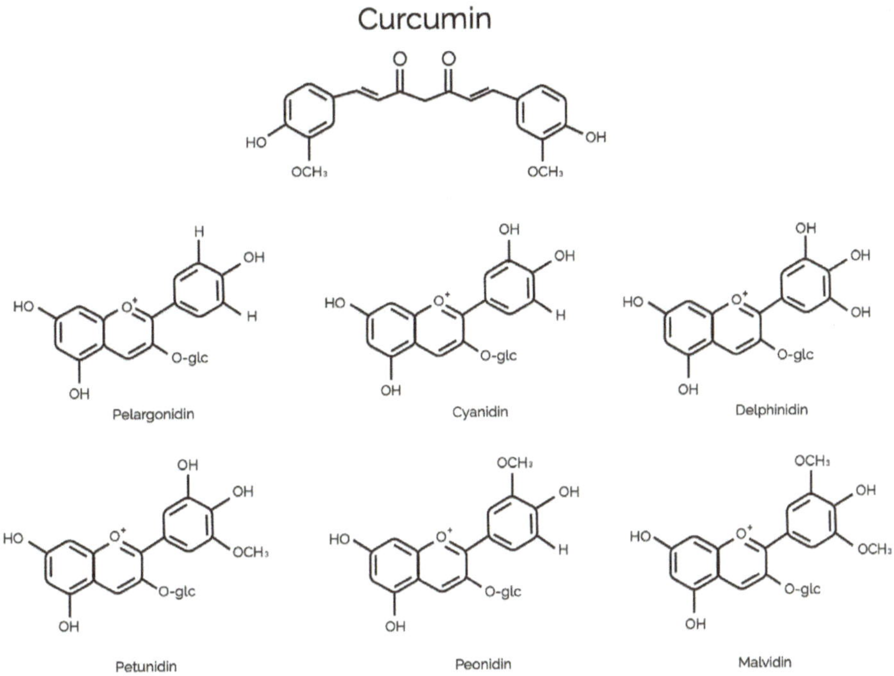

FIG. 6 Chemical structures of polyphenols.

et al., 2017). However, the impact in the microbiota may depend on the class or type of polyphenol, considering the differences between their metabolic paths (Kawabata et al., 2019).

Currently, one of the most studied phenolic compounds is curcumin (Fig. 6), a diferuloylmethane with yellow pigmentation belonging to the subclass of curcuminoids. Its main source is turmeric, the rhizome of *Curcuma longa,* a product very commonly consumed in India in the form of powder named Indian saffron or in association with other products like curry (Stanić, 2017). The percentage of curcumin usually found in dry spice powders of Indian saffron or curry is around 3% (Tayyem et al., 2006) (Fig. 1).

Curcumin is usually poorly bioavailable, meaning that it is very little metabolized and absorbed through the gastrointestinal tract and consequently not found in significant amounts in serum (Klickovic et al., 2014; Lao et al., 2006). Despite this potentially represents a problem in terms of drug delivery and efficacy, it also indicates that curcumin is present in large amounts in the colon and can be utilized as a substrate by the microbiota in order to increase gut health (Peterson et al., 2018) (Table 1). In addition, studies show that curcumin can act directly in colonic epithelial cells, modulating their integrity and permeability and therefore possibly contributing to the prevention of diseases related to the colon (Pluta et al., 2020; Wang et al., 2017a).

A recent double-blind, randomized, placebo-controlled pilot study attested the prebiotic-like effects of curcumin in the human gut. Peterson et al. (2018) gave tablets of turmeric, curcumin, or placebo to healthy individuals for 8 weeks at a dosage of 6 g/day and analyzed their microbiota by 16S rDNA sequencing. Individuals receiving curcumin presented a higher number of species and a tendency toward increased diversity of bacteria. Also, turmeric- or curcumin-treated groups had an increase in *Bacteroides* spp. (Peterson et al., 2018) (Table 1).

A decrease in microbial diversity is related to the worsening of several diseases, including colorectal cancer (Ai et al., 2019). In addition, a recent metaanalysis has shown that low levels of *Bacteroides* are related to the disease onset of patients with inflammatory bowel diseases, especially in the active phase (Zhou and Zhi, 2016). Lack of *Bacteroides* also seems to be closely related to an obesity profile, as children in that state present lower levels of this genus and obese adults have decreased populations of *Bacteroides faecichinchillae* and *Bacteroides thetaiotaomicron* (Castaner et al., 2018). Such findings suggest a relation between the consumption of curcumin and increased gut health. In fact, human and animal studies utilizing curcumin are increasingly proving that the microbiome-modulating effects of this polyphenol are related to a better outcome in chronic diseases, such as metabolic syndrome (du Preez et al., 2019), colorectal cancer, ulcerative colitis (Mcfadden et al., 2015; Ohno et al., 2017), and irritable bowel syndrome (Ng et al., 2018).

Besides curcuminoids, another subclass of polyphenols that may act as a prebiotic due to apparent limited absorption in the small intestine (Crozier et al., 2009) are the anthocyanins. These are parts of the flavonoids' family and are especially found in fruits with red, blue, and purple colors, such as grapes, blueberries, Brazilian berries (jaboticaba, açaí, and juçara), and pomegranate (Ramos et al., 2014; Rufino et al., 2010) (Fig. 1). Most common anthocyanins found in nature include cyanidin, delphinidin, malvidin, pelargonidin, peonidin, and petunidin (Fig. 6) (Ramos et al., 2014). Recent studies indicate that large amounts of these polyphenols may reach the large intestine and go through absorption and microbial fermentation, suggesting that the colon may be an important site of metabolization capable of improving the biological action of anthocyanins (Kamiloglu et al., 2015; Müller-Maatsch et al., 2016; Zhu et al., 2018).

Initial studies with animals indicate the potential of anthocyanins as future prebiotics for human health. According to Peng et al. (2019), the freeze-dried extract of black wolfberry (*Lycium ruthenicum*), a native fruit from Central Asia, is capable of increasing the health status of the microbiome of mice when given at a daily dosage of approximately 6 mg/animal for 12 weeks. Such a long-period intervention with anthocyanins is capable of increasing the concentration of total SCFAs in the feces from both cecum and colon and augmenting the abundance of *Coprobacter* and *Odoribacter*, genus inversely related to obesity and cardiovascular risk (Peng et al., 2020; Qian et al., 2018). In fact, anthocyanins may protect against the metabolic alterations derived from a high-fat regime. In a study by Wang et al. (2020a), for example, an anthocyanin-rich extract from black rice added to a high-fat and cholesterol diet for 12 weeks protected against increased body weight and accumulation of lipids in the serum on mice, probably because of its modulating effects in the microbiota, such as higher bacterial richness and diversity, and a decreased ratio of *Firmicutes/Bacteroidetes* (Wang et al., 2020a).

The *Firmicutes/Bacteroidetes* ratio has been associated with several gut diseases (Rinninella et al., 2019). For example, in a mice model of ulcerative colitis, this ratio was increased due to intestinal inflammation and regulated by the treatment with a monomer derived from the anthocyanin petunidin. Additionally, the short-period administration of this anthocyanin promoted antiinflammatory effects and improved epithelial barrier function (Peng et al., 2019). Finally, anthocyanins have been also associated with a better outcome in colorectal cancer, partially due to microbiota modulation from increases in probiotic *Eubacterium rectale* and *Faecalibacterium prausnitzii* bacteria species. A better microbiota profile is suggested to relate to improved epigenetic and DNA methylation process that may protect against colorectal cancer formation and progression (Lili et al., 2018).

Other specific polyphenols have been associated with microbiome health, such as caffeic acid from coffee, catechin and epicatechin from green tea, resveratrol from grapes and wine, and secoisolariciresinol from flaxseed products, for example (Wiciński et al., 2020). However, more studies are needed in order to confirm and deepen the discussion regarding (1) if polyphenols are entirely suitable as components of the prebiotic family and which classes should be considered as so; (2) the colonic metabolism and selective utilization of polyphenols by the microbiota; (3) the microbial changes that occur after the intake of polyphenols in healthy humans and its relation to the prevention of chronic diseases; and (4) the effects of the intake of polyphenols in clinical trials with chronic ill individuals, highlight the microbiota as a possible influencing factor.

Besides their most known health-related effects, polyphenols have been also implicated in technological and industrial usages, such as (1) enhancers of the organoleptic quality of foods, especially serving as natural coloring agents (most common: anthocyanins); (2) antimicrobial agents added to foods or packages to increase the shelf life of commercial products; (3) specific dietary supplements or products claiming to increase the health status of populations with or without chronic conditions (Baenas et al., 2018; Vinha et al., 2018); and (4) components of cosmetics capable of incorporating their antioxidant, antimicrobial, and antiinflammatory properties against ultraviolet damage and skin deterioration. Nowadays, it appears that the polyphenol utilization by the industry has been especially concentrated in the cosmetics field, with consumers expressing their will to purchase products with more natural and green composition, and with increased lasting efficacy. In addition to this, the increasing study and application of encapsulated polyphenols with improved bioaccessibility seems to be an important direction toward the future of cosmetology (Rodrigues et al., 2018).

RESISTANT STARCH

Resistant starch (RS) is a naturally occurring carbohydrate with a low molecular weight, linear in chemical structure, and formed by α-1,4-D-glucan usually derived from the retrograded amylose fraction. It represents the portion of the starch that resists digestion by enzymes of the gastrointestinal tract, mostly due to its compact molecular structure, but also by the formation of crystals during the cooling of starch gels previously heated (retrograded). By resisting digestion and absorption, RS can be fermentable by colonic bacteria, therefore indicating its prebiotic potential (Fuentes-Zaragoza et al., 2010; Zaman and Sarbini, 2016). However, proof of its selective utilization is still lacking.

As for its chemical structure and occurrence in food, RS can be divided into five types: RS1—natural starch physically protected in seed coats from whole-grain foods; RS2—natural and ungelatinized starches that resist digestion due to B-type crystallinity, usually found in amylose-rich starches (green banana) and tubers (potato); RS3—retrograded starch (cooked and cooled potatoes); RS4—experimentally or industrially modified starch by chemical agents (commercial breads and cakes); and RS5—mostly amylose-lipid complexes formed by food processing or created artificially. The last one is still in discussion if is a true RS or a slowly digested carbohydrate (DeMartino and Cockburn, 2020; Raigond et al., 2015).

Most rich food sources of RS in a dry weight matter are indicated as follows: green banana (*Musa acuminata*, variety Nanicão) flour at the first stage of maturation (49%) (Menezes et al., 2011), bean varieties raw (21%–37%) and boiled (2%–8%), rice varieties raw (12%–16%) and boiled (2%), and lentils raw or boiled (1.5%–5%) (Inan Eroglu and Buyuktuncer, 2017; Sagum and Arcot, 2000) (Fig. 1). The difference between the contents of RS between raw and cooked vegetables is related to the ability of heat to decompose such structures, forming a gelling aspect when in the presence of water, thus facilitating the acting of digestive enzymes. Some heating methods, however, may be preferable as they can maintain better the starch structure, avoiding a high loss of the structure of RS. These include boiling, steaming, and autoclave cooking (Inan Eroglu and Buyuktuncer, 2017). Also, the method of quantifying RS may influence on the final composition results, as McCleary (2007) found some important differences when comparing the classic "AOAC Official Method 2002.02 (RS)" with his proposed updated analysis, the "TDF/RS method" (McCleary, 2007).

Several clinical trials in the last years attested the potential of RS in different conditions, such as (1) a reducer of postprandial glucose and insulin responses (Stewart and Zimmer, 2018); (2) a factor capable of increasing fat oxidation in obese subjects when in combination with protein (Gentile et al., 2015); (3) an antiinflammatory and antioxidative product for renal patients (Esgalhado et al., 2018); (4) an ameliorative food for patients with diabetes (Costa et al., 2019); and (5) a microbiome-modulating and anticonstipation agent (Table 1), to be especially discussed next.

The capacity of RS to be fermented by intestinal bacteria and modulated the microbiome has been suggested. According to the metagenomic study of Vital et al. (2018), the natural form of RS, RS2, is largely degraded by the members of the *Firmicutes* phylum, mainly *Ruminococcus bromii, Faecalibacterium prausnitzii, Eubacterium rectale, and Roseburia* spp. Such bacteria are commonly known for their health-related effects, as they contribute to a butyrate-rich environment (Vital et al., 2018). Butyrate is a SCFA that has been mentioned in this chapter due to its preventive action toward inflammatory conditions.

In addition to the natural sources of RS, other types have implications in the human microbiome. For example, a recent study attested that different forms of RS (natural RS2 and manipulated RS4), given for 2 weeks at a dosage of 4–5 g, had similar positive effects in the microbiome of healthy Indian children (2–5 years old), as they significantly augmented the concentration of SCFAs (butyrate, acetate, and propionate) and decreased pH in feces, therefore indicating a better colonic fermentative capacity after the intake of RS. Additionally, from this study, the concentration of SCFAs in feces was also increased in stunted children, although less than in the healthy group. This situation may indicate a protective mechanism in the intestine of children with impaired growth, as SCFA absorption by the intestinal epithelium could have been increased as a protective mechanism, instead of largely excreted through feces (Balamurugan et al., 2019) (Table 1). The idea, however, that less SCFA in feces could relate to an increased absorption by intestine is not well defined, and in fact, this could also be associated with a disease outcome that limits the formation of such molecules.

RS appears to be interesting to children, as Cassettari et al. (2019) also attested the effects of the supplementation of 30 g of green banana biomass, source of RS (8% of fresh weight), for 8 weeks in constipated 5- to 15-year-old individuals. From this study, the authors found an improvement in straining and pain in defecation, in abdominal pain, and in the score of stool classification. These effects of the banana biomass were similar to those of the laxative drugs sodium picosulfate and polyethylene glycol 3350. However, when banana was combined with such drugs, the effects were amplified, as the interventions were also capable of increasing the defecation frequency for more than three bowel movements for a week and reducing the fecal incontinence. This study demonstrates that the combination of common laxative drugs with natural rich sources of RS may be an alternative to be considered for the treatment of functional constipation (Cassettari et al., 2019) (Table 1). More studies are needed in order to understand the relation between the constipation mechanisms with the profile of the microbiota before and after an intervention with RS or its food sources.

RS has also been used by the industry due to its unique and/or interesting properties. Evidence has shown that RS is able to improve the quality of foods, for example, by increasing crispiness, reducing oil in fried products, providing better mouthfeel, functioning as an important added material in microencapsulation and being included as a supplement in functional foods. The use of RS as a microencapsulation tool and as a supplement may be possible, since, in comparison with most parts of dietary fibers and some prebiotics, RS possesses more manageable characteristics, such as (1) fine particle size, (2) white color, (3) bland flavor, (4) lower water holding capacity, (5) higher water binding capacity, and (6) good extrusion qualities (Homayouni et al., 2014). Some of the foods that may benefit rheologically and nutritionally from the addition of RS are cheese, bread, pasta products, and probiotic particles (Ferrão et al., 2016; Homayouni et al., 2014).

Especially, the utilization of RS as a component of the probiotic microcapsule has been shown to be useful in improving the characteristics and function of such products. For example, in two studies by de Araújo Etchepare et al. (2016) and Etchepare et al. (2016), the addition of 1% RS to moist microcapsules of calcium alginate (1%) or calcium alginate (1%) plus chitosan (0.4%), containing *Lactobacillus acidophilus*, avoided the loss of viable microorganisms during long-period storage, in addition to providing thermal stability and protecting the bacteria against stimulated gastrointestinal stress (de Araújo Etchepare et al., 2016; Etchepare et al., 2016). Also, other studies using RS and calcium alginate microcapsules containing other probiotic bacteria (*Bifidobacterium bifidum*, *Lactobacillus casei*, and/or *Bifidobacterium lactis*) applied to cheese (Mirzaei et al., 2012), yogurt (Heydari et al., 2011; Kailasapathy, 2006), cake (Zanjani, 2012), ice cream (Homayouni et al., 2008), or mayonnaise (Maryam et al., 2013), showed the positive influence of such carbohydrate in improving the survival of microorganisms after extended storage. Overall, such results indicate the potential of RS to increase the shelf life of microencapsulated probiotics (isolated or applied to food products) and to provide an improved liberation and viability of bacteria during its gastrointestinal passage. As for the capacity of RS-added microcapsules to improve the sensorial properties of foods, some shreds of evidence have shown little or no influence on such aspect (Cruz et al., 2010; Heydari et al., 2018); however, more studies are needed.

SUGAR ALCOHOLS

Sugar alcohols (SAs) or polyols are hydrogenated carbohydrates that structurally resemble both sugar and alcohol, since they contain an alcohol group in the aldose and ketose portions of mono-, di-, and polysaccharides. They are mainly represented by xylitol, sorbitol, and mannitol, which are monosaccharides, but also lactitol, an example of disaccharide, and starch hydrolysates (Fig. 7) (Livesey, 2003). These carbohydrates are widely produced by the industry to work as sugar replacers, but also as flavor enhancers; stabilizers; emulsifiers; thickeners; sequestrants; and anticaking, bulking, and glazing agents (Grembecka, 2015). SAs can be found in several commercial high-sugar products, such as candies and gums, chocolate, ice cream, cookies, and cakes (Lenhart and Chey, 2017). Naturally, they are present in small amounts (0.2%–3.1% of fresh weight) in blackberry, black currant, and pear—sorbitol; cauliflower, celery bulb and leaves, and mushroom bottom—mannitol; and dandelion bulb, strawberry, and raspberry—xylitol (Jovanovic-Malinovska et al., 2014) (Fig. 1).

Most recently, SAs have been considered as emerging prebiotics (Míguez et al., 2016). They are known to be low-digestible carbohydrates, usually poorly absorbed in the small intestine and highly fermented by the colonic microbiota. Especially, in comparison with sorbitol, mannitol, and maltitol, lactitol appears to be the most fermented SA by the intestinal microbiota of healthy individuals, based on the breath hydrogen excretion (Wursch et al., 1989). By being fermented, SAs may act by altering the bowel function and modulating the microbiota environment, therefore potentially leading to preventive and healing health effects (Livesey, 2003; Míguez et al., 2016). However, so far, there is still inconclusive or little evidence regarding the selective microbial utilization of sugar alcohols.

Studies from the late 1990s and beginning of 2000 showed that the consumption of high amounts of SAs (20 g or more) can increase some unwanted gastrointestinal symptoms, such as abdominal pain,

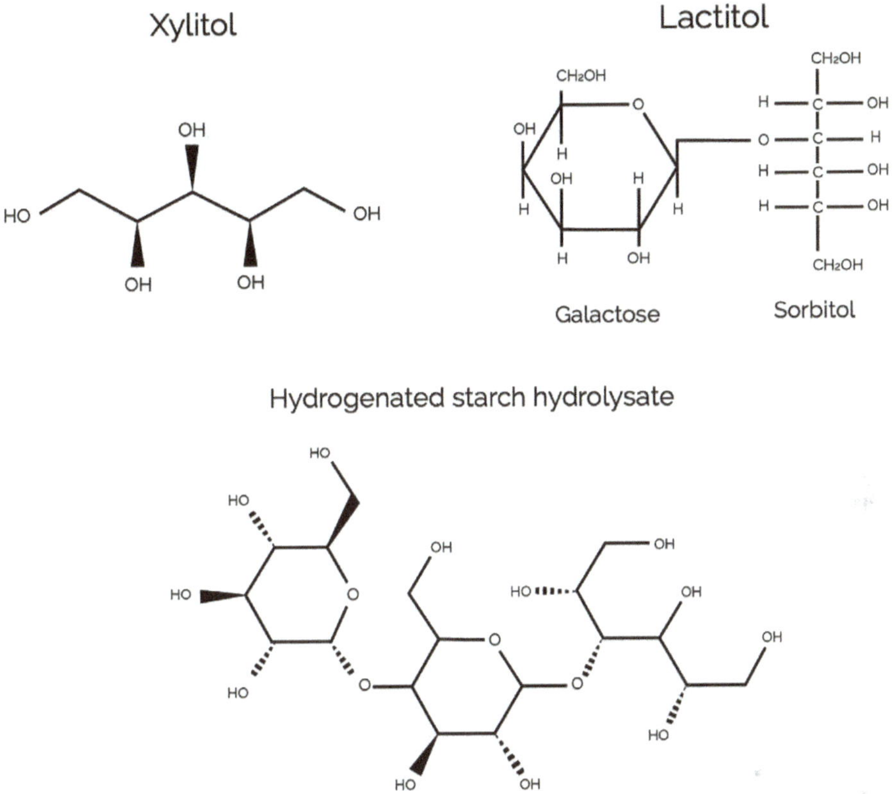

FIG. 7 Chemical structures of sugar alcohols.

excessive flatulence, and diarrhea, due to malabsorption that leads to increased osmosis and water retention (Hyams, 1983; Lenhart and Chey, 2017; Patil et al., 1987). Although there is evidence that the consumption of 20–40 g/day of lactitol or isomalt can produce important intestinal symptoms (Koutsou et al., 1996; Lee and Storey, 1999), it appears that such disaccharides may be more tolerable than the monosaccharides sorbitol and mannitol, for example, which usually produce higher laxative effects (Lenhart and Chey, 2017; Patil et al., 1987; Zumbé and Brinkworth, 1992). Additionally, erythritol, by being highly absorbed in the small intestine, can lead to fewer intestinal symptoms (Oku and Okazaki, 1996), but also has less potential to be fermented and produce modulatory microbiota effects. The potential of SAs to soft the stools may benefit chronic constipated individuals, as found by a recent clinical trial using 10 g/day of lactitol (Cheng et al., 2019) (Table 1). The knowledge of the human gastrointestinal tolerance for SAs may be valuable to researches that wish to test the effects of products in safe clinical trials.

The microbiota-modulatory effects of SAs, such as sorbitol, mannitol, isomalt, lactitol, and mannitol have been described by the literature (Míguez et al., 2016). Especially, the most numerous studies, especially clinical trials, are the ones with lactitol. The short or long intake of 10 g/day of such carbohydrate has been shown to be successful in increasing the abundance of beneficial *Bifidobacterium* and the concentration of SCFAs (Finney et al., 2007) (Table 1), in addition to promoting increased defecation frequency and relief in constipated individuals (Cheng et al., 2019). Interestingly, in individuals with chronic viral hepatitis, lactitol promoted decreased levels of endotoxin in plasma, which was theorized by the authors to be closely related to microbiota modulation through increases in *Bifidobacterium* and *Lactobacillus* and decreased in *Clostridium perfringens* (Chen et al., 2007; Table 1). *Clostridium perfringens* is a well-known toxin-producing bacterium associated with several infectious bowel diseases (Uzal et al., 2014); therefore, its blockage by a functional food, like lactitol, could represent an interesting nutritional approach to such conditions.

XYLOOLIGOSACCHARIDES

Xylooligosaccharides (XOSs) are short soluble sugars made of 2–10 xylose monomeric units, such as xylobiose and xylotriose, connected through β-(1,4) linkages (Fig. 8). Structurally, they also possess linked side groups of sugar acids, like acetyl groups or arabinofuranosyl residues, which may differentially interfere in the biological action of such prebiotics. Naturally, XOSs can be mostly found in bamboo shoots, milk, honey, and fruits (Fig. 1); however, characterization studies are still lacking (Aachary and Prapulla, 2011; Vázquez et al., 2000).

For industrial, technological, or experimental purposes, XOSs are usually obtained by chemical and/or enzymatic hydrolysis of hemicellulose from lignocellulosic materials, such as bagasses, corn cobs, hardwoods, hulls, malt cakes, and straws. Further, the hydrolytic degradation of the hemicellulose arabinoxylan, commonly found in cereal grains, also generates XOSs, in addition to manufacturing arabinoxylan-oligosaccharides, another recently considered type of emerging prebiotic (Broekaert et al., 2011; Saville and Saville, 2018).

In the last years, the utilization of agricultural waste to produce XOSs has been shown to be interesting when thinking in an economic and ecological perspective. By-products are challenged for the food and bioenergy industry as they represent a significant portion of the initial product, but are usually discarded in the process, impacting mostly the environment. Such materials, including the plant biomass, are rich sources of hemicellulose containing xylan, therefore, can be treated to produce an interesting food ingredient that may increase the quality of commercial products, and promote gastrointestinal and overall health when consumed (Jain et al., 2015).

A recent study by Ferrão et al. (2018) showed that the addition of XOSs to a popular commercial Brazilian cheese, *requeijão cremoso*, had a major impact in several aspects of the product. The carbohydrate was able to improve the rheological and physicochemical properties of the cheese by increasing its apparent viscosity, elasticity, and firmness, in addition to reducing the bitter taste of the product (Ferrão et al., 2018). In a more recent study, Dysvik et al. (2020) also found an interesting outcome from the addition of XOSs, this time in the production of sour beer. From this study, wood-derived XOSs provided a secondary fermentation mechanic from *Lactobacillus brevis* by reducing the fermentation time from 1 to 3 years to only 4 weeks, therefore suggesting a new approach to successfully producing this type of beer (Dysvik et al., 2020). A better technological and producing quality associated with the addition of a prebiotic with potential prohealth function may favor the yet-to-come growing recognition of XOSs as a food ingredient in different vehicles.

In addition to being a potential high-value product from discarded agroindustrial residues, XOSs have been showing promise results as a prebiotic in clinical trials. By being highly fermented by a selective microbiota, XOSs can increase the abundance of *Bifidobacterium* and *Lactobacillus,* in addition to elevating the levels of SCFAs in feces (Aachary and Prapulla, 2011; Childs et al., 2014; Finegold et al., 2014; Lin et al., 2016) (Table 1). Especially, when, in combination with inulin, the effects are extended beyond the usual prebiotic effects, as shown by the randomized, placebo-controlled, and double-blind study of Lecerf et al. (2012). From this study, the authors found that the consumption of 3 g of inulin plus 1 g of XOS/day for 4 weeks significantly reduced the levels in the plasma of circulating lipopolysaccharides and the gene expression

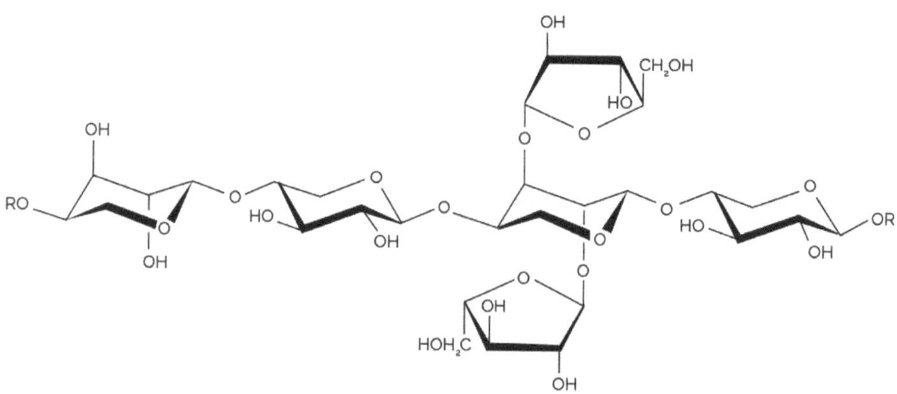

FIG. 8 Chemical structure of xylooligosaccharides.

of interleukins 1β and 13 in blood (Lecerf et al., 2012) (Table 1).

Some inflammatory markers, like lipopolysaccharides and proinflammatory interleukins, are closely related to chronic intestinal inflammatory diseases, especially ulcerative colitis. Interventions capable of decreasing such parameters in healthy individuals may indicate a possible target in chronic ill patients that possess high amounts of proinflammatory substances in the blood and/or colon (Guo et al., 2015; Ungaro et al., 2017; Vojdani et al., 2020). In fact, studies have shown that XOS, alone or in combination with a probiotic, given preventively, can alleviate against the poor outcome of chemically induced IBD-like colitis in mice or rat. Besides providing a reduction in inflammatory molecules, such as tumor necrosis factor alpha, interleukin 1β, and myeloperoxidase, XOSs can also fortify the epithelial barrier integrity and maintain the crypt cell integrity, therefore contributing to a healthier bowel structure (Grisham et al., 1996; Sheng et al., 2020). Although these studies did not perform any analysis regarding alterations in the microbiome, XOSs probably acted in an antiinflammatory manner by means of microbiota modulation, increased concentration of SCFAs, and/or improved bowel function.

The beneficial health and/or microbiota-related effects of XOSs have also been described recently in preclinical studies with obesity (Fei et al., 2020), diabetes (Khat-udomkiri et al., 2020), chronic kidney disease (Yang et al., 2018), and functional constipation (Fu et al., 2019). In comparison with other emerging prebiotics mentioned in this chapter, XOS appears to be one of the most successful ones attested by preclinical studies and clinical trials.

OTHER POTENTIAL EMERGING PREBIOTICS

Other new potential prebiotics mentioned by the literature include arabinoxylan-oligosaccharides, cyclodextrins, gentiooligosaccharides, glucomannan, glucooligosaccharides, human milk oligosaccharides, lactosucrose, *N*-acetylchitooligosaccharides, oligodextrans, polyunsaturated fatty acids, and soybean oligosaccharides. Glucomannan and soybean oligosaccharides will be especially mentioned next, as they are some of the most common studied ones regarding their prebiotic-related roles.

Glucomannan is a polysaccharide of 1,4-D-glucose and D-mannose (Fig. 9), uniquely found in the Japanese konjac plant *Amorphophallus konjac* (Fig. 1), which possess up to 60% (tuber) or 80% (flour) of the component in a dry weight basis (BeMiller, 2019). Besides being a fermentable product in the colon capable of providing beneficial effects in the microbiome, glucomannan can also act as a soluble fiber (Korczak et al., 2017). Studies in vitro, with animals and humans, are increasingly showing the beneficial effects of the consumption of konjac flour or hydrolyzed glucomannan in diverse situations, such as bowel function and microbiome modulation, diabetes, obesity, and inflammation. Especially, glucomannan is believed to form a protective cover in the intestinal epithelium (Devaraj et al., 2019), which could possibly provide a prevention mechanism against defective intestinal barrier-related diseases. Additionally, konjac glucomannan has been related to interesting technological usages, with drug delivery being probably the most studied. Due to its linear structure, ability to form gels, and high biodegradability, this carbohydrate has been used to target colonic delivery systems (Devaraj et al., 2019). It is important to notice, however, that konjac glucomannan, in addition to psyllium and guar gum, may produce esophagus obstruction when in large doses or if consumed for a long time, due to their high capacity to absorb water and great expansion rates. Therefore, individuals with esophageal pathology should probably avoid such products, especially the isolated commercial forms (Henry et al., 1986; Vanderbeek et al., 2007).

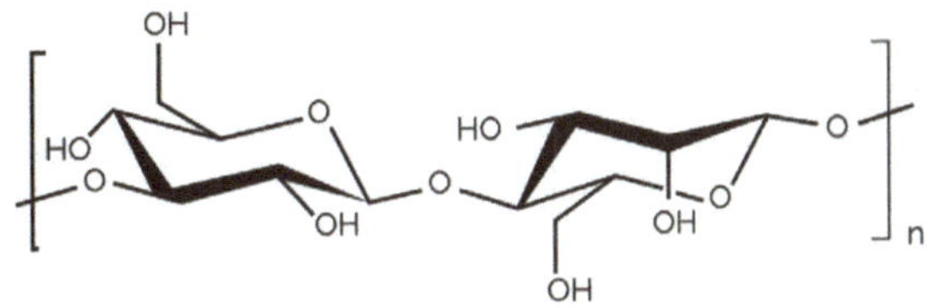

FIG. 9 Chemical structure of konjac glucomannan.

Soybean oligosaccharides are regarded as another group that has been identified as capable of modulating the microbiota (Fig. 1). Despite being rich in a series of interesting substances with health-associated effects, such as protein, lectins, saponins, trypsin inhibitors, genistein, and isoflavones (Csáky and Fekete, 2004), soybean also possess great amounts of oligosaccharides, mostly raffinose, stachyose, and verbascose. By not being digested, these reach the colon intact and are fermented by the microbiota (Guarino et al., 2020). Some of the positive effects of these substances have been shown by preclinical studies. For example, in a study by Pérez-López et al. (2016), native or treated Okara, a by-product from the processing of soybean, was included in batch cultures with inoculated human fecal slurries, which simulate the human gut.

TABLE 1
Clinical trials on the effects of emerging prebiotics in the bowel function and/or intestinal microbiome.

Prebiotic form, composition[a]	Dosage, period, administration	Subjects	Main results	Reference[b]	Observations
Algae[c]					
Lyophilized *Chlamydomonas reinhardtii* biomass, 40% carbohydrate and 5.9% fiber	1 or 3 g/day for 30 days	51 healthy adults and seniors with normal gut function or poor gut health	Overall positive results. Reduced bowel discomfort, diarrhea, bloating/gas, and increased defecation frequency and quality in individuals with poor gut health. No effects in the composition of the microbiota from the feces. No differences between the dosages	Fields et al. (2020)	No detailed analysis composition of the algae was performed. Not a blind trial
Water-extracted *Laminaria japonica* in an oval-shaped green tablet with 625 mg of the algae	1.25 g of algae (2 capsules)/day for 30 days	40 healthy adults and seniors	No effects in the composition of the microbiota or in bowel function. The microbiota-modulatory effects were achieved when subjects consumed the algae with a probiotic mix	Ko et al. (2014)	Algae composition analysis was not performed. Randomized, double-blind, placebo-controlled study
Bacterial exopolysaccharides[d]					
Sachets containing *Streptococcus thermophilus*-producing EPSs	1 billion viable cells/day for 30 days	25 healthy adults	Overall positive results. EPS concentration in feces was higher in the treated group after 30 days and even after 15 days of washout. EPSs provided improved colonic permeability	Del Piano et al. (2014)	Randomized, double-blind, placebo-controlled, pilot trial
Isomaltooligosaccharides					
Liquid base sachets, 56.1% IMOSs	30 g, divided into 2 doses a day, dissolved in warm water, for 28 days	20 elderly chronic constipated hemodialysis patients	Overall positive results. IMO was well tolerated and resulted in reduced constipation score, increased bowel frequency, and improved intestinal symptoms, mostly during the treatment and not after washout	Wang et al. (2001)	Not a randomized blind trial
IMOS supplementation, 42% of active component	Gradually increased from 8 g to 24 g/day during the first 10 days, for a total of 30 days of intervention. During the afternoon dessert	7 constipated elderly men	Overall positive results. Increased defecation frequency, wet stool output, and dry stool weight. Higher feces levels of acetate, propionate, and total SCFAs	Chen et al. (2001)	Not a randomized blind trial

IMOSs, 95%	Phase 1: 5 g after lunch and 5 g after dinner for 7 days. Phase 2: 2.5, 5.0, 7.5, or 10 g, divided between after lunch and after dinner for 7 days	Healthy adults Phase 1: 64 Phase 2: 136	IMO was not bifidogenic and had no effects in digestive symptoms in both phases. No differences were found between the dosages of Phase 2 in any aspect	Bouhnik et al. (2004)	Double-blind, randomized, placebo-controlled study
IMOSs, 45% of active component	Gradually increased from 11 to 22 g/day during the first 7 days, for a total of 8 weeks of intervention. Dissolved in drinking water as an afternoon drink within the 7-day cycle menu	13 chronic constipated elderly men (5) and women (8)	Overall positive results. Increased abundance in feces of *Bifidobacterium, Lactobacillus,* and *Bacteroides* and decreased abundance of *Clostridium.* Higher frequency of spontaneous defecation, wet fecal mass, and dry fecal mass. Higher concentration in feces of acetate, propionate, *n*-butyrate, and total SCFAs	Yen et al. (2011)	Double-blind, placebo-controlled, diet-controlled study
Pectin and pectic oligosaccharides[e]					
Formula with 0.2 g/100 mL of pectin-derived acidic oligosaccharides	Free, according to demand	82 healthy, full-term infants, ages of 1 week to 3 months, partially breast-feeding	Overall little positive results. Higher proportion of *Bifidobacterium.* No differences in other populations and in stool characteristics	Magne et al. (2008)	Double-blind, placebo-controlled, randomized trial
Pectin-based supplement	4 g/day for 14 days	6 adults and seniors with short bowel syndrome having jejunocolonic anastomosis	Overall little positive results. 80% of the pectin was fermented. Pectin increased the concentration of propionate, but not butyrate, acetate, or total SCFAs. Pectin resulted in nonsignificant or delayed prolongation of the gastric emptying time and orocolonic bowel transit time	Atia et al. (2011)	Not a randomized blind trial
Pectin, 50% weight/weight	20 g/day for 5 days following fecal microbiota transplantation	20 adults and seniors with active ulcerative colitis that went through fecal microbiota transplantation	Overall positive results. Clinical score was reduced after 4 and 12 weeks from the treatment. In addition, pectin helped delay the loss of bacterial diversity, decreased the Shiga toxin-producing *Escherichia coli,* and modulated the microbiota composition	Wei et al. (2016)	Randomized controlled trial, not blind

Continued

TABLE 1
Clinical trials on the effects of emerging prebiotics in the bowel function and/or intestinal microbiome—cont'd

Prebiotic form, composition	Dosage, period, administration	Subjects	Main results	Reference	Observations
Pectin supplement	8 g/day, after fecal microbiota transplantation. Period: not well defined, but long (up to 1 year)	29 adults and seniors (5 men, 24 women) with functional constipation and slow transit	Overall positive results. Increased complete spontaneous bowel movements per week and improved symptoms of constipation, with a cure rate of 69.0% after 4 weeks and 48.3% after 1 year	Zhang et al. (2018a)	Not a randomized controlled blind trial. Since this is not a controlled trial, the results found should not be attributed in totality to pectin
Sachets with dry powder GENU BETA high ester pectin, extracted from sugar beet pulp, with a degree of acetation of the homogalacturonan backbone of the pectin of 18%–26%, molecular weight > 60,000 Da	15 g, divided into 2 doses for a day for 4 weeks before breakfast and dinner, and mixed with flavored syrup and tap water	52 healthy young adults and 48 healthy elderly	No differences were found in the composition of the microbiota or the concentration of SCFAs in both populations	An et al. (2019)	Randomized, double-blind, placebo-controlled trial
Polyphenols[e]					
Hard-gelatin capsules with 450 mg of pomegranate extract, rich in punicalagins, ellagic acid and gallagic acid dilactone, and source of catequins	1 (450 mg extract, 164 mg phenolics) or 4 capsules (1.8 g, 656 mg phenolics)/day for 3 weeks	50 overweight-obese adults, BMI > 27 kg/m^2, over 40 years old	Overall positive results. The 1.8 g dosage decreased the levels of plasma lipopolysaccharide-binding protein and had a tendency to decrease C-reactive protein; increased *Bacteroides, Faecalibacterium, Butyricicoccus, Odoribacter,* and *Butyricimonas;* and decreased proinflammatory microorganisms *Parvimonas, Methanobrevibacter,* and *Methanosphaera*	González-Sarrías et al. (2018)	Randomized, double-blind, placebo-controlled study

Turmeric tablets or curcumin tablets with 1 g of active component plus 1.25 mg of black pepper extract	6 tablets (6000 mg) with food, divided into 2 meals/day, for 8 weeks	30 healthy adults	Overall inconclusive results. Treatment with curcumin highly increased the number of species and tended to increase alpha diversity ($P = -.08$). Microbial composition was highly personalized due to treatments. Commonly, there was increased *Clostridium* spp., *Bacteroides* spp., *Citrobacter* spp., *Cronobacter* spp., *Enterobacter* spp., *Enterococcus* spp., *Klebsiella* spp., *Parabacteroides* spp., and *Pseudomonas* spp., and reduced *Blautia* spp. and *Ruminococcus* spp.	Peterson et al. (2018)	Randomized, double-blind, placebo-controlled study
Olive pomace-enriched product PreBiÒ, composed of dehydrated food-grade olive powder, chestnut, peas, buckwheat flour, extra-virgin olive oil, salt, and sugar; 0.4% of total biophenols and 0.017% of tyrosol polyphenols	90 g/day for 8 weeks	73 mildly hypercholesterolemic adults	Overall inconclusive results. No differences in microbial diversity. Reduced *Lactobacilli* and *Ruminococcus* abundance. Tendency to increase *Bifidobacteria* abundance	Conterno et al. (2019)	Randomized, double-blind, placebo-controlled study. This study utilized an intervention with balanced amounts of fiber, carbohydrate, fat, and protein; therefore, the effects found may not be attributed only to the polyphenols
Montmorency cherry supplement	60 mL (296 mg of total anthocyanins and 1040 mg of total polyphenols)/day, divided into morning and evening, for 4 weeks	28 healthy middle-aged adults	No effects were found in microbiota richness and composition, or glycemic and inflammatory parameters	Lear et al. (2019)	Randomized, double-blind, placebo-controlled study
Aronia berry: (1) extract in capsule (75 g), 116 (HPLC) or 200 mg of total phenolics; (2) the whole fruit powder (10 g), 12 or 20 mg of total polyphenols	1 capsule or the whole fruit powder for 2 h (acute) and/or 12 weeks (chronic, every day)	66 healthy adult men	Overall positive results. Improvement in vascular function for two treatments in both acute and chronic interventions. No changes in gut microbiota diversity. The extract increased *Anaerostipes,* whereas the whole fruit, *Bacteroides*	Istas et al. (2019)	Randomized, double-blind, placebo-controlled study

Continued

TABLE 1
Clinical trials on the effects of emerging prebiotics in the bowel function and/or intestinal microbiome—cont'd

Prebiotic form, composition	Dosage, period, administration	Subjects	Main results	Reference	Observations
Resistant starch[e]					
Type 3 resistant starch within a controlled diet	25.4 g/day for 3 weeks	14 adult and senior men with metabolic syndrome and obesity (body mass index range: 27.9–51.3)	Overall negative results. Reduced microbiota diversity, richness, and evenness. Increased *Oscillospira guillermondii, Ruminococcus bromii, Sporobacter termitis, Clostridium leptum, Alistipes* spp., and *Clostridium cellulosi.* Decreased *Papillibacter cinnamivorans* and *Eubacterium rectale.* Decreased concentration of acetate, propionate, butyrate, and succinate. Increased concentration of isobutyrate and isovalerate	Salonen et al. (2014)	Randomized study, not double-blind or placebo-controlled
MSPrebiotic: unmodified RS derived from potatoes (*Solanum tuberosum*), 70% RS type 2, 20% amylose, and 80% amylopectin	30 g/day, 2 h before medication (if applicable), added to food or liquid not heated, for 12 weeks	84 healthy middle-aged adults or seniors	Overall positive results. Increased bacterial diversity and higher *Bifidobacterium* abundance for both adult and senior groups. Increased abundance of *Ruminococcus bromii* and decreased abundance of *Ruminococcus obeum* and *Ruminococcus torques* in the adults' group. Increased relative proportion of butyrate, reduced abundance of *Proteobacteria* and decreased use of stool softeners in the seniors' group	Alfa et al. (2018)	Randomized, placebo-controlled, double-blind study
Green banana biomass (*Musa* spp. AAA) from unripe bananas, 8% RS	30 g/day, mixed with the child's favorite foods, for 8 weeks	80 children and adolescents with functional constipation	Overall positive results. Improvement in the stool score and reduced straining and pain in defecation and abdominal pain. The combination of the biomass with laxative drugs had extended benefits: increased defecation frequency and fecal incontinence	Cassettari et al. (2019)	Randomized study, not placebo-controlled or double-blind

Biscuits with high amylose maize starch (42% RS2) or high amylose maize starch chemically acylated (50% RS4)	Sequential feeding: 10 g of RS2 for 2 weeks and 10 g of RS4 for 2 weeks. 2-week washout between the 2 feedings	20 healthy or stunted children (2–5 years old)	Overall positive results. RS2: decreased pH and increased levels of acetate, butyrate, and propionate in both healthy and stunted children. RS4: decreased pH, increased the levels of acetate and propionate in both healthy and stunted children, and increased the levels of butyrate in healthy children. Healthy children showed higher levels of acetate and butyrate in comparison with stunted when fed either type of RS	Balamurugan et al. (2019)	Not a randomized, placebo-controlled, double-blind study
Maize RS4 (VERSAFIBE 2470) or Potato RS4 (VERSAFIBE 1490), or tapioca RS4 (VERSAFIBE 3490)	Gradually week-increasing doses: 10, 20, 35, and 50 g/day, for 4 weeks. Doses were divided into 2–3 servings per day and incorporated in drinks or foods not heated	40 healthy adults	Overall positive results. Maize RS4: fermented by the colon; increased *Eubacterium rectale* abundance and concentration of butyrate in a dose-dependent manner and with the limit of 35 g/day. Tapioca RS4: fermented by the colon; increased *Parabacteroides distasonis* abundance and concentration of propionate in a dose-dependent manner. Potato RS4: not fermented by the colon, no effects in the microbiome	Deehan et al. (2020)	Randomized, placebo-controlled, double-blind study
Sugar alcohols[e]					
Isomalt	30 g/day added to sweet foods within a 7-day cycle menu, for 4 weeks	19 healthy adults (12 women and 7 men)	Overall inconclusive results. Increased counts or proportion of *Bifidobacteria, Actinobacteria,* and *Atopobium*. Reduced *Prevotella* and *Roseburia intestinalis*	Gostner et al. (2006)	Double-blind, placebo-controlled, diet-controlled study
Lactitol	15–45 g, divided into 3 times a day, after meals, for 3 weeks	60 adults and seniors (53 men, 7 women) with chronic viral hepatitis	Overall positive results. Increased *Bifidobacteria* and *Lactobacillus* and decreased *Clostridium perfringens* counts. Reduction in plasma endotoxin levels	Chen et al. (2007)	Not a randomized blind trial. This study did not establish a single dosage system for patients

Continued

TABLE 1
Clinical trials on the effects of emerging prebiotics in the bowel function and/or intestinal microbiome—cont'd

Prebiotic form, composition	Dosage, period, administration	Subjects	Main results	Reference	Observations
Lactitol in the form of chocolate bars	5 or 10 g/day, first half mid-morning and second mid-afternoon, for 7 days	75 healthy young adults	Overall positive results. Increased abundance of *Bifidobacterium* spp., reduced fecal pH, increased concentration of butyric and propionic acids and higher score of total symptom in individuals receiving 10 g. The 5 g dose had no effects	Finney et al. (2007)	Randomized, placebo-controlled, double-blind study
Milk chocolate bar with 22.8 g of mannitol	50 g/day of chocolate (22.8 g of mannitol), for 2 weeks, followed by 75 g (34.2 g of mannitol) for more 2 weeks, and 100 g (45.6 g of mannitol) for more 2 weeks	40 healthy adults (27 women and 13 men)	Overall positive results. Increased abundance of *Bifidobacteria, Bacteroides, Clostridium, Lactobacilli, Eubacteria, Atopobium, Fusobacterium prausnitzii, Ruminococcus flavefaciens,* and *Ruminococcus bromii*. Increased concentration of acetate, propionate, and butyrate	Beards et al. (2010)	Randomized, placebo-controlled, double-blind study
Powder sachets with 10 g of lactitol	10 g/day, mixed in 250 mL of water, preferably during breakfast, for 4 weeks	172 healthy adults with constipation	Overall positive results. Increased bowel movement frequency and defecation relief	Cheng et al. (2019)	Randomized, double-blind, placebo-controlled study
Xylooligosaccharides					
XOS-enriched compound derived from wheat arabinoxylans (Opti'flor; DF3 SAS), 80% of XOS content	6.64 g (5.3 g of XOSs), divided into 2 doses per day, 1 before breakfast and the other before the evening meal, dissolved in pure orange juice, for 4 weeks	59 healthy young adult students	Overall positive results. Increased levels of *Bifidobacterium* and a trend toward an increase in *Bacteroidetes* ($P = .064$). Increased concentration of acetic acid and in the profile of SCFAs. The combination of XOS and inulin had extended beneficial effects regarding inflammation and immunomodulation	Lecerf et al. (2012)	Randomized, placebo-controlled, double-blind study. Although this study possesses a control group, the use of orange juice may have interfered in the results found

XOS supplement in the form of powder sachets	8 g/day, dissolved in water, milk, or fruit juice, for 21 days	41 healthy adults	Overall positive results. Increased levels of *Bifidobacterium*. Increased number of bowel movements and higher vitality and happiness scores. Increased levels of HDL-cholesterol and immunomodulation	Childs et al. (2014)	Randomized, placebo-controlled, double-blind study
XOS supplement capsule, white or off-white crystalline, containing 12.5% of XOS	1.4 or 2.8 g (175 or 300 mg of XOS)/ day, for 8 weeks	32 healthy adults (21 women and 11 men)	Overall positive results. Increased abundance of *Bifidobacterium,* especially in the group with a higher dosage, and *Bacteroides fragilis* in the group taking 2.8 g/ day. No effects in *Lactobacillus* count	Finegold et al. (2014)	Randomized, placebo-controlled, double-blind study
XOS supplement, 70% XOS	2.8 g (2 g of XOS) daily for 8 weeks	16 healthy and 13 prediabetic adults with impaired fasting glucose and/ or glycosylated hemoglobin	Overall positive results. XOS decreased the abundance of *Howardella, Enterorhabdus,* and *Slackia* in both healthy and prediabetic, and decreased the abundance of *Blautia hydrogenotrophica* in prediabetic subjects. XOS reversed part of the profile of the microbiota of prediabetic subjects	Yang et al. (2015)	Randomized, placebo-controlled, double-blind study
Rice porridge containing 0.8% of XOS	150 g/day of rice porridge (1.2 g of XOS), during breakfast, for 6 weeks	20 healthy young adults	Overall positive results. Higher *Lactobacillus* spp. counts at weeks 4, 6, and washout. Higher *Bifidobacterium* counts at week 6. Lower *Clostridium perfringens* counts at week 6 and washout	Lin et al. (2016)	Randomized, placebo-controlled study, not blind

Abbreviations: *EPSs*, bacterial exopolysaccharides; *IMOSs*, isomaltooligosaccharides; *RS*, resistant starch; *SCFAs*, short-chain fatty acids; *XOSs*, xylooligosaccharides.

[a]Clinical trials using more than one potential prebiotic food or compound (e.g., mix of polyphenols, probiotic, and/or dietary fibers) were not included in this table, as the results found cannot be attributed to a single intervention component.

[b]Articles only in the English language and a limit of 5 per prebiotic were included.

[c]There are no studies with isolated carbohydrates from algae; therefore, the ones with the whole algae are presented in the table.

[d]Although there are some other studies attesting the health effects of lactic acid bacteria, it appears that there is only one study that related and analyzed the production of EPSs from such bacteria as a contributing factor, so that article was the only one considered for EPSs.

[e]When possible, the available clinical trials published in the latest 20 years for the main emerging prebiotics cited in this chapter are mentioned in this table, except for pectin, polyphenols, resistant starch, and sugar alcohols, which possess more numerous articles; therefore, the most interesting recent ones were presented.

The different forms of Okara, the native being especially rich in insoluble fibers, and the treated one, in soluble fibers, including oligosaccharides, had positive effects in the in vitro experiments. The treatments increased the levels of acetic, propionic, butyric, and lactic acids to concentrations higher than the control group and even than the group that received FOSs. In addition, Okara significantly increased the abundance of bifidobacteria, lactobacilli, and *Bacteroides* spp. (Pérez-López et al., 2016). Finally, a more recent study reported that mice receiving different doses of commercial soybean oligosaccharides for 22 days showed microbiota and immune-modulating effects. Especially, from this study, the intragastric high dosage of 4 g/kg of body weight increased the proliferation of bifidobacteria and decreased *Clostridium perfringens* (Ma et al., 2017). Such initial studies with soybean oligosaccharides indicate their potential to be future tested in clinical trials with individuals with dysbiosis and related immune disorders.

In conclusion, although some of the compounds mentioned in this chapter may be fermented by the microbiota and alter positively the overall health, not all of them fit the criteria to be a prebiotic (Gibson et al., 2017). Especially, the "selectively" criteria (the property of affecting or been utilized by some things—beneficial bacteria—and not others or all of them—overall bacteria population) appear to be the most difficult one to define, as in vitro studies with emerging prebiotics are lacking or can be sometimes very limited to a few species of bacteria. On the other side, studies with animals and humans that perform microbiota analyses may not identify all the microbiome changes or nuances that occur due to a prebiotic intervention. Some authors seem to question this criterion, as it is still to this day too subjective (Bindels et al., 2015; Hutkins et al., 2016). Finally, it appears that, despite some issues regarding concepts, emerging prebiotics are extremely relevant and more studies are needed to prove their diverse efficacy, as in the coming future they may represent potential tools capable of indirectly or directly increase the day-to-day facilities, health, and well-being of individuals.

ACKNOWLEDGMENTS

This chapter was supported by the Coordenação de Aperfeiçoamento de Pessoal de Nível Superior—Brasil (CAPES)—Finance Code 001; Conselho Nacional de Desenvolvimento Científico e Tecnológico (CNPq) (403328/2016-0; 301496/2019-6; 140812/2019-9); and Fundação de Amparo à Pesquisa do Estado de Paulo (FAPESP) (2015/50333-1; 2018/11069-5; 2015/13320-9). MRMJ acknowledges Red Iberomericana de Alimentos Autoctonos Subutilizados (ALSUB-CYTED, 118RT0543). Special thanks to Paula Januzzi Guedes, a self-employed design professional responsible for creating the original images of this chapter.

REFERENCES

Aachary, A.A., Prapulla, S.G., 2011. Xylooligosaccharides (XOS) as an emerging prebiotic: microbial synthesis, utilization, structural characterization, bioactive properties, and applications. Compr. Rev. Food Sci. Food Saf. https://doi.org/10.1111/j.1541-4337.2010.00135.x.

Ai, D., Pan, H., Li, X., Gao, Y., Liu, G., Xia, L.C., 2019. Identifying gut microbiota associated with colorectal cancer using a zero-inflated lognormal model. Front. Microbiol. https://doi.org/10.3389/fmicb.2019.00826.

Alba, K., Kontogiorgos, V., 2018. Seaweed polysaccharides (agar, alginate carrageenan). Encycl. Food Chem. https://doi.org/10.1016/B978-0-08-100596-5.21587-4.

Alfa, M.J., Strang, D., Tappia, P.S., et al., 2018. A randomized trial to determine the impact of a digestion resistant starch composition on the gut microbiome in older and mid-age adults. Clin. Nutr. 37 (3), 797–807. https://doi.org/10.1016/j.clnu.2017.03.025.

Al-Lahham, S.H., Peppelenbosch, M.P., Roelofsen, H., Vonk, R.J., Venema, K., 2010. Biological effects of propionic acid in humans; metabolism, potential applications and underlying mechanisms. Biochim. Biophys. Acta Mol. Cell Biol. Lipids. https://doi.org/10.1016/j.bbalip.2010.07.007.

Al-Tamimi, M.A.H.M., Palframan, R.J., Cooper, J.M., Gibson, G.R., Rastall, R.A., 2006. In vitro fermentation of sugar beet arabinan and arabino-oligosaccharides by the human gut microflora. J. Appl. Microbiol. https://doi.org/10.1111/j.1365-2672.2005.02780.x.

An, R., Wilms, E., Smolinska, A., et al., 2019. Sugar beet pectin supplementation did not alter profiles of fecal microbiota and exhaled breath in healthy young adults and healthy elderly. Nutrients 11, 2193. https://doi.org/10.3390/nu11092193.

Atia, A., Girard-Pipau, F., Hébuterne, X., et al., 2011. Macronutrient absorption characteristics in humans with short bowel syndrome and jejunocolonic anastomosis: starch is the most important carbohydrate substrate, although pectin supplementation may modestly enhance short chain fatty acid production and fluid absorption. J. Parenter. Enteral Nutr. 35 (2), 229–240. https://doi.org/10.1177/0148607110378410.

Babbar, N., Dejonghe, W., Gatti, M., Sforza, S., Elst, K., 2016. Pectic oligosaccharides from agricultural by-products: production, characterization and health benefits. Crit. Rev. Biotechnol. https://doi.org/10.3109/07388551.2014.996732.

Baenas, N., Abellán, Á., Rivera, S., Moreno, D.A., García-Viguera, C., Domínguez-Perles, R., 2018. Foods and Supplements. Elsevier, https://doi.org/10.1016/B978-0-12-813572-3.00010-5.

Balamurugan, R., Pugazhendhi, S., Balachander, G.M., Dharmalingam, T., Mortimer, E.K., Gopalsamy, G.L., et al., 2019. Effect of native and acetylated dietary resistant starches on intestinal fermentative capacity of normal and stunted children in southern India. Int. J. Environ. Res. Public Health. https://doi.org/10.3390/ijerph16203922.

Bastos, L.P.H., dos Santos, C.H.C., de Carvalho, M.G., Garcia-Rojas, E.E., 2020. Encapsulation of the black pepper (*Piper nigrum* L.) essential oil by lactoferrin-sodium alginate complex coacervates: structural characterization and simulated gastrointestinal conditions. Food Chem. https://doi.org/10.1016/j.foodchem.2020.126345.

Beards, E., Tuohy, K., Gibson, G., 2010. A human volunteer study to assess the impact of confectionery sweeteners on the gut microbiota composition. Br. J. Nutr. 104 (5), 701–708. https://doi.org/10.1017/S0007114510001078.

BeMiller, J.N., 2019. Inulin and konjac glucomannan. Carbohydr. Chem. Food Sci. https://doi.org/10.1016/b978-0-12-812069-9.00010-8.

Bindels, L.B., Delzenne, N.M., Cani, P.D., Walter, J., 2015. Opinion: towards a more comprehensive concept for prebiotics. Nat. Rev. Gastroenterol. Hepatol. https://doi.org/10.1038/nrgastro.2015.47.

Bouhnik, Y., Raskine, L., Simoneau, G., et al., 2004. The capacity of nondigestible carbohydrates to stimulate fecal bifidobacteria in healthy humans: a double-blind, randomized, placebo-controlled, parallel-group, dose-response relation study. Am. J. Clin. Nutr. 80 (6), 1658–1664. https://doi.org/10.1093/ajcn/80.6.1658.

Bouwhuis, M.A., McDonnell, M.J., Sweeney, T., Mukhopadhya, A., O'Shea, C.J., O'Doherty, J.V., 2017. Seaweed extracts and galacto-oligosaccharides improve intestinal health in pigs following Salmonella Typhimurium challenge. Animal. https://doi.org/10.1017/S1751731117000118.

Broekaert, W.F., Courtin, C.M., Verbeke, K., van de Wiele, T., Verstraete, W., Delcour, J.A., 2011. Prebiotic and other health-related effects of cereal-derived arabinoxylans, arabinoxylan-oligosaccharides, and xylooligosaccharides. Crit. Rev. Food Sci. Nutr. https://doi.org/10.1080/10408390903044768.

Camacho, F., Macedo, A., Malcata, F., 2019. Potential industrial applications and commercialization of microalgae in the functional food and feed industries: a short review. Mar. Drugs. https://doi.org/10.3390/md17060312.

Carlson, J.L., Erickson, J.M., Lloyd, B.B., Slavin, J.L., 2018. Health effects and sources of prebiotic dietary fiber. Curr. Dev. Nutr. https://doi.org/10.1093/cdn/nzy005.

Cassettari, V.M.G., Machado, N.C., de Arrudab Lourenção, P.L.T., Carvalho, M.A., Ortolan, E.V.P., 2019. Combinations of laxatives and green banana biomass on the treatment of functional constipation in children and adolescents: a randomized study. J. Pediatr. https://doi.org/10.1016/j.jped.2017.10.011.

Castaner, O., Goday, A., Park, Y.M., Lee, S.H., Magkos, F., Shiow, S.A.T.E., et al., 2018. The gut microbiome profile in obesity: a systematic review. Int. J. Endocrinol. https://doi.org/10.1155/2018/4095789.

Chee, S.Y., Wong, P.K., Wong, C.L., 2011. Extraction and characterisation of alginate from brown seaweeds (Fucales, Phaeophyceae) collected from Port Dickson, peninsular Malaysia. J. Appl. Phycol. https://doi.org/10.1007/s10811-010-9533-7.

Chen, H.L., Lu, Y.H., Lin, J.J., Ko, L.Y., 2001. Effects of isomalto-oligosaccharides on bowel functions and indicators of nutritional status in constipated elderly men. J. Am. Coll. Nutr. https://doi.org/10.1080/07315724.2001.10719013.

Chen, C., Li, L., Wu, Z., Chen, H., Fu, S., 2007. Effects of lactitol on intestinal microflora and plasma endotoxin in patients with chronic viral hepatitis. J. Inf. Secur. https://doi.org/10.1016/j.jinf.2005.11.013.

Chen, Q., Liu, M., Zhang, P., Fan, S., Huang, J., Yu, S., et al., 2019. Fucoidan and galactooligosaccharides ameliorate high-fat diet–induced dyslipidemia in rats by modulating the gut microbiota and bile acid metabolism. Nutrition. https://doi.org/10.1016/j.nut.2019.03.001.

Cheng, J., Tennilä, J., Stenman, L., Ibarra, A., Kumar, M., Gupta, K.K., et al., 2019. Influence of lactitol and psyllium on bowel function in constipated Indian volunteers: a randomized, controlled trial. Nutrients. https://doi.org/10.3390/nu11051130.

Cheong, K.L., Qiu, H.M., Du, H., Liu, Y., Khan, B.M., 2018. Oligosaccharides derived from red seaweed: production, properties, and potential health and cosmetic applications. Molecules. https://doi.org/10.3390/molecules23102451.

Cherry, P., Yadav, S., Strain, C.R., Allsopp, P.J., Mcsorley, E.M., Ross, R.P., et al., 2019. Prebiotics from seaweeds: an ocean of opportunity? Mar. Drugs 17. https://doi.org/10.3390/md17060327.

Childs, C.E., Röytiö, H., Alhoniemi, E., Fekete, A.A., Forssten, S.D., Hudjec, N., et al., 2014. Xylo-oligosaccharides alone or in synbiotic combination with Bifidobacterium animalis subsp. lactis induce bifidogenesis and modulate markers of immune function in healthy adults: a double-blind, placebo-controlled, randomised, factorial cross-over study. Br. J. Nutr. https://doi.org/10.1017/S0007114513004261.

Conterno, L., Martinelli, F., Tamburini, M., et al., 2019. Measuring the impact of olive pomace enriched biscuits on the gut microbiota and its metabolic activity in mildly hypercholesterolaemic subjects. Eur. J. Nutr. 58 (1), 63–81. https://doi.org/10.1007/s00394-017-1572-2.

Corzo, N., Alonso, J.L., Azpiroz, F., Calvo, M.A., Cirici, M., Leis, R., et al., 2015. Prebióticos; concepto, propiedades y efectos beneficiosos. Nutr. Hosp. https://doi.org/10.3305/nh.2015.31.sup1.8715.

Costa, N.E., Hannon, J.A., Guinee, T.P., Auty, M.A.E., McSweeney, P.L.H., Beresford, T.P., 2010. Effect of exopolysaccharide produced by isogenic strains of Lactococcus lactis on half-fat Cheddar cheese. J. Dairy Sci. https://doi.org/10.3168/jds.2009-3008.

Costa, E.S., França, C.N., Fonseca, F.A.H., Kato, J.T., Bianco, H.T., Freitas, T.T., et al., 2019. Beneficial effects of green banana biomass consumption in patients with pre-diabetes and type 2 diabetes: a randomised controlled trial. Br. J. Nutr. https://doi.org/10.1017/S0007114519000576.

Crozier, A., Jaganath, I.B., Clifford, M.N., 2009. Dietary phenolics: chemistry, bioavailability and effects on health. Nat. Prod. Rep. https://doi.org/10.1039/b802662a.

Cruz, A.G., Cadena, R.S., Walter, E.H.M., Mortazavian, A.M., Granato, D., Faria, J.A.F., et al., 2010. Sensory analysis: relevance for prebiotic, probiotic, and synbiotic product development. Compr. Rev. Food Sci. Food Saf. https://doi.org/10.1111/j.1541-4337.2010.00115.x.

Csáky, I., Fekete, S., 2004. Soybean: feed quality and safety part 1: biologically active components a review. Acta Vet. Hung. https://doi.org/10.1556/AVet.52.2004.3.6.

Davani-Davari, D., Negahdaripour, M., Karimzadeh, I., Seifan, M., Mohkam, M., Masoumi, S.J., et al., 2019. Prebiotics: definition, types, sources, mechanisms, and clinical applications. Foods. https://doi.org/10.3390/foods8030092.

de Araújo Etchepare, M., Raddatz, G.C., de Moraes Flores, É.M., Zepka, L.Q., Jacob-Lopes, E., Barin, J.S., et al., 2016. Effect of resistant starch and chitosan on survival of Lactobacillus acidophilus microencapsulated with sodium alginate. LWT - Food Sci. Technol. https://doi.org/10.1016/j.lwt.2015.08.039.

De Jesus Raposo, M.F., De Morais, A.M.M.B., De Morais, R.M.S.C., 2016. Emergent sources of prebiotics: seaweeds and microalgae. Mar. Drugs. https://doi.org/10.3390/md14020027.

Deehan, E.C., Yang, C., Perez-Muñoz, M.E., et al., 2020. Precision microbiome modulation with discrete dietary fiber structures directs short-chain fatty acid production. Cell Host Microbe 27 (3), 389–404.e6. https://doi.org/10.1016/j.chom.2020.01.006.

Del Piano, M., Balzarini, M., Carmagnola, S., Pagliarulo, M., Tari, R., Nicola, S., et al., 2014. Assessment of the capability of a gelling complex made of tara gum and the exopolysaccharides produced by the microorganism Streptococcus thermophilus ST10 to prospectively restore the gut physiological barrier: a pilot study. J. Clin. Gastroenterol. https://doi.org/10.1097/MCG.0000000000000254.

DeMartino, P., Cockburn, D.W., 2020. Resistant starch: impact on the gut microbiome and health. Curr. Opin. Biotechnol. https://doi.org/10.1016/j.copbio.2019.10.008.

Devaraj, R.D., Reddy, C.K., Xu, B., 2019. Health-promoting effects of konjac glucomannan and its practical applications: a critical review. Int. J. Biol. Macromol. https://doi.org/10.1016/j.ijbiomac.2018.12.203.

DeVries, J.W., Camire, M.E., Cho, S., Craig, S., Gordon, D., Jones, J.M., et al., 2001. The definition of dietary fiber. Cereal Foods World 46, 112–129.

Di, R., Vakkalanka, M.S., Onumpai, C., Chau, H.K., White, A., Rastall, R.A., et al., 2017. Pectic oligosaccharide structure-function relationships: prebiotics, inhibitors of *Escherichia coli* O157:H7 adhesion and reduction of Shiga toxin cytotoxicity in HT29 cells. Food Chem. https://doi.org/10.1016/j.foodchem.2017.01.100.

Dongowski, G., Lorenz, A., Anger, H., 2000. Degradation of pectins with different degrees of esterification by Bacteroides thetaiotaomicron isolated from human gut flora. Appl. Environ. Microbiol. https://doi.org/10.1128/AEM.66.4.1321-1327.2000.

du Preez, R., Pahl, J., Arora, M., Kumar, M.N.V.R., Brown, L., Panchal, S.K., 2019. Low-dose curcumin nanoparticles normalise blood pressure in male wistar rats with diet-induced metabolic syndrome. Nutrients. https://doi.org/10.3390/nu11071542.

du Preez, R., Paul, N., Mouatt, P., Majzoub, M.E., Thomas, T., Panchal, S.K., et al., 2020. Carrageenans from the red seaweed Sarconema filiforme attenuate symptoms of diet-induced metabolic syndrome in rats. Mar. Drugs. https://doi.org/10.3390/md18020097.

Dueñas, M., Muñoz-González, I., Cueva, C., Jiménez-Girón, A., Sánchez-Patán, F., Santos-Buelga, C., et al., 2015. A survey of modulation of gut microbiota by dietary polyphenols. Biomed. Res. Int. https://doi.org/10.1155/2015/850902.

Dysvik, A., La Rosa, S.L., Buffetto, F., Liland, K.H., Myhrer, K.S., Rukke, E.O., et al., 2020. Secondary lactic acid bacteria fermentation with wood-derived xylooligosaccharides as a tool to expedite sour beer production. J. Agric. Food Chem. https://doi.org/10.1021/acs.jafc.9b05459.

Ebrahimi-Mameghani, M., Sadeghi, Z., Abbasalizad Farhangi, M., Vaghef-Mehrabany, E., Aliashrafi, S., 2017. Glucose homeostasis, insulin resistance and inflammatory biomarkers in patients with non-alcoholic fatty liver disease: beneficial effects of supplementation with microalgae Chlorella vulgaris: a double-blind placebo-controlled randomized clinical trial. Clin. Nutr. https://doi.org/10.1016/j.clnu.2016.07.004.

El-Nawawi, S.A., Shehata, F.R., 1987. Extraction of pectin from Egyptian orange peel. Factors affecting the extraction. Biol. Wastes. https://doi.org/10.1016/0269-7483(87)90005-X.

Esgalhado, M., Kemp, J.A., Azevedo, R., Paiva, B.R., Stockler-Pinto, M.B., Dolenga, C.J., et al., 2018. Could resistant starch supplementation improve inflammatory and oxidative stress biomarkers and uremic toxins levels in hemodialysis patients? A pilot randomized controlled trial. Food Funct. https://doi.org/10.1039/c8fo01876f.

Etchepare, M.D.A., Raddatz, G.C., Cichoski, A.J., Flores, É.M.M., Barin, J.S., Queiroz Zepka, L., et al., 2016. Effect of resistant starch (Hi-maize) on the survival of Lactobacillus acidophilus microencapsulated with sodium alginate. J. Funct. Foods. https://doi.org/10.1016/j.jff.2015.12.025.

FAO, 2018. The State of World Fisheries and Aquaculture 2018—Meeting the Sustainable Development Goals. Rome., https://doi.org/10.1093/japr/3.1.101.

Farias, D.d.P., de Araújo, F.F., Neri-Numa, I.A., Pastore, G.M., 2019. Prebiotics: trends in food, health and technological applications. Trends Food Sci. Technol. 93, 23–35. https://doi.org/10.1016/J.TIFS.2019.09.004.

Fei, Y., Wang, Y., Pang, Y., Wang, W., Zhu, D., Xie, M., et al., 2020. Xylooligosaccharide modulates gut microbiota and alleviates colonic inflammation caused by high fat diet induced obesity. Front. Physiol. https://doi.org/10.3389/fphys.2019.01601.

Ferdouse, F., Holdt, S.L., Smith, R., Murúa, P., Yang, Z., 2018. The Global Status of Seaweed Production, Trade and Utilization: FAO Globefish Research Programme. vol. 124, p. 120.

Ferrão, L.L., Silva, E.B., Silva, H.L.A., Silva, R., Mollakhalili, N., Granato, D., et al., 2016. Strategies to develop healthier processed cheeses: reduction of sodium and fat contents and use of prebiotics. Food Res. Int. https://doi.org/10.1016/j.foodres.2016.04.034.

Ferrão, L.L., Ferreira, M.V.S., Cavalcanti, R.N., Carvalho, A.F.A., Pimentel, T.C., Silva, R., et al., 2018. The xylooligosaccharide addition and sodium reduction in requeijão cremoso processed cheese. Food Res. Int. https://doi.org/10.1016/j.foodres.2018.02.018.

Ferreira-Lazarte, A., Kachrimanidou, V., Villamiel, M., Rastall, R.A., Moreno, F.J., 2018. In vitro fermentation properties of pectins and enzymatic-modified pectins obtained from different renewable bioresources. Carbohydr. Polym. https://doi.org/10.1016/j.carbpol.2018.07.041.

Fields, F.J., Lejzerowicz, F., Schroeder, D., Ngoi, S.M., Tran, M., McDonald, D., et al., 2020. Effects of the microalgae Chlamydomonas on gastrointestinal health. J. Funct. Foods. https://doi.org/10.1016/j.jff.2019.103738.

Finegold, S.M., Li, Z., Summanen, P.H., Downes, J., Thames, G., Corbett, K., et al., 2014. Xylooligosaccharide increases bifidobacteria but not lactobacilli in human gut microbiota. Food Funct. https://doi.org/10.1039/c3fo60348b.

Finney, M., Smullen, J., Foster, H.A., Brokx, S., Storey, D.M., 2007. Effects of low doses of lactitol on faecal microflora, pH, short chain fatty acids and gastrointestinal symptomology. Eur. J. Nutr. https://doi.org/10.1007/s00394-007-0666-7.

Fu, X., Li, R., Zhang, T., Li, M., Mou, H., 2019. Study on the ability of partially hydrolyzed guar gum to modulate the gut microbiota and relieve constipation. J. Food Biochem. https://doi.org/10.1111/jfbc.12715.

Fuentes-Zaragoza, E., Riquelme-Navarrete, M.J., Sánchez-Zapata, E., Pérez-Álvarez, J.A., 2010. Resistant starch as functional ingredient: a review. Food Res. Int. https://doi.org/10.1016/j.foodres.2010.02.004.

Galle, S., Schwab, C., Dal Bello, F., Coffey, A., Gänzle, M.C., Arendt, E.K., 2012. Influence of in-situ synthesized exopolysaccharides on the quality of gluten-free sorghum sourdough bread. Int. J. Food Microbiol. https://doi.org/10.1016/j.ijfoodmicro.2012.01.009.

Gallego, C.G., Salminen, S., 2016. Novel probiotics and prebiotics: how can they help in human gut microbiota dysbiosis? Appl. Food Biotechnol. https://doi.org/10.22037/afb.v3i2.11276.

García-Ríos, V., Ríos-Leal, E., Robledo, D., Freile-Pelegrin, Y., 2012. Polysaccharides composition from tropical brown seaweeds. Phycol. Res. https://doi.org/10.1111/j.1440-1835.2012.00661.x.

Gaynor, P., 2016. Generally Recognized as Safe (GRAS) Notice for VF-DP3-IMO for Use in Conventional Foods.

Gentès, M.C., St-Gelais, D., Turgeon, S.L., 2011. Gel formation and rheological properties of fermented milk with in situ exopolysaccharide production by lactic acid bacteria. Dairy Sci. Technol. https://doi.org/10.1007/s13594-011-0039-0.

Gentile, C.L., Ward, E., Holst, J.J., Astrup, A., Ormsbee, M.J., Connelly, S., et al., 2015. Resistant starch and protein intake enhances fat oxidation and feelings of fullness in lean and overweight/obese women. Nutr. J. https://doi.org/10.1186/s12937-015-0104-2.

Gibson, G.R., Hutkins, R., Sanders, M.E., Prescott, S.L., Reimer, R.A., Salminen, S.J., et al., 2017. Expert consensus document: the international scientific association for probiotics and prebiotics (ISAPP) consensus statement on the definition and scope of prebiotics. Nat. Rev. Gastroenterol. Hepatol. 14, 491–502. https://doi.org/10.1038/nrgastro.2017.75.

Goffin, D., Delzenne, N., Blecker, C., Hanon, E., Deroanne, C., Paquot, M., 2011. Will isomalto-oligosaccharides, a well-established functional food in Asia, break through the European and American market? The status of knowledge on these prebiotics. Crit. Rev. Food Sci. Nutr. https://doi.org/10.1080/10408391003628955.

Gomez-Zavaglia, A., Prieto Lage, M.A., Jimenez-Lopez, C., Mejuto, J.C., Simal-Gandara, J., 2019. The potential of seaweeds as a source of functional ingredients of prebiotic and antioxidant value. Antioxidants. https://doi.org/10.3390/antiox8090406.

González-Sarrías, A., Romo-Vaquero, M., García-Villalba, R., et al., 2018. The endotoxemia marker lipopolysaccharide-binding protein is reduced in overweight-obese subjects consuming pomegranate extract by modulating the gut microbiota: a randomized clinical trial. Mol. Nutr. Food Res. 62 (11), e1800160. https://doi.org/10.1002/mnfr.201800160.

Gostner, A., Blaut, M., Schäffer, V., et al., 2006. Effect of isomalt consumption on faecal microflora and colonic metabolism in healthy volunteers. Br. J. Nutr. 95 (1), 40–50. https://doi.org/10.1079/bjn20051589.

Grand View Research, 2017. Polydextrose Market Size, Share & Trends Analysis Report By Product (Powder, Liquid), By Application (Nutritional Food, Bakery & Confectionary, Beverages), By Region, and Segment Forecasts, 2018–2025. Grand View Research.

Grand View Research, 2018. Prebiotics Market Size, Share & Trends Analysis By Ingredients (FOS, Inulin, GOS, MOS), By Application (Food and Beverages, Dietary Supplements, Animal Feed), By Region, and Segment Forecasts, 2014–2024. Grand View Research, p. 159.

Grembecka, M., 2015. Sugar alcohols—their role in the modern world of sweeteners: a review. Eur. Food Res. Technol. https://doi.org/10.1007/s00217-015-2437-7.

Grisham, M.B., DeMichele, S.J., Garleb, K.A., Specian, R.D., 1996. Sulfasalazine or enteral diets containing fish oil or oligosaccharides attenuate chronic colitis in rats. Inflamm. Bowel Dis. https://doi.org/10.1097/00054725-199609000-00004.

Guarino, M.P.L., Altomare, A., Emerenziani, S., Di Rosa, C., Ribolsi, M., Balestrieri, P., et al., 2020. Mechanisms of action of prebiotics and their effects on gastro-intestinal disorders in adults. Nutrients. https://doi.org/10.3390/nu12041037.

Gullón, B., Gómez, B., Martínez-Sabajanes, M., Yáñez, R., Parajó, J.C., Alonso, J.L., 2013. Pectic oligosaccharides: manufacture and functional properties. Trends Food Sci. Technol. https://doi.org/10.1016/j.tifs.2013.01.006.

Guo, Y., Zhou, G., He, C., Yang, W., He, Z., Liu, Z., 2015. Serum levels of lipopolysaccharide and 1,3-β-D-glucan refer to the severity in patients with Crohn's disease. Mediat. Inflamm. https://doi.org/10.1155/2015/843089.

Hamer, H.M., Jonkers, D., Venema, K., Vanhoutvin, S., Troost, F.J., Brummer, R.J., 2008. Review article: the role of butyrate on colonic function. Aliment. Pharmacol. Ther. https://doi.org/10.1111/j.1365-2036.2007.03562.x.

Hamet, M.F., Medrano, M., Pérez, P.F., Abraham, A.G., 2016. Oral administration of kefiran exerts a bifidogenic effect on BALB/c mice intestinal microbiota. Benef. Microbes. https://doi.org/10.3920/BM2015.0103.

Henry, D.A., Mitchell, A.S., Aylward, J., Fung, M.T., McEwen, J., Alain, R., 1986. Glucomannan and risk of oesophageal obstruction. Br. Med. J. (Clin. Res. Ed.). https://doi.org/10.1136/bmj.292.6520.591-a.

Hexa Research, 2018. Prebiotics Market Size, and Forecast By Ingredients (Inulin, FOS, MOS, GOS), By Application (Animal Feed, Dietary Supplements and, Food and Beverages), and Trend Analysis, 2014–2025. Hexa Research, p. 100. https://www.hexaresearch.com/research-report/prebiotics-industry?utm_source=Blogger&utm_medium=Referral&utm_campaign=Uday.

Heydari, S., Mortazavian, A.M., Ehsani, M.R., Mohammadifar, M.A., Ezzatpanah, H., 2011. Biochemical, microbiological and sensory characteristics of probiotic yogurt containing various prebiotic compounds. Ital. J. Food Sci. 23, 153–164.

Heydari, S., Amiri-Rigi, A., Ehsani, M.R., Mohammadifar, M.A., Khorshidian, N., Koushki, M.R., et al., 2018. Rheological behaviour, sensory properties and syneresis of probiotic yoghurt supplemented with various prebiotics. Int. J. Dairy Technol. https://doi.org/10.1111/1471-0307.12491.

Holloway, W.D., Tasman Jones, C., Maher, K., 1983. Pectin digestion in humans. Am. J. Clin. Nutr. https://doi.org/10.1093/ajcn/37.2.253.

Homayouni, A., Azizi, A., Ehsani, M.R., Yarmand, M.S., Razavi, S.H., 2008. Effect of microencapsulation and resistant starch on the probiotic survival and sensory properties of synbiotic ice cream. Food Chem. https://doi.org/10.1016/j.foodchem.2008.03.036.

Homayouni, A., Amini, A., Keshtiban, A.K., Mortazavian, A.M., Esazadeh, K., Pourmoradian, S., 2014. Resistant starch in food industry: a changing outlook for consumer and producer. Starch/Staerke. https://doi.org/10.1002/star.201300110.

Hondo, S., Mochizuki, T., 1979. Free sugars in miso. Nippon Shokuhin Kogyo Gakkaishi. https://doi.org/10.3136/nskkk1962.26.11_469.

Hongpattarakere, T., Cherntong, N., Wichienchot, S., Kolida, S., Rastall, R.A., 2012. In vitro prebiotic evaluation of exopolysaccharides produced by marine isolated lactic acid bacteria. Carbohydr. Polym. https://doi.org/10.1016/j.carbpol.2011.08.085.

Hosseini, S.S., Khodaiyan, F., Kazemi, M., Najari, Z., 2019. Optimization and characterization of pectin extracted from sour orange peel by ultrasound assisted method. Int. J. Biol. Macromol. https://doi.org/10.1016/j.ijbiomac.2018.12.096.

Hutkins, R.W., Krumbeck, J.A., Bindels, L.B., Cani, P.D., Fahey, G., Goh, Y.J., et al., 2016. Prebiotics: why definitions matter. Curr. Opin. Biotechnol. https://doi.org/10.1016/j.copbio.2015.09.001.

Hyams, J.S., 1983. Sorbitol intolerance: an unappreciated cause of functional gastrointestinal complaints. Gastroenterology. https://doi.org/10.1016/S0016-5085(83)80163-2.

Inan Eroglu, E., Buyuktuncer, Z., 2017. The effect of various cooking methods on resistant starch content of foods. Nutr. Food Sci. https://doi.org/10.1108/NFS-10-2016-0154.

Istas, G., Wood, E., Le Sayec, M., et al., 2019. Effects of aronia berry (poly)phenols on vascular function and gut microbiota: a double-blind randomized controlled trial in adult men. Am. J. Clin. Nutr. 110 (2), 316–329. https://doi.org/10.1093/ajcn/nqz075.

Jain, I., Kumar, V., Satyanarayana, T., 2015. Xylooligosaccharides: an economical prebiotic from agroresidues and their health benefits. Indian J. Exp. Biol.

Jovanovic-Malinovska, R., Kuzmanova, S., Winkelhausen, E., 2014. Oligosaccharide profile in fruits and vegetables as sources of prebiotics and functional foods. Int. J. Food Prop. https://doi.org/10.1080/10942912.2012.680221.

Judprasong, K., Tanjor, S., Puwastien, P., Sungpuag, P., 2011. Investigation of Thai plants for potential sources of inulin-type fructans. J. Food Compos. Anal. https://doi.org/10.1016/j.jfca.2010.12.001.

Jutur, P.P., Nesamma, A.A., Shaikh, K.M., 2016. Algae-derived marine oligosaccharides and their biological applications. Front. Mar. Sci. https://doi.org/10.3389/fmars.2016.00083.

Kailasapathy, K., 2006. Survival of free and encapsulated probiotic bacteria and their effect on the sensory properties of yoghurt. LWT Food Sci. Technol. https://doi.org/10.1016/j.lwt.2005.07.013.

Kamiloglu, S., Capanoglu, E., Grootaert, C., van Camp, J., 2015. Anthocyanin absorption and metabolism by human intestinal Caco-2 cells—a review. Int. J. Mol. Sci. https://doi.org/10.3390/ijms160921555.

Katina, K., Maina, N.H., Juvonen, R., Flander, L., Johansson, L., Virkki, L., et al., 2009. In situ production and analysis of Weissella confusa dextran in wheat sourdough. Food Microbiol. https://doi.org/10.1016/j.fm.2009.07.008.

Kawabata, K., Yoshioka, Y., Terao, J., 2019. Role of intestinal microbiota in the bioavailability and physiological functions of dietary polyphenols. Molecules. https://doi.org/10.3390/molecules24020370.

Khat-udomkiri, N., Toejing, P., Sirilun, S., Chaiyasut, C., Lailerd, N., 2020. Antihyperglycemic effect of rice husk derived xylooligosaccharides in high-fat diet and low-dose streptozotocin-induced type 2 diabetic rat model. Food Sci. Nutr. https://doi.org/10.1002/fsn3.1327.

Kim, C.O., Kim, Y.N., Lee, D.C., 2019. Effects of Gelidium elegans on weight and fat mass reduction and obesity biomarkers in overweight or obese adults: a randomized double-blinded study. Nutrients. https://doi.org/10.3390/nu11071513.

Klickovic, U., Doberer, D., Gouya, G., Aschauer, S., Weisshaar, S., Storka, A., et al., 2014. Human pharmacokinetics of high dose oral curcumin and its effect on heme oxygenase-1 expression in healthy male subjects. Biomed. Res. Int. https://doi.org/10.1155/2014/458592.

Ko, S.J., Kim, J., Han, G., Kim, S.K., Kim, H.G., Yeo, I., et al., 2014. Laminaria japonica combined with probiotics improves intestinal microbiota: a randomized clinical trial. J. Med. Food. https://doi.org/10.1089/jmf.2013.3054.

Kohmoto, T., Fukui, F., Takaku, H., Machida, Y., Arai, M., Mitsuoka, T., 1988. Effect of isomalto-oligosaccharides on human fecal flora. Bifidobact. Microflora. https://doi.org/10.12938/bifidus1982.7.2_61.

Korcz, E., Kerényi, Z., Varga, L., 2018. Dietary fibers, prebiotics, and exopolysaccharides produced by lactic acid bacteria: potential health benefits with special regard to cholesterol-lowering effects. Food Funct. https://doi.org/10.1039/c8fo00118a.

Korczak, R., Kamil, A., Fleige, L., Donovan, S.M., Slavin, J.L., 2017. Dietary fiber and digestive health in children. Nutr. Rev. https://doi.org/10.1093/nutrit/nuw068.

Koutsou, G.A., Storey, D.M., Lee, A., Zumbe, A., Flourie, B., Lebot, Y., et al., 1996. Dose-related gastrointestinal response to the ingestion of either isomalt, lactitol or maltitol in milk chocolate. Eur. J. Clin. Nutr. 50 (1), 17–21.

Kulshreshtha, G., Rathgeber, B., Stratton, G., Thomas, N., Evans, F., Critchley, A., et al., 2014. Immunology, health, and disease: feed supplementation with red seaweeds, *Chondrus crispus* and *Sarcodiotheca gaudichaudii*, affects performance, egg quality, and gut microbiota of layer hens. Poult. Sci. https://doi.org/10.3382/ps.2014-04200.

Kumar, A.S., Mody, K., Jha, B., 2007. Bacterial exopolysaccharides–a perception. J. Basic Microbiol. https://doi.org/10.1002/jobm.200610203.

Kumar, H., Salminen, S., Verhagen, H., Rowland, I., Heimbach, J., Bañares, S., et al., 2015. Novel probiotics and prebiotics: road to the market. Curr. Opin. Biotechnol. https://doi.org/10.1016/j.copbio.2014.11.021.

Kunal Ahuja, K.M., 2019. Prebiotics Market Size by Ingredient (Inulin, GOS, FOS, MOS), by Application (Animal Feed, Food & Beverages [Dairy, Cereals, Baked Goods, Fermented Meat, Dry Foods], Dietary Supplements [Food, Nutrition, Infant Formulations]), Regional Outlook, Industry An. Global Mark Insights, p. 235. https://www.gminsights.com/industry-analysis/prebiotics-market.

Lao, C.D., Ruffin IV, M.T., Normolle, D., Heath, D.D., Murray, S.I., Bailey, J.M., et al., 2006. Dose escalation of a curcuminoid formulation. BMC Complement. Altern. Med. https://doi.org/10.1186/1472-6882-6-10.

Lear, R., O'Leary, M., Andersen, L.O., et al., 2019. Tart cherry concentrate does not alter the gut microbiome, glycaemic control or systemic inflammation in a middle-aged population. Nutrients 11 (5), 1063. https://doi.org/10.3390/nu11051063.

Lecerf, J.M., Dépeint, F., Clerc, E., Dugenet, Y., Niamba, C.N., Rhazi, L., et al., 2012. Xylo-oligosaccharide (XOS) in combination with inulin modulates both the intestinal environment and immune status in healthy subjects, while XOS alone only shows prebiotic properties. Br. J. Nutr. https://doi.org/10.1017/S0007114511007252.

Lee, S.H., Jeon, Y.J., 2015. Efficacy and safety of a dieckol-rich extract (AG-dieckol) of brown algae, Ecklonia cava, in pre-diabetic individuals: a double-blind, randomized, placebo-controlled clinical trial. Food Funct. https://doi.org/10.1039/c4fo00940a.

Lee, A., Storey, D.M., 1999. Comparative gastrointestinal tolerance of sucrose, lactitol, or D-tagatose in chocolate. Regul. Toxicol. Pharmacol. https://doi.org/10.1006/rtph.1998.1255.

Lee, C.C., Wang, H.F., Lin, S.D., 2008. Effect of isomaltooligosaccharide syrup on quality characteristics of sponge cake. Cereal Chem. https://doi.org/10.1094/CCHEM-85-4-0515.

Leite-Legatti, A.V., Batista, A.G., Dragano, N.R.V., Marques, A.C., Malta, L.G., Riccio, M.F., et al., 2012. Jaboticaba peel: antioxidant compounds, antiproliferative and antimutagenic activities. Food Res. Int. 49, 596–603. https://doi.org/10.1016/j.foodres.2012.07.044.

Lenhart, A., Chey, W.D., 2017. A systematic review of the effects of polyols on gastrointestinal health and irritable bowel syndrome. Adv. Nutr. https://doi.org/10.3945/an.117.015560.

Li, T., Chen, L., Xiao, J., An, F., Wan, C., Song, H., 2018. Prebiotic effects of resistant starch from purple yam (*Dioscorea alata* L.) on the tolerance and proliferation ability of Bifidobacterium adolescentis in vitro. Food Funct. https://doi.org/10.1039/c7fo01919j.

Li, S., Wang, L., Liu, B., He, N., 2020. Unsaturated alginate oligosaccharides attenuated obesity-related metabolic abnormalities by modulating gut microbiota in high-fat-diet mice. Food Funct. https://doi.org/10.1039/c9fo02857a.

Ligen, C., Xu, W., Chen, D., Chen, G., Liu, J., Zeng, X., et al., 2018. Digestibility of sulfated polysaccharide from the brown seaweed *Ascophyllum nodosum* and its effect on the human gut microbiota in vitro. Int. J. Biol. Macromol. https://doi.org/10.1016/j.ijbiomac.2018.01.183.

Lili, C., Jiang, B., Zhong, C., Guo, J., Zhang, L., Mu, T., et al., 2018. Chemoprevention of colorectal cancer by black raspberry anthocyanins involved the modulation of gut microbiota and SFRP2 demethylation. Carcinogenesis. https://doi.org/10.1093/carcin/bgy009.

Lin, S.H., Chou, L.M., Chien, Y.W., Chang, J.S., Lin, C.I., 2016. Prebiotic effects of xylooligosaccharides on the improvement of microbiota balance in human subjects. Gastroenterol. Res. Pract. https://doi.org/10.1155/2016/5789232.

Liu, J., Wang, X., Pu, H., Liu, S., Kan, J., Jin, C., 2017. Recent advances in endophytic exopolysaccharides: production, structural characterization, physiological role and biological activity. Carbohydr. Polym. https://doi.org/10.1016/j.carbpol.2016.10.084.

Liu, G., Chen, H., Chen, J., Wang, X., Gu, Q., Yin, Y., 2019. Effects of bifidobacteria-produced exopolysaccharides on human gut microbiota in vitro. Appl. Microbiol. Biotechnol. https://doi.org/10.1007/s00253-018-9572-6.

Livesey, G., 2003. Health potential of polyols as sugar replacers, with emphasis on low glycaemic properties. Nutr. Res. Rev. https://doi.org/10.1079/nrr200371.

Lugani, Y., Sooch, S., 2017. Xylitol, an emerging prebiotic: a review. Int. J. Appl. Pharm. Biol. Res. 2 (2), 67–73.

Luo, X., Song, H., Yang, J., Han, B., Feng, Y., Leng, Y., et al., 2020. Encapsulation of *Escherichia coli* strain Nissle 1917 in a chitosan—alginate matrix by combining layer-by-layer assembly with CaCl2 cross-linking for an effective treatment of inflammatory bowel diseases. Colloids Surf. B Biointerfaces. https://doi.org/10.1016/j.colsurfb.2020.110818.

Lynch, K.M., Coffey, A., Arendt, E.K., 2018a. Exopolysaccharide producing lactic acid bacteria: their techno-functional role and potential application in gluten-free bread products. Food Res. Int. https://doi.org/10.1016/j.foodres.2017.03.012.

Lynch, K.M., Zannini, E., Coffey, A., Arendt, E.K., 2018b. Lactic acid Bacteria exopolysaccharides in foods and beverages: isolation, properties, characterization, and health benefits. Annu. Rev. Food Sci. Technol. https://doi.org/10.1146/annurev-food-030117-012537.

Ma, Y., Wu, X., Giovanni, V., Meng, X., 2017. Effects of soybean oligosaccharides on intestinal microbial communities and immune modulation in mice. Saudi J. Biol. Sci. https://doi.org/10.1016/j.sjbs.2016.09.004.

Madsen, L.R., Stanley, S., Swann, P., Oswald, J., 2017. A survey of commercially available Isomaltooligosaccharide-based food Ingredients. J. Food Sci. https://doi.org/10.1111/1750-3841.13623.

Magne, F., Hachelaf, W., Suau, A., et al., 2008. Effects on faecal microbiota of dietary and acidic oligosaccharides in children during partial formula feeding. J. Pediatr. Gastroenterol. Nutr. 46 (5), 580–588. https://doi.org/10.1097/MPG.0b013e318164d920.

Mallikarjuna, S.E., Dharmesh, S.M., 2018. Swallow root (Decalepis hamiltonii) pectic oligosaccharide (SRO1) induces cancer cell death via modulation of galectin-3 and survivin. Carbohydr. Polym. https://doi.org/10.1016/j.carbpol.2018.01.053.

Mano, M.C.R., Neri-Numa, I.A., da Silva, J.B., Paulino, B.N., Pessoa, M.G., Pastore, G.M., 2018. Oligosaccharide biotechnology: an approach of prebiotic revolution on the industry. Appl. Microbiol. Biotechnol. https://doi.org/10.1007/s00253-017-8564-2.

Maru, V., Hewale, S., Mantri, H., Ranade, V., 2015. Partial purification and characterization of mannan oligosaccharides from cell wall of *Saccharomyces cerevisiae*. Int. J. Curr. Microbiol. Appl. Sci. 4, 705–711.

Maryam, F., Nima, M., Hamed, A., Mohammad, A.Z., Fariba, Z.H., Kasra, B., 2013. Effect of microencapsulation plus resistant starch on survival of lactobacillus casei and Bifidobacterium bifidum in mayonnaise sauce. African J. Microbiol. Res. https://doi.org/10.5897/ajmr12.1240.

McCleary, B.V., 2007. An integrated procedure for the measurement of total dietary fibre (including resistant starch), non-digestible oligosaccharides and available carbohydrates. Anal. Bioanal. Chem. https://doi.org/10.1007/s00216-007-1389-6.

Mcfadden, R.M.T., Larmonier, C.B., Shehab, K.W., Midura-Kiela, M., Ramalingam, R., Harrison, C.A., et al., 2015. The role of curcumin in modulating colonic microbiota during colitis and colon cancer prevention. Inflamm. Bowel Dis. https://doi.org/10.1097/MIB.0000000000000522.

Menezes, E.W., Tadini, C.C., Tribess, T.B., Zuleta, A., Binaghi, J., Pak, N., et al., 2011. Chemical composition and nutritional value of unripe banana flour (*Musa acuminata*, var. Nanicão). Plant Foods Hum. Nutr. https://doi.org/10.1007/s11130-011-0238-0.

Míguez, B., Gómez, B., Gullón, P., Gullón, B., Alonso, J.L., 2016. Pectic oligosaccharides and other emerging prebiotics. Probiotics Prebiotics Hum. Nutr. Health. https://doi.org/10.5772/62830.

Mirzaei, H., Pourjafar, H., Homayouni, A., 2012. Effect of calcium alginate and resistant starch microencapsulation on the survival rate of lactobacillus acidophilus La5 and sensory properties in Iranian white brined cheese. Food Chem. https://doi.org/10.1016/j.foodchem.2011.12.033.

Mohammadi, R., Mortazavian, A.M., 2011. Review article: technological aspects of prebiotics in probiotic fermented milks. Food Rev. Int. https://doi.org/10.1080/87559129.2010.535235.

Moroeanu, V.I., Vamanu, E., Paun, G., Neagu, E., Ungureanu, O.R., Eremia, S.A.V., et al., 2015. Probiotic strains influence on infant microbiota in the in vitro colonic fermentation model GIS1. Indian J. Microbiol. https://doi.org/10.1007/s12088-015-0542-8.

Müller-Maatsch, J., Bencivenni, M., Caligiani, A., Tedeschi, T., Bruggeman, G., Bosch, M., et al., 2016. Pectin content and composition from different food waste streams. Food Chem. https://doi.org/10.1016/j.foodchem.2016.01.012.

Múzquiz De La Garza, A.R., Tapia-Salazar, M., Maldonado-Muñiz, M., De La Rosa-Millán, J., Gutiérrez-Uribe, J.A., Santos-Zea, L., et al., 2019. Nutraceutical potential of five Mexican brown seaweeds. Biomed. Res. Int. https://doi.org/10.1155/2019/3795160.

Ng, Q., Soh, A., Loke, W., Venkatanarayanan, N., Lim, D., Yeo, W.-S., 2018. A meta-analysis of the clinical use of Curcumin for irritable bowel syndrome (IBS). J. Clin. Med. https://doi.org/10.3390/jcm7100298.

Nguyen, S.G., Kim, J., Guevarra, R.B., Lee, J.H., Kim, E., Il, K.S., et al., 2016. Laminarin favorably modulates gut microbiota in mice fed a high-fat diet. Food Funct. https://doi.org/10.1039/c6fo00929h.

Nishino, R., Ozawa, Y., Yasuda, A., Sakasai, T., 1974. Oligosaccharides in soy sauce. Nippon Nå geikagaku Kaishi. https://doi.org/10.1271/nogeikagaku1924.48.291.

O'Sullivan, L., Murphy, B., McLoughlin, P., Duggan, P., Lawlor, P.G., Hughes, H., et al., 2010. Prebiotics from marine macroalgae for human and animal health applications. Mar. Drugs. https://doi.org/10.3390/md8072038.

Ochoa-Villarreal, M., Vargas-Arispuro, I., Islas-Osuna, M.A., González-Aguilar, G., Martínez-Téllez, M.Á., 2011. Pectin-derived oligosaccharides increase color and anthocyanin content in flame seedless grapes. J. Sci. Food Agric. https://doi.org/10.1002/jsfa.4412.

Ohno, M., Nishida, A., Sugitani, Y., Nishino, K., Inatomi, O., Sugimoto, M., et al., 2017. Nanoparticle curcumin ameliorates experimental colitis via modulation of gut microbiota and induction of regulatory T cells. PLoS One. https://doi.org/10.1371/journal.pone.0185999.

Oku, T., Nakamura, S., 2003. Comparison of digestibility and breath hydrogen gas excretion of fructo-oligosaccharide, galactosyl-sucrose, and isomalto-oligosaccharide in healthy human subjects. Eur. J. Clin. Nutr. https://doi.org/10.1038/sj.ejcn.1601666.

Oku, T., Okazaki, M., 1996. Laxative threshold of sugar alcohol erythritol in human subjects. Nutr. Res. https://doi.org/10.1016/0271-5317(96)00036-X.

Olano, A., Corzo, N., 2009. Lactulose as a food ingredient. J. Sci. Food Agric. https://doi.org/10.1002/jsfa.3694.

Olano-Martin, E., Rimbach, G.H., Gibson, G.R., Rastall, R.A., 2003. Pectin and pectic-oligosaccharides induce apoptosis in in vitro human colonic adenocarcinoma cells. Anticancer Res.

Onumpai, C., Kolida, S., Bonnin, E., Rastall, R.A., 2011. Microbial utilization and selectivity of pectin fractions with various structures. Appl. Environ. Microbiol. https://doi.org/10.1128/AEM.00179-11.

Pangestuti, R., Kim, S.K., 2015. An overview of phycocolloids: the principal commercial seaweed extracts. In: Marine Algae Extract: Processes Products and Applications. vol. 1–2, pp. 319–330, https://doi.org/10.1002/9783527679577.ch19.

Park, E.Y., Jang, S.B., Lim, S.T., 2016. Effect of fructo-oligosaccharide and isomalto-oligosaccharide addition on baking quality of frozen dough. Food Chem. https://doi.org/10.1016/j.foodchem.2016.06.067.

Patel, S., Goyal, A., 2012. The current trends and future perspectives of prebiotics research: a review. 3 Biotech. https://doi.org/10.1007/s13205-012-0044-x.

Patil, D.H., Grimble, G.K., Silk, D.B.A., 1987. Lactitol, a new hydrogenated lactose derivative: intestinal absorption and laxative threshold in normal human subjects. Br. J. Nutr. https://doi.org/10.1079/bjn19870025.

Peng, Y., Yan, Y., Wan, P., Chen, D., Ding, Y., Ran, L., et al., 2019. Gut microbiota modulation and anti-inflammatory properties of anthocyanins from the fruits of Lycium ruthenicum Murray in dextran sodium sulfate-induced colitis in mice. Free Radic. Biol. Med. 136, 96–107. https://doi.org/10.1016/j.freeradbiomed.2019.04.005.

Peng, Y., Yan, Y., Wan, P., Dong, W., Huang, K., Ran, L., et al., 2020. Effects of long-term intake of anthocyanins from Lycium ruthenicum Murray on the organism health and gut microbiota in vivo. Food Res. Int. 130, 108952. https://doi.org/10.1016/j.foodres.2019.108952.

Pérez-López, E., Cela, D., Costabile, A., Mateos-Aparicio, I., Rupérez, P., 2016. In vitro fermentability and prebiotic potential of soyabean Okara by human faecal microbiota. Br. J. Nutr. https://doi.org/10.1017/S0007114516002816.

Peterson, C.T., Vaughn, A.R., Sharma, V., Chopra, D., Mills, P.J., Peterson, S.N., et al., 2018. Effects of turmeric and Curcumin dietary supplementation on human gut microbiota: a double-blind, randomized, placebo-controlled pilot study. J. Evidence-Based Integr. Med. https://doi.org/10.1177/2515690X18790725.

Pluta, R., Januszewski, S., Ułamek-Kozioł, M., 2020. Mutual two-way interactions of Curcumin and gut microbiota. Int. J. Mol. Sci. 21. https://doi.org/10.3390/ijms21031055.

Qian, L., Gao, R., Hong, L., Pan, C., Li, H., Huang, J., et al., 2018. Association analysis of dietary habits with gut microbiota of a native Chinese community. Exp. Ther. Med. https://doi.org/10.3892/etm.2018.6249.

Rahbar Saadat, Y., Yari Khosroushahi, A., Pourghassem, G.B., 2019. A comprehensive review of anticancer, immunomodulatory and health beneficial effects of the lactic acid bacteria exopolysaccharides. Carbohydr. Polym. https://doi.org/10.1016/j.carbpol.2019.04.025.

Raigond, P., Ezekiel, R., Raigond, B., 2015. Resistant starch in food: a review. J. Sci. Food Agric. https://doi.org/10.1002/jsfa.6966.

Ramos, P., Herrera, R., Moya-león, M.A., 2014. Anthocyanins: food sources and benefits to consumer's health. In: Handbook of Anthocyanins.

Rinninella, E., Raoul, P., Cintoni, M., Franceschi, F., Miggiano, G.A.D., Gasbarrini, A., et al., 2019. What is the healthy gut microbiota composition? A changing ecosystem across age, environment, diet, and diseases. Microorganisms. https://doi.org/10.3390/microorganisms7010014.

Rodrigues, D., Freitas, A.C., Pereira, L., Rocha-Santos, T.A.P., Vasconcelos, M.W., Roriz, M., et al., 2015. Chemical composition of red, brown and green macroalgae from Buarcos bay in Central West Coast of Portugal. Food Chem. https://doi.org/10.1016/j.foodchem.2015.03.057.

Rodrigues, F., Cádiz-Gurrea, M.D.L.L., Nunes, M.A., Pinto, D., Vinha, A.F., Linares, I.B., et al., 2018. Cosmetics., https://doi.org/10.1016/B978-0-12-813572-3.00012-9.

Rolim, P.M., Hu, Y., Gänzle, M.G., 2019. Sensory analysis of juice blend containing isomalto-oligosaccharides produced by fermentation with Weissella cibaria. Food Res. Int. https://doi.org/10.1016/j.foodres.2018.08.089.

Rosales-Mendoza, S., Salazar-González, J.A., Decker, E.L., Reski, R., 2016. Implications of plant glycans in the development of innovative vaccines. Expert Rev. Vaccines 15, 915–925. https://doi.org/10.1586/14760584.2016.1155987.

Rufino, S.M., Alves, R.E., De Brito, E.S., Pérez-jiménez, J., Saura-calixto, F., Mancini-filho, J., 2010. Bioactive compounds and antioxidant capacities of 18 non-traditional tropical fruits from Brazil. Food Chem. 121, 996–1002. https://doi.org/10.1016/j.foodchem.2010.01.037.

Rycroft, C.E., Jones, M.R., Gibson, G.R., Rastall, R.A., 2001. A comparative in vitro evaluation of the fermentation properties of prebiotic oligosaccharides. J. Appl. Microbiol. https://doi.org/10.1046/j.1365-2672.2001.01446.x.

Sagum, R., Arcot, J., 2000. Effect of domestic processing methods on the starch, non-starch polysaccharides and in vitro starch and protein digestibility of three varieties of rice with varying levels of amylose. Food Chem. https://doi.org/10.1016/S0308-8146(00)00041-8.

Salazar, N., Ruas-Madiedo, P., Kolida, S., Collins, M., Rastall, R., Gibson, G., et al., 2009. Exopolysaccharides produced by Bifidobacterium longum IPLA E44 and Bifidobacterium animalis subsp. lactis IPLA R1 modify the composition and metabolic activity of human faecal microbiota in pH-controlled batch cultures. Int. J. Food Microbiol. https://doi.org/10.1016/j.ijfoodmicro.2009.08.017.

Salazar, N., Gueimonde, M., de los Reyes-Gavilán, C.G., Ruas-Madiedo, P., 2016. Exopolysaccharides produced by lactic acid bacteria and bifidobacteria as fermentable substrates by the intestinal microbiota. Crit. Rev. Food Sci. Nutr. https://doi.org/10.1080/10408398.2013.770728.

Salonen, A., Lahti, L., Salojärvi, J., et al., 2014. Impact of diet and individual variation on intestinal microbiota composition and fermentation products in obese men. ISME J. 8, 2218–2230. https://doi.org/10.1038/ismej.2014.63.

Saville, B.A., Saville, S., 2018. Xylooligosaccharidesand arabinoxylanoligosaccharides and their application as prebiotics. Appl. Food. Biotechnol. https://doi.org/10.22037/afb.v%vi%i.20212.

Schwab, C., Mastrangelo, M., Corsetti, A., Gänzle, M., 2008. Formation of oligosaccharides and polysaccharides by lactobacillus reuteri LTH5448 and Weissella cibaria 10M in sorghum sourdoughs. Cereal Chem. https://doi.org/10.1094/CCHEM-85-5-0679.

Sheng, K., He, S., Sun, M., Zhang, G., Kong, X.-W., Wang, J., et al., 2020. Synbiotic supplementation containing Bifidobacterium infantis and xylooligosaccharide alleviates dextran sulfate sodium-induced ulcerative colitis. Food Funct. https://doi.org/10.1039/d0fo00518e.

Singh, D.P., Khare, P., Bijalwan, V., Baboota, R.K., Singh, J., Kondepudi, K.K., et al., 2017a. Coadministration of isomalto-oligosaccharides augments metabolic health benefits of cinnamaldehyde in high fat diet fed mice. Biofactors. https://doi.org/10.1002/biof.1381.

Singh, D.P., Singh, J., Boparai, R.K., Zhu, J.H., Mantri, S., Khare, P., et al., 2017b. Isomalto-oligosaccharides, a prebiotic, functionally augment green tea effects against high fat diet-induced metabolic alterations via preventing gut dysbacteriosis in mice. Pharmacol. Res. https://doi.org/10.1016/j.phrs.2017.06.015.

Singh, D.P., Singh, S., Bijalwan, V., Kumar, V., Khare, P., Baboota, R.K., et al., 2018. Co-supplementation of isomalto-oligosaccharides potentiates metabolic health benefits of polyphenol-rich cranberry extract in high fat diet-fed mice via enhanced gut butyrate production. Eur. J. Nutr. https://doi.org/10.1007/s00394-017-1561-5.

Singh, A.K., Cabral, C., Kumar, R., Ganguly, R., Rana, H.K., Gupta, A., et al., 2019. Beneficial effects of dietary polyphenols on gut microbiota and strategies to improve delivery efficiency. Nutrients. https://doi.org/10.3390/nu11092216.

Sriamornsak, P., 2007. Chemistry of Pectin and Its Pharmaceutical Uses: A Review. pp. 207–228.

Stanić, Z., 2017. Curcumin, a compound from natural sources, a true scientific challenge—a review. Plant Foods Hum. Nutr. https://doi.org/10.1007/s11130-016-0590-1.

Stewart, M.L., Zimmer, J.P., 2018. Postprandial glucose and insulin response to a high-fiber muffin top containing resistant starch type 4 in healthy adults: a double-blind, randomized, controlled trial. Nutrition. https://doi.org/10.1016/j.nut.2018.01.002.

Strain, C.R., Collins, K.C., Naughton, V., McSorley, E.M., Stanton, C., Smyth, T.J., et al., 2020. Effects of a polysaccharide-rich extract derived from Irish-sourced Laminaria digitata on the composition and metabolic activity of the human gut microbiota using an in vitro colonic model. Eur. J. Nutr. https://doi.org/10.1007/s00394-019-01909-6.

Su, D.L., Li, P.J., Quek, S.Y., Huang, Z.Q., Yuan, Y.J., Li, G.Y., et al., 2019. Efficient extraction and characterization of pectin from orange peel by a combined surfactant and microwave assisted process. Food Chem. https://doi.org/10.1016/j.foodchem.2019.01.200.

Tayyem, R.F., Heath, D.D., Al-Delaimy, W.K., Rock, C.L., 2006. Curcumin content of turmeric and curry powders. Nutr. Cancer. https://doi.org/10.1207/s15327914nc5502_2.

Tingirikari, J.M.R., 2018. Microbiota-accessible pectic poly- and oligosaccharides in gut health. Food Funct. https://doi.org/10.1039/c8fo01296b.

Tribess, T.B., Hernández-Uribe, J.P., Méndez-Montealvo, M.G.C., Menezes, E.W., Bello-Perez, L.A., Tadini, C.C., 2009. Thermal properties and resistant starch content of green banana flour (Musa cavendishii) produced at different drying conditions. LWT - Food Sci. Technol. https://doi.org/10.1016/j.lwt.2008.12.017.

Tsao, R., 2010. Chemistry and biochemistry of dietary polyphenols. Nutrients. https://doi.org/10.3390/nu2121231.

Ungaro, R., Mehandru, S., Allen, P.B., Peyrin-biroulet, L., Colombel, J., 2017. Ulcerative colitis. Lancet 389. https://doi.org/10.1016/S0140-6736(16)32126-2.

Uzal, F.A., Freedman, J.C., Shrestha, A., Theoret, J.R., Garcia, J., Awad, M.M., et al., 2014. Towards an understanding of the role of Clostridium perfringens toxins in human and animal disease. Future Microbiol. https://doi.org/10.2217/fmb.13.168.

Vanderbeek, P.B., Fasano, C., O'Malley, G., Hornstein, J., 2007. Esophageal obstruction from a hygroscopic pharmacobezoar containing glucomannan. Clin. Toxicol. https://doi.org/10.1080/15563650601006215.

Vanitha, T., Khan, M., 2020. Role of pectin in food processing and food packaging. In: Pectins—Extraction Purification, Characterization and Applications., https://doi.org/10.5772/intechopen.83677.

Vázquez, M.J., Alonso, J.L., Domínguez, H., Parajó, J.C., 2000. Xylooligosaccharides: manufacture and applications. Trends Food Sci. Technol. https://doi.org/10.1016/S0924-2244(01)00031-0.

Vera-Guzmán, A.M., Aispuro-Hernández, E., Vargas-Arispuro, I., Islas-Osuna, M.A., Martínez-Téllez, M.Á., 2019. Expression of antioxidant-related genes in flavedo of cold-stored grapefruit (Citrus paradisi Macfad cv. Rio Red) treated with pectic oligosaccharides. Sci. Hortic. (Amsterdam). https://doi.org/10.1016/j.scienta.2018.08.035.

Vinha, A.F., Rodrigues, F., Nunes, M.A., MBPP, O., 2018. Natural Pigments and Colorants in Foods and Beverages. Elsevier Inc, https://doi.org/10.1016/B978-0-12-813572-3.00011-7.

Vinogradov, E., Bock, K., 1998. Structural determination of some new oligosaccharides and analysis of the branching pattern of isomaltooligosaccharides from beer. Carbohydr. Res. https://doi.org/10.1016/S0008-6215(98)00119-0.

Virgen-Ortiz, J.J., Morales-Ventura, J.M., Colín-Chávez, C., Esquivel-Chávez, F., Vargas-Arispuro, I., Aispuro-Hernández, E., et al., 2019. Postharvest application of pectic-oligosaccharides on quality attributes, activities of defense-related enzymes, and anthocyanin accumulation in strawberry. J. Sci. Food Agric. https://doi.org/10.1002/jsfa.10207.

Vital, M., Howe, A., Bergeron, N., Krauss, R.M., Jansson, J.K., Tiedje, J.M., 2018. Metagenomic insights into the degradation of resistant starch by human gut microbiota. Appl. Environ. Microbiol. https://doi.org/10.1128/AEM.01562-18.

Vojdani, A., Vojdani, E., Herbert, M., Kharrazian, D., 2020. Correlation between antibodies to bacterial lipopolysaccharides and barrier proteins in sera positive for ASCA and ANCA. Int. J. Mol. Sci. https://doi.org/10.3390/ijms21041381.

Wang, H.F., Lim, P.S., Kao, M.D., Chan, E.C., Lin, L.C., Wang, N.P., 2001. Use of isomalto-oligosaccharide in the treatment of lipid profiles and constipation in hemodialysis patients. J. Ren. Nutr. https://doi.org/10.1053/jren.2001.22485.

Wang, J., Ghosh, S.S., Ghosh, S., 2017a. Curcumin improves intestinal barrier function: modulation of intracellular signaling, and organization of tight junctions. Am. J. Physiol. Cell Physiol. https://doi.org/10.1152/ajpcell.00235.2016.

Wang, L., Hu, L., Yan, S., Jiang, T., Fang, S., Wang, G., et al., 2017b. Effects of different oligosaccharides at various dosages on the composition of gut microbiota and short-chain fatty acids in mice with constipation. Food Funct. https://doi.org/10.1039/c7fo00031f.

Wang, H., Liu, D., Ji, Y., Liu, Y., Xu, L., Guo, Y., 2020a. Dietary supplementation of black rice anthocyanin extract regulates cholesterol metabolism and improves gut microbiota dysbiosis in C57BL/6J mice fed a high-fat and cholesterol diet. Mol. Nutr. Food Res. 64 (8). https://doi.org/10.1002/mnfr.201900876, e1900876.

Wang, Y., Li, L., Ye, C., Yuan, J., Qin, S., 2020b. Alginate oligosaccharide improves lipid metabolism and inflammation by modulating gut microbiota in high-fat diet fed mice. Appl. Microbiol. Biotechnol. 104 (8), 3541–3554. https://doi.org/10.1007/s00253-020-10449-7.

Wei, Y., Gong, J., Zhu, W., Tian, H., Ding, C., Gu, L., et al., 2016. Pectin enhances the effect of fecal microbiota transplantation in ulcerative colitis by delaying the loss of diversity of gut flora. BMC Microbiol. https://doi.org/10.1186/s12866-016-0869-2.

Wells, M.L., Potin, P., Craigie, J.S., Raven, J.A., Merchant, S.S., Helliwell, K.E., et al., 2017. Algae as nutritional and functional food sources: revisiting our understanding. J. Appl. Phycol. https://doi.org/10.1007/s10811-016-0974-5.

Wiciński, M., Gębalski, J., Mazurek, E., Podhorecka, M., Śniegocki, M., Szychta, P., et al., 2020. The influence of polyphenol compounds on human gastrointestinal tract microbiota. Nutrients. https://doi.org/10.3390/nu12020350.

Wilkowska, A., Nowak, A., Antczak-Chrobot, A., Motyl, I., Czy, A., Paliwoda, A., 2019. Structurally different pectic oligosaccharides produced from apple pomace and their biological activity in vitro. Foods. https://doi.org/10.3390/foods8090365.

Wursch, P., Koellreutter, B., Schweizer, T.F., 1989. Hydrogen excretion after ingestion of five different sugar alcohols and lactulose. Eur. J. Clin. Nutr. 43, 819–825.

Yang, J., Li, Q., Henning, S.M., Zhong, J., Hsu, M., Lee, R., et al., 2018. Effects of prebiotic fiber xylooligosaccharide in adenine-induced nephropathy in mice. Mol. Nutr. Food Res. https://doi.org/10.1002/mnfr.201800014.

Yang, J., Summanen, P.H., Henning, S.M., et al., 2015. Xylooligosaccharide supplementation alters gut bacteria in both healthy and prediabetic adults: a pilot study. Front. Physiol. 6, 216. https://doi.org/10.3389/fphys.2015.00216.

Yen, C.H., Tseng, Y.H., Kuo, Y.W., Lee, M.C., Chen, H.L., 2011. Long-term supplementation of isomalto-oligosaccharides improved colonic microflora profile, bowel function, and blood cholesterol levels in constipated elderly people-a placebo-controlled, diet-controlled trial. Nutrition. https://doi.org/10.1016/j.nut.2010.05.012.

Zaman, S.A., Sarbini, S.R., 2016. The potential of resistant starch as a prebiotic. Crit. Rev. Biotechnol. https://doi.org/10.3109/07388551.2014.993590.

Zanjani, M.A.K., 2012. Microencapsulation of Lactobacillus casei with calcium alginate-resistant starch and evaluation of survival and sensory properties in cream-filled cake. African J. Microbiol. Res. https://doi.org/10.5897/ajmr12.972.

Zhang, H., Tsao, R., 2016. Dietary polyphenols, oxidative stress and antioxidant and anti-inflammatory effects. Curr. Opin. Food Sci. https://doi.org/10.1016/j.cofs.2016.02.002.

Zhang, X., Tian, H., Gu, L., Nie, Y., Ding, C., Ge, X., et al., 2018a. Long-term follow-up of the effects of fecal microbiota transplantation in combination with soluble dietary fiber as a therapeutic regimen in slow transit constipation. Sci. China Life Sci. https://doi.org/10.1007/s11427-017-9229-1.

Zhang, Z., Wang, X., Han, S., Liu, C., Liu, F., 2018b. Effect of two seaweed polysaccharides on intestinal microbiota in mice evaluated by illumina PE250 sequencing. Int. J. Biol. Macromol. https://doi.org/10.1016/j.ijbiomac.2018.01.192.

Zhong, W., Lu, X., Shi, H., Zhao, G., Song, Y., Wang, Y., et al., 2019. Distinct microbial populations exist in the mucosa-associated microbiota of diarrhea predominant irritable bowel syndrome and ulcerative colitis. J. Clin. Gastroenterol. 53, 660–672. https://doi.org/10.1097/MCG.0000000000000961.

Zhou, Y., Zhi, F., 2016. Lower level of Bacteroides in the gut microbiota is associated with inflammatory bowel disease: a meta-analysis. Biomed. Res. Int. https://doi.org/10.1155/2016/5828959.

Zhou, X., Qi, W., Hong, T., Xiong, T., Gong, D., Xie, M., et al., 2018. Exopolysaccharides from lactobacillus plantarum NCU116 regulate intestinal barrier function via STAT3 signaling pathway. J. Agric. Food Chem. https://doi.org/10.1021/acs.jafc.8b03340.

Zhou, Y., Cui, Y., Qu, X., 2019. Exopolysaccharides of lactic acid bacteria: structure, bioactivity and associations: a review. Carbohydr. Polym. https://doi.org/10.1016/j.carbpol.2018.11.093.

Zhu, Y., Sun, H., He, S., Lou, Q., Yu, M., Tang, M., et al., 2018. Metabolism and prebiotics activity of anthocyanins from black rice (*Oryza sativa* L.) in vitro. PLoS One. https://doi.org/10.1371/journal.pone.0195754.

Zumbé, A., Brinkworth, R.A., 1992. Comparative studies of gastrointestinal tolerance and acceptability of milk chocolate containing either sucrose, isomalt or sorbitol in healthy consumers and type II diabetics. Z. Ernahrungswiss. https://doi.org/10.1007/BF01612551.

CHAPTER 3

Probiotics and Prebiotics in Chronic Kidney Disease

MILENA BARCZA STOCKLER-PINTO[a,b] • JULIANA SARAIVA DOS ANJOS[a,b]
[a]Nutrition Faculty, Department of Nutrition and Dietary, Fluminense Federal University (UFF), Niterói, RJ, Brazil, [b]Medical School, Graduate Program in Cardiovascular Sciences, Fluminense Federal University (UFF), Niterói, RJ, Brazil

INTRODUCTION

According to the Improving Global Outcomes (KDIGO) foundation, chronic kidney disease (CKD) is defined as abnormalities of kidney structure or function present for > 3 months that have health implications. It is classified based on cause, the glomerular filtration rate (GFR) category, and the albuminuria category. The CKD severity is classified into five stages according to GFR. Stages one to four correspond to predialysis, and stage five is related to hemodialysis and peritoneal dialysis (Kidney Disease Improving Global Outcomes KDIGO, 2013). Recent publications showed that CKD affects 8%–16% of the world's population. For that reason, researchers state that CKD management includes cardiovascular risk reduction, treatment of albuminuria, avoidance of uremic toxins, and adjustments to drug dosage. Patients should be controlled with respect to hyperkalemia, metabolic acidosis, anemia, and metabolic abnormalities (Hill et al., 2016; Chen et al., 2019).

CKD triggers metabolic disturbances at all stages of disease, such as obesity, impaired immune system, oxidative stress, inflammation, and gut microbiota imbalance (Kanbay et al., 2018; Chen et al., 2019). CKD patients also have several risk factors for cardiovascular disease (CVD). Studies identified the imbalance of gut microbiota as a new factor that may contribute to inflammation and CVD in these patients (Li and Tang, 2018; Meijers et al., 2018).

Some uremic toxins produced by gut microbiota such as indoxyl sulfate (IS), *p*-cresyl sulfate (p-CS), indole-3-acetic acid (IAA), and trimethylamine N-oxide (TMAO) accumulate in CKD patients and can lead to conditions such as inflammation and oxidative stress, which are related in part to CVD risk in these conditions. In general, these toxins are metabolized in the liver and produced by some specific intestinal bacteria such as a metabolite of tryptophan, phenylalanine, tyrosine, choline, and L-carnitine derived from protein sources (Ito and Yoshida, 2014; Liu et al., 2014; Lau et al., 2018; Vahed et al., 2020).

Recently, studies proved that the gut-epithelial barrier is a CKD breakdown, leading to the translocation of bacteria-derived uremic toxins into systemic circulation. This induces inflammation and leukocyte influx, which causes endocytosis of transcellular tight junction proteins (Nishiyama et al., 2019). In dysbiosis conditions, intestinal permeability is increased. This leads to lipopolysaccharide (LPS) translocation. LPS promotes nuclear factor-kappa B (NF-κB) pathway activation, and this nuclear factor is responsible for cytokines production, such as tumor necrosis factor-alpha (TNF-α) and interleukin-6 (IL-6). (Mihai et al., 2018; Mikusic et al., 2020).

Furthermore, these uremic toxins exert a wide range of influence in endothelial dysfunction and vascular calcification while exacerbating reactive oxygen species (ROS), and progression of CKD and mortality (Opdebeeck et al., 2019). Studies show that IS exacerbates ROS production by nicotinamide adenine dinucleotide phosphate (NADPH) oxidase activation and promotes inflammation in those patients. Additionally, p-CS and IAA also lead to increased ROS production and are positively correlated with the C-reactive protein (CRP) (Lin et al., 2012; Shimazu et al., 2013; Lano et al., 2020). A recent study conducted by our group showed that nuclear factor-erythroid 2-related factor-2 (Nrf2) and NF-κB pathway stimulation (Stockler-Pinto et al., 2018) is a possible mechanism to induce oxidative stress and inflammation by uremic toxins. Moreover, TMAO can induce atherosclerosis by the inhibition of reverse cholesterol transport and bile acid synthesis (Bäckhed, 2013; Tang et al., 2013). In this sense, high levels of uremic toxins are considered a predictor of cardiovascular disease in CKD patients (Lin et al., 2014; Dou et al., 2015; Jing et al., 2016) (Fig. 1).

Probiotics and Prebiotics in Foods. https://doi.org/10.1016/B978-0-12-819662-5.00008-2

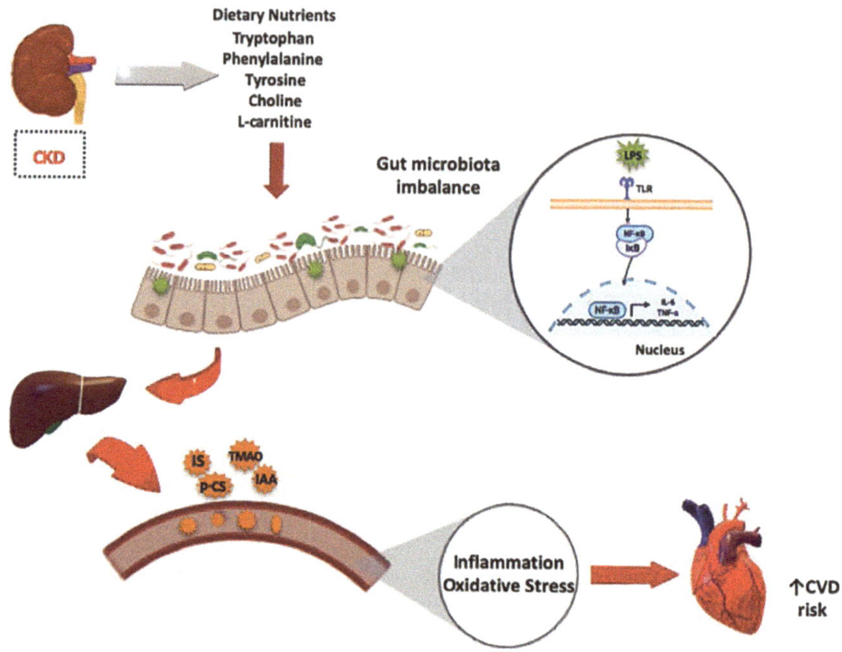

FIG. 1 **Crosstalk between gut microbiota and chronic kidney disease.**
CKD impairs the balance among bacteria. The consequences of specific dietary nutrient consumption are impairment of the intestinal barrier and increased intestinal permeability and translocation of bacteria and lipopolysaccharide (LPS). LPS could activate Toll-like receptor 4 (TLR4) and nuclear factor-kappa B (NF-κB) pathways, leading to cytokine production such as tumor necrosis factor-alpha (TNF-α) and interleukin-6 (IL-6); tryptophan, phenylalanine, tyrosine, choline, and L-carnitine are fermented by some specific intestinal bacteria, metabolized in the liver, and then converted to indoxyl sulfate (IS), p-cresyl sulfate (p-CS), indole-3-acetic acid (IAA), and trimethylamine N-oxide (TMAO); in the bloodstream, these toxins stimulate inflammation and oxidative stress while increasing the risk of cardiovascular disease (CVD).

Studies are in progress to reinforce the association between gut microbiota imbalance and inflammation in CKD patients. Recently, an advanced technique was applied to analyze fecal microbiota, a 16S ribosomal DNA pyrosequencing. The authors demonstrated that some genera such as *Parasutterella*, *Lactobacillus*, *Paraprevotella*, *Clostridium* sensu stricto, and *Desulfovibrio* are altered in CKD gut microbiota and correlated with disease-severity indicators such as the glomerular filtration rate. Beyond that, the decreased abundance of *Akkermansia* (an important probiotic) was negatively correlated with the production of interleukin-10, an antiinflammatory cytokine (Li et al., 2019).

CKD patients present other factors that can also impair maintenance of the gut microbiota and lead to a less-acid pH. It should be highlighted that consuming lower amounts of fiber, because restrictions of fruits and vegetables are necessary to control potassium levels in CKD patients contributes to lower production of short-chain fatty acids and a lack of pH control (Mafra et al., 2014; Vaziri et al., 2015).

Thereby, the microbiota remodeling appears to be a strategy to reduce the uremic toxins and their repercussions, including mainly oxidative stress, inflammation, and risk of CVD development. Thus, the use of dietary interventions as well as probiotics and prebiotics has been suggested in CKD (Vanholder et al., 2016; Ramezani and Raj, 2014; Koppe et al., 2015; Mafra et al., 2019).

EFFECTS OF PROBIOTICS IN CKD

Around the year 2000, researchers hypothesized that an increase in bifidobacteria produced compounds to inhibit potential pathogens, reducing blood ammonia levels by producing vitamins and digestive enzymes, and that that scenario could improve some conditions in CDK (Chow, 2000). Then, some experimental studies were conducted. In 2005, Ranganathan et al. (2005) published results from 5/6 nephrectomized rats as a chronic renal failure model randomized in different groups according to probiotic type: (1) *Bacillus pasteurii*; (2) *Lactobacillus sporogenes*; (3) *Lactobacillus acidophilus, Streptococcus thermophilus, Lactobacillus bulgaricus, Bifidus, Lactobacillus casei, Lactobacillus reuteri;* (4) *L. acidophilus, L. bulgaricus, Bifidus, S. thermophilus;* and (5) *Saccharomyces boulardii*, and compared that to the control group. After 16 weeks of treatment, the *B. pasteurii* and *L. sporogenes* groups significantly prolonged the lifespan of uremic rats and showed a reduction in blood urea nitrogen levels. The next year, the same group tested another probiotic type, a urease-positive bacterium *Sporosarcina pasteurii* (Sp), as a potential urea-targeted component in the same animal model. The authors demonstrated reduced blood urea-nitrogen levels and significantly prolonged the lifespan of uremic animals (Ranganathan et al., 2006).

Since 2005, clinical studies have been done to understand the impact of different strains in CKD patients. However, it is important to mention that a few animal studies were conducted before the clinical studies. Taki et al. (2005) orally administered *Bifidobacterium longum* to return the disturbed microflora to normal in hemodialysis patients and found that this probiotic was effective at reducing a uremic toxin (IS). Those results were corroborated with Eidi et al. (2018), who also observed a decrease in a uremic toxin (p-CS) in hemodialysis patients after administering *Lactobacillus rhamnosus*. On the other hand, after supplementing *S. thermophilus, L. acidophilus,* and *B. longum* in nondialysis (de Faria Barros et al., 2018) and in hemodialysis patients (Borges et al., 2018), researchers did not observe any changes in IS, p-CS, or IAA plasma levels.

To understand the effects of intestinal barrier translocation in CKD, Wei et al. (2014) supplemented a probiotic such as *Bifidobacterium animalis* subsp. lactis Bi-07 in uremic rats. This revealed that the probiotic decreases bacterial translocation and ameliorates microinflammation by the recovery of intestinal mucosal integrity. The main mechanisms associated with probiotic action in CKD are related to maintaining the integrity of the intestinal barrier, the immune modulation response, decreases of pathogenic bacteria, and changes in gut microbiota composition (Mafra et al., 2019).

In dysbiosis conditions, LPS promotes NF-κB pathway activation, which leads to cytokine production. Thus, probiotics could inhibit LPS translocation, restrain NF-κB activation, and decrease cytokine secretion (Koppe et al., 2015). In this sense, Wang et al. (2015) showed a reduction in proinflammatory cytokines (TNF-α and IL-6) after probiotic intervention in peritoneal dialysis; moreover, there were also increased serum levels of an antiinflammatory cytokine (interleukin-10). Nevertheless, an intervention with *S. thermophilus, L. acidophilus,* and *B. longum* failed to reduce uremic toxins, as demonstrated above, and an increase in IL-6 plasma levels was found in hemodialysis patients (Borges et al., 2018). Other studies did not demonstrate any significant effects of probiotics in inflammation (Natarajan et al., 2014; Soleimani et al., 2016; Shariaty et al., 2017).

The associated improvement in intestinal imbalance is still poorly defined and unclear. Few studies have assayed the impact of probiotics in CKD fecal microbiota. Borges et al. (2018) pointed any changes in the microbiota profile after probiotics supplementation. Recently, Liu et al. (2020) revealed that probiotics did not alter the species diversity of the fecal microbiome in HD patients. However, it restored the community composition, in particular in nondiabetic HD patients, and enhanced the proportions of the families *Bacteroidaceae* and *Enterococcaceae* while reducing *Ruminococcaceae, Halomonadaceae, Peptostreptococcaceae, Clostridiales Family XIII. Incertae Sedis,* and *Erysipelotrichaceae*; additionally, the authors showed that probiotics reduced several uremic toxins Table 1.

Despite the promising adjuvant therapy for CKD, as observed, clinical studies have reported discrepant results. According to Vaziri et al. (2015), one of the hypotheses for the discrepant results is related to some conditions presented in CKD (uremia, dietary and medicinal regimens) that promote an adverse milieu to probiotic bacteria therapy. In particular, patients who are undergoing dialysis have high levels of uremic toxins, which leads to raised urea influx into the intestinal lumen. That way, the urea generates a milieu for the expansion of urease-possessing bacteria, providing a greater amount of ammonia and ammonium hydroxide, which are detrimental to the intestinal barrier. Although the impact of probiotics in the gut microbiota is widely debated, insufficient studies have been published to understand this mechanism in CKD.

Because the beneficial effects of probiotics are not entirely elucidated so far, some systematic reviews and metaanalyses were conducted to guide probiotic prescriptions. In 2018, a manuscript included eight randomized controlled trials with nondialysis and hemodialysis patients to analyze a metaanalysis. It concluded that probiotics could reduce p-CS levels and increase IL-6 levels. The limited number of studies and small sample sizes were pointed out, outlining the need for more studies (Jia et al., 2018). Another metaanalysis was accomplished that considered 10 randomized controlled trials; the manuscript highlighted that probiotic supplements exerted an effect on urea levels in nondialysis CKD patients, although no evidence suggested outcomes in the reduction of uric acid, C-reactive protein, creatinine, and estimated glomerular filtration rate preservation in CKD (Tao et al., 2019).

TABLE 1
Summary of Studies Investigating the Relationship Between Probiotic Intake and CKD.

First Author and Year	Probiotics/Day	Study	Results
Studies in experimental models of CKD			
Ranganathan et al. (2005)	(1) 10^9 CFU—*B. pasteurii*; (2) 10^8 CFU—*L. sporogenes*; (3) 10^{10} CFU—*L. acidophilus*, *S. thermophilus*, *L. bulgaricus*, *Bifidus*, *L. casei*, *L. reuteri*; (4) 5×10^{10} CFU—*L. acidophilus*, *L. bulgaricus*, *Bifidus*, *S. thermophilus*; (5) 10^9 CFU—*S. boulardii*	5/6 nephrectomized Sprague-Dawley rats randomized in five groups, according to probiotic type (1–5), 16 weeks	↑ Lifespan ↓ blood urea nitrogen
Ranganathan et al. (2006)	*S. pasteurii* (Sp): 10^9 CFU	5/6 nephrectomized Sprague-Dawley rats, 16 weeks	↑ Lifespan ↓ blood urea nitrogen
Wei et al. (2014)	*B. animalis* subsp. lactis Bi-07: 10^9 CFU	5/6 nephrectomized Sprague-Dawley rats, 4 weeks	↓ Intestinal permeability ↓ IL-6, CRP, TNF-alpha ↓ endotoxin
Clinical studies			
Taki et al. (2005)	*B. longum:* 3×10^9 CFU—4 weeks 6×10^9 CFU—4th to 8th week 12×10^9 CFU—8th to 12th week	27 HD patients 3 months	↓ Plasma IS
Ranganathan et al. (2010)	*L. acidophilus*, *B. longum*, and *S. thermophilus*: 9×10^{10} CFU	46 nondialysis patients Randomized, double-blind, placebo-controlled, with crossover, 6 months	↑ Quality of life ↓ blood urea nitrogen
Alatriste et al. (2014)	*L. casei* Shirota: Dose A: 8×10^9 CFU Dose B: 16×10^9 CFU	30 nondialysis patients Randomized clinical trial, 2 months (15 in the probiotic group dose A and 15 in the probiotic group dose B)	↓ Urea in dose B
Natarajan et al. (2014)	*S. thermophilus*, *L. acidophilus*, and *B. longum*: 1.8×10^{11} CFU	22 HD patients Two 2-month treatment periods separated by 2 months washout and crossover	↔ Indoxyl glucuronide ↔ CRP
Wang et al. (2015)	*B. bifidum*, *B. catenulatum*, *B. longum*, and *L. plantarum*: 10^9 CFU of each strain	39 PD patients Randomized, double-blind, placebo-controlled, 6 months (21 in the probiotic group and 18 in the placebo group)	↓ Serum TNF-alpha, IL-5, IL-6 ↓ endotoxin ↑ serum IL-10
Soleimani et al. (2016)	*L. acidophilus*, *L. casei*, and *B. bifidum*: 2×10^9 CFU	60 diabetic patients on HD Randomized, double-blind placebo-controlled, 3 months (30 in the probiotic group and 30 in the placebo group)	↔ Hb ↔ CRP

TABLE 1
Summary of Studies Investigating the Relationship Between Probiotic Intake and CKD—cont'd

First Author and Year	Probiotics/Day	Study	Results
Shariaty et al. (2017)	3×10^{10} CFU—*L. acidophilus*, 3×10^{9} CFU—*L. casei*, 7×10^{9} CFU—*L. rhamnosus*, 5×10^{8} CFU—*L. bulgaricus* 2×10^{10} CFU—*B. breve*, 1×10^{9} CFU—*B. longum* and 3×10^{8} CFU—*S. thermophilus*	34 HD patients Randomized clinical trial, 3 months (17 in the probiotic group and 17 in the placebo group)	↔ CRP
Borges et al. (2018)	*S. thermophilus*, *L. acidophilus*, and *B. longum*: 9×10^{10} CFU	33 HD patients Randomized clinical trial, 3 months (16 in the probiotic group and 17 in the placebo group)	↑ Plasma IS ↔ p-CS, IAA ↔ CRP, IL-6
de Faria Barros et al. (2018)	*S. thermophilus*, *L. acidophilus*, and *B. longum*: 9×10^{10} CFU	22 nondialysis patients Randomized clinical trial, 3 months (12 in the probiotic group and 10 in the placebo group)	↑ IL-6 levels ↔ TMAO, IS, p-CS, IAA ↔ CRP
Eidi et al. (2018)	*L. Rhamnosus:* 1.6×10^{7} CFU	42 HD patients Randomized clinical trial, 1 month (21 in the probiotic group and 21 in the placebo group)	↓ Serum p-CS
Liu et al. (2020)	2.2×10^{9} CFU—*B. longum*, 0.53×10^{9} CFU—*L. acidophilus* 1.1×10^{9} CFU—*E. faecalis*	45 HD patients Randomized clinical trial, 6 months (22 in the probiotic group and 23 in the placebo group)	↓ Serum and fecal 1-methylinosine ↓ fecal 3-guanidinopropionic acid ↓ fecal phenol ↓ serum Indole-3-acetic acid-O-glucuronide

CFU, colony-forming units; *CRP*, C-reactive protein; *p-CS*, -*p*-cresyl sulfate; *Hb*, hemoglobin; *HD*, hemodialysis; *IAA*, indole-3-acetic acid; *IL-10*, interleukin-10; *IL-5*, interleukin-5; *Il-6*, interleukin-6; *IS*, indoxyl sulfate; *PD*, peritoneal dialysis; *TMAO*, trimethylamine N-oxide; *TNF-alpha*, tumor necrosis factor-alpha. ↔: no change; ↑: increase; ↓: decrease.

The main difficulty in comprehending the probiotic role in CKD is related to the diversity of strains, the heterogeneity of the populations, study sample sizes, and different experimental designs, which leads to tricky conclusions about the intervention. For that reason, experimental studies to advance and improve knowledge concerning the effects of probiotics in CKD are still greatly needed.

EFFECTS OF PREBIOTICS IN CKD

Roberfroid and Gibson created the term "prebiotic" in 1995 (Gibson and Roberfroid, 1995). However, in 2004 it was updated and is now defined as "selectively fermented ingredients that allow specific changes, both in the composition and/or activity in the gastrointestinal microflora, that confer benefits upon the host well-being and health" (Gibson et al., 2004). This is a well-known strategy for modulating the gut microbiota (Bindels et al., 2015; Holscher, 2017).

The types of fibers most used in the studies are oligosaccharides, resistant starch, chemically synthesized fibers, inulin, bran, and cereal fiber (Hijová et al., 2019). Prebiotics support the proliferation of bacteria such as bifidobacteria and lactobacilli, which are beneficial for increasing short-chain fatty acids (SCFAs) (Felizardo et al., 2019).

SCFA is the final product of nondigestible carbohydrate fermentation (Poul et al., 2003). The SCFA acts as an essential molecular signal between the microbiota and host health. It is known that prebiotics increase the production of SCFAs that lead to gut microbiota modulation, therefore increasing the intestinal barrier integrity, resulting in a decrease of uremic toxins, oxidative stress, and inflammation (de Lopes et al., 2018; Morrison and Preston, 2016). SCFAs can reduce the production of LPS, which activates the proinflammatory cascade of NF-κB (Hobby et al., 2019).

Wang et al. (2019) showed that there is a difference in SCFA levels between healthy people and CKD patients, where butyrate was almost three times higher in healthy individuals than in CKD patients. They performed a study with rats transplanted with microbiota obtained from CKD patients and supplemented with butyrate. This analysis showed a more significant progression of CKD by an increase in trimethylamine N-oxide (TMAO), which was able to be reversed after supplementation. Therefore, butyrate supplementation can slow CKD progression. The use of prebiotics is already consolidated as a factor that increases the intestinal production of SCFAs (Chiavaroli et al., 2015).

In 2014, Vaziri et al. (2014) published the first study that supplemented resistant starch with the aim of attenuating CKD progression in a CKD experimental model. According to the authors, rats were supplemented with amylopectin or high fermentable fiber (amylose maize resistant starch) for 3 weeks. It was demonstrated that a high resistant starch diet retards CKD progression and attenuates fibrosis, oxidative stress, and inflammation in rats. It also encourages the development of clinical studies with resistant starch and prebiotics in general.

Yang et al. (2018) conducted a study with male C57BL/6J mice. Three groups received adenine for 3 weeks to induce CKD. After this period, they were divided into six groups and received xylooligosaccharide supplementation (XOS). A decreased in Actinobacteria, and p-CS serum levels in all groups supplemented with XOS, control, and CKD was found. Additionally, all XOS-supplemented groups increased SCFAs cecum total, acetate, and butyrate concentrations. Those results show that dietary fiber has the potential to modify inflammation and oxidative stress in CKD.

Another prebiotic commonly used in clinical practice is lactulose, which is a disaccharide composed of galactose and fructose (Ruszkowski and Witkowski, 2019). In a study by Sueyoshi et al. (2019) Wistar rats with adenine-induced CKD received lactulose in different doses (3% and 7.5%). After the supplementation period, the groups that were supplemented with lactulose showed an improvement in serum creatinine and BUN levels. Interstitial fibrosis and serum IS levels were reduced. Thus, lactulose seems to modify the intestinal microbiota of the animals, delaying the progression of CKD and improving their uremic condition.

Researchers conducted a single-blind prospective randomized trial to evaluate the effects of resistant starch supplementation in patients with CKD in HD. At the end of the study, they proved that prebiotic fiber could decrease the plasma concentrations of uremic toxins in HD patients (Sirich et al., 2014). Those results corroborate with Esgalhado et al. (2018), where patients with CKD in HD received a supplementation of resistant starch (16 g) or a placebo (16 g of manioc flour) for 4 weeks and showed a decrease of IL-6, MDA, and IS plasma levels.

The best prebiotics are considered to be oligosaccharides, including fructooligosaccharide (FOS), which is found in foods such as onions, beets, and garlic (Slavin, 2013). To evaluate the benefits of this fiber, a double-blind, placebo-controlled, randomized trial with CKD patients was performed. However, the 3-month supplementation showed no significant difference in uremic toxin levels, only a trend toward decreased p-CS in nondiabetic patients (Ramos et al., 2019).

A single-center, nonrandomized study of patients in HD with a staggering dose of oligofructose-enriched inulin showed that supplementation was effective in promoting a decrease in p-CS concentration. These data are highly relevant because p-CS is related to the inflammatory process in patients with CKD (Meijers et al., 2010).

A recent metaanalysis showed the positive effects of prebiotics in promoting a decrease in oxidative stress and inflammatory markers in patients with CKD (Zheng et al., 2021). Another metaanalysis published in 2015 showed that a balanced diet associated with dietary fiber supplements decreased serum urea and creatinine concentrations, leading to improved renal function (Chiavaroli et al., 2015).

Patients with CKD receive nutritional guidance according to the stage of the disease. It is known that a lower intake of protein leads to a lower production of uremic toxins. However, those toxins in patients with CKD are already high. Patients in HD and PD receive a high-protein diet guideline, leading to a greater expression of uremic solutes. The use of prebiotics can reduce the production of solutes from the colon in these patients without harming protein intake. Studies with prebiotics show that supplementation does not present risks for patients with CKD, thus making them an excellent strategy to improve the conditions of these patients Table 2.

TABLE 2
Summary of Studies Investigating the Relationship Between Prebiotic Intake and CKD.

First Author and Year	Prebiotics/Day	Study	Results
Studies in experimental models of CKD			
Vaziri et al. (2014)	Resistant starch (59%)	Sprague-Dawley rats 0.7% adenine for 2 weeks to induce chronic interstitial nephropathy, high amylose maize resistant starch for 3 weeks	↓ Creatinine clearance ↓ activation of NF-κB ↑ nuclear translocation of Nrf2 ↑ nuclear translocation of p65
Yang et al. (2018)	XOS (2% and 7%)	C57BL/6J mice 0.7% adenine for 3 weeks to induce CKD, XOS 2% and 7% for 3 weeks	*XOS in CKD* ↓ interstitial fibrosis, IL-6 ↓ blood urea nitrogen ↓ creatinine *XOS in all supplemented groups* ↓ Actinobacteria ↓ serum p-CS ↑ SCFAs cecum total, acetate, and butyrate
Sueyoshi et al. (2019)	Lactulose (3% and 7%)	Wistar/ST rats 0.7% adenine for 3 weeks to induce CKD; 3% lactulose; 7% lactulose	↓ Creatinine ↓ blood urea nitrogen ↓ interstitial fibrosis ↓ serum IS
Clinical studies			
Meijers et al. (2010)	Oligofructose-enriched inulin (10 g)	22 HD patients Single-center, nonrandomized, open-label phase I/II study, 4 weeks	↓ Serum p-CS ↓ total p-CS solute removal ↔ serum IS ↔ total IS solute removal
Sirich et al. (2014)	Resistant starch (15 g)	56 HD patients Single blinded prospective randomized trial, 6 weeks (28 in the prebiotic group and 28 in the placebo group)	↓ Plasma p-CS ↓ plasma IS
Esgalhado et al. (2018)	Resistant starch (16 g)	31 HD patients Randomized, double-blind, placebo-controlled trial, 4 weeks (15 in the prebiotic group and 16 in the placebo group)	↓ IL-6 ↓ MDA ↓ plasma IS
Ramos et al. (2019)	FOS (12 g)	46 nondiabetic CKD Double-blind randomized controlled trial, 3 months (23 in the prebiotic group and 23 in the placebo group)	↓ Δ PCS; Δ%PCS

BUN, blood urea nitrogen; *FOS*, frutooligosaccharide; *HD*, hemodialysis; *IS*, indoxyl sulfate; *MDA*, malondialdehyde; *NF-κB*, nuclear factor-kappa B; *Nrf2*, nuclear factor-erythroid 2-related factor-2; *SCFA*, short chain fatty acid; *XOS*, xylooligosaccharide. ↔: no change; ↑: increase; ↓: decrease.

PERSPECTIVES

The gut microbiota role in CKD remains to be established to focus its efforts in understanding this cross-talk and establish efficient therapy for CKD patients. A recently published targeted metabolomics profiling of serum bile acids, short- and medium-chain fatty acids, and uremic solutes in CKD patients in different stages controls individuals (Wu et al., 2020). The dual-omics analyses divulged the association between gut microbial species to changes in circulating metabolites, particularly a set of species positively correlated with the levels of p-CS. Those findings are essential to guide how probiotics could control CKD progression and reduce some metabolites. The dual-omics analyses divulged the association between gut microbial species to changes in circulating metabolites, particularly a set of species positively correlated with the levels of p-CS. Those findings are essential to provide guidance as to what way probiotics could be acting to control CKD progression and reducing some metabolites.

Following the progress and challenges faced with probiotics and prebiotics, researchers have advanced toward synbiotics therapy in CKD. The recent use of prebiotics (glutamine, dietary fiber, and oligosaccharide) and probiotics (*B. longum* strain) in nephrectomized rats ameliorated kidney function and reduced creatinine and blood urea nitrogen, besides IS levels. According to the authors, these results could be attributed to an increase in the proportions of *Bifidobacterium* and Ruminococcus by the probiotic and prebiotic blend (Iwashita et al., 2018).

Clinical studies using synbiotics are still inconclusive and future protocols are demanded.

Concerning scientific developments, paraprobiotics and postbiotics seem to be promising therapies for CKD patients because they can reduce the risk of microbial translocation, infection, and inflammatory conditions. In addition, the use of foods as a vehicle for paraprobiotics and postbiotics has been suggested, although no experimental or clinical studies have been validated yet (Mafra et al., 2019).

CONCLUSIONS

Nowadays, with the rush of everyday life, a balanced diet is often challenging. Patients with CKD need a more balanced diet because of multiple factors, including their decreased renal function and several other disorders. Recently, nonpharmacological strategies have been adopted to improve the life conditions of CKD patients. Probiotics and prebiotics seem to be excellent alternatives. Thus, wide-ranging clinical randomized trials with probiotics, a selection of probiotic strains, and prebiotics should be developed to improve the statistical power and the detection accuracy involving changes in gut microbiota composition, uremic toxins, and inflammation. Indeed, probiotics and prebiotics may be promising therapies to prevent CKD progression and to act in controlling metabolic alterations.

REFERENCES

Alatriste, P.V.M., Arronte, R.U., Espinosa, C.O.G., Cuevas, M.E., 2014. Effect of probiotics on human blood urea levels in patients with chronic renal failure. Nutr. Hosp. 29, 582–590.

Bäckhed, F., 2013. Meat-metabolizing bacteria in atherosclerosis. Nat. Med. 19 (5), 533–534.

Bindels, L.B., Delzenne, N.M., Cani, P.D., Walter, J., 2015. Towards a more comprehensive concept for prebiotics. Nat. Rev. Gastroenterol. Hepatol. 12, 303–310.

Borges, N.A., Carmo, F.L., Stockler-Pinto, M.B., de Brito, J.S., Dolenga, C.J., Ferreira, D.C., Nakao, L.S., Rosado, A., Fouque, D., Mafra, D., 2018. Probiotic supplementation in chronic kidney disease: a double-blind, randomized, placebo-controlled trial. J. Ren. Nutr. 28 (1), 28–36.

Chen, T.K., Knicely, D.H., Grams, M.E., 2019. Chronic kidney disease diagnosis and management: a review. JAMA 322 (13), 1294–1304.

Chiavaroli, L., Mirrahimi, A., Sievenpiper, J.L., Jenkins, D.J.A., Darling, P.B., 2015. Dietary fiber effects in chronic kidney disease: a systematic review and meta-analysis of controlled feeding trials. Eur. J. Clin. Nutr. 69, 761–768.

Chow, J., 2000. Probiotics and prebiotics: a brief overview. J. Ren. Nutr. 12 (2), 76–86.

Li, D.Y., Tang, W.H.W., 2018. Contributory role of gut microbiota and their metabolites towards cardiovascular complications in chronic kidney disease. Semin. Nephrol. 38 (2), 193–205.

de Faria Barros, A., Borges, N.A., Nakao, L.S., Dolenga, C.J., do Carmo, F.L., de Carvalho Ferreira, D., Stenvinkel, P., Bergman, P., Lindholm, B., Mafra, D., 2018. Effects of probiotic supplementation on inflammatory biomarkers and uremic toxins in non-dialysis chronic kidney patients: a double-blind, randomized, placebo-controlled trial. J. Funct. Foods 46, 378–383.

Dou, L., Sallée, M., Cerini, C., Poitevin, S., Gondouin, B., Jourde-Chiche, N., Fallague, K., Brunet, P., Calaf, R., Dussol, B., Mallet, B., Dignat-George, F., Burtey, S., 2015. The cardiovascular effect of the uremic solute indole-3 acetic acid. J. Am. Soc. Nephrol. 26 (4), 876–887.

Eidi, F., Poor-reza Gholi, F., Ostadrahimi, A., Dalili, N., Samadian, F., Barzegari, A., 2018. Effect of *Lactobacillus rhamnosus* on serum uremic toxins (phenol and p-cresol) in hemodialysis patients: a double blind randomized clinical trial. Clin. Nutr. ESPEN 28, 158–164.

Esgalhado, M., Kemp, J.A., Azevedo, R., Paiva, B.R., Stockler-Pinto, M.B., Dolenga, C.J., Borges, N.A., Nakao, L.S., Mafra, D., 2018. Could resistant starch supplementation improve inflammatory and oxidative stress biomarkers and uremic toxins levels in hemodialysis patients? A pilot randomized controlled trial. Food Funct. 9, 6508–6516.

Felizardo, R.J.F., Watanabe, I.K.M., Dardi, P., Rossoni, L.V., Câmara, N.O.S., 2019. The interplay among gut microbiota, hypertension and kidney diseases: the role of short-chain fatty acids. Pharmacol. Res. 141, 366–377.

Gibson, G.R., Probert, H.M., Loo, J.V., Rastall, R.A., Roberfroid, M.B., 2004. Dietary modulation of the human colonic microbiota: updating the concept of prebiotics. Nutr. Res. Rev. 17, 259–275.

Gibson, G.R., Roberfroid, M.B., 1995. Dietary modulation of the human colonic microbiota: introducing the concept of prebiotics. J. Nutr. 125, 1401–1412.

Hijová, E., Bertková, I., Štofilová, J., 2019. Dietary fibre as prebiotics in nutrition. Cent. Eur. J. Public Health 27, 251–255.

Hill, N.R., Fatoba, S.T., Oke, J.L., Hirst, J.A., O'Callaghan, C.A., Lasserson, D.S., Hobbs, F.D., 2016. Global prevalence of chronic kidney disease—a systematic review and meta-analysis. PLoS One 11 (7), e0158765.

Holscher, H.D., 2017. Dietary fiber and prebiotics and the gastrointestinal microbiota. Gut Microbes 8, 172–184.

Hobby, G.P., Karaduta, O., Dusio, G.F., Singh, M., Zybailov, B.L., Arthur, J.M., 2019. Chronic kidney disease and the gut microbiome. Am. J. Physiol. Renal Physiol. 316, F1211–F1217.

Ito, S., Yoshida, M., 2014. Protein-bound uremic toxins: new culprits of cardiovascular events in chronic kidney disease patients. Toxins (Basel) 6 (2), 665–678.

Iwashita, Y., Ohya, M., Yashiro, M., Sonou, T., Kawakami, K., Nakashima, Y., Yano, T., Iwashita, Y., Mima, T., Negi, S., Kubo, K., Tomoda, K., Odamaki, T., Shigematsu, T., 2018. Dietary changes involving *Bifidobacterium longum* and other nutrients delays chronic kidney disease progression. Am. J. Nephrol. 47, 325–332.

Jia, L., Jia, Q., Yang, J., Jia, R., Zhang, H., 2018. Efficacy of probiotics supplementation on chronic kidney disease: a systematic review and meta-analysis. Kidney Blood Press. Res. 43, 1623–1635.

Jing, Y.J., Ni, J.W., Ding, F.H., Fang, Y.H., Wang, X.Q., Wang, H.B., Chen, X.N., Chen, N., Zhan, W.W., Lu, L., Zhang, R.Y., 2016. p-Cresyl sulfate is associated with carotid arteriosclerosis in hemodialysis patients and promotes atherogenesis in apoE −/− mice. Kidney Int. 89 (2), 439–449.

Kanbay, M., Onal, E., Afsar, B., Dagel, T., Yerlikaya, A., Covic, A., Vaziri, N.D., 2018. The crosstalk of gut microbiota and chronic kidney disease: role of inflammation, proteinuria, hypertension, and diabetes mellitus. Int. Urol. Nephrol. 50 (8), 1453–1466.

Kidney Disease Improving Global Outcomes (KDIGO), 2013. KDIGO 2012 clinical practice guideline for the evaluation and management of chronic kidney disease. Kidney Int. Suppl. 3 (1), 1–150.

Koppe, L., Mafra, D., Fouque, D., 2015. Probiotics and chronic kidney disease. Kidney Int. 88, 958–966.

Lano, G., Burtey, S., Sallée, M., 2020. Indoxyl sulfate, a uremic endotheliotoxin. Toxins (Basel) 12 (4), 1–14. https://doi.org/10.3390/toxins12040229. 229.

Lau, W.L., Savoj, J., Nakata, M.B., Vaziri, N.D., 2018. Altered microbiome in chronic kidney disease: systemic effects of gut-derived uremic toxins. Clin. Sci. (Lond.) 132 (5), 509–522.

Li, F., Wang, M., Wang, J., Li, R., Zhang, Y., 2019. Alterations to the gut microbiota and their correlation with inflammatory factors in chronic kidney disease. Front. Cell. Infect. Microbiol. 9, 206.

Lin, C.J., Liu, H.L., Pan, C.F., Chuang, C.K., Jayakumar, T., Wang, T.J., Chen, H.H., Wu, C.J., 2012. Indoxyl sulfate predicts cardiovascular disease and renal function deterioration in advanced chronic kidney disease. Arch. Med. Res. 43, 451–456.

Lin, C.J., Pan, C.F., Chuang, C.K., Sun, F.J., Wang, D.J., Chen, H.H., Liu, H.L., Wu, C.J., 2014. P-cresyl sulfate is a valuable predictor of clinical outcomes in pre-ESRD patients. Biomed. Res. Int., 526932.

Liu, M., Li, X.-C., Lu, L., Cao, Y., Sun, R.-R., Chen, S., Zhang, P.-Y., 2014. Cardiovascular disease and its relationship with chronic kidney disease. Eur. Rev. Med. Pharmacol. Sci. 18 (19), 2918–2926.

Liu, S., Liu, H., Chen, L., Liang, S.S., Shi, K., Meng, W., Xue, J., He, Q., Jiang, H., 2020. Effect of probiotics on the intestinal microbiota of hemodialysis patients: a randomized trial. Eur. J. Nutr. 59 (8), 3755–3766. https://doi.org/10.1007/s00394-020-02207-2.

de Lopes, R.C.S.O., Balbino, K.P., Jorge, M.D.P., Ribeiro, A.Q., Martino, H.S.D., Alfenas, R.D.C.G., 2018. Modulation of intestinal microbiota, control of nitrogen products and inflammation by pre/probiotics in chronic kidney disease: a systematic review. Nutr. Hosp. 35, 722–730. https://doi.org/10.20960/nh.1642.

Mafra, D., Borges, N., Alvarenga, L., Esgalhado, M., Cardozo, L., Lindholm, B., Stenvinkel, P., 2019. Dietary components that may influence the disturbed gut microbiota in chronic kidney disease. Nutrients 11 (3), 496.

Mafra, D., Lobo, J.C., Barros, A.F., Koppe, L., Vaziri, N.D., Fouque, D., 2014. Role of altered intestinal microbiota in systemic inflammation and cardiovascular disease in chronic kidney disease. Future Microbiol. 9, 399–410.

Meijers, B., Jouret, F., Evenepoel, P., 2018. Linking gut microbiota to cardiovascular disease and hypertension: lessons from chronic kidney disease. Pharmacol. Res. 133, 101–107.

Meijers, B.K.I., De Preter, V., Verbeke, K., Vanrenterghem, Y., Evenepoel, P., 2010. p-Cresyl sulfate serum concentrations in haemodialysis patients are reduced by the prebiotic oligofructose-enriched inulin. Nephrol. Dial. Transplant. Off. Publ. Eur. Dial. Transpl. Assoc. Eur. Ren. Assoc. 25, 219–224.

Mihai, S., Codrici, E., Daniela Popescu, D.I., Enciu, A.M., Albulescu, L., Necula, L.G., Mambet, C., Anton, G., Tanase, C., 2018. Inflammation-related mechanisms in chronic kidney disease prediction, progression, and outcome. J. Immunol. Res. 2018, 2180373. https://doi.org/10.1155/2018/2180373. eCollection 2018.

Mikusic, N.L.R., Kouyoumdzian, N.M., Choi, M.R., 2020. Gut microbiota and chronic kidney disease: evidences and mechanisms that mediate a new communication in the gastrointestinal-renal axis. Pflugers Arch. 472 (3), 303–320.

Morrison, D.J., Preston, T., 2016. Formation of short chain fatty acids by the gut microbiota and their impact on human metabolism. Gut Microbes 7, 189–200.

Natarajan, R., Pechenyak, B., Vyas, U., Ranganathan, P., Weinberg, A., Liang, P., Mallappallil, M.C., Norin, A.J., Friedman, E.A., Saggi, S.J., 2014. Randomized controlled trial of strain-specific probiotic formulation (renadyl) in dialysis patients. Biomed. Res. Int. 2014, 568571. https://doi.org/10.1155/2014/568571.

Nishiyama, K., Aono, K., Fujimoto, Y., Kuwamura, M., Okada, T., Tokumoto, H., Izawa, T., Okano, R., Nakajima, H., Takeuchi, T., Azuma, Y.T., 2019. Chronic kidney disease after 5/6 nephrectomy disturbs the intestinal microbiota and alters intestinal motility. Cell. Physiol. 234 (5), 6667–6678.

Opdebeeck, B., Maudsley, S., Azmi, A., De Maré, A., De Leger, W., Meijers, B., Verhulst, A., Evenepoel, P., D'Haese, P.C., Neven, E., 2019. Indoxyl sulfate and p-cresyl sulfate promote vascular calcification and associate with glucose intolerance. J. Am. Soc. Nephrol. 30 (5), 751–766.

Poul, E.L., Loison, C., Struyf, S., Springael, J.-Y., Lannoy, V., Decobecq, M.-E., Brezillon, S., Dupriez, V., Vassart, G., Damme, J.V., Parmentier, M., Detheux, M., 2003. Functional characterization of human receptors for short chain fatty acids and their role in polymorphonuclear cell activation. J. Biol. Chem. 278, 25481–25489.

Ramezani, A., Raj, S., 2014. The gut microbiome, kidney disease, and targeted interventions. J. Am. Soc. Nephrol. 25 (4), 657–670.

Ramos, C.I., Armani, R.G., Canziani, M.E.F., Dalboni, M.A., Dolenga, C.J.R., Nakao, L.S., Campbell, K.L., Cuppari, L., 2019. Effect of prebiotic (fructooligosaccharide) on uremic toxins of chronic kidney disease patients: a randomized controlled trial. Nephrol. Dial. Transplant. Off. Publ. Eur. Dial. Transpl. Assoc. Eur. Ren. Assoc. 34, 1876–1884.

Ranganathan, N., Patel, B., Ranganathan, P., Marczely, J., Dheer, R., Chordia, T., Dunn, S.R., Friedman, E.A., 2005. Probiotic amelioration of azotemia in 5/6th nephrectomized Sprague-Dawley rats. Sci. World J. 5, 652–660.

Ranganathan, N., Patel, B.G., Ranganathan, P., Marczely, J., Dheer, R., Pechenyak, B., Dunn, S.R., Verstraete, W., Decroos, K., Mehta, R., Friedman, E.A., 2006. In vitro and in vivo assessment of intraintestinal bacteriotherapy in chronic kidney disease. ASAIO J. 52 (1), 70–79.

Ranganathan, N., Ranganathan, P., Friedman, E.A., Joseph, A., Delano, B., Goldfarb, D.S., Tam, P., Rao, A.V., Anteyi, E., Musso, C.G., 2010. Pilot study of probiotic dietary supplementation for promoting healthy kidney function in patients with chronic kidney disease. Adv. Ther. 27, 634–647.

Ruszkowski, J., Witkowski, J.M., 2019. Lactulose: patient- and dose-dependent prebiotic properties in humans. Anaerobe 59, 100–106.

Shariaty, Z., Shan, G.R.M., Farajollahi, M., Amerian, M., Pour, N.B., 2017. The effects of probiotic supplement on hemoglobin in chronic renal failure patients under hemodialysis: a randomized clinical trial. J. Res. Med. Sci. 22, 1–5.

Shimazu, S., Hirashiki, A., Okumura, T., Yamada, T., Okamoto, R., Shinoda, N., Takeshita, K., Kondo, T., Nida, T., Murohara, T., 2013. Association between indoxyl sulfate and cardiac dysfunction and prognosis in patients with dilated cardiomyopathy. Circ. J. 77, 390–396.

Sirich, T.L., Plummer, N.S., Gardner, C.D., Hostetter, T.H., Meyer, T.W., 2014. Effect of increasing dietary fiber on plasma levels of colon-derived solutes in hemodialysis patients. Clin. J. Am. Soc. Nephrol. 9, 1603–1610.

Slavin, J., 2013. Fiber and prebiotics: mechanisms and health benefits. Nutrients 5, 1417–1435.

Soleimani, A., Mojarrad, M.Z., Bahmani, F., Taghizadeh, M., Ramezani, M., Tajabadi-Ebrahimi, M., Jafari, P., Esmaillzadeh, A., Asemi, Z., 2016. Probiotic supplementation in diabetic hemodialysis patients has beneficial metabolic effects. Kidney Int. 91, 435–442.

Stockler-Pinto, M.B., Soulage, C.O., Borges, N.A., Cardozo, L.F.M.F., Dolenga, C.J., Nakao, L.S., Pecoits-Filho, R., Fouque, D., Mafra, D., 2018. From bench to the hemodialysis clinic: protein-bound uremic toxins modulate NF-κB/Nrf2 expression. Int. Urol. Nephrol. 50, 347–354.

Sueyoshi, M., Fukunaga, M., Mei, M., Nakajima, A., Tanaka, G., Murase, T., Narita, Y., Hirata, S., Kadowaki, D., 2019. Effects of lactulose on renal function and gut microbiota in adenine-induced chronic kidney disease rats. Clin. Exp. Nephrol. 23, 908–919.

Taki, K., Takayama, F., Niwa, T., 2005. Beneficial effects of Bifidobacteria in a gastroresistant seamless capsule on hyperhomocysteinemia in hemodialysis patients. J. Ren. Nutr. 15 (1), 77–80.

Tang, W.H.W., Wang, Z., Levison, B.S., Koeth, R.A., Britt, E.B., Fu, X., Wu, Y., Hazen, S.L., 2013. Intestinal microbial metabolism of phosphatidylcholine and cardiovascular risk. N. Engl. J. Med. 368 (17), 1575–1584.

Tao, S., Tao, S., Cheng, Y., Liu, J., Ma, L., Fu, P., 2019. Effects of probiotic supplements on the progression of chronic kidney disease: a meta-analysis. Nephrology 24, 1122–1130.

Vahed, S.Z., Mostafavi, S., Khatibi, S.M.H., Shoja, M.M., Ardalan, M., 2020. Vascular calcification: an important understanding in nephrology. Vasc. Health Risk Manag. 16, 167–180.

Vanholder, R.C., Eloot, S., Glorieux, G.L., 2016. Future avenues to decrease uremic toxin concentration. Am. J. Kidney Dis. 67 (4), 664–676.

Vaziri, N.D., Liu, S.M., Lau, W.L., Khazaeli, M., Nazertehrani, S., Farzaneh, S.H., Kieffer, D.A., Adams, S.H., Martin, R.J., 2014. High amylose resistant starch diet ameliorates oxidative stress, inflammation, and progression of chronic kidney disease. PLoS One 9 (12), e114881.

Vaziri, N.D., Zhao, Y.Y., Pahl, M.V., 2015. Altered intestinal microbial flora and impaired epithelial barrier structure and function in CKD: the nature, mechanisms, consequences and potential treatment. Nephrol. Dial. Transplant. 31, 737–746.

Wang, I.K., Wu, Y.Y., Yang, Y.F., Ting, I.W., Lin, C.C., Yen, T.H., Chen, J.H., Wang, C.H., Huang, C.C., Lin, H.C., 2015. The effect of probiotics on serum levels of cytokine and endotoxin in peritoneal dialysis patients: a randomised, double-blind, placebo-controlled trial. Benefic. Microbes 6, 423–430.

Wang, S., Lv, D., Jiang, S., Jiang, J., Liang, M., Hou, F., Chen, Y., 2019. Quantitative reduction in short-chain fatty acids, especially butyrate, contributes to the progression of chronic kidney disease. Clin. Sci. 133, 1857–1870.

Wei, M., Wang, Z., Liu, H., Jiang, H., Wang, M., Liang, S., Shi, K., Feng, J., 2014. Probiotic *Bifidobacterium animalis* subsp. lactis Bi-07 alleviates bacterial translocation and ameliorates microinflammation in experimental uraemia. Nephrology (Carlton) 19 (8), 500–506.

Wu, I.W., Gao, S.S., Chou, H.C., Yang, H.Y., Chang, L.C., Kuo, Y.L., Dinh, M.C.V., Chung, W.H., Yang, C.W., Lai, H.C., Hsieh, W.P., Su, S.C., 2020. Integrative metagenomic and metabolomic analyses reveal severity-specific signatures of gut microbiota in chronic kidney disease. Theranostics 10 (12), 5398–5411.

Yang, J., Li, Q., Henning, S.M., Zhong, J., Hsu, M., Lee, R., Long, J., Chan, B., Nagami, G.T., Heber, D., Li, Z., 2018. Effects of prebiotic fiber xylooligosaccharide in adenine-induced nephropathy in mice. Mol. Nutr. Food Res. 62 (15), e1800014. https://doi.org/10.1002/mnfr.201800014.

Zheng, H.J., Guo, J., Wang, Q., Wang, L., Wang, Y., Zhang, F., Huang, W.-J., Zhang, W., Liu, W.J., Wang, Y., 2021. Probiotics, prebiotics, and synbiotics for the improvement of metabolic profiles in patients with chronic kidney disease: a systematic review and meta-analysis of randomized controlled trials. Crit. Rev. Food Sci. Nutr. 61 (4), 577–598. https://doi.org/10.1080/10408398.2020.1740645.

CHAPTER 4

Probiotics and Prebiotic in Oral Health

MICHEL REIS MESSORA[a] • PEDRO HENRIQUE FELIX SILVA[a] • LUCIANA PRADO MAIA[b] • SÉRGIO SALVADOR[c] • ARTHUR C. OUWEHAND[d] • FLÁVIA FURLANETO[a]
[a]Department of Oral and Maxillofacial Surgery and Periodontology, School of Dentistry of Ribeirao Preto, University of Sao Paulo—USP, Ribeirao Preto, SP, Brazil, [b]Graduate Program in Dentistry (GPD-Master's Degree), University of Western Sao Paulo-UNOESTE, Presidente Prudente, SP, Brazil, [c]Department of Clinical Analyses, School of Pharmaceutical Sciences of Ribeirao Preto, University of Sao Paulo—USP, Ribeirao Preto, SP, Brazil, [d]Global Health & Nutrition Sciences, International Flavors & Fragrances Inc., Kantvik, Finland

INTRODUCTION

In the oral cavity, numerous bacteria adhere to dental surfaces and form microbial communities or biofilms (Costerton et al., 1978), with a predominance of commensal, nonpathogenic microorganisms, which have an intimate but harmonious relationship with the host tissues, contributing to the development of a normal immune system (Seneviratne et al., 2011). However, changes in the environment can lead to dental diseases, such as dental caries and periodontal disease (Hajishengallis and Lamont, 2012; Kristofersson et al., 1985; Marsh, 2003; Parahitiyawa et al., 2009). Thus, in general, oral health can be achieved with good oral hygiene, which is directly related to the control of biofilm in the oral cavity. There are different methods of biofilm control, among which mechanical control, through brushing and scaling and root planing (SRP), plays an important role, with chemical methods being indicated as complementary (Sugano, 2012). Chemical control of dental biofilm includes antibiotics, enzymes, biguanides, and metal salts. Although these methods are effective, they fail to prevent microbial recolonization, and their continued and prolonged use can do more harm than good. These deficiencies paved the way for the development of less aggressive but more efficient agents.

Numerous beneficial health effects have been documented with the use of probiotics and prebiotics, and their antimicrobial and immune response modulation capacities constitute a new therapeutic approach in the prevention of infectious diseases. Their proven effects on general health have led to increased research in the field of oral health. In this context, isolated probiotic species and different combinations have been investigated for their effects on the treatment and prevention of periodontal diseases, peri-implant diseases, and cavities.

ORAL CAVITY AND MICROBIOME

Within the oral cavity, the microbial community represents a well-characterized interaction between microorganisms and host. These bacteria originate from distinct and highly specialized communities that reside in a variety of environmental niches in the human mouth, such as the mucous surfaces of the tongue, cheek mucosa, palate, tonsils, and in the biofilms that accumulate on the surfaces of teeth (Curtis et al., 2011), and there is a correlation between the qualitative composition of the oral microbiota and the state of health/disease (Curtis et al., 2011).

Oral bacteria developed highly specific adhesions for proteins and carbohydrates from the acquired film, which are quite different from the adhesion mechanisms of other commensal and pathogenic bacteria that manifest tropisms for highly specific tissues. In the formation process of the oral biofilm, the initial colonizers establish themselves on the tooth surface through interactions with the acquired film from the host and serve as additional binding sites for intermediate and late colonizers. This process demonstrates that each stage in the formation of dental biofilm is highly specific and represents coevolution between the different oral bacterial species and the host (Socransky and Haffajee, 2005; Curtis et al., 2011).

Recent studies examining the host-bacterial interaction have revealed that commensal bacteria not only protect the host from pathogenic bacteria by occupying niches, but can also promote the development of adequate structure and function in tissues. These data indicate that polymicrobial communities associated with the host, such as those found in the oral cavity, evolved together with the host and became an integral part of the same organism (Ley et al., 2008; Roberts and Darveau, 2015).

Probiotics and Prebiotics in Foods. https://doi.org/10.1016/B978-0-12-819662-5.00015-X

The complex polymicrobial communities of the oral biofilm, which interact with the host, are influenced by diet and immunity and explain symbiosis and/or dysbiosis. The interruption in the beneficial relationship between the oral microbiota and the host (a state of dysbiosis) is associated with inflammation of the periodontal tissues and may contribute to more severe systemic conditions, via bacteremia (Han and Wang, 2013). Recent studies emphasize that the oral microbiome interacts with all other bodily microbiomes and is extensively involved in inflammatory diseases (Caballero and Pamer, 2015; Pickard et al., 2017; Gao et al., 2018). Thus, the microorganisms present in the oral cavity, due to their easy dissemination, interfere with the intestinal microbiota and, due to this change, an imbalance in the intestine-brain axis may occur. Further, the immune changes mediated by the intestine, contribute to the induction of a process of disease with systemic involvement (Hajishengallis, 2015; Gao et al., 2018).

PROBIOTICS AS AN ANTIMICROBIAL STRATEGY

In vitro studies demonstrate the potential of probiotic bacteria of the *Lactobacillus* and *Bifidobacterium* genera in the microbial modulation of biofilms that develop in the oral environment (Vuotto et al., 2014), as some probiotic strains form a biofilm on the teeth and gums, preventing the adhesion of pathogenic bacteria, which denotes a mechanism of competition for binding sites. The mechanisms of antimicrobial action of the probiotics are listed in Box 1.

BOX 1
Mechanisms of antimicrobial action of probiotics, which inhibit the adhesion of pathogenic bacteria.

Mechanism of antimicrobial action of probiotics
Competition for binding sites
Modification of the protein composition of the salivary film
Adhesion to gingival epithelial cells
Co-aggregation capacity
Secretion of active molecules (bacteriocins, teichoic acid, free fatty acids, hydrogen peroxide, and lactic acid)
Reduced expression of virulence factors by pathogenic bacteria

Haukioja et al. (2006) demonstrated that *L. rhamnosus* GG and *L. casei* can modify the protein composition of the salivary film acquired by binding and/or degrading salivary proteins, preventing the adhesion of other bacteria.

Mendi et al. (2016) also demonstrated that *L. rhamnosus* have good adherence capacity in gingival mesenchymal stem cells, being able to protect them from invasion by *P. gingivalis*.

Another microbiological mechanism is the ability to co-aggregate, through which probiotic microorganisms integrate with oral microbial communities and prevent the colonization of other pathogenic bacteria. Some species of *Lactobacillus* and *Bifidobacterium* have the ability to co-aggregate with *F. nucleatum* (Stamatova et al., 2008; Nagaoka et al., 2008; Ledder et al., 2008), a species that adheres to almost all oral bacteria (Rickard et al., 2003). Jasberg et al. (2016) also demonstrated that integration of the *B. animalis* subsp. *lactis* BB-12 probiotic and of *B. dentium* and *B. longum* isolates with *P. gingivalis*, *Actinomyces naeslundii*, and *F. nucleatum*. However, it is important to consider that the co-aggregation with free oral pathogens may be greater than that on pathogens integrated in an already established biofilm (Marttinen et al., 2013). Therefore, removal of the existing oral biofilm may be necessary before administration of a probiotic (Teughels et al., 2013) in order to enhance its antimicrobial effects and colonization capacity.

Additionally, probiotic microorganisms can secrete active molecules, such as bacteriocins, teichoic acid, free fatty acids, hydrogen peroxide, and lactic acid, which are used to control the growth and exchange of nutrients (Wohlgemuth et al., 2010; Travers et al., 2011). Bacteriocins are peptides capable of causing the death of bacteria by permeabilizing their membranes or by interfering with essential enzymes. Many of these peptides are produced by probiotic strains of the *Lactobacillus* (Travers et al., 2011; Vuotto et al., 2014). Lactic acid decreases local pH, affecting the growth of several microorganisms sensitive to this change in the microenvironment (Wohlgemuth et al., 2010; Travers et al., 2011), including some periodontal pathogens (Koll-Klais et al., 2005).

Some probiotic strains cannot only quantitatively affect the composition of a pathogenic biofilm, but also interfere with the virulence of the periodontopathogens present. Nissen et al. (2014) investigated the effect of *Lactobacillus* on the expression of the two main virulence factors of *A. actinomycetemcomitans*, leukotoxin (LtxA), and cytotoxic toxin (CdtB) and showed that *L. salivarius* and *L. gasseri* greatly attenuated the expressions of CdtB and LtxA.

PROBIOTICS AND THE HOST'S IMMUNE-INFLAMMATORY RESPONSE

With regard to the immunological mechanisms, probiotic microorganisms can induce the secretion of antimicrobial peptides or anti-inflammatory molecules through an interaction with several cell populations (Vuotto et al., 2014b; Teitelbaum and Walker, 2002; Sartor, 2004; Patel and Denning, 2013). These extracellular proteins of probiotic bacteria can diffuse through the mucosa and interact with epithelial and immunological cells, modulating the host's immune response and potentiating the epithelial and mucosal barrier function, by rearranging the cellular junctions (Sánchez et al., 2010). However, the exact mechanisms by which probiotic bacteria promote their effects are not fully understood. Probiotics have the ability to attenuate the activation of the nuclear kappa B factor (NF-κB), to promote the overexpression of cytoprotective genes, to prevent apoptosis and cell death, to generate reactive oxygen species important for cell and cell signaling, and to induce the expression of proteins necessary for mucosal barrier function (Patel and Denning, 2013; Bron et al., 2011). In general, it is believed that the beneficial effects of the probiotics are based, in part, on the ability to regulate the production of inflammatory cytokines and to balance the types of cellular immune responses (Cross et al., 2004; Ghadimi et al., 2008).

Considering that similar immunoinflammatory mechanisms occur in the periodontal tissues and in the intestinal mucosa, it is believed that the immunoinflammatory action of the probiotics in the oral cavity is analogous to that described in the intestinal mucosa. Through analysis of saliva, gingival crevicular fluid, and gingival biopsies, some clinical and preclinical studies have demonstrated that the ingestion of probiotics can interfere with the levels of several inflammatory and anti-inflammatory markers, as detailed later (Stamatova and Meurman, 2009a,b; Oliveira et al., 2017; Riccia et al., 2007; Shimauchi et al., 2008; Szkaradkiewicz et al., 2014; Maekawa and Hajishengallis, 2014; Foureaux et al., 2014; Messora et al., 2016).

TOOTH DECAY

Dental caries is one of the most common multifactorial chronic diseases; individuals remain susceptible to this disease throughout their lives. Environmental aspects and diverse risk factors for caries include nutrition, oral hygiene, fluoride intake and degree of colonization of cariogenic bacteria (which reflects food consumption), cultural characteristics, lifestyle, and socioeconomic status in relation to oral odor and caries dentistry (Hoceini et al., 2016). In addition, there are risk factors related to the host itself, such as salivary secretion, salivary damping capacity, position of teeth in juxtaposition with each other, characteristics of the tooth enamel surface, and the size of occlusal cracks in the posterior teeth. Other factors related to the host may increase the risk of caries, such as hereditary changes (Featherstone et al., 2003; Marsh et al., 2012; Struzycka, 2014). Several epidemiological research studies have identified dental caries and periodontal diseases as the most important burden of global oral health historically (WHO, 2003).

The onset of tooth decay is due to a bacterial infection that leads to demineralization and destruction of the tissues that make up the tooth. The progression process is slow, and the action of mainly acidogenic bacteria within the biofilm and accumulation of food residues on the tooth surface leads to damage, infection, and tooth loss (Chen and Wang, 2010; Marsh and Martin, 1999). These bacteria commonly exist within bacterial ecosystems, forming dental plaques that accumulate on the surface of tooth enamel. In the biofilm, some bacteria, such as *Streptococcus mutans*, *S. sobrinus*, and *Lactobacillus* spp., produce organic acids as metabolic by-products of fermentable carbohydrates (Conrads et al., 2014; Simon, 2007). These compounds result in low pH levels, causing demineralization of the dental tissue (Kidd and Fejerskov, 2004). Oral bacteria are considered an etiological factor in the development of caries and several studies have tried to elucidate the relationship of dental caries and to explain the disease process (Keyes and Jordan, 1964).

In the etiology of dental caries, the involvement of the *Streptococcus* genus was first reported by Clarke (1924). The first isolated species of a carious lesion was *S. mutans*, later recognized as the primary microorganism causing tooth decay, as several studies have reported an association between the presence of *S. mutans* and the risk of caries (Bartold and Van Dyke, 2013; Marsh, 2004). This bacterium, according to the ecological plaque hypothesis, represents an essential cariogenic organism (Marsh, 1994). When dental caries sets in, environmental changes occur in the oral cavity that has adverse effects on oral health due to the production of acids from fermentable carbohydrates. Acidogenic and acid-tolerant species, such as *S. mutans* and *Lactobacillus* survive under low pH conditions (Caufield et al., 2000). Recent studies show two important points in the appearance of caries lesions, the first one is that *S. mutans* is associated with the onset of tooth decay in human beings, and the second one would be that bacteria of the *Lactobacillus* genus are important for the development/progression of caries, especially in dentin (Karpiński and Szkaradkiewicz, 2013).

In the oral microbiota, the most frequently isolated species are *S. salivarius*, *S. sanguis*, *S. mitis*, and *S. mutans* (Aas et al., 2005). *S. mutans*, an indigenous bacterium in the mouth, is the bacterial agent that activates caries lesions in humans. This cariogenic microorganism exhibits the ability to colonize teeth, reduce the pH in the oral cavity, and synthesize extracellular polysaccharides from sucrose (Koo et al., 2010). A study shows that most of the carious lesions reflect infection by sucrose-dependent *S. mutans*. Eliminating or suppressing the growth of this organism on the dental surfaces offers a potentially effective approach to preventing tooth decay (Keyes and Jordan, 1964; Chen and Wang, 2010; Tokunaga et al., 1989).

Evidence Regarding the Use of Probiotics in the Prevention and Treatment of Dental Caries

Various studies have shown new approaches in the treatment of caries and focused on the effect of probiotics on oral health (Table 1). A number of studies suggest that the consumption of products containing probiotics reduces the number of *S. mutans* in saliva (Näse et al., 2001; Lin et al., 2017). Like bacteria that influence their environment, probiotics are considered capable of interacting synergistically and antagonistically with bacteria in the dental biofilm. The ecological plaque hypothesis suggests that selective pressure via changes in the environmental conditions can alter the balance between health and diseases (Marsh, 2004).

In vitro studies showed that some probiotic strains can survive in human saliva (Hasslöf et al., 2010). In addition, it was discovered that the *Lactobacillus* ssp. co-aggregate with *S. mutans* and other strains associated with caries and inhibit their growth (Hasslöf et al., 2010; Lodi et al., 2010). The bactericidal and bacteriostatic effects of probiotic strains such as *Lactobacillus* are attributable to the production of antimicrobial agents, such as organic acids, hydrogen peroxide, and antifungal compounds, such as fatty acids and bacteriocins (Twetman et al., 2009).

To discover the effects on antibacterial and anticariogenic activity, a study administered the *L. paracasei* subsp. *NTU paracasei 101* strain and skimmed soy milk fermented in Sprague Dawley rats. In addition, caries scores were assessed to evaluate the process of caries formation. The results showed that there was a suppression of bacteriological factors involved in caries and a reduction in the caries index in rats colonized by *S. mutans* and associated with the application of NTU 101. The probiotic clearly attenuates bacterial levels in the teeth, formation of lesions, and penetration of *S. mutans* in dental tissues, thus suggesting that NTU 101 and its fermented products are protective factors that have potential for use in preventing tooth decay in rats (Lin and Pan, 2014).

Yakult is a known probiotic drink consisting of a single live bacterial species, *L. casei* Shirota. Lin et al. (2017) analyzed the potential cariogenic/cariostatic effects of Yakult intake in children, investigating the effects of short-term ingestion on acidogenicity of oral biofilm and cariogenic bacterial count. The ingestion of probiotics resulted in a significant increase in the pH value and in significant cariostatic effects on the acidogenicity of the oral biofilm (Lin et al., 2017). These findings suggest that Yakult intake is beneficial in promoting oral health; however, these findings should be supported by additional studies.

A clinical study (Lodi et al., 2010) evaluated the effect of two fermented milk-based drinks, Yakult and Batavito, and of a 20% sucrose solution (control) on the development of caries, through the analysis of ionic concentration, polysaccharide insoluble extracellular tissue and degree of mineral loss from the biofilm collected from the dental surface. The results revealed that the daily administration of the drinks decreased the pH value of the dental biofilm and reduced the enamel demineralization (Lodi et al., 2010). Another double-blind, randomized, controlled study demonstrated that the *L. paracasei* GMNL-33 probiotic may reduce the count of *S. mutans* in the saliva of healthy adults (Chuang et al., 2011).

PERIODONTAL DISEASES

Periodontal disease is a chronic inflammatory disease that affects the supporting tissues of the teeth and can lead to tooth loss (Stamatova et al., 2008), and both the bacterial challenge and the host response are key factors in its development (Teughels et al., 2011). It is a prevalent disease both in developed and in developing countries, affecting about 20%–50% of the global population (Nazir, 2017). Periodontitis is an important public health problem due to its high prevalence, as well as because it can lead to tooth loss and disability, negatively affecting chewing and aesthetics, being a source of social inequality, and can also impair quality of life (Papapanou et al., 2018), and its association with systemic diseases has made the prevention of periodontal diseases important for maintaining general health (Beck et al., 1996; Ishihara et al., 2004; Iwamoto et al., 2001).

TABLE 1
Descriptive characteristics and individual results of the clinical studies that evaluated the effects of probiotic therapy in individuals at risk for dental caries[a].

Reference	N	Population	Intervention	Results
Villavicencio et al. (2018)	363	Children (3–4 years old)	200 mL of milk with *Lactobacillus rhamnosus* 5×10^6 and *Bifidobacterium longum* 3×10^6 CFU/g, 5 days a week for 9 months	↓ *S. mutans*, ↓ *Lactobacillus* spp. and ↑ salivary buffering capacity after 9 months
Alamoudi et al. (2018)	178	Healthy children (3–6 years old)	Tablets containing a minimum of 200 million *L. reuteri* DSM17938 and ATCC PTA5289 (≈ 2×10^8 UFC/800 mg) for 28 days, twice a day	↓ *S. mutans*, ↓ lactobacilli
Pahumunto et al. (2018)	124	Children (1.5–5 years old)	50 mL of milk with 5 g of *L. paracasei* SD1 10^7 CFU/g once a day for 3 months	The children in the probiotic group had a significantly lower risk of an increase in the level of *S. mutans*, in addition to significantly reducing the risk of caries compared to the control group
Ghasemi et al. (2017)	50	Women (19–27 years old) with *S. mutans* greater than 10^5	Yogurt containing 1.5×10^8 CFUs/ 200 g *Lactobacillus acidophilus* ATCC 4356 and *Bifidobacterium bifidum* ATCC 29521 for three weeks once a day	Reduction in the *S. mutans* count with results similar to xylitol gum
Rodríguez et al. (2016)	261	Children (2–3 years old)	150 mL of milk supplemented with *Lactobacillus rhamnosus* SP1 10^7 (CFU/mL) 5 days a week for 10 months	Decrease in the prevalence of caries, reduction in the percentage of new individuals who developed cavitated lesions. The increase in dental caries presented an odds ratio of 0.35 in favor of the probiotic group. At the level of the cavitated lesions, the probiotic group showed an increase of 0.58 ± 1.17 new lesions compared to 1.08 ± 1.70 new lesions observed in the control group
Stensson et al. (2014)	113	Children (9 years old)	Five drops of daily oil containing *L. reuteri* ATCC 55730 (≈ 10^8 CFU/5 drops) for 4 weeks prior to the expected date of delivery (mother) and in the child until 1 year of birth	At 9 years old, the group that took probiotics in the first year of life was 82% free of caries when compared to 58% in the control group. In addition to presenting a lower prevalence of interproximal caries in the probiotic group

[a] ↑ = Increase, ↓ = Reduction, ↔ = No difference, ≈ Approximately.

The main periodontal diseases are gingivitis and periodontitis. Biofilm-induced gingivitis is defined as an inflammatory lesion resulting from interactions between dental biofilm and the host's immune-inflammatory response, which remains contained in the gum, and is not understood by the supporting periodontium (cementum, periodontal ligament, and alveolar bone) (Chapple et al., 2018). This inflammation is reversible by reducing the levels of dental biofilm (Chapple et al., 2018). Periodontitis, on the contrary, is characterized by the progressive destruction of the dental support apparatus, having the following as primary characteristics: loss of clinical attachment and alveolar bone loss, presence of periodontal pocket, and bleeding on probing (BoP) (Papapanou et al., 2018).

The progression of periodontal disease is site-specific and varies from individual to individual, depending on the host's resistance and on the pathogenicity of the bacterial plaque (Kornman, 2001). *Porphyromonas gingivalis*, *Tannerella forsythia*, and *Treponema denticola* are widely regarded as major periodontal pathogens (Socransky et al., 1998). *Aggregatibacter actinomycetemcomitans* and *Prevotella intermedia* have also been found in subgingival lesions in patients with periodontitis (Ashimoto et al., 1996; Zambon, 1996; Loomer, 2004). Thus, suppressing the growth of these bacteria plays an important role in preventing periodontal diseases. Additionally, considering the importance of the host's response in the pathogenesis of periodontal disease, several studies have been evaluating the use of therapies to modulate the host's response.

Evidence Regarding the Use of Probiotics in the Prevention and Treatment of Periodontal Diseases

Many types of probiotic bacteria have been explored for the prevention and treatment of periodontal diseases, but the most widely studied are those belonging to the *Lactobacillus* and *Bifidobacterium* genera.

Their mechanism of action includes the antimicrobial effect and the modulation of the host's response. In vitro studies demonstrated that some probiotic strains are capable of inhibiting the growth of periodontopathogenic bacteria (Cornacchione et al., 2019; Esteban-Fernández et al., 2019; Higuchia et al., 2019) and of altering the immune response of gingival epithelial cells to microbial aggression (Albuquerque-Souza et al., 2019; Esteban-Fernández et al., 2019).

Gingivitis

Considering the different types of bacteria used as probiotics in periodontal healthy individuals or with gingivitis, the most studied are those of the *Lactobacillus* genus, and different species were investigated (Table 2).

In this context, it has been demonstrated that the administration of L. *reuteri* reduces the clinical signs and alters the microbiota in individuals with gingivitis. Krasse et al. (2006) observed a reduction in the plaque (PI) and gingival (GI) indexes; Twetman et al. (2009) demonstrated a reduction in BoP, crevicular fluid volume, and expression of the IL-8 and TNF-α proinflammatory cytokines; while Iniesta et al. (2012) report a reduction in the total count of anaerobic bacteria and *P. intermedia* in saliva and of *P. gingivalis* in the subgingival biofilm. Several studies have also shown that this therapy results in the oral colonization of this bacterium (Iniesta et al., 2012; Krasse et al., 2006), with a higher proportion in saliva, but with a longer duration in the subgingival microbiota (Iniesta et al., 2012).

Staab et al. (2009) observed a significant decrease in the elastase activity and in the amount of MMP-3 after the consumption of *L. casei* for 8 weeks, followed by interruption of oral hygiene for 96 h, but without clinical differences regarding the PI and the papillary bleeding index.

Oral administration of *L. salivarius* in adults without serious periodontitis did not generate differences in the clinical signs of gingivitis; however, when the individuals were grouped into smokers and nonsmokers, the smokers who received the probiotic showed a significant improvement in the (PI) and probing depth (PD), when compared to smokers in the placebo group (Shimauchi et al., 2008), indicating that the probiotic therapy can be a useful tool in the treatment of clinical symptoms of periodontitis, especially in high-risk individuals (Shimauchi et al., 2008). A significant suppressive effect was also observed in the numerical sum of *A. actinomycetemcomitans*, *P. intermedia*, *P. gingivalis*, *T. denticola*, and *T. forsythia* in all the individuals, regardless of their tobacco status (Mayanagi et al., 2009).

Regarding the effects of the *L. crispatus* in healthy adults, Tobita et al. (2018) observed a greater reduction in the number of *P. gingivalis* in saliva and in the PI and GI of the female participants, with no significant clinical differences between groups when considering all the participants or only the male ones, showing that there may be differences in the host's response against gender-dependent microorganisms.

In turn, Lee et al. (2015) demonstrated that the daily consumption of *L. brevis* can delay the development of gingivitis by regulating the inflammatory cascade in adults with gingival health but who were instructed to abstain from oral hygiene for 2 weeks. In this study,

TABLE 2
Descriptive characteristics and individual results of the clinical studies that evaluated the effects of probiotic therapy on individuals with periodontal health or gingivitis.

Reference	N	Population	Intervention	Results
Alanzi et al. (2018)	108	Teenagers without periodontitis	1 Tablet containing 4.4×10^8 CFU *L. rhamnosus* GG and 4.8×10^8 CFU *B. lactis* BB-12, 2 times a day, for 4 weeks	↓ PI, GI, levels of *A. actinomycetemcomitans* and *F. nucleatum* in saliva and biofilm, levels of *P. gingivalis* in biofilm, and total count of salivary bacteria
Harini and Anegundi (2010)	45	Healthy children	Daily mouthwash with probiotic solution, for 14 days	↓ PI, GI
Iniesta et al. (2012)	40	Young individuals with gingivitis	1 Tablet containing 2×10^8 CFU *L. reuteri* DSM-17938 and ATCC PTA 5289, once a day, for 28 days	↓ Total anaerobic bacteria count and *P. intermedia* in saliva and *P. gingivalis* in subgingival biofilm ↑ *L. reuteri* in saliva and subgingival biofilm
Karuppaiah et al. (2013)	216	Adolescents	Daily curd consumption for 30 days	↓ PI ↔ GI
Krasse et al. (2006)	59	Moderate to severe gingivitis	Daily consumption of 2×108 CFU of *L. reuteri*	↓ PI, GI
Kuru et al. (2017)	51	Young individuals and adults with gingival health	Daily consumption of 110 g of yogurt with $\geq10^8$ CFU *B. lactis* WB21, 3 times a day, for 28 days	↔ PI, GI, BoP, crevicular fluid volume, and IL-1β
Lee et al. (2015)	34	Adults with gingival health	1 Tablet with 1×109 CFU *L. brevis* CD2, 3 times a day, for 2 weeks, interruption of oral hygiene during the study	↔ BoP ↑ PI, GI
Mayanagi et al, (2009)	66	Adults without periodontitis	1 Tablet containing 6.7×10^8 CFU *L. salivarius* WB21, 3 times a day, for 8 weeks	↓ Sum of *A. actinomycetemcomitans*, *P. intermedia*, *P. gingivalis*, *T. denticola*, and *T. forsythia* in subgingival biofilm after 4 weeks
Shimauchi et al. (2008)	66	Adults without periodontitis	1 Tablet containing 6.7×10^8 CFUs *L. salivarius* WB21, 3 times a day, for 8 weeks	↔ PD, GI, BoP, PI ↑ *L. salivarius* in saliva
Staab et al. (2009)	50	Dentistry and medical students with gingival health	Daily consumption of fermented milk containing *L. casei* Shirota, for 8 weeks, followed by interruption of mechanical plate control for 96 h	↔ PI and BoP ↓ Elastase activity, MMP-3
Tobita et al. (2018)	16	Adults without periodontitis	1 Tablet containing 400×10^8 CFU *L. crispatus* KT-11, 3 times a day, for 4 weeks	↓ PI, GI, *P. gingivalis* of saliva in women
Toiviainen et al. (2015)	60	Adults with gingival health	1 Chewing gum containing 4.4×10^8 CFU *L. rhamnosus* GG and 4.8×10^8 CFUs *B. lactis* BB-12, 4 times a day, for 4 weeks	↓ PI and GI ↔ Microbial composition of saliva

Continued

TABLE 2
Descriptive characteristics and individual results of the clinical studies that evaluated the effects of probiotic therapy on individuals with periodontal health or gingivitis—cont'd

Reference	N	Population	Intervention	Results
Twetman et al. (2009)	42	Adults with moderate gingivitis	1 to 2 Chewing gums with 1×10^8 CFU *L. reuteri* ATCC 55730 and ATCC PTA 5289 daily, for 2 weeks	↓ BoP, crevicular fluid volume, IL-8, and TNF-α in the crevicular fluid

CFU, colony-forming units; *PI*, plaque index; *GI*, gingival index; *PD*, probing depth; BoP, bleeding on probing; *IL*, interleukin; *PBI*, papillary bleeding index; ↑ = Increase, ↓ = Reduction, ↔ = No difference.

while the placebo group showed a significant increase in PB, no significant differences were observed in the probiotic group, despite the fact that the two groups showed an increase in PI and GI, with no difference between the groups. The measurements of inflammatory markers in the gingival crevicular fluid also revealed an increase in the production of nitric oxide in the placebo group.

Considering the *Bifidobacterium* genus, it has been demonstrated that the use of *B. lactis* may have a positive effect on the PI and on the (GI) after abstaining from oral hygiene, in addition to decreasing levels of IL-1β in the crevicular gingival fluid (Kuru et al., 2017).

Additionally, the combination of *L. rhamnosus* and *B. lactis* seems to have favorable results in periodontally healthy adolescents (Alanzi et al., 2018) and adults (Toiviainen et al., 2015), leading to a significant reduction in PI and GI (Alanzi et al., 2018; Toiviainen et al., 2015) and in the levels of *A. actinomycetemcomitans* and *F. nucleatum* in saliva and biofilm, in the levels of *P. gingivalis* in biofilm, and in the total count of salivary bacteria (Alanzi et al., 2018). However, the results of the study by Toiviainen et al. (2015) suggest poor retention of bacteria from this probiotic combination in the oral cavity.

It was also demonstrated that the daily mouthwash with probiotic solution in healthy children showed results similar to the mouthwash with chlorhexidine in terms of PI and superior in terms of GI (Harini and Anegundi, 2010). Considering that chlorhexidine is the most effective antiplaque chemical agent, these results suggest a potential therapeutic value of the mouthwash with probiotics.

Periodontitis

Validated preclinical studies, such as the use of animal models with experimental periodontitis, may provide important new data on the safety and efficacy of new therapeutic approaches, and in this context, beneficial metabolic and periodontal effects of probiotics have been reported in the prevention and treatment of experimental periodontitis in rats using different probiotic strains and administration protocols. These results are illustrated in Table 3.

Considering the use of *Lactobacillus*, the following strains were investigated in rats or mice: *L. brevis* (Maekawa and Hajishengallis, 2014), *L. gasseri* (Kobayashi et al., 2017), *L. paracasei* (Liu et al., 2018), and *L. rhamnosus* (Gatej et al., 2018). The studies have shown that this therapy decreased gingival inflammation, bone loss, and disorganization and detachment of the fibers of the periodontal ligament. There was also a reduction in the expression of proinflammatory cytokines and in the total number of bacteria and anaerobic bacteria, while the number of aerobic bacteria increased.

Regarding the effects of bacteria of the *Bifidobacterium* genus, it has been shown that *B. lactis* promotes a protective effect against alveolar and attachment bone loss, in addition to modifying the subgingival microbiota and to modulating the inflammatory response in rats (Oliveira et al., 2017; Ricoldi et al., 2017).

Bacillus subtilis has also been extensively studied in the prevention of experimental periodontitis and, when used as a single strain (Foureaux et al., 2014; Messora et al., 2013) or combined with the *B. licheniformis* strain (Messora et al., 2016), reduced alveolar bone and attachment loss in rats, in addition to reducing the expression of proinflammatory cytokines and of markers related to bone resorption (Foureaux et al., 2014; Messora et al., 2016).

Similar results were observed after the administration of *Saccharomyces cerevisiae* in rats (Garcia et al., 2016) and the combination of *S. sanguinis*, *S. salivarius*, and *S. mitis* (Teughels et al., 2007).

In addition, both *B. lactis* (Ricoldi et al., 2017) and the combination of *Bacillus subtilis* and *B. licheniformis*

TABLE 3
Descriptive characteristics and individual results of the preclinical studies that evaluated the effects of probiotic therapy on experimental periodontitis.

Reference	Intervention	Results
Foureaux et al. (2014)	Oral administration of 1.5×10^8 CFU *B. subtilis* CH201 for 44 days, starting 30 days before EP induction in rats	↓ Bone loss, OPG in the gingival tissues, and number of inflammatory cells in the interproximal area
Garcia et al. (2016)	Topical application of *S. cerevisiae* at 0, 48, and 96 h alone or associated with SRP, after EP induction in rats	↓ Alveolar bone loss when associated with SRP, TNF-α, IL-1β ↑ IL-10 when associated with SRP
Gatej et al. (2018)	Daily administration for 44 days, via oral inoculation or gavage, of 200 μL of 2–9 × 10^9 CFU *L. rhamnosus* GG, starting 3 days before EP induction in rats	↓ Alveolar bone loss and gingival inflammation
Kobayashi et al. (2017)	Gastric incubation of 1×10^9 CFU *L. gasseri* SBT2055, once a day for 5 weeks, starting 3 weeks before EP induction in mice	↓ Alveolar bone loss, disorganization, and detachment of the periodontal ligament, IL-6, and IL-10 ↑ β-Defensin-14
Liu et al. (2018)	Topical application of an ethanol extract obtained from fermented skimmed milk containing *L. paracasei* subsp. *paracasei* NTU 101 at concentrations of 0.83, 1.66, and 3.32 mg/mL, daily, for 4 weeks, after EP induction in rats	↓ Alveolar bone loss, number of bacteria in gingival tissue, TNF-α, IL-1β, IL-6, and IL-17 in gingival tissue and blood
Maekawa and Hajishengallis (2014)	Topical application of 8.2×10^5 CFU *L. brevis* CD2, 3 times a day, for 5 days, after EP induction in mice	↓ Bone loss, TNF, IL-1β, IL-6, and IL-17A, anaerobic bacteria ↑ Aerobic bacteria
Messora et al. (2013)	Oral administration of 1.5×10^8 CFU *B. subtilis* CH201 for 44 days, starting 30 days before EP induction in rats	↓ Bone and attachment loss
Messora et al. (2016)	Oral administration of 1.5×10^8 CFU *B. subtilis*-DSM 5750 and *B. licheniformis*-DSM 5749 for 44 days, starting 30 days before EP induction in rats	↓ Bone loss, eosinophil count ↔ RANKL/OPG
	Oral administration of 1.5×10^8 CFU *B. subtilis*-DSM 5750 and *B. licheniformis*-DSM 5749 for 15 days, starting after SRP in rats	↓ Bone and attachment loss, osteoclasts ↑ IL-10
Oliveira et al. (2017)	Topical administration of 0.5×10^8 CFU *B. lactis* HN019 on days 0, 3 and 7 after EP induction in rats	↓ Bone and attachment loss, *V. parvula, C. sputigena, E. corrodens, P. intermedia*, IL-1b, and RANKL ↑ *Actinomyces* and *Streptococcus*, OPG and β-defensin
Ricoldi et al. (2017)	Oral administration of 1×10^9 CFU *B. lactis* HN019 for 15 days, starting after SRP in rats	↓ Bone and attachment loss, osteoclasts,IL-1β, CINC ↑ IL-10, TGF-β, aerobic/anaerobic bacteria ratio
Teughels et al. (2007)	Single or multiple topical application (weekly for 4 weeks) of 11×10^{10} CFUs *S. sanguinis*, 10.7×10^{10} CFUs *S. salivarius*, and 9.6×10^{10} CFU *S. mitis*, after SRP, in EP in dogs	↓ BoP, anaerobic species and chromogenic bacteria, *P. gulae, P. intermedia, C. rectus, F. nucleatum, A. actinomycetemcomitans and T. forsythia,* and its recolonization, with better results for multiple applications

CFU, colony-forming units; *EP*, experimental periodontitis; *OPG*, osteoprotegerin; *SRP*, scaling and root planing; *TNF*, tumor necrosis factor; *IL*, interleukin; *RANKL*, bonding agent of the receptor of the nuclear kappa B factor receptor, *CINC*, goat anticytokine-induced neutrophil chemoattractant; *BoP*, bleeding on probing; ↑ = Increase; ↓ = Reduction; ↔ = No difference.

(Messora et al., 2016) potentiated the clinical and immunological effects of scaling and root planing.

In randomized clinical studies evaluating the effects of probiotic therapy as an adjunct to nonsurgical periodontal therapy (NSPT) in individuals with periodontitis, the results are conflicting due to different strains, dosage, frequency, and method of administration used (Table 4).

In this context, the *Lactobacillus* genus bacteria are also the most studied, with the majority of clinical studies evaluating the effects of the *L. reuteri* strain, demonstrating better results regarding PD, clinical attachment level (CAL), PI, GI, and BoP when probiotic therapy was associated with NSPT (Ince et al., 2015; Szkaradkiewicz et al., 2014; Tekce et al., 2015; Teughels et al., 2013; Vivekananda et al., 2010). Teughels et al. (2013) also found that the use of L. *reuteri* as an adjunct to SRP reduces the risk of progression of periodontitis and the number of patients, teeth, and sites with indication for additional periodontal therapy. Tekce et al. (2015) also noted the persistence of this microorganism in the oral cavity 9 weeks after the discontinuation of probiotic therapy, suggesting that it is incorporated into the subgingival biofilm. Regarding the immune response, Szkaradkiewicz et al. (2014) observed a significant decrease in the levels of proinflammatory cytokines, while Ince et al. (2015) reported a significant decrease in the levels of matrix 8 metalloproteinases (MMP-8) and increased levels of tissue inhibitors of matrix 1 metalloproteinases (TIMP-1) when probiotic therapy was associated with SRT. Therapy with *L. reuteri* without previous NSPT also resulted in a significant reduction in all the periodontal clinical parameters (Vicario et al., 2013). However, more recently, Pelekos et al. (2019) did not observe any additional clinical effects of this therapy in the treatment of periodontitis.

Other species of lactobacilli also evaluated were *L. rhamnosus*, *L. plantarum*, *L. salivarius*, *L. brevis*, and *L. casei*. The consumption of the *L. plantarum* L-137 probiotic significantly reduced the PD of sites with PD $\geq$ 4 mm in patients undergoing periodontal support therapy (Iwasaki et al., 2016). In turn, the combination of *L. salivarius* and *L. reuteri* used in the form of daily mouthwash for 15 days after SRP and weekly subgingival irrigation for 30 days led to a significant reduction in the PD of moderate bags, and an improvement in PI, GI, BoP, and halitosis (Penala et al., 2016). Regarding the consumption of *L. brevis*, Shah et al. (2013) observed a significant clinical improvement, similar to that observed with the use of antibiotic therapy with doxycycline. These data reinforce the use of probiotics as an alternative to antibiotics, which can lead to the emergence of drug-resistant microorganisms. The authors also observed an increase in *Lactobacillus* in saliva, indicating the colonization of the probiotic in the oral cavity. Already Imran et al. (2015) did not observe any additional clinical benefits with the consumption of fermented milk containing *L. casei* Shirota. However, there was a significant reduction in the number of periodontopathogenic bacteria in the subgingival biofilm. On the contrary, the consumption of *L. rhamnosus* did not generate clinical and microbiological results additional to NSPT (Morales et al., 2016, 2018).

Invernici et al. (2018) evaluated the effects of *B. lactis* (Invernici et al., 2018) and reported a significant decrease in PD and a gain in clinical attachment, in addition to a smaller number of moderate and deep residual bags and a lower percentage of patients in need of additional periodontal treatment. In addition, lower levels of periodontopathogens of the red and orange complexes were observed 90 days after the beginning of the study, suggesting that the probiotics may delay the recolonization of periodontal pockets by these bacteria. Regarding the inflammatory response, the authors observed a decrease in the rates of the IL-1β and IL-8 proinflammatory cytokines and an increase in the levels of the IL-10 anti-inflammatory cytokine in the crevicular gingival fluid. Additionally, the group that received the probiotic therapy showed an increase in the levels of *B. lactis* HN019 in the subgingival biofilm even 60 days after the discontinuation of the therapy, which suggests the exogenous translocation of the probiotic to the subgingival biofilm, showing that the microorganism must have temporarily integrated the subgingival biofilm.

It has also been demonstrated that the use of a rinse containing *B. subtilis* reduced the scores of the BANA test and the GI (Tsubura et al., 2009), while the administration of probiotic tablets containing *S. oralis*, *S. uberis*, and *S. rattus* significantly reduces PI and salivary *P. intermedia* counts, but does not result in any improvement in PD, BoP, or CAL (Laleman et al., 2015).

Considering the great heterogeneity in the results obtained in these clinical studies, diverse meta-analyses point to discrete clinical benefits in the adjunct use of probiotics in reducing PD and gaining clinical attachment in patients with periodontitis (Donos et al., 2019; Ikram et al., 2018), suggesting the need for high quality randomized clinical studies, with immunological and microbiological results, in order to draw strong conclusions regarding this therapy.

TABLE 4
Descriptive characteristics and individual results of the randomized clinical studies that evaluated the effects of probiotic therapy in individuals with periodontitis.

Reference	N	Intervention	Results
Ikram et al. (2018)	30	NSPT + 1 sachet containing 1.2×10^9 CFU *L. reuteri*, twice a day for 3 months	↔ PI, BoP, PD, and CAL
Imran et al. (2015)	42	Daily consumption of probiotic drink containing *L. casei* for 1 month	↓ *P. gingivalis*, *A. actinomycetemcomitans*, and *P. intermedia*
Ince et al. (2015)	30	NSPT + 1 tablet containing *L. reuteri*, twice a day, for 3 weeks	↓ PI, GI, BoP, CAL, and MMP-8 ↑ TIMP-1
Invernici et al. (2018)	41	NSPT + 1 insert containing 10^9 CFU *B. lactis* HN019, twice a day for 30 days	↓ BoP, CAL, red and orange complex periodontopathogens, and proinflammatory cytokines ↑ *B. lactis* HN019 in subgingival biofilm
Iwasaki et al. (2016)	36	SPT + 1 capsule containing 10 mg of *L. plantarum* L-137, once a day, for 12 weeks	↔ BoP and number of teeth or sites with PD ≥ 4 mm ↓ PD in sites with PD ≥ 4 mm
Laleman et al. (2015)	48	NSPT + 1 tablet containing at least 10^8 CFU *S. oralis* KJ3, *S. uberis* KJ2, and *S. rattus* JH145, twice a day for 24 weeks	↓ PI and salivary *P. intermedia* ↔ BoP, PD, and CAL
Morales et al. (2016)	28	NSPT + 1 sachet containing 2×10^7 CFU *L. rhamnosus* SP1, once a day, for 3 months	↓ PD, number of participants with PD ≥ 6 mm ↔ PI, PD, and CAL
Morales et al. (2018)	47	NSPT + 1 sachet containing 2×10^7 CFU *L. rhamnosus*, once a day, for 3 months	↔ *P. gingivalis*, *T. forsythias*, and *A. actinomycetemcomitans*
Pelekos et al. (2019)	41	NSPT + 1 insert containing 10^8 CFU *L. reuteri* DSM17938 and ATCC PTA5289, twice a day, for 28 days	↔ PI, BoP, PD, and CAL
Penala et al. (2016)	32	NSPT + daily mouthwash for 15 days and weekly subgingival irrigation for 30 days with 2×10^9 CFU *L. salivarius* and *L. reuteri*	↓ PI, GI, BoP, and PD of moderate bags and halitosis
Shah et al. (2013)	30	NSPT + 1 insert containing 10^8 CFU, twice a day, for 14 days	↓ PI, GI, BoP, CAL, and *A. actinomycetemcomitans*
Szkaradkiewicz et al. (2014)	38	NSPT + 1 tablet containing 10^8 CFU *L. reuteri* ATCC PTA5289, twice a day, for 2 weeks	↓ BoP, PD, CAL, TNF-α, IL-1β, and IL-17
Tekce et al. (2015)	40	NSPT + 1 tablet containing L. *reuteri*, twice a day, for 3 weeks	↓ PI, GI, BoP, PD, total number of bacteria, and anaerobes
Teughels et al. (2013)	30	NSPT + 1 capsule containing 10^8 CFU *L. reuteri* DSM17938 and ATCC PTA5289, twice a day, for 12 weeks	↔ PI, GI, BoP, PD, and GR ↓ *P. gingivalis*
Tsubura et al. (2009)	54	NSPT + 2 mouthwashes daily for 1 month with rinse containing *B. subtilis*	↓ GI, BANA test scores
Vicario et al. (2013)	20	1 Tablet containing Prodentis *L. reuteri*, once a day, for 30 days	↓ PI, BoP, and PD
Vivekananda et al. (2010)	30	NSPT + 1 insert containing 10^8 CFU *L. reuteri* DSM17938 and ATCC PTA5289, twice a day, for 20 days	↓ PI, GI, BoP, CAL, *A. actinomycetemcomitans*, *P. gingivalis*, and *P. intermedia*

NSPT, nonsurgical periodontal therapy; *SPT*, support periodontal therapy; *CFUs*, colony-forming units; *PI*, plaque index; *GI*, gingival index; *PD*, probing depth; *BoP*, bleeding on probing; *CAL*, clinical attachment level; *GR*, gingival retraction; *MMP-8*, matrix 8 metalloproteinases; *TIMP-1*, tissue inhibitors of matrix 1 metalloproteinases; *TNF*, tumor necrosis factor; ↑ = Increase; ↓ = Reduction; ↔ = No difference.

PERI-IMPLANT DISEASES

When a tooth is replaced by a dental implant, the tissues around the implant remodel in a different configuration than what is seen around natural teeth, which makes these tissues more susceptible to disease processes. Peri-implant diseases are divided into mucositis and peri-implantitis, and mucositis is inflammation of the peri-implant mucosa around a functional implant, but without signs of bone loss, while the presence of inflammation with marginal peri-implant bone loss after biological bone remodeling is described as peri-implantitis (Caton et al., 2018; Zitzmann and Berglundh, 2008). The prevalence of mucositis is high, affecting 19% to 65% of the patients, with a mean of 42.9%, while the prevalence of peri-implantitis is 1% to 47%, with a mean of 21.7% (Derks and Tomasi, 2015). The microbiota associated with peri-implant diseases seems to be more complex than in individuals with healthy peri-implant conditions, with the main flora consisting of gram-negative anaerobic bacteria (Ata-Ali et al., 2011), as in periodontitis.

Usually, the treatment of peri-implant diseases is performed through mechanical debridement (MD) of the implant surface and adjacent tissues (Flichy-Fernandez et al., 2015), being an efficient therapy in the treatment of mucositis (Schwarz et al., 2015); however, nonsurgical therapy has limited efficacy and low predictability in peri-implantitis, and the use of local and/or systemic antiseptics or antibiotics can improve clinical parameters, but does not completely resolve the peri-implant injury, there is also no consensus regarding the ideal protocol for the surgical treatment of peri-implantitis (Schwarz et al., 2015).

Evidence Regarding the Use of Probiotics in the Treatment of Peri-Implant Diseases

To date, few clinical studies have evaluated the effects of the probiotic therapy on the prevention and treatment of peri-implant diseases, and the results are controversial (Table 5).

With regard to peri-implant mucositis, while some studies have observed an improvement in the clinical and microbiological parameters and in the levels of proinflammatory cytokines only with the administration of *L. reuteri* (Flichy-Fernandez et al., 2015) or when this therapy was used as an adjunct to MF (Alqahtani et al., 2019; Galofre et al., 2018), other studies (Hallström et al., 2016; Peña et al., 2019) did not observe additional clinical and microbiological effects with the use of the same probiotic strain in the adjunctive treatment of mucositis. The administration of *L. plantarum* and *L. brevis* in combination with the mechanical therapy, in turn, did not result in significant clinical differences in mucositis implants (Mongardini et al., 2017).

In the treatment of peri-implantitis, Galofre et al. (2018) and Tada et al. (2018) reported that the probiotic therapy with *L. reuteri* associated with mechanical debridement provides better clinical results (Galofre et al., 2018; Tada et al., 2018).

Despite promising results regarding the use of adjunct probiotic therapy in the treatment of mucositis and peri-implantitis, studies in this field are still scarce and do not make it possible to establish the role of probiotics in this context or to justify their clinical use. More high-quality randomized clinical trials are needed, as the current evidence is still weak.

PREBIOTICS AND ORAL DISEASES

A limitation of the probiotic therapy is the fact that the presence of probiotic bacteria in the oral cavity is only temporary, since they are normally not part of the endogenous microbiota in that region. Eventually, these probiotic bacteria will be eliminated from the oral cavity, reducing their effects. This limitation led to the search for more functional and ecological alternatives to manipulate the dental biofilm. An interesting approach is the nutritional stimulation of beneficial endogenous bacteria, with the use of prebiotics. Prebiotics allow for changes in the composition and microbial activity of the resident microflora, conferring benefits, such as well-being and health (Pawar et al., 2017; Roberfroid, 2007). However, the research studies regarding the application of prebiotics in the oral cavity are still limited.

Slomka et al. (2017) demonstrated in an in vitro study that nutritional stimulation with prebiotics restores the balance of the oral microbiota, with a predominance of beneficial bacteria and consequent suppression of pathogens.

Considering the effects of the use of prebiotics in the prevention of EP, it has been demonstrated that the administration of B-glucans isolated from *Saccharomyces cerevisiae* reduces alveolar bone loss in rats with EP with and without diabetes (Silva et al., 2015). In diabetic animals with EP, it also reduced the levels of glucose, cholesterol, and triglycerides and the expression of RANKL and cycloxygenase-2 (COX-2), an important inflammatory marker, and increased the expression of OPG (Silva et al., 2015). These results suggest that the prebiotic therapy may be particularly interesting in the treatment of individuals with systemically compromised periodontitis.

Levi et al. (2018) also demonstrated that the mannanoligosaccharide prebiotic, also isolated from

TABLE 5
Descriptive characteristics and individual results of the clinical studies that evaluated the effects of probiotic therapy in individuals with mucositis or peri-implantitis.

Reference	N	Population	Intervention	Results
Alqahtani et al. (2019)	80	Individuals with mucositis	MD + 1 tablet containing 1×10^8 CFU *L. reuteri* ATCC PTA 5289 and DSM 17938, twice a day, for 3 weeks	↓ PD, PI, and BoP in smokers after 3 months
Galofre et al. (2018)	44	Individuals with mucositis or peri-implantitis	MD + 1 pellet containing 1×10^8 CFU *L. reuteri* ATCC PTA 5289 and DSM 17938, once a day, for 30 days	↓ PD, BoP, and *P. gingivalis*
Flichy-Fernandez et al. (2015)	34	Individuals with mucositis	1 Tablet containing 1×10^8 CFU *L. reuteri* ATCC PTA 5289 and DSM 17938, once a day, for 30 days	↓ PD, PI, and crevicular fluid volume
Hallström et al. (2016)	49	Individuals with mucositis	MD + 1 pellet containing 2×10^7 CFU *L. reuteri* ATCC PTA 5289 and DSM 17938, twice a day, for 3 months	↔ PD, PI, BoP, crevicular fluid volume, IL-1β, IL-1RA, IL-4, IL-6, IL-8, IL-17A, CCL5, TNF-α, IFN-g, GM-CSF, and microbial profile
Mongardini et al. (2017)	20	Individuals with mucositis	MD + 1 tablet containing *L. plantarum* and *L. brevis*, once a day, for 14 days	↔ PI and BoP
Peña et al. (2019)	50	Individuals with mucositis	MD + 1 tablet containing 1×10^8 CFU *L. reuteri* ATCC PTA 5289 and DSM 17938, twice a day, for 3 weeks	↔ PI, BoP, and microbial profile
Tada et al. (2018)	30	Individuals with peri-implantitis	MD + 1 tablet containing 1×10^8 CFU *L. reuteri* ATCC PTA 5289 and DSM 17938, once a day, for 6 months	↓ GI, PD ↔ PI, BoP, and number of bacteria

MD, mechanical debridement; *CFUs*, colony-forming units; *PI*, plaque index; *GI*, gingival index; *PD*, probing depth; *BoP*, bleeding on probing; *IL*, Interleukin; *TNF*, tumor necrosis factor; *IFN*, interferon; *CCL5*, protein encoded by the CCL5 gene; *GM-CSF*, macrophage and granulocyte colony-stimulating factor, ↑ = Increase, ↓ = Reduction, ↔ = No difference.

isolates of *S. cerevisiae*, prevented bone loss, decreased the expression of TNF-a and IL-1b, and increased immunostaining for TGF-β in rats submitted to EP, reinforcing its therapeutic potential in preventing EP. However, future studies are indispensable to confirm these findings in more advanced experimental models and in controlled clinical studies.

COMMENSAL BACTERIA AND PROBIOTICS

Probiotic bacteria, as well as oral commensal bacteria, are able to increase the host's immunotolerance toward pathogen agents. The interaction of the surface molecules of the probiotic bacterium with host's pattern recognition receptors (PRRs) and associated coreceptors plays a key role in the health benefits observed by the probiotic bacteria (Lebeer et al., 2010). Flagella, fimbriae, secreted proteins, cell wall-associated polysaccharides (CPSs), lipoteichoic acid (LTA), lipopolysaccharide (LPS), and peptidoglycan (PG) are important probiotic surface molecules that contain molecular patterns associated with microorganisms (MAMPs), recognized by PRRs, which can transduce signals in the interaction with probiotic surface molecules, including the Toll-like 2 (TLR2) receptor (in heterodimers with TLR1 or TLR6), TLR4 and TLR5, the type C-specific intercellular cell adhesion molecule nongrabbing integrin (DC-SIGN) and protein-like receptors containing the nucleotide-binding oligomerization domain (NOD) (Lebeer et al., 2010). These receivers do not work in isolation, but cooperate with several coreceptors.

The PRRs transduce these signals to the cell nucleus, through associated adapter proteins that are linked to κB factor (NF-κB) and mitogen-activated protein kinase (MAPK) signaling cascades, to modulate the expression of response genes that include genes for cytokines, chemokines, antimicrobial peptides, cytoprotective factors, and costimulatory molecules.

The nonphagocytic epithelial cells and the phagocytic DCs differ in their ability to respond to the probiotic MAMPs. This is especially relevant for LTA, LPS, and PG, as their MAMPs are normally embedded in the cell walls of entire bacterial cells.

Some parallels can be drawn between the interactions of the isolated surface molecules of probiotics, commensals, and pathogens. However, it is the coordinated sum of the interactions of the various effector molecules of the bacterial surface with the PRRs and coreceptors that determine the final response of the host cell against a given microorganism. The surface molecules of the probiotic bacteria and their host receptors are subject to dynamic regulation as a function of time and space. Glycans are important in this dynamics, possibly modulating ligand-receptor interactions and providing an additional opportunity for discrimination between probiotic, commensal, and pathogen MAMPs (Lebeer et al., 2010).

In the last decade, there has been a growing awareness of the role played by the commensal bacteria in modulating mucosal immune responses and, as a consequence, there is now a great deal of interest in the therapeutic potential of probiotics and other bacterial-based strategies for a range of immune disorders, since these bacteria have a high ability to interact with epithelial cells, dendritic cells, and regulatory T cells (Forsythe and Bienenstock, 2010). Thus, it has been proposed that commensal bacteria, through their ability to interact with the host's immune system, are essential for the maintenance of healthy tissues, having multiple roles, including primary immune responses to ensure rapid and efficient defenses against pathogens (Devine et al., 2015). Currently, it is known that the standard state of the oral tissues, as well as of the intestine's, is an inflammatory process that can be balanced by regulatory mechanisms and by the activities of resident bacteria, which can be commensal and anti-inflammatory, modulating the TLR signaling or the activation of NF-κB, or influencing the development and activity of the immune cells. However, the widespread ability of normal resident organisms to suppress inflammation could impose an unsustainable burden on the immune system and compromise responses to the pathogens.

Immunosuppressive resident bacteria have been isolated from the mouth and may constitute 30% of the *Streptococcus* residing in the plaque or in the tongue, for example. Their roles in oral health and dysbiosis have yet to be determined. A wide range of bacterial components and/or products can mediate the immunomodulatory activity, increasing the possibility of developing alternative strategies for therapy and health promotion using probiotics, prebiotics, or immunomodulatory molecules derived from the commensals (Devine et al., 2015).

Based on the effects that commensal bacteria can have on the host's immune system, Zhang et al. (2008) investigated the ability of *Streptococcus cristatus*, considered a commensal bacteria, to invoke the production of IL-8 by oral epithelial cells, compared to the *Fusobacterium nucleatum* pathogen. The results showed that *F. nucleatum* alone evoked a potent response to IL-8, while *S. cristatus* alone did not induce any significant expression of IL-8. When present together, *S. cristatus* attenuated the production of IL-8 induced by *F. nucleatum* in four oral epithelial cell lines to varying degrees. The inhibitory effect of *S. cristatus* was independent of its viability, and of its co-aggregation with *F. nucleatum*, was not related to soluble bacterial products, and seemed to require bacterial contact with epithelial cells (Zhang et al., 2008).

Another study using transcriptional tools to define the biological pathways of gingival epithelial cell modulation by the coculture with the *S. gordonii* oral commensal microorganism and the *F. nucleatum* pathogen showed that *F. nucleatum* and *S. gordonii* disrupted the gingival epithelial cell transcriptome much less significantly than *P. gingivalis* and then, *A. actinomycetemcomitans*, oral pathogens with a high virulence profile, indicating that there was a greater degree of adaptation of the host by the commensal species (Handfield et al., 2005). As demonstrated by Hasegawa et al. (2007), the biological pathways significantly affected by *F. nucleatum* and by *S. gordonii* included mitogen-activated protein kinase (MAPK) and the Toll-like receptor signaling pathways. The differential regulation of GADD45 and DUSP4, key components of the MAPK pathway, has been confirmed at the protein level. Modulation of the MAPK pathway probably affects the proliferation and differentiation of host cells. In addition, both MAPK receptor and Toll-like pathways eventually converge in the expression of the cytokine gene. An immunosorbent assay linked to the secreted interleukin-6 (IL-6) and IL-8 enzymes demonstrated that *F. nucleatum* induced the production of these cytokines, while *S. gordonii* inhibited the secretion of the epithelial cells. Stimulating the secretion of proinflammatory cytokines of the epithelial cells may reflect the invasive phenotype of *F. nucleatum* and contribute to the pathogenic potential of *F. nucleatum* being greater than that of *S. gordonii*.

The junctional epithelium, comprising the gingival connection to the tooth, acts as an important barrier against the microorganisms of the pathogenic subgingival biofilm and its products. There is evidence that pathogenic bacteria have the potential to impair epithelial integrity, contributing to the breakdown of the junctional epithelium characteristic of the immunopathological response of periodontal disease. In a study, Ye et al. (2013) investigated the capacity of the *S. gordonii* oral commensal bacteria in increasing the epithelial barrier function and in maintaining the epithelial integrity of the healthy tissue. The function of the oral epithelial barrier was measured by a permeability test. The data showed an increase in the expression of genes that encode the ZO-1, ZO-2, JAM-A, and occludin tight junction components, in the proportion of 100 bacterial colony-forming units per epithelial cell. This was associated with an increase in the expression at the protein level of ZO-1, ZO-2, and JAM-A. The reduction in permeability to fluorochrome-labeled dextran accompanied these changes. The data support the hypothesis that commensal bacteria have a beneficial effect on the oral epithelium (Ye et al., 2013).

After these studies showing the positive effects of commensal bacteria, which have also been demonstrated in the intestine, over time research studies have emerged in order to induce a favorable environment for them. Thus, Slomka et al. (2017) used Phenotype MicroArrays as a high-throughput method to assess the change in the respiratory activity of 16 oral bacteria in response to 742 nutritional compounds. The most promising prebiotic compounds were selected and applied in tests of growth of single species and biofilm formation, as well as competition tests of dual species (beneficial-pathogen). The results were that the increase in the respiratory activity cannot always be related to an increase in growth or biofilm formation. Six compounds were used in dual-species competition trials to directly monitor whether the selective nutritional stimulation of the beneficial bacteria results in the suppression of the pathogenic bacteria. Two compounds, beta-methyl-d-galactoside and *N*-acetyl-d-mannosamine, can be identified as potential oral prebiotic compounds, triggering the growth of selectively beneficial oral bacteria throughout the experiments and displacing biofilm communities from two species to one beneficial dominant in vitro composition. These data suggest that prebiotic compounds can favor commensal bacteria and inhibit possible pathogens. Associated with the other research results described earlier, these effects occur through the bacterium-host interaction (state of immunotolerance activated by commensal bacteria) and commensal-pathogen (inhibiting the growth and disfavoring the adhesion of certain, as well as inactivating, products).

FUTURE PERSPECTIVES

The research field has been working to bring new safe and effective therapies to the clinical practice. The use of bacteriotherapy has been increasingly inviting in these promicrobial approaches and without generating resistant microorganisms. Recently, a study evaluated the use of predatory bacteria in the treatment of periodontitis in rats (Silva et al., 2019), and the group treated with predatory bacteria showed greater bone volume and less loss of connective tissue attachment than the untreated group. Furthermore, bacteriotherapy was able to modulate the rat microbiota and the expression of IL-17, IL-10, and the FOXP3 nuclear factor. These data suggest a promising field for the use of predatory bacteria as an antimicrobial agent in the treatment of periodontitis; however, these data need to be validated in clinical studies.

Another point would be the use of probiotics as epigenetic modulators in the treatment of periodontal disease. Currently, there is strong evidence of the epigenome as an important marker of disease susceptibility (Larsson et al., 2015). Research studies involving gene activation and silencing through probiotic therapy have been carried out and may open a new field in disease prevention, since epigenetic modulation can be carried out since intrauterine life (Vähämiko et al., 2018). Thus, new therapeutic approaches may emerge in the field of oral disease treatment.

CONCLUSION

The increase in resistant pathogens is already a public health issue, and this concern has led to the need to search for safe, economical, and creative alternatives to the traditional use of antibiotic therapy. In this context, interest in the use of probiotics and prebiotics in oral health has gradually evolved, and this therapy has shown promising results. However, it is important to emphasize that the results obtained with this emerging therapy cannot yet be generalized, since they depend on the strain, dose, frequency, and form of administration. The performance of more high-quality preclinical studies and controlled clinical studies is essential to elucidate the mechanisms of action and to establish protocols for the clinical use of this therapy.

REFERENCES

Aas, J.A., Paster, B.J., Stokes, L.N., Olsen, I., Dewhirst, F.E., 2005. Defining the normal bacterial flora of the oral cavity. J. Clin. Microbiol. 43 (11), 5721–5732. https://doi.org/10.1128/JCM.43.11.5721-5732.2005.

Alamoudi, N.M., Almabadi, E.S., El Ashiry, E.A., El Derwi, D.A., 2018. Effect of probiotic lactobacillus reuteri on salivary cariogenic bacterial counts among groups of preschool children in Jeddah, Saudi Arabia: a randomized clinical trial. J. Clin. Pediatr. Dent. 42 (5), 331–338. https://doi.org/10.17796/1053-4625-42.5.2.

Alanzi, A., Honkala, S., Honkala, E., Varghese, A., Tolvanen, M., Soderling, E., 2018. Effect of *Lactobacillus rhamnosus* and Bifidobacterium lactis on gingival health, dental plaque, and periodontopathogens in adolescents: a randomised placebo-controlled clinical trial. Benefic. Microbes 9, 593–602.

Albuquerque-Souza, E., Balzarini, D., Ando-Suguimoto, E.S., Ishikawa, K.H., Simionato, M.R.L., Holzhausen, M., Mayer, M.P.A., 2019. Probiotics alter the immune response of gingival epithelial cells challenged by *Porphyromonas gingivalis*. J. Periodontal Res. 54, 115–127.

Alqahtani, F., Alqahtani, M., Shafqat, S.S., Akram, Z., Al-Kheraif, A.A., Javed, F., 2019. Efficacy of mechanical debridement with adjunctive probiotic therapy in the treatment of peri-implant mucositis in cigarette-smokers and never-smokers. Clin. Implant. Dent. Relat. Res. 21 (4), 734–740. https://doi.org/10.1111/cid.12795.

Ashimoto, A., Chen, C., Bakker, I., Slots, J., 1996. Polymerase chain reaction detection of 8 putative periodontal pathogens in sub gingival plaque of gingivitis and advanced periodontitis lesions. Oral Microbiol. Immunol. 11, 266–273.

Ata-Ali, J., Candel-Marti, M.E., Flichy-Fernández, A.J., Peñarrocha-Oltra, D., Balaguer-Martinez, J.F., Peñarrocha Diago, M., 2011. Peri-implantitis: associated microbiota and treatment. Med. Oral Patol. Oral Cir. Bucal. 16, e937–e943.

Bartold, P.M., Van Dyke, T.E., 2013. Periodontitis: a host-mediated disruption of microbial homeostasis. Periodontol. 62 (1), 203–217. https://doi.org/10.1111/j.1600-0757.2012.00450.x.

Beck, J., Garcia, R., Heiss, G., Vokonas, P.S., Offenbacher, S., 1996. Periodontal disease and cardiovascular disease. J. Periodontol. 67, 1123–1137.

Bron, P.A., van Baarlen, P., Kleerebezem, M., 2011. Emerging molecular insights into the interaction between probiotics and the host intestinal mucosa. Nat. Rev. Microbiol. 10 (1), 66–78.

Caballero, S., Pamer, E.G., 2015. Microbiota-mediated inflammation and antimicrobial defense in the intestine. Annu. Rev. Immunol. 33, 227–256.

Caton, G.J., Armitage, G., Berglundh, T., Chapple, I.L.C., Jepsen, S., Kornman, K.S., et al., 2018. A new classification scheme for periodontal and peri-implant diseases and conditions—introduction and key changes from the 1999 classification. J. Clin. Periodontol. 45 (Suppl. 20), S1–S8. https://doi.org/10.1111/jcpe.12935.

Caufield, P.W., Dasanayake, A.P., Li, Y., Pan, Y., Hsu, J., Hardin, J.M., 2000. Natural history of *Streptococcus sanguinis* in the oral cavity of infants: evidence for a discrete window of infectivity. Infect. Immun. 68 (7), 4018–4023. https://doi.org/10.1128/IAI.68.7.4018-4023.2000.

Chapple, I.L.C., Mealey, B.L., Van Dyke, T.E., Bartold, P.M., Dommisch, H., Eickholz, P., Geisinger, M.L., Genco, R.J., Glogauer, M., Goldstein, M., Griffin, T.J., Holmstrup, P., Johnson, G.K., Kapila, Y., Lang, N.P., Meyle, J., Murakami, S., Plemons, J., Romito, G.A., Shapira, L., Tatakis, D.N., Teughels, W., Trombelli, L., Walter, C., Wimmer, G., Xenoudi, P., Yoshie, H., 2018. Periodontal health and gingival diseases and conditions on an intact and a reduced periodontium: consensus report of workgroup 1 of the 2017 world workshop on the classification of periodontaland peri-implant diseases and conditions. J. Clin. Periodontol. 45 (Suppl. 20), S68–S77. https://doi.org/10.1111/jcpe.12940.

Chen, F., Wang, D., 2010. Novel technologies for the prevention and treatment of dental caries: a patent survey. Expert Opin. Ther. Pat. 20 (5), 681–694. https://doi.org/10.1517/13543771003720491.

Chuang, L.C., Huang, C.S., Ou-Yang, L.W., Lin, S.Y., 2011. Probiotic *Lactobacillus paracasei* effect on cariogenic bacterial flora. Clin. Oral Investig. 15 (4), 471–476. https://doi.org/10.1007/s00784-010-0423-9.

Clarke, J.K., 1924. On the bacterial factor in the etiology of dental caries. Br. J. Exp. Pathol. 5, 141–147.

Conrads, G., de Soet, J.J., Song, J., Henne, K., Sztajer, H., Wagner-Döbler, I., Zeng, A.P., 2014. Comparing the cariogenic species *Streptococcus sobrinus* and *S. mutans* on whole genome level. J. Oral Microbiol. 6 (1), 26189. https://doi.org/10.3402/jom.v6.26189.

Cornacchione, L.P., Klein, B.A., Duncan, M.J., Hua, L.T., 2019. Interspecies inhibition of *Porphyromonas gingivalis* by yogurt-derived *Lactobacillus delbrueckii* requires active pyruvate oxidase. Appl. Environ. Microbiol. 85, e01271-19.

Costerton, J.W., Geesey, G.G., Cheng, K.J., 1978. How bacteria stick. Sci. Am. 238, 86–95.

Cross, M.L., Ganner, A., Teilab, D., Fray, L.M., 2004. Patterns of cytokine induction by gram-positive and gram-negative probiotic bacteria. FEMS Immunol. Med. Microbiol. 42 (2), 173–180.

Curtis, M.A., Zenobia, C., Darveau, R.P., 2011. The relationship of the oral microbiota to periodontal health and disease. Cell Host Microbe 10 (4), 302–306.

Derks, J., Tomasi, C., 2015. Peri-implant health and disease. A systematic review of current epidemiology. J. Clin. Periodontol. 42 (Suppl. 16), 158–171.

Devine, D.A., Marsh, P.D., Meade, J., 2015. Modulation of host responses by oral commensal bacteria. J. Oral Microbiol. 7 (1), 26941.

Donos, N., Calciolari, E., Brusselaers, N., Goldoni, M., Bostanci, N., Belibasakis, G.N., 2019. The adjunctive use of host modulators in nonsurgical periodontal therapy. A systematic review of randomized, placebo-controlled clinical studies. J. Clin. Periodontol. 47, 199–238. https://doi.org/10.1111/jcpe.13232.

Esteban-Fernández, A., Ferrer, M.D., Zorraquín-Peña, I., López-López, A., Moreno-Arribas, M.V., Mira, A., 2019. In vitro beneficial effects of *Streptococcus dentisani* as potential oral probiotic for periodontal diseases. J. Periodontol. 90, 1346–1355.

Featherstone, J.D., Adair, S.M., Anderson, M.H., Berkowitz, R.J., Bird, W.F., Crall, J.J., Den Besten, P.K., Donly, K.J., Glassman, P., Milgrom, P., Roth, J.R., Snow, R., Stewart, R.E., 2003. Caries management by risk assessment: consensus statement. J. Calif. Dent. Assoc. 31, 257–269.

Flichy-Fernandez, A.J., Ata-Ali, J., Alegre-Domingo, T., et al., 2015. The effect of orally administered probiotic *Lactobacillus reuteri*-containing tablets in peri-implant mucositis: a double-blind randomized controlled trial. J. Periodontal Res. 50 (6), 775–785.

Forsythe, P., Bienenstock, J., 2010. Immunomodulation by commensal and probiotic bacteria. Immunol. Investig. 39 (4–5), 429–448.

Foureaux, R.C., Messora, M.R., de Oliveira, L.F., Napimoga, M.H., Pereira, A.N., Ferreira, M.S., Pereira, L.J., 2014. Effects of probiotic therapy on metabolic and inflammatory parameters of rats with ligature-induced periodontitis associated with restraint stress. J. Periodontol. 85 (7), 975–983. https://doi.org/10.1902/jop.2013.130356.

Galofre, M., Palao, D., Vicario, M., Nart, J., Violant, D., 2018. Clinical and microbiological evaluation of the effect of *Lactobacillus reuteri* in the treatment of mucositis and peri-implantitis: a triple-blind randomized clinical trial. J. Periodontal Res. 53 (3), 378–390.

Gao, L., et al., 2018. Oral microbiomes: more and more importance in oral cavity and whole body. Protein Cell 9, 488–500.

Garcia, V.G., Knoll, L.R., Longo, M., Novaes, V.C., Assem, N.Z., Ervolino, E., de Toledo, B.E., Theodoro, L.H., 2016. Effect of the probiotic *Saccharomyces cerevisiae* on ligature-induced periodontitis in rats. J. Periodontal Res. 51 (1), 26–37. https://doi.org/10.1111/jre.12274.

Gatej, S.M., Marino, V., Bright, R., Fitzsimmons, T.R., Gully, N., Zilm, P., Gibson, R.J., Edwards, S., Bartold, P.M., 2018. Probiotic *Lactobacillus rhamnosus* GG prevents alveolar bone loss in a mouse model of experimental periodontitis. J. Clin. Periodontol. 45 (2), 204–212. https://doi.org/10.1111/jcpe.12838.

Ghadimi, D., Folster-Holst, R., de Vrese, M., Winkler, P., Heller, K.J., Schrezenmeir, J., 2008. Effects of probiotic bacteria and their genomic DNA on TH1/TH2-cytokine production by peripheral blood mononuclear cells (PBMCs) of healthy and allergic subjects. Immunobiology 213 (8), 677–692.

Ghasemi, E., Mazaheri, R., Tahmourespour, A., 2017. Effect of probiotic yogurt and xylitol-containing chewing gums on salivary S mutans count. J. Clin. Pediatr. Dent. 41 (4), 257–263. https://doi.org/10.17796/1053-4628-41.4.257.

Hajishengallis, G., 2015. Periodontitis: from microbial immune subversion to systemic inflammation. Nat. Rev. Immunol. 15 (1), 30.

Hajishengallis, G., Lamont, R.J., 2012. Beyond the red complex and into more complexity: the polymicrobial synergy and dysbiosis (PSD) model of periodontal disease etiology. Mol Oral Microbiol 27 (6), 409–419.

Hallström, H., Lindgren, S., Widén, C., Renvert, S., Twetman, S., 2016. Probiotic supplements and debridement of peri-implant mucositis: a randomized controlled trial. Acta Odontol. Scand. 74, 60–66.

Han, Y.W., Wang, X., 2013. Mobile microbiome: oral bacteria in extra-oral infections and inflammation. J. Dent. Res. 92 (6), 485–491.

Handfield, M., Mans, J.J., Zheng, G., Lopez, M.C., Mao, S., Progulske-Fox, A., et al., 2005. Distinct transcriptional profiles characterize oral epithelium-microbiota interactions. Cell. Microbiol. 7 (6), 811–823.

Harini, P.M., Anegundi, R.T., 2010. Efficacy of a probiotic and chlorhexidine mouth rinses: a short-term clinical study. J. Indian Soc. Pedod. Prev. Dent. 28 (3), 179–182.

Hasegawa, Y., Mans, J.J., Mao, S., Lopez, M.C., Baker, H.V., Handfield, M., Lamont, R.J., 2007. Gingival epithelial cell transcriptional responses to commensal and opportunistic oral microbial species. Infect. Immun. 75 (5), 2540–2547.

Hasslöf, P., Hedberg, M., Twetman, S., Stecksén-Blicks, C., 2010. Growth inhibition of oral mutans streptococci and candida by commercial probiotic lactobacilli—an in vitro study. BMC Oral Health 10 (1), 18–23. https://doi.org/10.1186/1472-6831-10-18.

Haukioja, A., Yli-Knuuttila, H., Loimaranta, V., Kari, K., Ouwehand, A.C., Meurman, J.H., Tenovuo, J., 2006. Oral adhesion and survival of probiotic and other lactobacilli and bifidobacteria in vitro. Oral Microbiol. Immunol. 21, 326–332. https://doi.org/10.1111/j.1399-302X.2006.00299.x.

Higuchia, T., Suzukib, N., Nakayac, S., Omagaria, S., Yonedaa, M., Haniokab, T., Hirofujia, T., 2019. Effects of *Lactobacillus salivarius* WB21 combined with green tea catechins on dental caries, periodontitis, and oral malodor. Arch. Oral Biol. 98, 243–247.

Hoceini, A., Khelil, N.K., Ben-Yelles, I., Mesli, A., Ziouani, S., Ghellai, L., Aissaoui, N., Nas, F., Arab, M., 2016. Caries-related factors and bacterial composition of supragingival plaques in caries free and caries active Algerian adults. Asian Pac. J. Trop. Biomed. 6 (8), 720–726. https://doi.org/10.1016/j.apjtb.2016.06.011.

Ikram, S., Hassan, N., Raffat, M.A., Mirza, S., Akram, Z., 2018 Aug. Systematic review and meta-analysis of double-blind, placebo-controlled, randomized clinical trials using probiotics in chronic periodontitis. J. Investig. Clin. Dent. 9 (3). https://doi.org/10.1111/jicd.12338, e12338 (Epub 2018 Mar 31).

Imran, F., Das, S., Padmanabhan, S., Rao, R., Suresh, A., Bharath, D., 2015. Evaluation of the efficacy of a probiotic drink containing *Lactobacillus casei* on the levels of periodontopathic bacteria in periodontitis: a clinico-microbiologic study. Indian J. Dent. Res. 26 (5), 462–468.

Ince, G., G€ursoy, H., Ipci, S.D., Cakar, G., Emekli-Alturfan, E., Yılmaz, S., 2015. Clinical and biochemical evaluation of *Lactobacillus reuteri* containing lozenges as an adjunct to non-surgical periodontal therapy in chronic periodontitis. J. Periodontol. 86, 746–754.

Iniesta, M., Herrera, D., Montero, E., Zurbriggen, M., Matos, A.R., Marin, M.J., Sánchez-Beltrán, M.C., Llama-Palacio, A., Sanz, M., 2012. Probiotic effects of orally administered *Lactobacillus reuteri*-containing tablets on the subgingival and salivary microbiota in patients with gingivitis. A randomized clinical trial. J. Clin. Periodontol. 39, 736–744.

Invernici, M.M., Salvador, S.L., Silva, P.H.F., Soares, M.S.M., Casarin, R., Palioto, D.B., Souza, S.L.S., Taba Jr., M., Novaes Jr., A.B., Furlaneto, F.A.C., Messora, M.R., 2018 Oct. Effects of bifidobacterium probiotic on the treatment of chronic periodontitis: a randomized clinical trial. J. Clin. Periodontol. 45 (10), 1198–1210. https://doi.org/10.1111/jcpe.12995.

Ishihara, K., Nabuchi, A., Ito, R., Miyachi, K., Kuramitsu, H.K., Okuda, K., 2004. Correlation between detection rates of periodontopathic bacterial DNA in carotid coronary stenotic artery plaque and in dental plaque samples. J. Clin. Microbiol. 42, 1313–1315.

Iwamoto, Y., Nishimura, F., Nakagawa, M., Sugimoto, H., Shikata, K., Makino, H., Fukuda, T., Tsuji, T., Iwamoto, M., Murayama, Y., 2001. The effect of antimicrobial periodontal treatment on circulating tumor necrosis factor-alpha and glycated hemoglobin level in patients with type 2 diabetes. J. Periodontol. 72, 774–778.

Iwasaki, K., Maeda, K., Hidaka, K., Nemoto, K., Hirose, Y., Deguchi, S., 2016. Daily intake of heat-killed *Lactobacillus plantarum* L-137 decreases the probing depth in patients undergoing supportive periodontal therapy. Oral Health Prev. Dent. 14 (3), 207–214. https://doi.org/10.3290/j.ohpd.a36099.

Jasberg, H., Soderling, E., Endo, A., Beighton, D., Haukioja, A., 2016. Bifidobacteria inhibit the growth of *Porphyromonas gingivalis* but not of Streptococcus mutans in an in vitro biofilm model. Eur. J. Oral Sci. 124 (3), 251–258.

Karpiński, T.M., Szkaradkiewicz, A.K., 2013. Microbiology of dental caries. J. Biol. Earth Sci. 3, M21–M24.

Karuppaiah, R.M., Shankar, S., Raj, S.K., Ramesh, K., Prakash, R., Kruthika, M., 2013. Evaluation of the efficacy of probiotics in plaque reduction and gingival health maintenance among school children—a randomized control trial. J. Int. Oral Health 5, 33–37.

Keyes, P.H., Jordan, H.V., 1964. Periodontal lesions in the Syrian hamster III. Findings related to an infectious and transmissible component. Arch. Oral Biol. 9 (4), 377–400. https://doi.org/10.1016/0003-9969(64)90024-X.

Kidd, E.A., Fejerskov, O., 2004. What constitutes dental caries? Histopathology of carious enamel and dentin related to the action of cariogenic biofilms. J. Dent. Res. 83 (Spec C), C35–C38.

Kobayashi, R., Kobayashi, T., Sakai, F., Hosoya, T., Yamamoto, M., Kurita-Ochiai, T., 2017. Oral administration of *Lactobacillus gasseri* SBT2055 is effective in preventing *Porphyromonas gingivalis*-accelerated periodontal disease. Sci. Rep. 7 (1), 545. https://doi.org/10.1038/s41598-017-00623-9.

Koll-Klais, P., Mandar, R., Leibur, E., Marcotte, H., Hammarstrom, L., Mikelsaar, M., 2005. Oral Lactobacilli in chronic periodontitis and periodontal health: species composition and antimicrobial activity. Oral Microbiol. Immunol. 20, 354–361.

Koo, H., Xiao, J., Klein, M.I., Jeon, J.G., 2010. Exopolysaccharides produced by Streptococcus mutans glucosyltransferases modulate the establishment of microcolonies within multispecies biofilms. J. Bacteriol. 192 (12), 3024–3032. https://doi.org/10.1128/JB.01649-09.

Kornman, K.S., 2001. Patients are not equally susceptible to periodontitis: does this change dental practice and the dental curriculum? J. Dent. Educ. 65, 777–784.

Krasse, P., Carlsson, B., Dahl, C., Paulsson, A., Nilsson, A., Sinkiewicz, G., 2006. Decreased gum bleeding and reduced gingivitis by the probiotic *Lactobacillus reuteri*. Swed. Dent. J. 30 (2), 55–60.

Kristofersson, K., Grondahl, J.G., Bratthall, D., 1985. The more Streptococcus mutans, the more caries on approximal surfaces. J. Dent. Res. 64 (1), 58–61. https://doi.org/10.1177/00220345850640011101.

Kuru, B.E., Laleman, I., Yalnizoglu, T., Kuru, L., Teughels, W., 2017. The influence of a bifidobacterium animalis probiotic on gingival health: a randomized controlled clinical trial. J. Periodontol. 88, 1115–1123. https://doi.org/10.1902/jop.2017.170213.

Laleman, I., Yilmaz, E., Ozcelik, O., Haytac, C., Pauwels, M., Herrero, E.R., Slomka, V., Quirynen, M., Alkaya, B., Teughels, W., 2015. The effect of a streptococci containing probiotic in periodontal therapy: a randomized controlled trial. J. Clin. Periodontol. 42, 1032–1041. https://doi.org/10.1111/jcpe.12464.

Larsson, L., Castilho, R.M., Giannobile, W.V., 2015. Epigenetics and its role in periodontal diseases: a state-of-the-art review. J. Periodontol. 86 (4), 556–568. https://doi.org/10.1902/jop.2014.140559.

Lebeer, S., Vanderleyden, J., De Keersmaecker, S.C., 2010. Host interactions of probiotic bacterial surface molecules: comparison with commensals and pathogens. Nat. Rev. Microbiol. 8 (3), 171–184.

Ledder, R.G., Timperley, A.S., Friswell, M.K., Macfarlane, S., McBain, A.J., 2008. Coaggregation between and among human intestinal and oral bacteria. FEMS Microbiol. Ecol. 66 (3), 630–636.

Lee, J.K., Kim, S.J., Ko, S.H., Ouwehand, A.C., Ma, D.S., 2015. Modulation of the host response by probiotic *Lactobacillus brevis* CD2 in experimental gingivitis. Oral Dis. 21 (6), 705–712.

Levi, Y.L.A.S., Novais, G.S., Dias, R.B., Andraus, R.A.C., Messora, M.R., Neto, H.B., Ervolino, E., Santinoni, C.S., Maia, L.P., 2018. Effects of the prebiotic mannan oligosaccharide on the experimental periodontitis in rats. J. Clin. Periodontol. 45 (9), 1078–1089. https://doi.org/10.1111/jcpe.12987.

Ley, R.E., Hamady, M., Lozupone, C., et al., 2008. Evolution of mammals and their gut microbes. Science 320 (5883), 1647–1651.

Lin, T.H., Pan, T.M., 2014. Inhibitory effect of *Lactobacillus paracasei* subsp. paracasei NTU 101 on rat dental caries. J. Funct. Foods 10, 223–231. https://doi.org/10.1016/j.jff.2014.06.015.

Lin, T.-H., Lin, C.-H., Pan, T.-M., 2017. The implication of probiotics in the prevention of dental caries. Appl. Microbiol. Biotechnol. 102 (2), 577–586. https://doi.org/10.1007/s00253-017-8664-z.

Liu, T.H., Tsai, T.Y., Pan, T.M., 2018. Effects of an ethanol extract from *Lactobacillus paracasei* subsp. paracasei NTU 101 fermented skimmed milk on lipopolysaccharide-induced periodontal inflammation in rats. Food Funct. 9 (9), 4916–4925. https://doi.org/10.1039/c8fo01303a.

Lodi, C.S., Manarelli, M.M., Sassaki, K.T., Fraiz, F.C., Delbem, A.C.B., Martinhon, C.C.R., 2010. Evaluation of fermented milk containing probiotic on dental enamel and biofilm: in situ study. Arch. Oral Biol. 55 (1), 29–33. https://doi.org/10.1016/j.archoralbio.2009.10.009.

Loomer, P.M., 2004. Microbiological diagnostic testing in the treatment of periodontal diseases. Periodontology 2000 34, 49–56.

Maekawa, T., Hajishengallis, G., 2014. Topical treatment with probiotic *Lactobacillus brevis* CD2 inhibits experimental periodontal inflammation and bone loss. J. Periodontal Res. 49 (6), 785–791.

Marsh, P.D., 1994. Microbial ecology of dental plaque and its significance in health and disease. Adv. Dent. Res. 8 (2), 263–271.

Marsh, P.D., 2003. Are dental diseases examples of ecological catastrophes? Microbiology 149, 279–294.

Marsh, P.D., 2004. Dental plaque as a microbial biofilm. Caries Res. 38 (3), 204–211. https://doi.org/10.1159/000077756.

Marsh, P.D., Martin, M.V., 1999. Oral Microbiology, fourth ed. Wright, Oxford.

Marsh, P.D., Martin, M.V., Lewis, M.A., Williams, D.W., Marsh, P.D., Martin, M.V., 2012. Oral Microbiology, fifth ed. Elsevier, Edinburgh.

Marttinen, A.M., Haukioja, A.L., Keskin, M., Soderling, E.M., 2013. Effects of *Lactobacillus reuteri* PTA 5289 and *L. paracasei* DSMZ16671 on the adhesion and biofilm formation of Streptococcus mutans. Curr. Microbiol. 67 (2), 193–199.

Mayanagi, G., Kimura, M., Nakaya, S., Hirata, H., Sakamoto, M., Benno, Y., Shimauchi, H., 2009. Probiotic effects of orally administered *Lactobacillus salivarius* WB21-containing tablets on periodontopathic bacteria: a double-blinded, placebo-controlled, randomized clinical trial. J. Clin. Periodontol. 36, 506–513.

Mendi, A., Kose, S., Uckan, D., Akca, G., Yilmaz, D., Aral, L., et al., 2016. Lactobacillus rhamnosus could inhibit *Porphyromonas gingivalis* derived CXCL8 attenuation. J. Appl. Oral Sci. 24 (1), 67–75.

Messora, M.R., Oliveira, L.F., Foureaux, R.C., Taba Jr., M., Zangerônimo, M.G., Furlaneto, F.A., Pereira, L.J., 2013. Probiotic therapy reduces periodontal tissue destruction and improves the intestinal morphology in rats with ligature-induced periodontitis. J. Periodontol. 84 (12), 1818–1826. https://doi.org/10.1902/jop.2013.120644.

Messora, M.R., Pereira, L.J., Foureaux, R., Oliveira, L.F., Sordi, C.G., Alves, A.J., Napimoga, M.H., Nagata, M.J., Ervolino, E., Furlaneto, F.A., 2016. Favourable effects of *Bacillus subtilis* and *Bacillus licheniformis* on experimental periodontitis in rats. Arch. Oral Biol. 66, 108–119. https://doi.org/10.1016/j.archoralbio.2016.02.014.

Mongardini, C., Pilloni, A., Farina, R., Di Tanna, G., Adjunctive, Z.B., Farina, R., 2017. Adjunctive efficacy of probiotics in the treatment of experimental peri-implant mucositis with mechanical and photodynamic therapy: a randomized, cross-over clinical trial. J. Clin. Periodontol. 44, 410–417.

Morales, A., Carvajal, P., Silva, N., Hernandez, M., Godoy, C., Rodriguez, G., Cabello, R., Garcia-Sesnich, J., Hoare, A., Diaz, P.I., et al., 2016. Clinical effects of *Lactobacillus rhamnosus* in non-surgical treatment of chronic periodontitis: a randomized placebo-controlled trial with 1-year follow-up. J. Periodontol. 87, 944–952. https://doi.org/10.1902/jop.2016.150665.

Morales, A., Gandolfo, A., Bravo, J., Carvajal, P., Silva, N., Godoy, C., Gamonal, J., 2018. Microbiological and clinical effects of probiotics and antibiotics on nonsurgical treatment of chronic periodontitis: a randomized placebo-controlled trial with 9-month follow-up. J. Appl. Oral Sci. 26. https://doi.org/10.1590/1678-7757-2017-0075.

Nagaoka, S., Hojo, K., Murata, S., Mori, T., Ohshima, T., Maeda, N., 2008. Interactions between salivary *Bifidobacterium adolescentis* and other oral bacteria: in vitro coaggregation and coadhesion assays. FEMS Microbiol. Lett. 281 (2), 183–189.

Nazir, M.A., 2017. Prevalence of periodontal disease, its association with systemic diseases and prevention. Int. J. Health Sci. 11, 72–80.

Nissen, L., Sgorbati, B., Biavati, B., Belibasakis, G.N., 2014. *Lactobacillus salivarius* and *L. gasseri* down-regulate *Aggregatibacter actinomycetemcomitans* exotoxins expression. Ann. Microbiol. 64, 611–617.

Näse, L., Hatakka, K., Savilahti, E., Saxelin, M., Pönkä, A., Poussa, T., Korpela, R., Meurman, J.H., 2001. Effect of long-term consumption of a probiotic bacterium, *Lactobacillus rhamnosus* GG, in milk on dental caries and caries risk in children. Caries Res. 35 (6), 412–420. https://doi.org/10.1159/000047484.

Oliveira, L.F., Salvador, S.L., Silva, P.H., Furlaneto, F.A., Figueiredo, L., Casarin, R., Ervolino, E., Palioto, D.B., Souza, S.L., Taba Jr., M., Novaes Jr., A.B., Messora, M.R., 2017. Benefits of *Bifidobacterium animalis* subsp. lactis probiotic in experimental periodontitis. J. Periodontol. 88 (2), 197–208. https://doi.org/10.1902/jop.2016.160217.

Pahumunto, N., Piwat, S., Chankanka, O., Akkarachaneeyakorn, N., Rangsitsathian, K., Teanpaisan, R., 2018. Reducing mutans streptococci and caries development by *Lactobacillus paracasei* SD1 in preschool children: a randomized placebo-controlled trial. Acta Odontol. Scand. 76 (5), 331–337. https://doi.org/10.1080/00016357.2018.1453083.

Papapanou, P.N., Sanz, M., Buduneli, N., Dietrich, T., Feres, M., Fine, D.H., Flemmig, T.F., Garcia, R., Giannobile, W.V., Graziani, F., Greenwell, H., Herrera, D., Kao, R.T., Kebschull, M., Kinane, D.F., Kirkwood, K.L., Kocher, T., Kornman, K.S., Kumar, P.S., Loos, B.G., Machtei, E., Meng, H., Mombelli, A., Needleman, I., Offenbacher, S., Seymour, G.J., Teles, R., Tonetti, M.S., 2018. Periodontitis: consensus report of workgroup 2 of the 2017 world workshop on the classification of periodontal and peri-implant diseases and conditions. J. Periodontol. 89 (Suppl 1), S173–S182. https://doi.org/10.1002/JPER.17-0721.

Parahitiyawa, N.B., Jun, L.J., Leung, W.K., Yam, W.X., Samaranayake, L.P., 2009. Microbiology of odontogenic bacteremia: beyond endocarditis. Clin. Microbiol. Rev. 22, 46–64.

Patel, R.M., Denning, P.W., 2013. Therapeutic use of prebiotics, probiotics, and postbiotics to prevent necrotizing enterocolitis: what is the current evidence? Clin. Perinatol. 40 (1), 11–25.

Pawar, M.M., Pattanaik, A.K., Sinha, D.K., Goswami, T.K., Sharma, K., 2017. Effect of dietary mannanoligosaccharide supplementation on nutrient digestibility, hindgut fermentation, immune response and antioxidant indices in dogs. J. Anim. Sci. Technol. 59 (1), 11. https://doi.org/10.1186/s40781-017-0136-6.

Pelekos, F., Ho, S.N., Acharya, A., Leung, W.K., McGrath, C., 2019. A double-blind, paralleled-arm, placebo-controlled and randomized clinical trial of the effectiveness of probiotics as an adjunct in periodontal care. J. Clin. Periodontol. 46, 1217–1227.

Penala, S., Kalakonda, B., Pathakota, K.R., et al., 2016. Efficacy of local use of probiotics as an adjunct to scaling and root planing in chronic periodontitis and halitosis: a randomized controlled trial. J. Res. Pharm. Pract. 5, 86.

Peña, M., Barallat, L., Vilarrasa, J., Vicario, M., Violant, D., Nart, J., 2019 Apr. Evaluation of the effect of probiotics in the treatment of peri-implant mucositis: a triple-blind randomized clinical trial. Clin. Oral Investig. 23 (4), 1673–1683. https://doi.org/10.1007/s00784-018-2578-8.

Pickard, J.M., Zeng, M.Y., Caruso, R., Núñez, G., 2017. Gut microbiota: role in pathogen colonization, immune responses and inflammatory disease. Immunol. Rev. 279, 70–89.

Riccia, D.N., Bizzini, F., Perilli, M.G., Polimeni, A., Trinchieri, V., Amicosante, G., et al., 2007. Anti-inflammatory effects of *Lactobacillus brevis* (CD2) on periodontal disease. Oral Dis. 13 (4), 376–385.

Rickard, A.H., Gilbert, P., High, N.J., Kolenbrander, P.E., Handley, P.S., 2003. Bacterial coaggregation: an integral process in the development of multi-species biofilms. Trends Microbiol. 11 (2), 94–100.

Ricoldi, M.S.T., Furlaneto, F.A.C., Oliveira, L.F.F., Teixeira, G.C., Pischiotini, J.P., Moreira, A.L.G., Ervolino, E., de Oliveira, M.N., Bogsan, C.S.B., Salvador, S.L., Messora, M.R., 2017. Effects of the probiotic *Bifidobacterium animalis* subsp. lactis on the non-surgical treatment of periodontitis. A histomorphometric, microtomographic and immunohistochemical study in rats. PLoS One 12 (6). https://doi.org/10.1371/journal.pone.0179946, e0179946.

Roberfroid, M., 2007. Prebiotics: the concept revisited. J. Nutr. 137, 830S–837S.

Roberts, F.A., Darveau, R.P., 2015. Microbial protection and virulence in periodontal tissue as a function of polymicrobial communities: symbiosis and dysbiosis. Periodontol. 2000 69 (1), 18–27.

Rodríguez, G., Ruiz, B., Faleiros, S., et al., 2016. Probiotic compared with standard milk for high-caries children: a cluster randomized trial. J. Dent. Res. 95 (4), 402–407. https://doi.org/10.1177/0022034515623935.

Sartor, R.B., 2004. Therapeutic manipulation of the enteric microflora in inflammatory bowel diseases: antibiotics, probiotics, and prebiotics. Gastroenterology 126 (6), 1620–1633.

Schwarz, F., Schmucker, A., Becker, J., 2015. Efficacy of alternative or adjunctive measures to conventional treatment of peri-implant mucositis and peri-implantitis: a systematic review and meta-analysis. Int. J. Implant Dent. 1, 22.

Seneviratne, C.J., Zhang, C.F., Samaranayake, L.P., 2011. Dental plaque biofilm in oral health and disease. Chin. J. Dent. Res. 14 (2), 87–94.

Shah, M.P., Gujjari, S.K., Chandrasekhar, V.S., 2013. Evaluation of the effect of probiotic (Inersan®) alone, combination of probiotic with doxycycline and doxycycline alone on aggressive periodontitis—a clinical and microbiological study. J. Clin. Diagn. Res. 7 (3), 595–600.

Shimauchi, H., Mayanagi, G., Nakaya, S., Minamibuchi, M., Ito, Y., Yamaki, K., Hirata, H., 2008. Improvement of periodontal condition by probiotics with *Lactobacillus salivarius* WB21: a randomized, double-blind, placebo-controlled study. J. Clin. Periodontol. 35, 897–905.

Silva, V.O., Lobato, R.V., Andrade, E.F., de Macedo, C.G., Napimoga, J.T., Napimoga, M.H., Messora, M.R., Murata, R.M., Pereira, L.J., 2015. β-Glucans (*Saccharomyces cerevisiae*) reduce glucose levels and attenuate alveolar bone loss in diabetic rats with periodontal disease. PLoS One 10 (8). https://doi.org/10.1371/journal.pone.0134742, e0134742.

Silva, P.H.F., Oliveira, L.F.F., Cardoso, R.S., et al., 2019. The impact of predatory bacteria on experimental periodontitis. J. Periodontol. 90 (9), 1053–1063. https://doi.org/10.1002/JPER.18-0485.

Simon, L., 2007. The role of Streptococcus mutans and oral ecology in the formation of dental caries. Lethbridge Undergraduate Res. J. 2, 1–6.

Slomka, V., Hernandez-Sanabria, E., Herrero, E.R., Zaidel, L., Bernaerts, K., Boon, N., et al., 2017. Nutritional stimulation of commensal oral bacteria suppresses pathogens: the prebiotic concept. J. Clin. Periodontol. 44 (4), 344–352. https://doi.org/10.1111/jcpe.12700.

Socransky, S.S., Haffajee, A.D., 2005. Periodontal microbial ecology. Periodontol. 2000 38, 135–187.

Socransky, S.S., Haffajee, A.D., Cugini, M.A., Smith, C., Kent Jr., R.L., 1998. Microbial complexes in subgingival plaque. J. Clin. Periodontol. 25, 134–144.

Staab, B., Eick, S., Knofler, G., Jentsch, H., 2009. The influence of a probiotic milk drink on the development of gingivitis: a pilot study. J. Clin. Periodontol. 36 (10), 850–856.

Stamatova, I., Meurman, J.H., 2009a. Probiotics and periodontal disease. Periodontol 2000 51, 141–151.

Stamatova, I., Meurman, J.H., 2009b. Probiotics and periodontal disease. Periodontology 2000 51, 141–151.

Stamatova, I., Kari, K., Meurman, J.H., 2008. In vitro evaluation of antimicrobial activity of putative probiotic lactobacilli against oral pathogens. Int. J. Probiotics Prebiotics 2, 225–232.

Stensson, M., Koch, G., Coric, S., et al., 2014. Oral administration of *Lactobacillus reuteri* during the first year of life reduces caries prevalence in the primary dentition at 9 years of age. Caries Res. 48 (2), 111–117. https://doi.org/10.1159/000354412.

Struzycka, I., 2014. The oral microbiome in dental caries. Pol. J. Microbiol. 63 (2), 127–135.

Sugano, N., 2012. Biological plaque control: novel therapeutic approach to periodontal disease. J. Oral Sci. 54 (1), 1–5.

Szkaradkiewicz, A.K., Stopa, J., Karpinski, T.M., 2014. Effect of oral administration involving a probiotic strain of *Lactobacillus reuteri* on proinflammatory cytokine response in patients with chronic periodontitis. Arch. Immunol. Ther. Exp. (Warsz.) 62, 495–500. https://doi.org/10.1007/s00005-014-0277-y.

Sánchez, B., Urdaci, M.C., Margolles, A., 2010. Extracellular proteins secreted by probiotic bacteria as mediators of effects that promote mucosa-bacteria interactions. Microbiology (Reading) 156 (11), 3232–3242. https://doi.org/10.1099/mic.0.044057-0.

Tada, H., Masaki, C., Tsuka, S., Mukaibo, T., Kondo, Y., Hosokawa, R., 2018. The effects of *Lactobacillus reuteri* probiotics combined with azithromycin on peri-implantitis: a randomized placebo-controlled study. J. Prosthodont. Res. 62 (1), 89–96. https://doi.org/10.1016/j.jpor.2017.06.006.

Teitelbaum, J.E., Walker, W.A., 2002. Nutritional impact of pre- and probiotics as protective gastrointestinal organisms. Annu. Rev. Nutr. 22, 107–138.

Tekce, M., Ince, G., Gursoy, H., Dirikan Ipci, S., Cakar, G., Kadir, T., Yılmaz, S., 2015. Clinical and microbiological effects of probiotic lozenges in the treatment of chronic periodontitis: a 1-year follow-up study. J. Clin. Periodontol. 42, 363–372. https://doi.org/10.1111/jcpe.12387.

Teughels, W., Newman, M.G., Coucke, W., Haffajee, A.D., Van Der Mei, H.C., Haake, S.K., Schepers, E., Cassiman, J.J., Van Eldere, J., van Steenberghe, D., Quirynen, M., 2007. Guiding periodontal pocket recolonization: a proof of concept. J. Dent. Res. 86 (11), 1078–1082. https://doi.org/10.1177/154405910708601111.

Teughels, W., Loozen, G., Quirynen, M., 2011. Do probiotics offer opportunities to manipulate the periodontal oral microbiota? J. Clin. Periodontol. 38 (Suppl. 11), 159–177.

Teughels, W., Durukan, A., Ozcelik, O., Pauwels, M., Quirynen, M., Haytac, M.C., 2013. Clinical and microbiological effects of *Lactobacillus reuteri* probiotics in the treatment of chronic periodontitis: a randomized placebo-controlled study. J. Clin. Periodontol. 40 (11), 1025–1035. https://doi.org/10.1111/jcpe.12155.

Tobita, K., Watanabe, I., Tomokiyo, M., Saito, M., 2018. Effects of heat-treated *Lactobacillus crispatus* KT-11 strain consumption on improvement of oral cavity environment: a randomized double-blind clinical trial. Benef. Microbes. 9 (4), 585–592. https://doi.org/10.3920/BM2017.0137.

Toiviainen, A., Jalasvuori, H., Lahti, E., Gursoy, U., Salminen, S., Fontana, M., Flannagan, S., Eckert, G., Kokaras, A., Paster, B., Söderling, E., 2015. Impact of orally administered lozenges with *Lactobacillus rhamnosus* GG and *Bifidobacterium animalis* subsp. lactis BB-12 on the number of salivary mutans streptococci, amount of plaque, gingival inflammation and the oral microbiome in healthy adults. Clin. Oral Investig. 19, 77–83.

Tokunaga, T., Oku, T., Hosoya, N., 1989. Utilization and excretion of a new sweetener, fructooligosaccharide (Neosugar), in rats. J. Nutr. 119 (4), 553–559.

Travers, M.A., Florent, I., Kohl, L., Grellier, P., 2011. Probiotics for the control of parasites: an overview. J. Parasitol. Res. 2011, 610769.

Tsubura, S., Mizunuma, H., Ishikawa, S., Oyake, I., Okabayashi, M., Katoh, K., et al., 2009. The effect of *Bacillus subtilis* mouth rinsing in patients with periodontitis. Eur. J. Clin. Microbiol. Infect. Dis. 28 (11), 1353–1356.

Twetman, L., Larsen, U., Fiehn, N.E., Stecksén-Blicks, C., Twetman, S., 2009. Coaggregation between probiotic bacteria and caries-associated strains: an in vitro study. Acta Odontol. Scand. 67 (5), 284–288. https://doi.org/10.1080/00016350902984237.

Vicario, M., Santos, A., Violant, D., Nart, J., Ginger, L., 2013. Clinical changes in periodontal subjects with the probiotic *Lactobacillus reuteri* Prodentis: a preliminary randomized clinical trial. Acta Odontol. Scand. 71 (3–4), 813–819. https://doi.org/10.3109/00016357.2012.734404.

Vicario, M., Santos, A., Violant, D., Nart, J., Giner, L., 2013. Clinical changes in periodontal subjects with the probiotic *Lactobacillus reuteri* prodentis: a preliminary randomized clinical trial. Acta Odontol. Scand. 71 (3–4), 813–819. https://doi.org/10.3109/00016357.2012.734404.

Villavicencio, J., Villegas, L.M., Arango, M.C., Arias, S., Triana, F., 2018. Effects of a food enriched with probiotics on *Streptococcus* mutans and *Lactobacillus* spp. salivary counts in preschool children: a cluster randomized trial. J. Appl. Oral Sci. 26. https://doi.org/10.1590/1678-7757-2017-0318, e20170318.

Vivekananda, M., Vandana, K., Bhat, K., 2010. Effect of the probiotic *Lactobacilli reuteri* (Prodentis) in the management of periodontal disease: a preliminary randomized clinical trial. J. Oral Microbiol. 2, 5344.

Vuotto, C., Longo, F., Donelli, G., 2014. Probiotics to counteract biofilm-associated infections: promising and conflicting data. Int. J. Oral Sci. 6 (4), 189–194.

Vuotto, C., Barbanti, F., Mastrantonio, P., Donelli, G., 2014b. Lactobacillus brevis CD2 inhibits *Prevotella melaninogenica* biofilm. Oral Dis. 20 (7), 668–674.

Vähämiko, S., Laiho, A., Lund, R., et al., 2018. The impact of probiotic supplementation during pregnancy on DNA methylation of obesity-related genes in mothers and their children. Eur. J. Nutr. 58, 367–377. https://doi.org/10.1007/s00394-017-1601-1.

WHO, 2003. The world oral health report. Continuous improvement of oral health in the 21st century—the approach of the WHO Global Oral Health Programme. http://www.who.int/oral_health.

Wohlgemuth, S., Loh, G., Blaut, M., 2010. Recent developments and perspectives in the investigation of probiotic effects. Int. J. Med. Microbiol. 300 (1), 3–10.

Ye, P., Harty, D., Commandeur, Z., Hunter, N., 2013. Binding of *Streptococcus gordonii* to oral epithelial monolayers increases paracellular barrier function. Microb. Pathog. 56, 53–59.

Zambon, J.J., 1996. American academy of periodontology consensus report periodontal disease. Microbial factors. Ann. Periodontol. 67, 879–925.

Zhang, G., Chen, R., Rudney, J.D., 2008. Streptococcus cristatus attenuates *Fusobacterium nucleatum*-induced interleukin-8 expression in oral epithelial cells. J. Periodontal Res. 43 (4), 408–416.

Zitzmann, N.U., Berglundh, T., 2008. Definition and prevalence of peri-implant diseases. J. Clin. Periodontol. 35 (Suppl. 8), 286–291.

CHAPTER 5

Multistrain Probiotics and Benefits to Consumer's Health

ARTHUR C. OUWEHAND[a] • MARCOS INVERNICI[b] •
MICHEL REIS MESSORA[b] • FLÁVIA FURLANETO[b]
[a]International Flavors & Fragrances Inc., Kantvik, Finland, [b]Department of Oral and Maxillofacial Surgery and Periodontology, School of Dentistry of Ribeirao Preto, University of Sao Paulo—USP, Ribeirao Preto, SP, Brazil

INTRODUCTION

Probiotics are commonly defined as "Live microbes that, when administered in adequate amounts, confer a health benefit on the host" (Hill et al., 2014). This definition is valid for both single- and multistrain probiotics. For the potential health benefits of probiotics in general and specific strain (combinations) in particular, the reader is referred to other chapters in this book or to general reviews on this topic (e.g., Khalesi et al., 2019).

Probiotic products can come in many different formats and compositions. In their most basic form, probiotics consist of only one strain: a single-strain probiotic. However, a probiotic product can also be composed of combinations of different species and/or different strains: a multistrain probiotic. Different strains in a product may belong to the same species; this would still be a multistrain probiotic (Chapman et al., 2011).

In the market, there is a trend to commercialize products with ever more probiotics. There are a few thoughts behind this. More strains means more chances for success through a broader spectrum of activity. More strains provide a greater diversity with the potential to influence more potential niches. Many multistrain products, however, hope or assume an additive or synergistic effect through the creation of a mutually favorable environment or reduced antagonism of the endogenous microbiota. Whether this is the case or just wishful thinking remains to be determined. Interestingly, potential antagonism with multistrain probiotics is rarely considered.

An earlier study concluded that most of the studies reviewed (75%) showed a greater efficacy for multistrain probiotics compared with single-strain probiotics for a single health benefit (Forssten and Ouwehand, 2017; Piyadeatsoontorn et al., 2019). However, a more recent review did not find the human data convincing enough to conclude that multistrain probiotics are indeed more efficacious on a single health benefit (Ouwehand et al., 2018).

Investigating whether there is a difference in efficacy of single- vs. multistrain probiotics is more challenging than it may appear. The simplest multistrain probiotic consists of two strains. To test if it has a different efficacy as compared with the single strains, it requires a four-arm study: the two probiotic strains separately, their combination, and a placebo, as is schematically represented in Fig. 1.

For every additional strain in the combination, an additional study arm is required. This also influences the study power and requires larger groups in each study arm. Further, there is the question of dose. If single strains are tested at, e.g., 10^9 colony-forming units (CFU), is the combination then to be tested at 10^9 CFU as well, which implies that the components are at a fraction of that, or are components to be tested at 10^9 CFU and is the total count of the combination several times that of the single strains? Especially with a high number of strains, the dose difference may become substantial which may have implications just by itself (Ouwehand, 2017).

When composing multistrain probiotics, the cell counts known for the single strains can be used to calculate the amount needed for the formulation. However, once blended, it is difficult to determine the viable counts of the single components in the multistrain product. The organisms tend to be very similar and are difficult to separate by selective culturing. Molecular techniques such as digital PCR can help, and when combining such technique with propidium monoazide as an extracellular DNA intercalator, in inhibiting the amplification of DNA associated with dead probiotic bacteria; thus, live bacteria can be distinguished from damaged/dead bacteria (Hansen et al., 2018).

Probiotics and Prebiotics in Foods. https://doi.org/10.1016/B978-0-12-819662-5.00002-1

FIG. 1 Schematic representation of testing requirements for a two-strain probiotic (the simplest multistrain probiotic).

The aim of the current chapter is to provide an updated review of the current literature comparing single- and multistrain probiotic efficacy, in vitro, in animal and human studies. To this end, PubMed was searched for "probiotic" AND ("multi" OR "combination" OR "blend") AND "single." The latest search was performed on the January 16th, 2020.

IN VITRO STUDIES

Because of the complexity of studying the effects of single-strain and multistrain probiotics, and the observation that probiotic effects are usually subtle, in vitro studies provide a convenient and clearer-cut setup for this type of studies. The most commonly investigated endpoints are antimicrobial activity, adhesion, and immune modulation (Table 1).

For antimicrobial activity, we see that there is a tendency for multistrain products to be more efficacious (Table 1). Interestingly, this may be influenced by the methodology used as was shown by Fijan et al. (2018); agar well diffusion indicated superiority for multistrain probiotics, while coculture indicated superiority for single-strain probiotics. Further, as pointed out earlier, multistrain products will have a higher total count especially with a large number of strains. This may explain the observation by Chapman et al. (2012) on a higher efficacy for the multistrain product. Despite the

TABLE 1
Nonexhaustive Overview of In Vitro Studies Comparing Single- vs. Multistrain Probiotics.

Health Target	Strains	Model	Outcome	Reference
Antimicrobial activity	*B. longum* BB536, *L. rhamnosus* HN001	Agar well diffusion and broth dilution	S<M	Inturri et al. (2019)
Antimicrobial activity	5 single strains, 4 multistrain products	Agar well diffusion and coculture	S<M S>M	Fijan et al. (2018)
Antimicrobial activity against *E. coli* and *S. aureus*	*B. bifidum, B. longum, B. longum, L. acidophilus, L. fermentum, L. plantarum*	Coculture in broth	S<M	Mathipa and Thantsha (2015)
Intestinal pathogen growth inhibition	*L. acidophilus, L. bulgaricus, L. casei, L. plantarum, L. rhamnosus, L. salivarius, L. helveticus, B. bifidum, B. breve, B. infantis, B. longum, S. thermophilus, Lc. lactis, Bc. subtilis*	Cross-streak and agar spot assay	S<M	Chapman et al. (2012)
Urinary pathogen growth inhibition	*L. acidophilus, L. bulgaricus, L. casei, L. plantarum, L. rhamnosus, L. salivarius, L. helveticus, B. bifidum, B. breve, B. infantis, B. longum, S. thermophilus, Lc. lactis, B. subtilis*	Agar spot test	S=M	Chapman et al. (2013)
Pathogen adhesion inhibition and coaggregation	*L. gasseri* (seven strains)	Mucus adhesion and coaggregation	S=M	Ferreira et al. (2011)
Biofilm formation and adhesion	*L. acidophilus, L. bulgaricus, L. casei, L. plantarum, L. rhamnosus, L. salivarius, L. helveticus, B. bifidum, B. breve, B. infantis, B. longum, S. thermophilus, Lc. lactis, B. subtilis*	Bladder cell model	S=M	Chapman et al. (2014)
Adhesion	*L. rhamnosus* GG, *L. bulgaricus, B. lactis* Bb-12	Mucus adhesion	S<M	Ouwehand et al. (2000)

TABLE 1
Nonexhaustive Overview of In Vitro Studies Comparing Single- vs. Multistrain Probiotics—cont'd

Health Target	Strains	Model	Outcome	Reference
Anti-inflammatory response	*L. acidophilus*, *B. lactis*	HT-29 cells, LPS, and TNF-α challenge	S<M	Li et al. (2019)
Immune modulation	*L. helveticus* R0052, *B. infantis* R0033, *B. bifidum* R0071	HT-29 tissue culture cell gene expression	S<M	MacPherson et al. (2017)
Immune modulation	*L. rhamnosus* ATCC 7469, *L. acidophilus* La-14, *L. paracasei* 19.1	Macrophages	S>M	de Oliveira et al. (2017)
Growth yield	*L. paracasei* F8, *L. rhamnosus* GG, *B. breve* 46	Coculture	S<M	Ambalam et al. (2015)
Digestibility and antioxidant activity	*T. delbrueckii* CCMA 0235, *L. acidophilus* LAC-04, *L. plantarum* CCMA 0743	Cassava-rice blend fermentation	S<M	Freire et al. (2017)
Gastrointestinal survival	*L. acidophilus* NCFM, *L. paracasei* Lpc-37, *B. lactis* Bi-07, *B. lactis* Bl-04	In vitro gastrointestinal tract simulation	S=M	Forssten and Ouwehand (2017)
Coexistence	*L. plantarum* L21, *L. plantarum* L80, *L. paraplantarum* L103	Cross-streak assay	S=M	Piyadeatsoontorn et al. (2019)

B., *Bifidobacterium*; *Bc.*, *Bacillus*; *L.*, *Lactobacillus* (*sensu lato*); *Lc.*, *Lactococcus*; *S.*, *Streptococcus*; *T.* = *Torulaspora*.

ability to produce antimicrobial components, probiotics do not seem to exert an antagonistic activity against each other: neither during simulated gastrointestinal transit (Forssten and Ouwehand, 2017) nor in coculture (Piyadeatsoontorn et al., 2019).

Pathogen adhesion and biofilm formation do not appear to be different between single- and multistrain probiotics (Table 1). There is a strain-dependent effect here. The effect is mainly driven by the most active strain in the combination with no further synergistic or additive effect of the other strains, and no antagonistic effects either. For probiotics, an improved adhesion has been reported for a combination of two strains (Ouwehand et al., 2000).

Immune modulation is broad and diverse, and many models exist. Here, two studies utilized tissue culture cells and observed a greater effect with multistrain probiotics, while the study using macrophages observed the opposite (Table 1). What would deserve more attention here is what the interactions between the so-called proinflammatory and anti-inflammatory strains would result in. Are they canceling each other's effect or do they influence different arms of the immune system?

Growth yield can be improved upon the combination of strains; this most likely happens through cross-feeding and is well known, for instance, in yogurt production (Moon and Reinbold, 1976); this is typical symbiosis. Not surprisingly, it is strain dependent (Ambalam et al., 2015). Such cross-feeding and symbiosis may also improve the digestibility and availability of nutrients in fermented foods (Freire et al., 2017).

The above-mentioned fact indicates there is no consistent picture in comparing single- and multistrain probiotics. Following the same strategy as Chapman et al. (2011) in their comparison of single- and multistrain probiotics, 56% of the study outcomes listed in Table 1 indicate a greater effect for multistrain probiotics, followed by 31% no difference and 13% a greater effect for single-strain probiotics.

ANIMAL STUDIES

Animal studies comparing single- and multistrain probiotics have, in general, been performed for two main purposes: to study performance of production animals or as animal models. Some of the latter studies do also serve the study of production animals as with specific infection models.

In aquaculture, the studies comparing single- and multistrain probiotics have given inconsistent results. Two studies reported no difference between single- and multistrain probiotics (Table 2); the effect is driven by one effective strain. There is no additional benefit by adding other strains to the effective strain, nor is there a detrimental effect (Vazirzadeh et al., 2019).

TABLE 2
Nonexhaustive Overview of Studies With Production Animals Comparing Single- vs. Multistrain Probiotics.

Health Target	Strains	Model	Outcome	Reference
Growth and immune function	*Sc. cerevisiae*, *L. fermentum* ATCC 14931, *L. buchneri* PRM 205	Rainbow trout	S=M	Vazirzadeh et al. (2019)
Gut microbiota modulation and growth	*Bacillus* sp. *P. acidilactici*, *Enterococcus* sp., *Lactobacillus* sp.	Juvenile rainbow trout	S<M (only at lower dose tested)	Ramos et al. (2013)
Growth performance, immune modulation, disease resistance	*Bc. subtilis*, *Bc. licheniformis*, *Pb. polymyxa*, *A. oryzae*, *Ss. cerevisiae*	Starry flounder infection model	S=M	Park et al. (2016)
Body weight, cecal coliform numbers, and immune response	*L. crispatus*, *L. johnsonii*	Broilers	S=M	Taheri et al. (2010)
Animal performance	*L. acidophilus* + *L. casei* + *B. thermophilum* + *E. faecium*, *P. acidilactici*	Chicken	S=M	Aalaei et al. (2018)
Gut beneficial bacterial populations, nutrient digestibility, cecal fermentation, feed efficiency, growth performance	*Bc. subtilis*, *L. acidophilus*	Rabbits	S=M	Phuoc and Jamikorn (2017)
Mucosal colonization	*L. plantarum* 423, *L. salivarius* 241	Piglets	S=M	Mare et al. (2006)
Immune modulation	*L. zeae*, *L. plantarum*, *L. reuteri*	Chicken salmonella infection model	S<M	Hu et al. (2015)
Meta-analysis		Lactobacilli counts, *E. coli* counts	S>M S=M	Heak et al. (2018)

A., Aspergillus; B., Bifidobacterium; Bc., Bacillus; E., Enterococcus; L., Lactobacillus (sensu lato); P., Propionibacterium; Pb., Paenibacillus; Sc., Saccharomyces.

While Ramos et al. (2013) observed the improved performance with a combination of probiotics, this effect was only observed in the lower dose tested, not in the higher dose. As slightly higher doses (here twice as much) are not known to have negative effects, the observation raises the question about the reproducibility of the observed effect. Also in chickens, rabbits, and piglets, no difference was noted between single- and multistrain probiotics (Table 2). Effects appear to be often driven by one active strain without any influence from additional strains (Phuoc and Jamikorn, 2017; Taheri et al., 2010).

Mucosal colonization was not found to be influenced whether strains are single or in combination. This may be explained by the observation that the tested strains preferable colonize different parts of the piglet intestine (Mare et al., 2006). A combination of probiotics was, however, found to be more efficacious in reducing the *Salmonella*-induced inflammatory response in chickens than the single strains (Hu et al., 2015).

One meta-analysis assessed the influence of probiotics on microbiota composition in the chicken intestine (Table 2). The study reported higher intestinal *Lactobacillus* (*sensu lato*) counts for single-strain probiotics than for multistrain probiotics. This difference can be explained by the generally higher *Lactobacillus* counts in single-strain products than in multistrain products (Heak et al., 2018). There was no difference between single- and multistrain probiotics for intestinal *Escherichia coli* counts.

Of the eight studies reported in Table 2, six did not observe a difference in outcome between single- and multistrain probiotics, while two studies reported multistrain probiotics to be superior. The meta-analysis given in Table 2, however, reported single strains to be superior for one parameter (lactobacilli counts), but not for another (*E. coli* counts). Overall, this suggests there is no difference in efficacy between single- and multistrain probiotics in production animals.

Experimental animal models that have compared single- and multistrain probiotics have investigated various outcomes. Experimental colitis models have been used in three studies and report a better protection from colitis and inflammation with multistrain probiotics (Table 3). However, in one case (Foligne et al., 2016), this effect is driven by an effective strain and is therefore higher as compared with a single less-effective strain.

Interestingly, both in a high-fat diet mouse and in an unchallenged Kunming mouse model, multistrain probiotics were reported to be more efficacious in managing blood lipid levels and other markers related to metabolic health (Table 3).

Also, visceral hypersensitivity was reported to be more improved with a multistrain probiotic (Table 3).

Modulation of the microbiota also indicated a superior effect for multistrain probiotics; however, this was again related to a strain-specific response of the tested probiotics (Dicks and ten Doeschate, 2010). In suppressing *Salmonella* infection in rats, multistrain probiotics were more efficacious (Table 3). In an antibiotic-induced dysbiosis in rats, Ermolenko et al. (2018) showed that in a three-strain probiotic combination, the effects on microbiota were mainly driven by *B. longum* GT15 and *E. faecium* L3 and cytokine induction was mainly driven by *B. longum* GT15 and *L. rhamnosus* K32. Intestinal mucosal ultrastructure was equally protected by all three strains and their combination. There does not seem to be an additive, synergistic, or average effect; the effects are driven by the most active strain in combination.

Interestingly, a combination of a *L. paracasei* and *B. breve* strain induced fewer changes in mucosal gene

TABLE 3
Nonexhaustive Overview of Animal Model Studies Comparing Single- vs. Multistrain Probiotics.

Health Target	Strains	Model	Outcome	Reference
Colitis and inflammation	*L. acidophilus* CGMCC 7282, *C. butyricum* CGMCC 7281	TNBS colitis mouse model	S<M	Wang et al. (2018)
Anti-inflammatory response, colitis	*L. plantarum, L. fermentum*	Experimental colitis	S<M	Dos Santos et al. (2016)
Anti-inflammatory response, colitis	*Lc. lactis* ssp. *lactis* CB 460, *Lc. lactis* ssp. *cremoris* CB 461, *S. thermophilus, P. freudenreichii* CB129	Experimental colitis	S≤M	Foligne et al. (2016)
Antioxidant activity, serum lipids, fecal SCFA	*C. butyricum* Sx-01, *L. salivarius* C-1-3	Mouse model	S<M	Long et al. (2018)
Lipid metabolism, body weight gain	*L. plantarum* KY1032, *L. curvatus* HY7601	Diet-induced obesity	S<M	Yoo et al. (2013)
Visceral hypersensitivity	*L. helveticus* R0052, *B. longum* R0175	Stressed mouse model	S<M	Ait-Belgnaoui et al. (2018)
Salmonella infection	*E. mundtii* ST4SA, *L. plantarum* 423	Experimental salmonellosis in rats	S<M	Dicks and ten Doeschate (2010)
Gene expression	*L. paracasei* CNCM I-4034, *B. breve* CNCM I-4035, *L. rhamnosus* CNCM I-4036	Obese Zucker rats	S<M	Plaza-Diaz et al. (2017)
Colonic microbiota (*Lactococcus lactis* colonization)	*L. garvieae, Co. variabile, Bc. coagulans*	Frogs	S<M	Weng et al. (2017)
Meta-analysis		Anxiety/stress/depression Cognition/social behavior	S=M S<M	Joseph and Law (2019)

B., Bifidobacterium; Bc., Bacillus; C., Clostridium; Co., Corynebacterium; E., Enterococcus; L., Lactobacillus (*sensu lato*); *Lc., Lactococcus.*

expression as compared with the strains separately or a strain of *L. rhamnosus*, in an obese rat model. It is not clear if there is up- or down-regulation of the same genes by the *L. paracasei* and *B. breve* strains that may have canceled each other. The same model also reported less lipopolysaccharide-binding protein concentrations in the serum of the rats, when fed the two-strain combination as compared with the single strains, which indicates an additive effect (Plaza-Diaz et al., 2017) (Table 3).

A meta-analysis investigating the neuropsychiatric effect of probiotics reported that studies with single-strain probiotics reported a similar frequency that often changes in stress/anxiety/depression as studies with multistrain probiotics. Though studies with multistrains reported more often changes in cognition and social behaviors. No statistical evidence for these conclusions was reported (Joseph and Law, 2019).

In contrast to production animals, animal models were reported in most cases; in eight out of nine models, there is a superior benefit for multistrain probiotics as compared with single-strain probiotics. Only one study reported an equal or better effect for multistrain probiotics. While this might suggest that multistrain probiotics perform better, the only meta-analysis is less clear-cut with one equal comparison (stress/anxiety/depression) and one superior comparison (cognition and social behaviors) for multistrain probiotics.

HUMAN STUDIES

Strongest evidence for probiotics intended for human use obviously comes from human intervention studies. Similarly, the strongest evidence for a comparison of single- and multistrain probiotics comes from human studies making a direct comparison. Such studies are unfortunately few, and many of them do not compare the same strains (Table 4). Further, some articles are actually reporting on different outcomes from the same

TABLE 4
Clinical Studies That Compared the Efficacy of Single- and Multistrain Probiotics to Each Other.

Objective	Strains	Dose (10^9 CFU)	Subjects age (years) *N*	Single × Multistrain	Comparison	Reference
Recovery of *Lactobacillus rhamnosus* GG	*L. rhamnosus* GG	20	24.4 years *N*=20	Recovered from tonsil tissue 40% subjects Fecal recovery from 95% of subjects	S=M	Kumpu et al. (2013)
	L. rhamnosus GG *L. rhamnosus* LC705 *P. freudenreichii* subsp. *shermanii* JS *B. lactis* Bb12	14	25.5 years *N*=17	Recovered from tonsil tissue 42% subjects Fecal recovery from 100% of subjects		
Correcting antibiotic-induced dysbiosis and stool frequency	*Sc. boulardii*	10	6.8 years *N*=6	NR	S≤M	Zoppi et al. (2001)
	E. faecium SF68	0.075	3.4 years *N*=7	NR		
	L. rhamnosus GG	6	4.0 years *N*=7	NR		
	L. rhamnosus "*L. bifidus*" *L. acidophilus*	15.4	6.3 years *N*=7	NR		
	B. bifidum *L. acidophilus*	4	5.3 years *N*=7	"Fewer stools" (no *P*-value)		

TABLE 4
Clinical Studies That Compared the Efficacy of Single- and Multistrain Probiotics to Each Other—cont'd

Objective	Strains	Dose (10^9 CFU)	Subjects age (years) *N*	Single × Multistrain	Comparison	Reference
	S. thermophilus *B. breve* *B. infantis* *B. longum* *L. acidophilus* *L. plantarum* *L. casei* *L. bulgaricus* *E. faecium*	485.67	7.4 years *N*=5	Fewer stools ($P \leq 0.01$) Most significant impact on changes in the microflora composition caused by ceftriaxone treatment (no p-value)		
Respiratory tract infections	*L. acidophilus* NCFM	10	3.7 years *N*=110	Fever OR=0.74 CI 0.38–1.33 *P*=0.33 Cough OR=0.84 CI 0.46–1.40 *P*=0.47 Rhinorrhea OR=0.64 CI 0.34–1.11 *P*=0.089 Any symptom OR=0.50 CI 0.26–0.98 *P*=0.04 Antibiotics OR=0.66 CI 0.29–1.49 *P*=0.29 Symptom duration mean=−1.06±0.81 *P*=0.195	S≤M	Leyer et al. (2009)
	L. acidophilus NCFM+*B. lactis* Bi-07	10	3.8 years *N*=112			
Respiratory tract infections	*B. lactis* Bl-04	2	36 years *N*=161	Disease duration 6.3 day Disease risk 0.73 (0.55±0.95)	S=M	West et al. (2014b)
	L. acidophilus NCFM+*B. lactis* Bi-07	10	36 years *N*=155	Disease duration 7.0 days Disease risk 0.81 (0.62±1.08)		
Diarrhea	*L. rhamnosus* GG	6	20 Mo *N*=100	−32 h CI −41 to −23 *P*<0.001	S=M	Canani et al. (2007)
	Sc. boulardii	5	18 Mo *N*=91	−5 h CI −13 - 5 *P*>0.05		
	Bc. clausii O/C84, N/R84, T84, SIN84	1	19 Mo *N*=100	1 h CI −7 - 8 *P*>0.05		
	L. bulgaricus LMG-P17550, *L. acidophilus* LMG-P 17549, *S. thermophilus* LMG-P 17503, *B. bifidum* LMG-P 17500	3.5	16 Mo *N*=97	−37 h CI −47 to −25 *P*<0.001		
	E. faecium SF68	0.075	15 Mo *N*=91	2 h CI −5 to 11 *P*>0.05		

Continued

TABLE 4
Clinical Studies That Compared the Efficacy of Single- and Multistrain Probiotics to Each Other—cont'd

Objective	Strains	Dose (10^9 CFU)	Subjects age (years) *N*	Single × Multistrain	Comparison	Reference
Immune markers	*B. lactis* Bl-04	2	50 years *N*=46	No change in most of the measured cytokines and white blood cell differentials. No difference between probiotic groups	S=M	West et al. (2014a)
	L. acidophilus NCFM+*B. lactis* Bi-07	10	51 years *N*=47			
Routine hematology and clinical chemistry measures	*B. lactis* Bl-04	2	42.2 years *N*=39	Few changes in routine hematology and clinical chemistry measures Serum calcium −0.01±0.11 mmol/L Serum urea −0.07±0.93 mmol/L	S≤M	Cox et al. (2014)
	L. acidophilus NCFM+*B. lactis* Bi-07	10	37.3 years *N*=45	Few changes in routine hematology and clinical chemistry measures Serum calcium 0.01±0.09 mmol/L Serum urea −0.55±1.03 mmol/L		
Peripheral blood regulatory	*B. lactis* Bl-04	2	39 years *N*=30	No change in peripheral blood regulatory T cells	S=M	West et al. (2016)
	L. acidophilus NCFM+*B. lactis* Bi-07	10	37 years *N*=32	No change in peripheral blood regulatory T cells		
Analgesic receptor expression	*L. acidophilus* NCFM	20	18–70 years *N*=13–14	μ-opioid receptor: 39.9-fold increase Cannabinoid receptor-2 4.15-fold decrease	S≥M	Ringel-Kulka et al. (2014)
	L. acidophilus NCFM+*B. lactis* Bi-07	20	18–70 years *N*=13–14	μ-opioid receptor no significant change Cannabinoid receptor-2 no change		

NR, not reported.
B., *Bifidobacterium*; *Bc.*, *Bacillus*; *E.*, *Enterococcus*; *L.*, *Lactobacillus* (sensu lato); *P.*, *Propionibacterium*; *S.*, *Streptococcus*; *Sc.*, *Saccharomyces*.

study (Cox et al., 2014; West et al., 2014a, b, 2016). Despite these limitations, five of the nine studies in Table 4 report no difference between single- and multistrain probiotics, three report equal or better, and one reports equal or worse outcome for multistrain probiotics. There is thus no convincing evidence suggesting that multistrain probiotics are different in efficacy compared with single-strain probiotics.

Meta-analyses can be used as an additional source of information to gain insight into the matter of single- vs. multistrain efficacy. However, as is common with meta-analyses on probiotic efficacy, they combine the results of different strains for a given endpoint. Further, studies with multistrain products are most often not split out according to the number of strains. Nevertheless, despite these shortcomings, meta-analyses may provide some insights, as there is a larger number of meta-analyses. Moreover, some of these meta-analyses are based on many individual primary studies, as with, e.g., necrotizing enterocolitis (NEC), and subgroup analysis is performed for single- vs. multistrain probiotics (Table 5).

TABLE 5
Nonexhaustive Overview of Meta-Analyses That Compared the Efficacy of Single- and Multistrain Probiotics to Each Other in Meta-Analyses of Human Studies.

Objective	Efficacy Single-Strain Probiotic	Single-Strain Number of Trials	Efficacy Multistrain Probiotic	Multistrain Number of Trials	Comparison	References
Prevention of postoperative infections		Not reported		Not reported	S = M	Lytvyn et al. (2016)
Prevention of necrotizing enterocolitis		8		7	No conclusion	Mihatsch et al. (2012)
Prevention of necrotizing enterocolitis, sepsis, and mortality	NEC *Lactobacillus* species RR = 0. 45, CI 0.27–0.75 *P* = 0.0023	5	NEC RR = 0.37, CI 0.25–0.54 *P* < 0.00001	9	*P* = 0.48 M = S	AlFaleh and Anabrees (2014)
	Bifidobacterium species RR = 0. 48, CI 0.16–1.47 *P* = 0.20	4				
	Saccharomyces boulardii RR = 0. 72, CI 0.34–1.55 *P* = 0.40	2				
	Sepsis *Lactobacillus* species RR = 0.91, CI 0. 71–1.16 *P* = 0.44	5	Sepsis RR = 0.91, CI 0. 78–1.06 *P* = 0.23	9	*P* = 1.00 M = S	
	Bifidobacterium species RR = 0.88, CI 0. 58–1.34 *P* = 0.56	3				
	Saccharomyces boulardii RR = 0.92, CI 0. 54–1.57 *P* = 0.77	2				
	Mortality *Lactobacillus* species RR 0.72, CI 0.47–1.10 *P* = 0.13	4	Mortality RR = 0.62, CI 0.47 to 0.81 *P* = 0.00057	9	*P* = 0.84 M = S	
	Bifidobactirium species RR 0.71, CI 0.14–3.60 *p* = 0.68	2				
	Saccharomyces boulardii RR 1.01, CI 0.30–3.4 *P* = 0.99	2				

Continued

TABLE 5
Nonexhaustive Overview of Meta-Analyses That Compared the Efficacy of Single- and Multistrain Probiotics to Each Other in Meta-Analyses of Human Studies—cont'd

Objective	Efficacy Single-Strain Probiotic	Single-Strain Number of Trials	Efficacy Multistrain Probiotic	Multistrain Number of Trials	Comparison	References
Prevention of necrotizing enterocolitis, sepsis, and mortality	NEC RR 0.36 CI 0.12–1.13	3	NEC 2 strains = RR 0.26 CI 0.11–0.62 ≥3 strains = RR 0.41 CI 0.23–0.73	4 2	$P > 0.05$ S = 2M = 3M	Baucells et al. (2016)
	Sepsis RR 0.99 CI 0.55–1.79	3	Sepsis 2 strains = RR 1.05 CI 0.76–1.46 ≥3 strains = RR 0.73 CI 0.57–0.95	4 2	$P = 0.016$ S < 2M $P = 0.004$ 2M > 3M	
	Mortality RR 0.99 CI 0.32–3.06	3	Mortality 2 strains = RR 0.32 CI 0.15–0.66 ≥3 strains = RR 0.74 CI 0.47–1.17	4 2	$P = 0.845$ S = 2M $P = 0.028$ 2M < 3M	
Prevention of necrotizing enterocolitis and mortality	NEC *Lactobacillus* OR = 0.60 CI 0.36–1.0 $P = 0.05$ *Saccharomyces* OR = 0.80 CI 0.41–1.56 $P = 0.52$ *Bifidobacterium* OR = 0.85 CI 0.61–1.20 $P = 0.37$	5 3 6	NEC OR = 0.36 CI 0.24–0.53 $P < 0.00001$	11	$P = 0.01$ S < M	(Chang et al., 2017)
	Mortality *Lactobacillus* OR = 0.94 CI 0.24–3.08 $P = 0.89$ *Saccharomyces* OR = 0.90 CI 0.36–2.26 $p = 0.82$ *Bifidobacterium* OR 0.95 CI 0.68 to 1.33 $P = 0.76$	3 2 6	Mortality OR = 0.58, CI 0.43–0.79 $P = 0.0006$	10	$P = 0.18$ M = S	

Prevention of necrotizing enterocolitis and sepsis	NEC RR = 0.74 CI 0.56–0.97 Sepsis RR = 0.79 CI 0.60–1.03	11 11	NEC RR = 0.37 CI 0.26–0.54 Sepsis RR = 0.74 CI = 0.52–1.04	14 12	$P < 0.01$ S < M $P = 0.77$ S = M	Sun et al. (2017)
Prevention of necrotizing enterocolitis	NEC RR = 0.36 CI 0.23–0.53		NEC 2 strains RR = 0.26 CI 0.12–0.51 3 or more strains RR = 0.029 CI 0.15–0.51		$P > 0.05$ S = M S = M	van den Akker et al. (2020)
Time to full enteral feeds	Mean – 1.34d CI – 2.10 to – 0.58 $P = 0.0006$	11	Mean – 1.74d CI – 4.22 to 0.73 $P = 0.17$	8	$P = 0.76$ M = S	Athalye-Jape et al. (2014)
Correction of dysbiosis	Normal microbiota restoration after disruption 5 out of 6 restore Disrupted normal microbiota at baseline 3 out of 10 change Undisrupted normal microbiota studies 2 out of 18 change	6 10 18	Normal microbiota restoration after disruption 7 out of 8 restore Disrupted normal microbiota at baseline 9 out of 14 Undisrupted normal microbiota studies 2 out of 11 change	8 14 11	$P = 0.5692$ S = M $P = 0.1107$ S = M $P = 0.6221$ S = M	McFarland (2014)
Irritable bowel syndrome	13 out of 29 "positive"	29	18 out of 27 "positive"	27	$P = 0.1159$ S = M	Mazurak et al. (2015)
Intestinal transit time	SMD 0.49 CI 0.32–0.66 $P = 0.001$	10	SMD 0.23 CI – 0.01 to 0.47 $P = 0.06$	7	$P = 0.09$ S = M	Miller et al. (2016)
Prevention of antibiotic-associated diarrhea	RR 0.41 CI 0.28 to 0.62 $P = 0.000016$	11	RR 0.52 CI 0.35 to 0.77 $P = 0.0010$	11	$P = 0.44$ S = M	Goldenberg et al. (2015)

Continued

TABLE 5
Nonexhaustive Overview of Meta-Analyses That Compared the Efficacy of Single- and Multistrain Probiotics to Each Other in Meta-Analyses of Human Studies—cont'd

Objective	Efficacy Single-Strain Probiotic	Single-Strain Number of Trials	Efficacy Multistrain Probiotic	Multistrain Number of Trials	Comparison	References
Prevention of pediatric antibiotic-associated diarrhea	RR 0.42 CI 0.32 to 0.56; $P<0.00001$	20	RR 0.53 CI 0.37 to 0.75; $P=0.0003$	13	$P=0.34$ S=M	Guo et al. (2019)
Gastrointestinal disease	RR=0.73 CI 0.68 to 0.79	51	RR=0.63 CI 0.53 to 0.76	33	$P>0.05$ S=M	Ritchie and Romanuk (2012)
Central nervous system function	7 out of 8 change	8	3 out of 7 change	7	$P=0.0932$ S=M	Wang et al. (2016)
Adverse events	Gastrointestinal disorders RR=0.99, CI 0.93–1.05, $P=0.30$	2	Gastrointestinal disorders RR=1.00, CI 0.94–1.06	2	$P=0.30$ S=M	Tapiovaara et al. (2016)
	Respiratory, thoracic, and mediastinal disorders RR=0.98, CI 0.92–1.06, $P=0.35$	2	Respiratory, thoracic, and mediastinal disorders RR=0.99, CI 0.97–1.01	2	$P=0.35$ S=M	
	Infections and infestations RR=1.09, CI 0.95–1.24, $P=0.62$	2	Infections and infestations RR=1.00, CI 0.81–1.23	2	$P=0.62$ S=M	
Fasting blood glucose control	−0.01 (−0.16, 0.14; $\rho=0.89$)	7	−0.31 (−0.58, −0.03; $\rho=0.03$)	11	$P=0.07$ S=M	Nikbakht et al. (2018)

Weight management/metabolic health	BMI −0.36 (−0.52, −0.20 kg/m^2; $P < 0.05$)	4	BMI −0.15 (−0.39, 0.10 kg/m^2; $P > 0.05$)	6	$P > 0.05$ S = M	Wang et al. (2019)
	Waist circumference −1.69 (−3.04, −0.33 cm; $P < 0.05$)	4	Waist circumference −0.42 (−0.91, 0.07 cm; $P > 0.05$)	4	$P > 0.05$ S = M	
	Fat mass −1.15 (−1.28, −1.02 kg; $P < 0.05$)	5	Fat mass −0.60 (−0.94, −0.26 kg; $P < 0.05$)	3	$P < 0.05$ S > M	
	Total cholesterol −0.61 (−1.54, −0.32; $P < 0.05$)	3	Total cholesterol −0.39 (−0.66, −0.13; $P < 0.05$)	4	$P < 0.05$ S > M	
	Triglycerides 0.45 (−0.23, 1.13: $P > 0.05$)	3	Triglycerides −0.10 (−0.43, 0.22; $P > 0.05$)	4	$P > 0.05$ S = M	
	LDL cholesterol −0. 64 (−1.66, 0.37; $P > 0.05$)	3	LDL cholesterol −0.33 (−0.57, −0.09; $P < 0.05$)	4	$P > 0.05$ S = M	
Vulvovaginal candidiasis	Intravaginal 1.11 (1.02, 1.21; $P = 0.01$)	4	Oral 1.41 (1.07, 1.84; $P = 0.001$)	1	$P = 0.35$ S = M	Xie et al. (2017)

Some meta-analyses attempted to perform a subgroup analysis for single- vs. multistrain probiotics but concluded there were too few observations to make a meaningful analysis. As with Feng et al. (2017) for probiotic supplementation to triple therapy for *Helicobacter pylori* eradication.

For NEC, older studies draw no conclusion (Mihatsch et al., 2012) or state no difference between single- and multistrain probiotics (AlFaleh and Anabrees, 2014; Baucells et al., 2016). More recent meta-analyses, however, conclude that multistrain probiotics are superior in reducing NEC risk as compared with single-strain probiotics (Chang et al., 2017; Sun et al., 2017). However, other NEC-related outcomes such as sepsis and mortality were not found to be different. Interestingly, Baucells et al. (2016) evaluated the difference between probiotic products with two or at least three strains. No consistent outcome was observed.

Of the other reported endpoints reported in Table 5, only in the case of weight management, reductions in fat mass and total serum cholesterol levels were different, and greater reductions with single-strain probiotics were also reported (Wang et al., 2019).

Fattorusso et al. (2019) conclude that probiotics (mostly a mixture of bifidobacteria, streptococci, and lactobacilli) are the most promising treatment for neurobehavioral symptoms and bowel dysfunction. However, this conclusion is based on an unbalanced data set with few single-strain probiotics, not because multistrain probiotics are more efficacious. This is a general shortcoming in subgroup analyses; while the original meta-analysis may be based on a substantial number of primary studies, the subgroups may be small and not warrant the conclusions that are presented.

Meta-analyses combine the results from studies with different strains (combinations). If there is sufficient primary data, a conclusion can be drawn on a specific strain (combination) as was done by Partty et al. (2018) for pediatric functional gastrointestinal disorders. The authors concluded that there was insufficient data for both single- and multistrain probiotics, possibly with the exception of *L. reuteri* DSM 17938 for the management of breast-fed colic infants.

Most meta-analyses in Table 5 do not report a difference in efficacy between single- and multistrain probiotics. However, the subgroups reported in these analyses are in many cases small, making it difficult to draw conclusions. When focusing on analyses with a relatively large number of primary studies included, such as "gastrointestinal disease" with 84 primary studies (Ritchie and Romanuk, 2012), reducing risk for pediatric antibiotic associated diarrhea with 33 primary studies (Guo et al., 2019) and irritable bowel syndrome (IBS) with 56 primary studies (Mazurak et al., 2015), none of these report a difference in efficacy between single- and multistrain probiotics.

DISCUSSION

Although some health benefits appear to be relatively common among probiotics, the general assumption in the field of probiotics is that health benefits are strain specific (Hill et al., 2014). It is therefore reasonable to assume the same holds true for probiotic combinations, which makes it challenging to draw conclusion on differences in efficacy between single and multistrains. Earlier communications suggested, more on a theoretical basis, that multistrain probiotics are more effective than single-strain products (Chapman et al., 2011). In an earlier communication (Ouwehand et al., 2018), we concluded that while multistrain probiotics can be efficacious, we could not find convincing evidence that multistrain probiotic products would be more efficacious than single-strain products. Nor did we find evidence that multistrain probiotics would be less effective; there does not seem to be an antagonistic activity between strains in a multistrain combination. Also, Korada et al. (2018) concluded that there is currently insufficient data to draw conclusions on the efficacy of single- vs. multistrain probiotics. In the current report, we did not observe a clear trend for difference in efficacy between single- and multistrain probiotics in vitro and in experimental animal models (Tables 1 and 2). Studies in production animals, however, gave a clear indication of superiority for multistrain probiotics (Table 3). Human studies are again less clear: no difference when compared directly (Table 4) and no difference when compared in meta-analyses (Table 5). Meta-analyses suffer in most cases from low study numbers. In this perspective, it may be interesting to look at the results related to NEC, one of the best-documented health benefits for probiotics (Underwood, 2019). There are indications that with an increasing number of studies, multistrain probiotics do indeed appear to be more efficacious (Table 5). Also, some studies that have not made head-on comparisons tend to conclude that multistrain probiotics are more efficacious in reducing NEC risk (Jiang et al., 2019), while others caution and require more studies (Aceti et al., 2018). However, the most recent analysis by the ESPGHAN Committee on Nutrition and the ESPGHAN Working Group for Probiotics and Prebiotics concludes that there is no a priori advantage of administering multiple strains versus a singlestrain (van den Akker et al., 2020).

One potential advantage of multistrain probiotics remains uninvestigated: a broader spectrum of activity. This is a an even more complex question as it not only compares single- vs. multistrain probiotics but would need to be investigated over multiple health endpoints.

RECOMMENDATIONS

To better answer the question if there is a difference in efficacy between single- and multistrain probiotics, and in particular if multistrain probiotics are superior, structured research will have to be performed as recommended earlier (Ouwehand et al., 2018).

(1) In meta-analyses comparing the efficacy of single- and multistrain probiotics, the results should be interpreted in the light of the number of included studies.
(2) Human intervention studies should be performed comparing a multistrain product with its single-strain components and placebo. The challenge here is that the dose will be different, as the combination will have a higher total count than the single strains. Thus, future studies should be conducted for the following reasons.
    (a) To determine whether a multistrain probiotic has a bigger effect size for a given health target as compared with a single-strain probiotic.
    (b) To determine whether a multistrain probiotic has a broader efficacy panel, as compared with a single-strain probiotic. This will be particularly challenging as it will require studies on multiple health targets.
(3) In vitro and animal studies will not answer the question but will be able to suggest potential mechanisms by which probiotics in a multistrain preparation interact.
(4) It would be recommended to use the simplest possible setup for such studies as follows:
    (a) 2-strain combination (a+b).
    (b) Single strain (a).
    (c) Single strain (b).
    (d) Placebo.
(5) In vitro studies could also be used to investigate the potential of a broader spectrum of activity for multistrain probiotics, following a similar study setup as suggested above under point (4).

CONCLUSIONS

It is currently not possible to make generalized recommendation as to whether single-strain and multistrain probiotics have a difference in efficacy. Both have been shown to be efficacious in multiple health endpoints. If there is a difference in efficacy, this will need to be documented in carefully setup studies as outlined earlier. However, risk for antagonistic activities between strains in a multistrain product appears to be small or nonexistent. The best recommendation therefore is to use strains and strain combinations that have documented efficacy for the health benefits that are being sought.

CONFLICT OF INTEREST

ACO is an employee of IFF, which manufactures and markets probiotics. The other authors report no conflict of interest.

ACKNOWLEDGMENT

The authors thank Nicholas Kruczynski and Dr. Johanna Maukonen for critically reading the manuscript.

REFERENCES

Aalaei, M., Khatibjoo, A., Zaghari, M., et al., 2018. Comparison of single- and multi-strain probiotics effects on broiler breeder performance, egg production, egg quality and hatchability. Br. Poult. Sci. 59, 531–538.

Aceti, A., Beghetti, I., Maggio, L., et al., 2018. Filling the gaps: current research directions for a rational use of probiotics in preterm infants. Nutrients 10, 1472.

Ait-Belgnaoui, A., Payard, I., Rolland, C., et al., 2018. Bifidobacterium longum and Lactobacillus helveticus synergistically suppress stress-related visceral hypersensitivity through hypothalamic-pituitary-adrenal axis modulation. J. Neurogastroenterol. Motil. 24, 138–146.

AlFaleh, K., Anabrees, J., 2014. Probiotics for prevention of necrotizing enterocolitis in preterm infants. Cochrane Database Syst. Rev. 4, CD005496.

Ambalam, P., Kondepudi, K.K., Balusupati, P., et al., 2015. Prebiotic preferences of human lactobacilli strains in co-culture with bifidobacteria and antimicrobial activity against *Clostridium difficile*. J. Appl. Microbiol. 119, 1672–1682.

Athalye-Jape, G., Deshpande, G., Rao, S., et al., 2014. Benefits of probiotics on enteral nutrition in preterm neonates: a systematic review. Am. J. Clin. Nutr. 100, 1508–1519.

Baucells, B.J., Mercadal Hally, M., Alvarez Sanchez, A.T., et al., 2016. Probiotic associations in the prevention of necrotising enterocolitis and the reduction of late-onset sepsis and neonatal mortality in preterm infants under 1,500g: a systematic review. An. Pediatr. (Barc.) 85, 247–255.

Canani, R.B., Cirillo, P., Terrin, G., et al., 2007. Probiotics for treatment of acute diarrhoea in children: randomised clinical trial of five different preparations. BMJ 335, 340.

Chang, H.Y., Chen, J.H., Chang, J.H., et al., 2017. Multiple strains probiotics appear to be the most effective probiotics in the prevention of necrotizing enterocolitis and mortality: an updated meta-analysis. PLoS One 12, e0171579.

Chapman, C.M., Gibson, G.R., Rowland, I., 2011. Health benefits of probiotics: are mixtures more effective than single strains? Eur. J. Nutr. 50, 1–17.

Chapman, C.M., Gibson, G.R., Rowland, I., 2012. In vitro evaluation of single- and multi-strain probiotics: inter-species inhibition between probiotic strains, and inhibition of pathogens. Anaerobe 18, 405–413.

Chapman, C.M., Gibson, G.R., Todd, S., et al., 2013. Comparative in vitro inhibition of urinary tract pathogens by single- and multi-strain probiotics. Eur. J. Nutr. 52, 1669–1677.

Chapman, C.M., Gibson, G.R., Rowland, I., 2014. Effects of single- and multi-strain probiotics on biofilm formation and in vitro adhesion to bladder cells by urinary tract pathogens. Anaerobe 27, 71–76.

Cox, A.J., West, N.P., Horn, P.L., et al., 2014. Effects of probiotic supplementation over 5 months on routine haematology and clinical chemistry measures in healthy active adults. Eur. J. Clin. Nutr. 68, 1255–1257.

de Oliveira, F.E., Rossoni, R.D., de Barros, P.P., et al., 2017. Immunomodulatory effects and anti-Candida activity of lactobacilli in macrophages and in invertebrate model of Galleria mellonella. Microb. Pathog. 110, 603–611.

Dicks, L.M., ten Doeschate, K., 2010. Enterococcus mundtii ST4SA and *Lactobacillus plantarum* 423 alleviated symptoms of Salmonella infection, as determined in Wistar rats challenged with *Salmonella enterica* serovar typhimurium. Curr. Microbiol. 61, 184–189.

Dos Santos, T.F., Melo, T.A., Santos, D.S., et al., 2016. Efficacy of oral administration of lactic acid bacteria isolated from cocoa in a fermented milk preparation: reduction of colitis in an experimental rat model. Genet. Mol. Res. 15.

Ermolenko, E., Rybalchenko, O., Borshev, Y., et al., 2018. Influence of monostrain and multistrain probiotics on immunity, intestinal ultrastructure and microbiota in experimental dysbiosis. Benefic. Microbes 9, 937–949.

Fattorusso, A., Di Genova, L., Dell'Isola, G.B., et al., 2019. Autism spectrum disorders and the gut microbiota. Nutrients 11, 521.

Feng, J.R., Wang, F., Qiu, X., et al., 2017. Efficacy and safety of probiotic-supplemented triple therapy for eradication of helicobacter pylori in children: a systematic review and network meta-analysis. Eur. J. Clin. Pharmacol. 73, 1199–1208.

Ferreira, C.L., Grzeskowiak, L., Collado, M.C., et al., 2011. In vitro evaluation of Lactobacillus gasseri strains of infant origin on adhesion and aggregation of specific pathogens. J. Food Prot. 74, 1482–1487.

Fijan, S., Sulc, D., Steyer, A., 2018. Study of the in vitro antagonistic activity of various single-strain and multi-strain probiotics against *Escherichia coli*. Int. J. Environ. Res. Public Health 15, 1539.

Foligne, B., Parayre, S., Cheddani, R., et al., 2016. Immunomodulation properties of multi-species fermented milks. Food Microbiol. 53, 60–69.

Forssten, S.D., Ouwehand, A.C., 2017. Simulating colonic survival of probiotics in single-strain products compared to multi-strain products. Microb. Ecol. Health Dis. 28, 1378061.

Freire, A.L., Ramos, C.L., da Costa Souza, P.N., et al., 2017. Nondairy beverage produced by controlled fermentation with potential probiotic starter cultures of lactic acid bacteria and yeast. Int. J. Food Microbiol. 248, 39–46.

Goldenberg, J.Z., Lytvyn, L., Steurich, J., et al., 2015. Probiotics for the prevention of pediatric antibiotic-associated diarrhea. Cochrane Database Syst. Rev. 4, CD004827.

Guo, Q., Goldenberg, J.Z., Humphrey, C., et al., 2019. Probiotics for the prevention of pediatric antibiotic-associated diarrhea. Cochrane Database Syst. Rev. 4, CD004827.

Hansen, S.J.Z., Morovic, W., DeMeules, M., et al., 2018. Absolute enumeration of probiotic strains Lactobacillus acidophilus NCFM((R)) and Bifidobacterium animalis subsp. lactis Bl-04 ((R)) via chip-based digital PCR. Front. Microbiol. 9, 704.

Heak, C., Sukon, P., Sornplang, P., 2018. Effect of direct-fed microbials on culturable gut microbiotas in broiler chickens: a meta-analysis of controlled trials. Asian Australas. J. Anim. Sci. 31, 1781–1794.

Hill, C., Guarner, F., Reid, G., et al., 2014. Expert consensus document. The International Scientific Association for Probiotics and Prebiotics consensus statement on the scope and appropriate use of the term probiotic. Nat. Rev. Gastroenterol. Hepatol. 11, 506–514.

Hu, J.L., Yu, H., Kulkarni, R.R., et al., 2015. Modulation of cytokine gene expression by selected Lactobacillus isolates in the ileum, caecal tonsils and spleen of Salmonella-challenged broilers. Avian Pathol. 44, 463–469.

Inturri, R., Trovato, L., Volti, G.L., et al., 2019. In vitro inhibitory activity of Bifidobacterium longum BB536 and Lactobacillus rhamnosus HN001 alone or in combination against bacterial and Candida reference strains and clinical isolates. Heliyon 5, e02891.

Jiang, T., Zhang, H., Xu, X., et al., 2019. Mixed probiotics decrease the incidence of stage II-III necrotizing enterocolitis and death: a systematic review and meta-analysis. Microb. Pathog. 138, 103794.

Joseph, J.M., Law, C., 2019. Cross-species examination of single- and multi-strain probiotic treatment effects on neuropsychiatric outcomes. Neurosci. Biobehav. Rev. 99, 160–197.

Khalesi, S., Bellissimo, N., Vandelanotte, C., et al., 2019. A review of probiotic supplementation in healthy adults: helpful or hype? Eur. J. Clin. Nutr. 73, 24–37.

Korada, S.K., Yarla, N.S., Mishra, V., et al., 2018. Single probiotic versus multiple probiotics - a debate on current scenario for alleviating health benefits. Curr. Pharm. Des. 24, 4150–4153.

Kumpu, M., Swanljung, E., Tynkkynen, S., et al., 2013. Recovery of probiotic Lactobacillus rhamnosus GG in tonsil tissue after oral administration: randomised, placebo-controlled, double-blind clinical trial. Br. J. Nutr. 109, 2240–2246.

Leyer, G.J., Li, S., Mubasher, M.E., et al., 2009. Probiotic effects on cold and influenza-like symptom incidence and duration in children. Pediatrics 124, e172–e179.

Li, S.C., Hsu, W.F., Chang, J.S., et al., 2019. Combination of Lactobacillus acidophilus and Bifidobacterium animalis subsp. lactis shows a stronger anti-inflammatory effect than individual strains in HT-29 cells. Nutrients 11, 969.

Long, M., Yang, S., Li, P., et al., 2018. Combined use of C. butyricum Sx-01 and L. salivarius C-1-3 improves intestinal health and reduces the amount of lipids in serum via modulation of gut microbiota in mice. Nutrients 10, 810.

Lytvyn, L., Quach, K., Banfield, L., et al., 2016. Probiotics and synbiotics for the prevention of postoperative infections following abdominal surgery: a systematic review and meta-analysis of randomized controlled trials. J. Hosp. Infect. 92, 130–139.

MacPherson, C.W., Shastri, P., Mathieu, O., et al., 2017. Genome-wide immune modulation of TLR3-mediated inflammation in intestinal epithelial cells differs between single and multi-strain probiotic combination. PLoS One 12, e0169847.

Mare, L., Wolfaardt, G.M., Dicks, L.M., 2006. Adhesion of Lactobacillus plantarum 423 and *Lactobacillus salivarius* 241 to the intestinal tract of piglets, as recorded with fluorescent in situ hybridization (FISH), and production of plantaricin 423 by cells colonized to the ileum. J. Appl. Microbiol. 100, 838–845.

Mathipa, M.G., Thantsha, M.S., 2015. Cocktails of probiotics pre-adapted to multiple stress factors are more robust under simulated gastrointestinal conditions than their parental counterparts and exhibit enhanced antagonistic capabilities against *Escherichia coli* and *Staphylococcus aureus*. Gut Pathog. 7, 5.

Mazurak, N., Broelz, E., Storr, M., et al., 2015. Probiotic therapy of the irritable bowel syndrome: why is the evidence still poor and what can be done about it? J. Neurogastroenterol. Motil. 21, 471–485.

McFarland, L.V., 2014. Use of probiotics to correct dysbiosis of normal microbiota following disease or disruptive events: a systematic review. BMJ Open 4, e005047.

Mihatsch, W.A., Braegger, C.P., Decsi, T., et al., 2012. Critical systematic review of the level of evidence for routine use of probiotics for reduction of mortality and prevention of necrotizing enterocolitis and sepsis in preterm infants. Clin. Nutr. 31, 6–15.

Miller, L.E., Zimmermann, A.K., Ouwehand, A.C., 2016. Contemporary meta-analysis of short-term probiotic consumption on gastrointestinal transit. World J. Gastroenterol. 22, 5122–5131.

Moon, N.J., Reinbold, G.W., 1976. Commensalism and competition in mixed cultures of *Lactobacillus bulgaricus* and *Streptococcus thermophilus*. J. Milk Food Technol. 39, 337–341.

Nikbakht, E., Khalesi, S., Singh, I., et al., 2018. Effect of probiotics and synbiotics on blood glucose: a systematic review and meta-analysis of controlled trials. Eur. J. Nutr. 57, 95–106.

Ouwehand, A.C., 2017. A review of dose-responses of probiotics in human studies. Benefic. Microbes 8, 143–151.

Ouwehand, A.C., Isolauri, E., Kirjavainen, P.V., et al., 2000. The mucus binding of Bifidobacterium lactis Bb12 is enhanced in the presence of Lactobacillus GG and Lact. delbrueckii subsp. bulgaricus. Lett. Appl. Microbiol. 30, 10–13.

Ouwehand, A.C., Invernici, M.M., Furlaneto, F.A.C., et al., 2018. Effectiveness of multistrain versus single-strain probiotics: current status and recommendations for the future. J. Clin. Gastroenterol. 52, S35–S40.

Park, Y., Moniruzzaman, M., Lee, S., et al., 2016. Comparison of the effects of dietary single and multi-probiotics on growth, non-specific immune responses and disease resistance in starry flounder, *Platichthys stellatus*. Fish Shellfish Immunol. 59, 351–357.

Partty, A., Rautava, S., Kalliomaki, M., 2018. Probiotics on pediatric functional gastrointestinal disorders. Nutrients 10, 1836.

Phuoc, T.L., Jamikorn, U., 2017. Effects of probiotic supplement (Bacillus subtilis and Lactobacillus acidophilus) on feed efficiency, growth performance, and microbial population of weaning rabbits. Asian Australas. J. Anim. Sci. 30, 198–205.

Piyadeatsoontorn, S., Taharnklaew, R., Upathanpreecha, T., et al., 2019. Encapsulating viability of multi-strain lactobacilli as potential probiotic in pigs. Probiotics Antimicrob. Proteins 11, 438–446.

Plaza-Diaz, J., Robles-Sanchez, C., Abadia-Molina, F., et al., 2017. Adamdec1, Ednrb and Ptgs1/Cox1, inflammation genes upregulated in the intestinal mucosa of obese rats, are downregulated by three probiotic strains. Sci. Rep. 7, 1939.

Ramos, M.A., Weber, B., Goncalves, J.F., et al., 2013. Dietary probiotic supplementation modulated gut microbiota and improved growth of juvenile rainbow trout (*Oncorhynchus mykiss*). Comp. Biochem. Physiol. A Mol. Integr. Physiol. 166, 302–307.

Ringel-Kulka, T., Goldsmith, J.R., Carroll, I.M., et al., 2014. Lactobacillus acidophilus NCFM affects colonic mucosal opioid receptor expression in patients with functional abdominal pain—a randomised clinical study. Aliment. Pharmacol. Ther. 40, 200–207.

Ritchie, M.L., Romanuk, T.N., 2012. A meta-analysis of probiotic efficacy for gastrointestinal diseases. PLoS One 7, e34938.

Sun, J., Marwah, G., Westgarth, M., et al., 2017. Effects of probiotics on necrotizing enterocolitis, sepsis, intraventricular hemorrhage, mortality, length of hospital stay, and weight gain in very preterm infants: a meta-analysis. Adv. Nutr. 8, 749–763.

Taheri, H.R., Moravej, H., Tabandeh, F., et al., 2010. Efficacy of combined or single use of Lactobacillus crispatus LT116 and L. johnsonii LT171 on broiler performance. Br. Poult. Sci. 51, 580–585.

Tapiovaara, L., Lehtoranta, L., Poussa, T., et al., 2016. Absence of adverse events in healthy individuals using probiotics—analysis of six randomised studies by one study group. Benefic. Microbes 7, 161–169.

Underwood, M.A., 2019. Probiotics and the prevention of necrotizing enterocolitis. J. Pediatr. Surg. 54, 405–412.

van den Akker, C.H.P., van Goudoever, J.B., Shamir, R., et al., 2020. Probiotics and preterm infants: a position paper by the ESPGHAN Committee on Nutrition and the ESPGHAN Working Group for Probiotics and Prebiotics. J. Pediatr. Gastroenterol. Nutr. 70, 664–680.

Vazirzadeh, A., Roosta, H., Masoumi, H., et al., 2019. Long-term effects of three probiotics, singular or combined, on serum innate immune parameters and expressions of cytokine genes in rainbow trout during grow-out. Fish Shellfish Immunol. 98, 748–757.

Wang, H., Lee, I.S., Braun, C., et al., 2016. Effect of probiotics on central nervous system functions in animals and humans: a systematic review. J. Neurogastroenterol. Motil. 22, 589–605.

Wang, Y., Gu, Y., Fang, K., et al., 2018. Lactobacillus acidophilus and Clostridium butyricum ameliorate colitis in murine by strengthening the gut barrier function and decreasing inflammatory factors. Benefic. Microbes 9, 775–787.

Wang, Z.B., Xin, S.S., Ding, L.N., et al., 2019. The potential role of probiotics in controlling overweight/obesity and associated metabolic parameters in adults: a systematic review and meta-analysis. Evid. Based Complement. Alternat. Med. 2019, 3862971.

Weng, F.C., Shaw, G.T., Weng, C.Y., et al., 2017. Inferring microbial interactions in the gut of the Hong Kong whipping frog (*Polypedates megacephalus*) and a validation using probiotics. Front. Microbiol. 8, 525.

West, N.P., Horn, P.L., Barrett, S., et al., 2014a. Supplementation with a single and double strain probiotic on the innate immune system for respiratory illness. e-SPEN J. 9, e178–e184.

West, N.P., Horn, P.L., Pyne, D.B., et al., 2014b. Probiotic supplementation for respiratory and gastrointestinal illness symptoms in healthy physically active individuals. Clin. Nutr. 33, 581–587.

West, N.P., Horn, P.L., Pyne, D.B., et al., 2016. Probiotic supplementation has little effect on peripheral blood regulatory T-cells. J. Allergy Clin. Immunol. 138, 1749–1752.

Xie, H.Y., Feng, D., Wei, D.M., et al., 2017. Probiotics for vulvovaginal candidiasis in non-pregnant women. Cochrane Database Syst. Rev. 11, CD010496.

Yoo, S.R., Kim, Y.J., Park, D.Y., et al., 2013. Probiotics L. plantarum and L. curvatus in combination alter hepatic lipid metabolism and suppress diet-induced obesity. Obesity (Silver Spring) 21, 2571–2578.

Zoppi, G., Cinquetti, M., Benini, A., et al., 2001. Modulation of the intestinal ecosystem by probiotics and lactulose in children during treatment with ceftriaxone. Curr. Ther. Res. Clin. Exp. 62, 418–435.

CHAPTER 6

Milk Quality for Development of Probiotic and Prebiotic Dairy Foods

ERHAN KEYVAN
Faculty of Veterinary Medicine, Department of Food Hygiene and Technology, Burdur Mehmet Akif Ersoy University, Burdur, Turkey

INTRODUCTION

Milk is one of the most important natural protein sources for all mammalians. In particular, it is crucial for infants in the first years of life to keep milk in their diets in an adequate level for appropriate and balanced nutrition (Park et al., 2013). Cow's milk is the main source of dairy production. However, milk of other animals such as goats, sheep, camels, and buffaloes can also be consumed. Due to its health benefits, world milk production has raised over the past decade (Table 1) (FAOSTAT, 2020).

Foodborne pathogens are a major concern and represent a public health problem globally. Each year worldwide, it is estimated that pathogens of food origin cause 600 million cases of foodborne diseases and approximately 420,000 deaths (WHO, 2015). Milk is a suitable medium for foodborne pathogens, which can be a major public health risk (Ding et al., 2016). Although raw milk is considered sterile during secretion, raw milk can be contaminated by microorganisms during milk handling, storage, and processing (De Silva et al., 2016). In this context, foodborne outbreaks caused by milk and dairy products have led to hospitalizations and deaths (Painter et al., 2013).

Considering the nutritional values and the new trends, it is reported that there has been an increase in the consumption of raw milk due to the increasing interest of consumers for untreated natural foods like fresh, organic, or raw in recent years (Verraes et al., 2014). Because of this interest, raw milk producers use various methods to sell milk to consumers such as vending machines and via the internet with strict regulations in Europe (EFSA, 2015). In the United States, 3.2% of the population consumes raw milk, and they are 800 times more susceptible to illnesses, which raises public health issue (Costard et al., 2017). While the sale of raw milk is legal in various states in the United States, it is prohibited or restricted in other states. In some countries, raw milk is completely prohibited, but certified milk can be sold in countries such as New Zealand, England, and Wales (Ijaz, 2014; Baars, 2019).

However, many pathogenic microorganisms isolated from the raw cow's milk such as *Listeria monocytogenes*, *Brucella* spp., Shiga toxin-producing *Escherichia coli*, *Yersinia enterocolitica*, *Campylobacter jejuni*, enterotoxigenic *Staphylococcus aureus*, and methicillin-resistant *S. aureus* (MRSA) can adversely affect human health (Jayarao et al., 2006; Oliver et al., 2009; Bianchi et al., 2013; Giacometti et al., 2013; Quigley et al., 2013; Keyvan et al., 2020). On the other hand, parasites like *Toxoplasma gondii*, *Cryptosporidium parvum*, and tick-borne encephalitis virus are also potential hazards originating from the raw milk (Holzmann et al., 2009; Ng et al., 2012; Rosenthal et al., 2015; Saad et al., 2018). Apart from the above-mentioned, there are some assertions about the possible positive effects of raw milk consumption on human health. However, according to the latest evaluations, these views were reported to have no scientific basis (Lucey, 2015). This chapter aims to provide information about reported studies on the benefits of raw milk consumption, its potential hazards and milk quality for probiotic and prebiotic dairy foods.

BENEFITS OF RAW MILK CONSUMPTION

Raw milk is defined by the European Commission (EC) (Regulation (EC) No 853/2004) as: "Milk which has not been heated beyond 40°C or undergone any treatment that has an equivalent effect" (EC, 2004). A glass of milk can provide 48% of the protein requirements of a 5-year-old child and 9% of calories, and plays a key role especially in infancy nutrition. Also, milk is a source of vitamins (riboflavin, vitamin B_{12}, and vitamin B_5) and minerals (including trace elements) for adults (FAO, 2020).

Before the application of pasteurization in the dairy industry, milk was consumed as unpasteurized or raw product. Some claims have been proposed about the advantages of raw milk consumption compared with

Probiotics and Prebiotics in Foods. https://doi.org/10.1016/B978-0-12-819662-5.00003-3

TABLE 1
World milk production status from different animals.

	WORLD MILK PRODUCTION (MILLION TONNES)				
Year	Cow	Goat	Sheep	Buffalo	Camel
2008	586	15	9	8	2
2009	590	15	9	8	2
2010	601	16	10	9	2
2011	616	16	9	9	2
2012	630	17	9	9	3
2013	635	16	10	10	2
2014	655	17	10	10	2
2015	661	17	10	10	2
2016	665	18	10	11	2
2017	667	18	10	12	2
2018	682	18	10	12	3

Source: FAOSTAT.

those of the pasteurized or heat-treated milk. Raw milk consumption is recommended due to its nutritional properties, and possible advantages such as reduced lactose intolerance incidence and the presence of good bacteria (O'Callaghan et al., 2019). An evidence has been suggested that the immunomodulatory effect has changed significantly as a result of heat treatment in the raw milk (McCarthy et al., 2015). Also, some epidemiological studies have been reported on the positive effects of raw milk consumption (Riedler et al., 2001; Perkin and Strachan, 2006; Waser et al., 2007; Bieli et al., 2007; Loss et al., 2011; Sozańska et al., 2013).

Milk, raw or processed, is a highly nutritious food. Protein, lipid, mineral, and vitamin compositions of milk are essential especially for infants. In recent years, the benefits of raw milk consumption have been expressed by some researchers. However, the consumption of raw milk is a potential health risk for young, immunocompromised, and elderly individuals. Pasteurization of milk at 72°C for 15 s inactivates all of the pathogens in milk, and it does not have any significant negative effects on nutritional and sensory values of milk. For this reason, it is recommended to consume milk after pasteurization (O'Callaghan et al., 2019).

General Nutritional Contents of Milk

Milk is an emulsion with fat particles (globules) dispersed in an aqueous (watery) environment. Complex proteins like enzymes, phospholipids, and other minor components are found in the milk fat globule membranes (MacGibbon and Taylor, 2006). Among the cow's milk components, water, lactose, fat, protein, and ash (mineral substances and other ingredients) are approximately 87%, 4.6%, 4%, 3.3%, and 1%, respectively (Walstra et al., 2006). The chemical composition profile of milk changes among animals (Rafiq et al., 2016).

Milk is a source of high-quality protein. Proteins are one of the main contents of cow's milk with an average content of approximately 3.5%. This rate covers 38% of the total nonfat dry matter content and on average 21% of the energy provided by milk (Jensen, 1995). Casein is the most essential protein of milk. It consists of four different fractions: α_{S1}-, α_{S2}-, β-, and κ-caseins. Serum proteins are more heterogeneous than casein and consist largely of β-lactoglobulin and α-lactalbumin. The amount of tryptophan (amino acid) is higher in α-lactalbumin. Milk is an important source of niacin because tryptophan is a precursor of niacin. Serum proteins represent a small fraction and can be found as serum albumin, immunoglobulins (IgA, IgG, IgM), protease peptone, lactoferrin, and transferrin (Pihlanto and Korhonen, 2003; Wedholm et al., 2006).

Lactose is another component present (2.0%–10%) in almost all mammalian milk (Gänzle et al., 2008). Lactose is the unique carbohydrate of milk, and it is synthesized in the Golgi vesicles of the lactating cells (Walstra et al., 2006). It is the main bovine milk solid and composes around 4.8% of the total solids (Gänzle et al., 2008). This rate covers 54% of the total nonfat dry matter content and 30% of the energy provided by milk (Jensen and Kroger, 2006). Udder health, cow's energy balance, and metabolism affect lactose synthesis and concentration in milk. In addition, lactose is an essential ingredient for the food and pharmaceutical industries (Ohlsson et al., 2017).

Milk fat is a source of energy, fat-soluble vitamins, essential fatty acids, and other health components as well as giving dairy products its characteristic features such as appearance, taste, flavor, and texture (German and Dillard, 1998). Milk fat consists of microscopic globules in a structure defined as an oil-in-water emulsion. Milk fat mainly contains triglycerides, phospholipids, cholesterol-cholesterol esters, diacylglycerols, monoacylglycerols, and free fatty acids, and fat-soluble vitamins A, D, E, and K. Milk fat contains 49% of the total energy supplied by milk. The fat concentration in milk is affected by cow's calving number, lactation period, feeding status, season, and geographic location (German and Dillard, 1998; Parodi, 2004).

Most of the vitamins present in milk are essential for humans. Vitamin A plays an important role in vision, gene expression, cellular differentiation, embryonic development, growth, proliferation, and immunity. Vitamin A and its precursors known as carotenoids are found in varying proportions in milk fat (Jensen, 1995). The major biological function of vitamin D is to maintain normal blood levels of calcium and phosphorus. As a result of inadequate mineralization in the bones, vitamin D deficiency diseases like rickets and osteomalacia are seen in children and adults (Laurent et al., 2018). Milk also contains vitamin E but in low amounts. Vitamin E is an antioxidant and protects cell membrane and lipoproteins from oxidative damage caused by free radicals. Some studies have reported that vitamin E has protective effects on cardiovascular diseases and cancer (Peh et al., 2016; Grażyna et al., 2017). Vitamin K is found in low concentrations in milk and is necessary for blood clotting. Vitamin K has also a protective effect on bones (Parodi, 2004).

Allergy and Asthma

Allergy is an immune response by the body to a substance, especially a particular food, pollen, fur, or dust, to which it has become hypersensitive. Allergy originated from cow's milk is considered as one of the most important food allergies (El-Agamy, 2007). Cow's milk contains α_{s1}-casein, β-lactoglobulin, and α-lactalbumin, which cause allergy or hypersensitivity with a prevalence of 2%–3% in infants and 0.5%–3% in adults. This allergy in infants and children can last from a few months to several years. Later, the tolerance may develop against milk allergy (Bindels and Hoijer, 2000). Gastrointestinal, respiratory, cutaneous, and systemic anaphylactic symptoms are commonly involved in milk allergies (El-Agamy, 2007).

Despite the hazards of unprocessed milk consumption, epidemiological studies suggest that the consumption of raw milk decreases the risk of asthma, hay fever, and allergy (Braun-Fahrländer and Von Mutius, 2011; Brick et al., 2019). In a study from village and town inhabitants in Poland, the researchers claimed that unpasteurized milk consumption may protect against atopy and asthma (Sozańska et al., 2013). In ALEX study, the consumption of farm milk decreases the risk of allergy, asthma, and hay fever in children younger than 1 year (Riedler et al., 2001). Waser et al. (2007)'s PARSIFAL study suggested that the consumption of farm milk can protect against asthma and allergy in rural and suburban populations across Europe. In the GABRIELA study, Loss et al. (2011) reported an inverse relationship of raw milk consumption with asthma. Despite the positive results of these studies, raw milk consumption poses unpredictable health risk in humans (Claeys et al., 2013).

Lactose Intolerance

Mammalian milk, including human milk, contains lactose (Gänzle et al., 2008). Lactose is hydrolyzed by the lactase enzyme (β-galactosidase) into glucose and galactose, and absorbed in the small intestines. The β-galactosidase can be genetically inadequate, or lactose digestibility may decrease due to aging. Lactose intolerance appears with digestive system symptoms such as bloating, diarrhea, abdominal pain, and gas as a result of the inability to digest lactose (Di Stefano et al., 2001; Swagerty et al., 2002; Lucey, 2015). Lactose intolerance has been surveyed in people in many countries and ethnic groups (Madsen, 1997; Yang et al., 2000; Keith et al., 2011; Latorre et al., 2014). There are some claims that the raw milk consumption reduces lactose intolerance (Lucey, 2015). However, in a study, it has been reported that the consumption of raw milk does not decrease lactose digestibility compared with pasteurized milk (Mummah et al., 2014). In dairy products, lactose is hydrolyzed by lactic acid bacteria, and it is the healthiest suggestion to be consumed by individuals with lactose intolerance.

Beneficial Microflora

Microflora in the raw milk derived from cattle, sheep, goats, and buffalo generally contains lactic acid bacteria prior to pasteurization (Aziz et al., 2009; Quigley et al., 2011; Tormo et al., 2015). Lactic acid bacteria in raw milk can inhibit the growth of pathogenic bacteria. However, this feature is very limited in the cold storage of raw milk. In addition, the inhibitory effect against some bacteria that can show pathogenic properties even at low doses may be insufficient (Claeys et al., 2013). Pasteurization inactivates pathogenic bacteria along with good bacteria in the microflora of raw milk. Inactivation of beneficial bacteria is a negligible loss compared with the potential hazards of raw milk consumption.

Antimicrobial Systems

Raw cow's milk contains various systems with antimicrobial properties, which contribute to the immunity of young calf (Claeys et al., 2013). Various proteins and enzymes are responsible for an antimicrobial effect. Immunoglobulins, lactoperoxidase, lysozyme, lactoferrin, xanthine oxidase, bacteriocins, oligosaccharides, and lipid fragments are the antimicrobial components of milk (Expósito and Recio, 2006). Besides, milk peptides have antibacterial and antitumoral properties (Lopez-Exposito and Recio, 2008). Lactoferrin is an iron-binding protein, and it has bacteriostatic effects on microorganisms (Sanchez et al., 1992). While pasteurized and unheated lactoferrin have similar antimicrobial effects, ultra-high-temperature (UHT)-treated milk does not contain lactoferrin (Wang et al., 2019). Immunoglobulins are essential components of the immunity present in milk and colostrum (Hurley and Theil, 2011). Colostrum is not recommended for human consumption because it contains the necessary components for calve's immunity (Stelwagen et al., 2009). Lactoperoxidase is one of the major enzymes in bovine milk, and it has inhibitory effects with lactoperoxidase system on a wide range of microorganisms. Thermal denaturation of lactoperoxidase starts at 70°C (Kussendrager and Van Hooijdonk, 2000). Although there are various factors that prevent the development of pathogenic microorganisms in raw milk, it does not completely prevent the growth of pathogenic microorganism. Therefore, the presence of antimicrobial systems in milk is not a sufficient feature for the consumption of raw milk.

NUTRITIONAL STATUS OF HEAT TREATED MILK

Heat treatment is the best way to ensure milk safety. The objectives in applying heat treatment to milk are to inactivate pathogenic microorganisms, to inactivate most microorganisms that cause degradation, and to stop the enzymatic activity leading to milk quality losses. By applying heat treatment types such as pasteurization and UHT sterilization, milk's shelf life is extended and it is made safe for human consumption (Melini et al., 2017).

Pasteurization is a process applied to milk in order to prolong the storage period by inactivating vegetative forms of microbial agents that may pose a risk for human health. Pasteurization can be done for milk at 63°C for 30 min or at 68°C for 10 min as low-temperature long-time (LTLT) and at 72°C for 15 s as high-temperature short-time (HTST). Pasteurization has a low impact on the nutritional value, flavor, casein structure, and whey protein structure, and no effect on mineral contents, fatty acid profile, and bioavailability. Pasteurization decreases vitamin B_{12}, A, D, and E contents of milk. Also, pre- and postpasteurization contamination with various mycotoxins, antibiotic residues, and disinfectant residues can be found in milk, since the contaminants are not affected by heat treatment, thus posing a risk to human health (Walstra et al., 2006; MacDonald et al., 2011; Cappozzo et al., 2015; Melini et al., 2017).

UHT sterilization process is defined as the treatment of milk at 135–150°C for 1–4 s, thus inactivating pathogenic and nonpathogenic microorganisms as well as spores in milk. UHT processing of milk causes denaturation of whey protein structure and loss of lysine but it has a little or no effect on the nutritional quality of protein (MacDonald et al., 2011; Melini et al., 2017). Compared with raw milk, UHT-treated milk has the similar chemical composition characteristics but with lower levels of vitamins D and A (Cappozzo et al., 2015). UHT may cause some organoleptic effects such as cooked and ketone flavor (MacDonald et al., 2011). UHT processing of milk has many advantages like long shelf life, storage at ambient temperature, and microbiological safety.

POTENTIAL HAZARDS OF RAW MILK

Despite the improvement of hygienic conditions in milk production and processing, pathogenic microorganisms in raw milk cannot be completely destroyed. In recent studies, the presence of foodborne pathogens in raw milk was reported from various countries (Table 2). Foodborne diseases pose an important threat to human health; outbreaks originating from the consumption of raw milk and products are presented in Table 3. Hazards from raw milk and its products can be defined as physical hazards, microbiological hazards, mycotoxins, hormones, chemical contaminants, and veterinary drugs.

Physical Hazards

Milk and dairy products should be produced in accordance with hygiene and sanitation conditions. Otherwise, unwanted physical hazards such as insects, hair, fabrics, metals, plastics, and other foreign bodies can be found in milk and dairy products. Physical hazards can cause damage to the digestive system and other organs in humans (Aguiar et al., 2018; Singhal et al., 2020). Processed milk and dairy products are produced by using filters and foreign matter detection systems. For this reason, compared with raw milk, the risk of physical hazards is very low in processed milk and dairy products. However, there are studies reporting that physical hazards can be found in processed milk and dairy products (Van Asselt et al., 2017; Aguiar et al., 2018).

TABLE 2
Prevalence of foodborne pathogens in raw milk.

Pathogen	Occurrence in raw milk (%)	Number of samples	Country	References
Brucella spp.	1.07	5211	China	Ning et al. (2013)
Listeria spp.	18.6	446	Iran	Jamali et al. (2013)
Mycobacterium bovis	0.25	401	Brazil	Da Silva Cezar et al. (2016)
S. aureus	16.3	141	Egypt	Mansour et al. (2017)
Campylobacter spp.	24.9	234	United States	Del Collo et al. (2017)
Yersinia enterocolitica	3.1	509	Italy	Bonardi et al. (2018)
Salmonella enterica *L. monocytogenes* *Escherichia coli* O157:H7	1.6 7.5 3.3	305	Jordan	Obaidat and Stringer (2019)

TABLE 3
Selected outbreaks due to consumption of raw milk or dairy products made from raw milk.

Pathogenic agent	Origin of outbreak	Year	Country	Status of outbreak	References
Campylobacter jejuni	Raw milk	1981	United States	25 cases, 22 hospitalized	Korlath et al. (1985)
Campylobacter jejuni	Raw milk	1994	United Kingdom	23 cases	Evans et al. (1996)
Staphylococcal enterotoxins	Raw milk	1999	Brazil	328 cases	Do Carmo et al. (2002)
Mycobacterium bovis	Raw milk	2005	Ireland	5 cases	Doran et al. (2009)
Brucella melitensis	Raw milk	2006	Spain	11 cases (*Brucella melitensis* positive in 5 cases), 11 hospitalized	Ramos et al. (2008)
Salmonella Typhimurium	Raw milk	2007	United States	29 cases	CDC (2007)
Tick-borne encephalitis virus	Raw goat milk	2007	Hungary	25 cases	Balogh et al. (2010)
Escherichia coli O157	Raw milk	2008	United States	14 cases, 5 hospitalized	Guh et al. (2010)
Brucella	Raw goat milk	2011–2012	Malaysia	80 cases	Leong et al. (2015)
Listeria monocytogenes	Raw milk	2014	United States	2 cases, 2 hospitalized, 1 dead	CDC (2020)

Continued

TABLE 3
Selected outbreaks due to consumption of raw milk or dairy products made from raw milk—cont'd

Pathogenic agent	Origin of outbreak	Year	Country	Status of outbreak	References
Escherichia coli O157: H7	Raw milk Gouda cheese	2013	Canada	29 cases, 5 hospitalized, 1 dead	Currie et al. (2018)
Yersinia pseudotuberculosis O:1	Raw milk	2014	Finland	55 cases	Pärn et al. (2015).
Staphylococcal enterotoxins	Soft raw milk cheese	2014	Switzerland	14 cases	Johler et al. (2015)
Salmonella Stanley	Raw cashew cheese	2014	United States	17 cases, 3 hospitalized	CDC (2020)
Tick-borne encephalitis virus	Raw goat milk or raw milk cheese	2015	Croatia	7 cases	Markovinović et al. (2016)
Listeria monocytogenes	Soft raw milk cheese	2017	United States	8 cases, 8 hospitalized, 2 dead	CDC (2020)

Microbiological Hazards

Milk has a complex microbial flora that affects the quality and safety of dairy products. Pathogenic bacteria can be found in this flora along with some beneficial bacteria (Porcellato et al., 2018). Raw milk can be contaminated with foodborne pathogens in a variety of ways. Among these contamination sources are dairy animals (even if they appear to be clinically healthy), environmental contamination at the stage of collecting milk or storage, contamination during transportation, poor hygiene, udder infection, and cross-contamination of raw milk (Verraes et al., 2014). Milk is an organic compound and a suitable medium for the growth of bacteria. Since milk is a good medium for bacterial growth, contamination even in low levels is dangerous because in this suitable condition bacteria can multiply fastly (O'Mahony et al., 2009). Foodborne pathogens such as *E. coli* O157:H7, *Listeria* spp., and *Salmonella* in raw cow's milk present a major public health concern, and the consumption of raw milk without heat treatment is not recommended for humans (Chye et al., 2004).

Pathogenic microorganisms can contaminate milk in different ways. Due to diseases such as tuberculosis and brucellosis, which cause systemic diseases in dairy animals, milk secretion may carry the pathogens (Leite et al., 2003; Firdessa et al., 2012). *Mycobacterium bovis* and *Brucella abortus* are important pathogens that cause disease in humans (Müller et al., 2013; Dadar et al., 2019). Raw milk and products are the main source of these pathogens causing disease in humans (Jay et al., 2012). Pathogen microorganisms such as *L. monocytogenes*, *Brucella* spp., Shiga toxin-producing *E. coli* (STEC), *Y. enterocolitica*, *C. jejuni*, and enterotoxigenic *S. aureus* appear due to deficiencies in hygienic conditions (Perin et al., 2019).

S. aureus is a foodborne pathogen that can be found naturally in humans and animals. *S. aureus* may cause foodborne intoxications with heat-resistant enterotoxins. It is also one of the most important agents of mastitis in dairy animals. Overuse of antibiotics for prophylactic and therapeutic purposes has led to the development of resistance to certain antibiotics in bacteria. The presence of methicillin-resistant *S. aureus* (MRSA) in milk may create a risk for public health (Le Loir et al., 2003; Sawant et al., 2005; Haran et al., 2012; Jamali et al., 2015; Gomes and Henriques, 2016).

L. monocytogenes is a ubiquitous bacteria commonly found in nature, and its psychotrophic feature and biofilm formation represent the main problems in the dairy industry (Oliver et al., 2005; Latorre et al., 2010). Raw milk and products are potentially hazardous foods in terms of the existence of *Listeria* spp. (Rahimi et al., 2010; EFSA, 2015; CDC, 2020). Although pasteurization has a lethal effect on *L. monocytogenes*, postpasteurization contamination of milk may cause listeriosis (Jordan et al., 2016). Illnesses due to *L. monocytogenes* are more frequently caused by the consumption of cheese made of unpasteurized milk (Costard et al., 2017). Pregnant women, neonates, the immunocompromised, and the elderly are defined as high-risk groups for infections caused by *Listeria* spp. (McLauchlin et al., 2014).

E. coli is a normal commensal of the human gut; its presence in foods is an indicator of fecal contamination and poor hygiene practices (Sugrue et al., 2019). Some strains of *E. coli* can cause a public health hazard by causing intestinal and extraintestinal infections. STEC is an important pathogenic group of *E. coli*, which causes serious symptoms like diarrhea, hemorrhagic colitis, hemolytic uremic syndrome (HUS), and deaths in humans. Ruminants are a significant reservoir of STEC and spread this bacteria with their feces. Milk obtained under insufficient hygiene conditions may contain STEC (Rios et al., 2019). STEC can cause outbreaks due to the consumption of unpasteurized milk and dairy products (Guh et al., 2010; Currie et al., 2018).

In recent years, *Campylobacter* spp. have been identified as the main cause of bacterial gastroenteritis in humans (Platts-Mills and Kosek, 2014). Campylobacteriosis is a zoonotic disease, which was reported by European Food Safety Authority (EFSA) and the European Center for Disease Prevention and Control (ECDC) as the most common disease in the European Union with 246,571 human cases in 2018 (EFSA and ECDC, 2019). *C. jejuni* is one of the most widespread *Campylobacter* species in milk and dairy products (Christidis et al., 2016; El-Zamkan and Hameed, 2016; Del Collo et al., 2017). Contamination of *Campylobacter* spp. in milk and dairy products occurs with fecal shedding of dairy cattle (Wesley et al., 2000). Pasteurization is the most effective inactivation method for this bacteria (D'Aoust et al., 1988).

Mycotoxins

Mycotoxins are produced by yeast and fungi, and have carcinogenic, genotoxic, teratogenic, mutagenic, nephrotoxic, hepatotoxic, and immunotoxic effects (Becker-Algeri et al., 2016). Animal feed preparation and preservation methods can cause mycotoxin contamination. Mycotoxins can be metabolized and biotransformed in foods such as milk and meat as a result of the consumption of contaminated feed by animals (Yiannikouris and Jouany, 2002). Molds such as *Fusarium*, *Aspergillus*, and *Penicillium* are the species that produce the most mycotoxins in human food and animal feed under improper storage conditions (Binder, 2007). The exposure of humans to mycotoxins can be directly through the consumption of contaminated agricultural products (cereals, corn, fruit, etc.) or products such as meat, milk obtained from animals that consume contaminated feed (Capriotti et al., 2012). The harmful effects and consequent reduction productivity caused by mycotoxins create economic losses in animal husbandry (Flores-Flores et al., 2015).

Ruminants are less sensitive to mycotoxins than other animal species (Kiessling et al., 1984). Mycotoxins are converted into less effective or biologically inactive substances with the effect of rumen flora. For example, ochratoxin A (OTA) is quickly converted to the less harmful metabolite (ochratoxin α) by the rumen flora and a small amount of OTA remains (Ringot et al., 2006). Some of the aflatoxins (AF) are inactivated by the rumen flora and converted into a secondary metabolite, aflatoxicols. In dairy animals, continuous exposure to aflatoxins results in a liver damage. It can be said that aflatoxin contaminations occurring especially in milk and dairy products are caused by liver-related damage (Miller and Wilson, 1994). In order to increase milk yield, energy-rich rations are needed in dairy cows. For this reason, complex rations have been created for feeding dairy cows. Different ingredients in the feeding regime may also increase the risk of mycotoxins (Fink-Gremmels, 2008).

Mycotoxin concerns in food and feed started as an outbreak of turkeys due to the contamination of turkey's feed with aflatoxins in the 1960s (Eaton and Groopman, 2013). Mycotoxins such as ochratoxin, zearalenone, fumonisin, T-2 toxin, and deoxynivalenol present in milk and dairy products are the potential hazards for humans (Coffey et al., 2009; Turkoglu and Keyvan, 2019). Milk pasteurization is the efficient process for the inactivation of most molds and spores. However, some mold types such as milk-sourced *Byssochlamys fulva* are resistant to heat treatment at 90°C for 5 min (Sant'Ana et al., 2009). The presence of this species in cheese milk may also cause the production of mycotoxins in cheese. *B. fulva* can synthesize many mycotoxin types. These mycotoxins have not yet been detected in milk and dairy products (Benkerroum, 2016).

Aflatoxins have harmful effects on the body like immunosuppression, hepatotoxic, and carcinogenic. They are usually produced by *Aspergillus flavus*, *Aspergillus parasiticus*, and *Aspergillus nomius* (Prandini et al., 2009). International Cancer Research Agency (IARC) classified aflatoxins as a carcinogen (Group 1) and a potential carcinogen in humans based on epidemiological and laboratory studies (Group 2B) (IARC, 1993). Within the aflatoxin types (B_1, B_2, G_1, G_2, M_1, M_2), AFM_1 poses a risk on the health arising from milk and dairy products (Veldman et al., 1992; Gourama and Bullerman,

1995). Aflatoxin M_1 is resistant to thermal processes like pasteurization and sterilization, which are not sufficient for the inactivation of AFM_1 (Assem and Mohamad, 2011). For this reason, monitoring the presence of AFM_1 in milk should be ensured continuously (Becker-Algeri et al., 2016).

Different strategies can be used to prevent the presence of mycotoxins in milk and dairy products, such as the prevention of toxin production or inactivation of toxins. Even as a result of the application of protective measures, complete protection may not be provided. In order to prevent the presence of mycotoxins in milk, production should be made in accordance with hygiene rules during the product processing stage. In addition, feed contaminated with yeast and molds should not be used in the feeding of dairy animals (Benkerroum, 2016).

Hormones

Milk is an important source of steroid hormones. Progesterone and estrone may increase depending on the amount of lipid in milk and dairy products. Many of the hormones are not affected by processes such as heating and cooling (Khaniki, 2007). The occurrence of hormones in milk can occur without endogenous origin (Matraszek-Zuchowska et al., 2017). Drug applications in dairy animals in order to increase milk yield may cause abnormal hormone secretions and increase hormone residues in milk (Tan et al., 2016). The presence of hormones in milk can cause hormone-dependent diseases, allergic reactions, and some types of cancer in humans (Khaniki, 2007; Tate et al., 2011). Also, most types of cancer have been reported to be closely related to foods (Ganmaa and Sato, 2005). For healthy and reliable milk production, like all other contaminants, hormone residues should be closely monitored and controlled in milk.

Chemical Contaminants in Milk

Milk and dairy products can be contaminated with certain chemicals such as pesticides, antiparasitic drugs, heavy metals, detergents, and disinfectants at various stages of the production chain. Chemical contaminations may occur in milk as a result of contaminated feed, contaminated soilborne plants, and veterinary practices. Chemical residues in milk and dairy products pose a major concern for public health (Khaniki, 2007; Nag, 2010).

Pesticides can be found in milk as a result of application in milk processing units and equipment for insect control and contamination of feed and drinking water. Pyrethroids, carbamates, organophosphate, and organochlorine are the main classes of pesticide used in cattle management (Khaniki, 2007). The presence of organophosphates and organochlorine has been reported from several countries (Melgar et al., 2010; Kampire et al., 2011; Kotinagu and Krishnaiah, 2015). Extensive and unnecessary use of pesticides may exhibit harmful effects on humans and the environment (Andreescu and Marty, 2006).

Antiparasitic drugs are used in food-producing animals for the treatment and control of parasitic diseases. Avermectins ivermectin (IVR), abamectin, doramectin, eprinomectin, and the milbemycin moxidectin are examples of the most commonly used antiparasitic drugs in cattle (Barreto et al., 2019). Antiparasitic residues in milk pose a risk for human health (Imperiale et al., 2002).

As a result of some industrial activities, the air is contaminated with some toxic elements such as As, Cd, Cu, Hg, and Pb, and environmental contamination may occur as a result of its transportation to other regions (WHO, 2007; Sanchez de la Campa et al., 2008). Heavy metals are resistant to heat applications as opposed to microorganisms. Milk and dairy products are not an adequate source of nutrients in terms of some essential elements (iron, copper, and zinc). However, some harmful elements such as lead and cadmium can be found in milk. Contaminated feed and grazing in pasture can be effective in contaminating milk with heavy metals (Bilandžić et al., 2011).

Detergents and disinfectants are widely used in dairy industry for cleaning of the milking equipment. It is a critical requirement for preventing contamination in milk. Chlorine compounds have rapid decomposition properties which are not dangerous for health. However, quaternary compounds are stable in milk, and these compounds may pose a risk for human health (Salomskiene et al., 2013).

Veterinary Drugs

Veterinary drugs are used in the treatment of many diseases such as mastitis, respiratory diseases, and diarrhea in dairy animals. Antimicrobials are used in food-producing animals for prophylactic and therapeutic purposes. Gentamicin, erythromycin, penicillin, ciprofloxacin, tetracycline, virginiamycin, florfenicol, sulfonamides, and colistin are the examples of antimicrobials utilized in these treatments. Overuse of these drugs in dairy animals may cause residues in milk and dairy products (Kabir et al., 2004; Barreto et al., 2019; Van et al., 2020).

The presence of antibiotic residues in raw milk creates disorders in the reproduction of starter culture bacteria, and this creates problems in the production of yogurt, cheese, and other fermented products. It can also cause undesirable flavor in butter and cream. Pasteurization has little or no effect on the antibiotic residue in milk (Barreto et al., 2019). Antimicrobials in milk may cause some adverse effects on health like unbalanced intestinal flora, allergic reactions, and antibiotic resistance in humans (Stolker et al., 2008). Studies have reported the presence of antibiotic residues in milk and dairy products (Darko et al., 2017; Wang et al., 2017). In order to ensure food safety and protect public health, excessive antibiotics should not be used in food-producing animals.

The detection and control of chemical contaminants in milk can be carried out with high-precision detection methods. In order to prevent veterinary drug residues in milk, dairy farmers should pay attention to waiting period in dairy animals after drug application (Rana et al., 2019).

MILK QUALITY FOR DEVELOPMENT OF PROBIOTIC AND PREBIOTIC DAIRY FOODS

Probiotics are living microorganisms that have positive effects on health when taken in sufficient amounts (McFarland, 2015). These microorganisms regulate gut microbes and create a protective effect against diseases (Sarowska et al., 2013). Due to the beneficial effects of probiotic products on human health, consumers' interest in these products has also increased. The raised interest toward functional foods has also led to increased researches in this field (Granato et al., 2010; Kumar et al., 2015). Nondairy probiotic products have also been developed due to vegan preferences, lactose intolerance, allergy, and the cholesterol content of dairy products (Prado et al., 2008). However, dairy products are the most important source of probiotic activity (Tamime et al., 2005).

It increases milk yield and some milk components (immunoglobulin G (IgG), lactoferrin (LTF), lysozyme (LYS) and lactoperoxidase (LP)) by giving some probiotics (*Lactobacillus casei* Zhang and *Lactobacillus plantarum* P-8) to dairy cows. A significant decrease has been reported in the rate of somatic cell counts (SCC) (Xu et al., 2017). Milk obtained from different animal species is also effective in the production of probiotic fermented products. It has been determined that the fermented product produced using goat's milk has a lower pH value and higher acidity than that produced using cow's milk (Dimitrellou et al., 2019).

CONCLUSION

Milk is a food of animal origin with a high nutritional quality. In recent years, there are various claims regarding the benefits from raw milk consumption. These claims include antimicrobial agents present in milk, the presence of beneficial microorganisms, and effects on allergies and asthma. However, it is reported that the consumption of raw milk can be quite dangerous for human health due to the possible pathogen and other hazards in milk. There is no significant difference between raw milk and heat-treated milk in terms of nutritional values. The benefits obtained by applying heat treatment to milk are more important than the potential benefits of raw milk consumption. For this reason, consumption of raw milk is not recommended as it poses significant risks for human health. In order to obtain quality probiotic dairy products, it is recommended to use chemically and microbiologically superior quality milk in production.

REFERENCES

Aguiar, R.S., Esmerino, E.A., Rocha, R.S., Pimentel, T.C., Alvarenga, V.O., Freitas, M.Q., Silva, M.C., Sant'Ana, A.S., Silva, A.C.O., Cruz, A.G., 2018. Physical hazards in dairy products: incidence in a consumer complaint website in Brazil. Food Control 86, 66–70.

Andreescu, S., Marty, J.L., 2006. Twenty years research in cholinesterase biosensors: from basic research to practical applications. Biomol. Eng. 23 (1), 1–15.

Assem, E., Mohamad, A., 2011. A survey on the occurrence of aflatoxin M1 in raw and processed milk samples marketed in Lebanon. Food Control 22 (12), 1856–1858.

Aziz, T., Khan, H., Bakhtair, S.M., Naurin, M., 2009. Incidence and relative abundance of lactic acid bacteria in raw milk of buffalo, cow and sheep. J. Anim. Plant Sci. 19 (4), 168–173.

Baars, T., 2019. Regulations and production of raw milk. In: Nero, L.A., De Carvalho, A.F. (Eds.), Raw Milk: Balance Between Hazards and Benefits. Academic Press, pp. 65–89.

Balogh, Z., Ferenczi, E., Szeles, K., Stefanoff, P., Gut, W., Szomor, K.N., Takacs, M., Berencsi, G., 2010. Tick-borne encephalitis outbreak in Hungary due to consumption of raw goat milk. J. Virol. Methods 163 (2), 481–485.

Barreto, F., Jank, L., Castilhos, T., Rau, R.B., Tomaszewski, C.A., Ribeiro, C., Hillesheim, D.R., 2019. Chemical residues and mycotoxins in raw milk. In: Nero, L.A., De Carvalho, A.F. (Eds.), Raw Milk: Balance Between Hazards and Benefits. Academic Press, pp. 273–293.

Becker-Algeri, T.A., Castagnaro, D., de Bortoli, K., de Souza, C., Drunkler, D.A., Badiale-Furlong, E., 2016. Mycotoxins in bovine milk and dairy products: a review. J. Food Sci. 81 (3), 544–552.

Benkerroum, N., 2016. Mycotoxins in dairy products: a review. Int. Dairy J. 62, 63–75.

Bianchi, D.M., Barbaro, A., Gallina, S., Vitale, N., Chiavacci, L., Caramelli, M., Decastelli, L., 2013. Monitoring of food-borne pathogenic bacteria in vending machine raw milk in Piedmont, Italy. Food Control 32 (2), 435–439.

Bieli, C., Eder, W., Frei, R., Braun-Fahrländer, C., Klimecki, W., Waser, M., Doekes, G., Launer, R., Martinez, F.D., 2007. A polymorphism in CD14 modifies the effect of farm milk consumption on allergic diseases and CD14 gene expression. J. Allergy Clin. Immunol. 120 (6), 1308–1315.

Bilandžić, N., Đokić, M., Sedak, M., Solomun, B., Varenina, I., Knežević, Z., Benić, M., 2011. Trace element levels in raw milk from northern and southern regions of Croatia. Food Chem. 127 (1), 63–66.

Bindels, J.G., Hoijer, M., 2000. Allergens: latest developments, newest techniques. Bull. Int. Dairy Fed. 351, 31–32.

Binder, E.M., 2007. Managing the risk of mycotoxins in modern feed production. Anim. Feed Sci. Technol. 133 (1–2), 149–166.

Bonardi, S., Le Guern, A.S., Savin, C., Pupillo, G., Bolzoni, L., Cavalca, M., Pongolini, S., 2018. Detection, virulence and antimicrobial resistance of *Yersinia enterocolitica* in bulk tank milk in Italy. Int. Dairy J. 84, 46–53.

Braun-Fahrländer, C., Von Mutius, E., 2011. Can farm milk consumption prevent allergic diseases? Clin. Exp. Allergy 41 (1), 29–35.

Brick, T., Hettinga, K., Kirchner, B., Pfaffl, M.W., Ege, M.J., 2019. The beneficial effect of farm milk consumption on asthma, allergies, and infections: from meta-analysis of evidence to clinical trial. J. Allergy Clin. Immunol. 8 (3), 878–889.

Cappozzo, J.C., Koutchma, T., Barnes, G., 2015. Chemical characterization of milk after treatment with thermal (HTST and UHT) and nonthermal (turbulent flow ultraviolet) processing technologies. J. Dairy Sci. 98 (8), 5068–5079.

Capriotti, A.L., Caruso, G., Cavaliere, C., Foglia, P., Samperi, R., Laganà, A., 2012. Multiclass mycotoxin analysis in food, environmental and biological matrices with chromatography/mass spectrometry. Mass Spectrom. Rev. 31 (4), 466–503.

CDC, 2007. *Salmonella* Typhimurium infection associated with raw milk and cheese consumption—Pennsylvania, 2007. MMWR 56 (44), 1161.

CDC, 2020. Foodborne Outbreak. Available from: https://www.cdc.gov/foodsafety/outbreaks/multistate-outbreaks/outbreaks-list.html. (Accessed 15 February 2020).

Christidis, T., Pintar, K.D.M., Butler, A.J., Nesbitt, A., Thomas, M.K., Marshall, B., Pollari, F., 2016. *Campylobacter* spp. prevalence and levels in raw milk: a systematic review and meta-analysis. J. Food Prot. 79 (10), 1775–1783.

Chye, F.Y., Abdullah, A., Ayob, M.K., 2004. Bacteriological quality and safety of raw milk in Malaysia. Food Microbiol. 21 (5), 535–541.

Claeys, W.L., Cardoen, S., Daube, G., De Block, J., Dewettinck, K., Dierick, K., Vandenplas, Y., Herman, L., 2013. Raw or heated cow milk consumption: review of risks and benefits. Food Control 31 (1), 251–262.

Coffey, R., Cummins, E., Ward, S., 2009. Exposure assessment of mycotoxins in dairy milk. Food Control 20 (3), 239–249.

Costard, S., Espejo, L., Groenendaal, H., Zagmutt, F.J., 2017. Outbreak-related disease burden associated with consumption of unpasteurized cow's milk and cheese, United States, 2009–2014. Emerg. Infect. Dis. 23 (6), 957–964.

Currie, A., Galanis, E., Chacon, P.A., Murray, R., Wilcott, L., Kirkby, P., Honish, L., Franklin, K., Shyng, S., Sharma, D., Tschetter, L., Hoang, L., Chui, L., Pacagnella, A., Wong, J., Pritchard, J., Kerr, A., Taylor, M., Mah, V., Flint, J., 2018. Outbreak of *Escherichia coli* O157: H7 infections linked to aged raw milk gouda cheese, Canada, 2013. J. Food Prot. 81 (2), 325–331.

D'Aoust, J.Y., Park, C.E., Szabo, R.A., Todd, E.C.D., Emmons, D.B., McKellar, R.C., 1988. Thermal inactivation of *Campylobacter* species, *Yersinia enterocolitica*, and hemorrhagic *Escherichia coli* 0157: H7 in fluid milk. Int. J. Dairy Sci. 71 (12), 3230–3236.

Da Silva Cezar, R.D., Lucena-Silva, N., Batista Filho, A.F.B., de Melo Borges, J., de Oliveira, P.R.F., Lúcio, É.C., Junior, J.W.P., 2016. Molecular detection of *Mycobacterium bovis* in cattle herds of the state of Pernambuco, Brazil. BMC Vet. Res. 12 (1), 31.

Dadar, M., Shahali, Y., Whatmore, A.M., 2019. Human brucellosis caused by raw dairy products: a review on the occurrence, major risk factors and prevention. Int. J. Food Microbiol. 292, 39–47.

Darko, G., Borquaye, L.S., Acheampong, A., Oppong, K., 2017. Veterinary antibiotics in dairy products from Kumasi, Ghana. Cogent Chem. 3 (1), 1343636.

De Silva, S.A.S.D., Kanugala, K.A.N.P., Weerakkody, N.S., 2016. Microbiological quality of raw milk and effect on quality by implementing good management practices. Procedia Food Sci. 6, 92–96.

Del Collo, L.P., Karns, J.S., Biswas, D., Lombard, J.E., Haley, B.J., Kristensen, R.C., Van Kessel, J.A.S., 2017. Prevalence, antimicrobial resistance, and molecular characterization of Campylobacter spp. in bulk tank milk and milk filters from US dairies. Int. J. Dairy Sci. 100 (5), 3470–3479.

Di Stefano, G., Veneto, S., Malservisi, A., Strocchi, G.R., Corazza, M., 2001. Lactose malabsorption and intolerance in the elderly. Scand. J. Gastroenterol. 36 (12), 1274–1278.

Dimitrellou, D., Salamoura, C., Kontogianni, A., Katsipi, D., Kandylis, P., Zakynthinos, G., Varzakas, T., 2019. Effect of milk type on the microbiological, physicochemical and sensory characteristics of probiotic fermented milk. Microorganisms 7 (9), 274.

Ding, T., Yu, Y.Y., Schaffner, D.W., Chen, S.G., Ye, X.Q., Liu, D.H., 2016. Farm to consumption risk assessment for *Staphylococcus aureus* and staphylococcal enterotoxins in fluid milk in China. Food Control 59, 636–643.

Do Carmo, L.S., Dias, R.S., Linardi, V.R., de Sena, M.J., dos Santos, D.A., de Faria, M.E., Pena, E.C., Heneine, L.G., 2002. Food poisoning due to enterotoxigenic strains of *Staphylococcus* present in Minas cheese and raw milk in Brazil. Food Microbiol. 19 (1), 9–14.

Doran, P., Carson, J., Costello, E., More, S.J., 2009. An outbreak of tuberculosis affecting cattle and people on an Irish dairy farm, following the consumption of raw milk. Ir. Vet. J. 62 (6), 390–397.

Eaton, D.L., Groopman, J.D. (Eds.), 2013. The Toxicology of Aflatoxins: Human Health, Veterinary, and Agricultural Significance. Elsevier.

EFSA Panel on Biological Hazards (BIOHAZ), 2015. Scientific opinion on the public health risks related to the consumption of raw drinking milk. EFSA J. 13 (1), 3940.

El-Agamy, E.I., 2007. The challenge of cow milk protein allergy. Small Rumin. Res. 68 (1–2), 64–72.

El-Zamkan, M.A., Hameed, K.G.A., 2016. Prevalence of *Campylobacter jejuni* and *Campylobacter coli* in raw milk and some dairy products. Vet. World 9 (10), 1147.

European Commission (EC), 2004. 853/2004 of the European Parliament and of the Council of 29 April 2004 laying down specific hygiene rules for food of animal origin. Off. J. Eur. Union 30, 151.

European Food Safety Authority and European Centre for Disease Prevention and Control (EFSA and ECDC), 2019. The European Union One Health 2018 Zoonoses report. EFSA J. 17 (12), e05926.

Evans, M.R., Roberts, R.J., Ribeiro, C.D., Gardner, D., Kembrey, D., 1996. A milk-borne *Campylobacter* outbreak following an educational farm visit. Epidemiol. Infect. 117 (3), 457–462.

Expósito, I.L., Recio, I., 2006. Antibacterial activity of peptides and folding variants from milk proteins. Int. Dairy J. 16 (11), 1294–1305.

FAO, 2020. Milk Facts. http://www.fao.org/resources/infographics/infographics-details/en/c/273893/. (Accessed 15 April 2020).

FAOSTAT, 2020. Livestock Primary. http://www.fao.org/faostat/en/#data/QL. (Accessed 15 April 2020).

Fink-Gremmels, J., 2008. Mycotoxins in cattle feeds and carry-over to dairy milk: a review. Food Addit. Contam. 25 (2), 172–180.

Firdessa, R., Tschopp, R., Wubete, A., Sombo, M., Hailu, E., Erenso, G., Young, D., Gordon, S.V., Sahile, M., Aseffa, A., Berg, S., 2012. High prevalence of bovine tuberculosis in dairy cattle in central Ethiopia: implications for the dairy industry and public health. PLoS One 7 (12), 1–12.

Flores-Flores, M.E., Lizarraga, E., de Cerain, A.L., González-Peñas, E., 2015. Presence of mycotoxins in animal milk: a review. Food Control 53, 163–176.

Ganmaa, D., Sato, A., 2005. The possible role of female sex hormones in milk from pregnant cows in the development of breast, ovarian and corpus uteri cancers. Med. Hypotheses 65 (6), 1028–1037.

Gänzle, M.G., Haase, G., Jelen, P., 2008. Lactose: crystallization, hydrolysis and value-added derivatives. Int. Dairy J. 18 (7), 685–694.

German, J.B., Dillard, C.J., 1998. Fractionated milk fat: composition, structure, and functional properties. Food Technol. 52, 33.

Giacometti, F., Bonilauri, P., Serraino, A., Peli, A., Amatiste, S., Arrigoni, N., Bianchi, M., Bilei, S., Cascone, G., Comin, D., Daminelli, P., Decastelli, L., Fustini, M., Mion, R., Petruzzelli, A., Rosmini, R., Rugna, G., Tamba, M., Tonucci, F., Bolzoni, G., 2013. Four-year monitoring of foodborne pathogens in raw milk sold by vending machines in Italy. J. Food Prot. 76 (11), 1902–1907.

Gomes, F., Henriques, M., 2016. Control of bovine mastitis: old and recent therapeutic approaches. Curr. Microbiol. 72 (4), 377–382.

Gourama, H., Bullerman, L.B., 1995. *Aspergillus flavus* and *Aspergillus parasiticus*: Aflatoxigenic fungi of concern in foods and feeds: a review. J. Food Prot. 58 (12), 1395–1404.

Granato, D., Branco, G.F., Cruz, A.G., Faria, J.D.A.F., Shah, N.P., 2010. Probiotic dairy products as functional foods. Compr. Rev. Food Sci. Food Saf. 9 (5), 455–470.

Grażyna, C., Hanna, C., Adam, A., Magdalena, B.M., 2017. Natural antioxidants in milk and dairy products. Int. J. Dairy Technol. 70 (2), 165–178.

Guh, A., Phan, Q., Randall, N., Purviance, K., Milardo, E., Kinney, S., Mshar, P., Kasacek, W., Cartter, M., 2010. Outbreak of *Escherichia coli* O157 associated with raw milk, Connecticut, 2008. Clin. Infect. Dis. 51 (12), 1411–1417.

Haran, K.P., Godden, S.M., Boxrud, D., Jawahir, S., Bender, J.B., Sreevatsan, S., 2012. Prevalence and characterization of *Staphylococcus aureus*, including methicillin-resistant *Staphylococcus aureus*, isolated from bulk tank milk from Minnesota dairy farms. J. Clin. Microbiol. 50 (3), 688–695.

Holzmann, H., Aberle, S.W., Stiasny, K., Werner, P., Mischak, A., Zainer, B., Netzer, M., Koppi, S., Bechter, A., Heinz, F.X., 2009. Tick-borne encephalitis from eating goat cheese in a mountain region of Austria. Emerg. Infect. Dis. 15 (10), 1671–1673.

Hurley, W.L., Theil, P.K., 2011. Perspectives on immunoglobulins in colostrum and milk. Nutrients 3 (4), 442–474.

Ijaz, N., 2014. Canada's 'other' illegal white substance: evidence, economics and raw milk policy. Health Law Rev. 22 (1), 26–39.

Imperiale, F., Sallovitz, J., Lifschitz, A., Lanusse, C., 2002. Determination of ivermectin and moxidecin residues in bovine milk and examination of the effects of these residues on acid fermentation of milk. Food Addit. Contam. 19 (9), 810–818.

International Agency Research of Cancer-IARC, 1993. Some naturally occurring substances: food items and constituents, heterocyclic aromatic amines and mycotoxins. In: IARC Monographs on the Evaluation of Carcinogenic Risks to Humans, Lyon. vol. 56. IARC Press, pp. 489–521.

Jamali, H., Radmehr, B., Thong, K.L., 2013. Prevalence, characterisation, and antimicrobial resistance of *Listeria* species and *Listeria monocytogenes* isolates from raw milk in farm bulk tanks. Food Control 34 (1), 121–125.

Jamali, H., Paydar, M., Radmehr, B., Ismail, S., Dadrasnia, A., 2015. Prevalence and antimicrobial resistance of *Staphylococcus aureus* isolated from raw milk and dairy products. Food Control 54, 383–388.

Jay, J.M., Loessner, M.J., Golden, D.A. (Eds.), 2012. Modern Food Microbiology. Springer, New York.

Jayarao, B.M., Donaldson, S.C., Straley, B.A., Sawant, A.A., Hegde, N.V., Brown, J.L., 2006. A survey of foodborne pathogens in bulk tank milk and raw milk consumption among farm families in Pennsylvania. J. Dairy Sci. 89 (7), 2451–2458.

Jensen, R.G., 1995. Handbook of Milk Composition. Academic Press, New York.

Jensen, R.G., Kroger, M., 2006. The importance of milk and milk products in the diet. In: Miller, G.D., Jarvis, J.K., McBean, L.D. (Eds.), Handbook of Dairy Foods and Nutrition. vol. 3. CRC Press, Boca Raton, FL, pp. 1–53.

Johler, S., Weder, D., Bridy, C., Huguenin, M.C., Robert, L., Hummerjohann, J., Stephan, R., 2015. Outbreak of staphylococcal food poisoning among children and staff at a Swiss boarding school due to soft cheese made from raw milk. J. Dairy Sci. 98 (5), 2944–2948.

Jordan, K., Hunt, K., Dalmasso, M., 2016. *Listeria monocytogenes* in milk products. In: Garg, N., Abdel-Aziz, S., Aeron, A. (Eds.), Microbes in Food and Health. Springer, New York, pp. 289–315.

Kabir, J., Umoh, V.J., Audu-Okoh, E., Umoh, J.U., Kwaga, J.K.P., 2004. Veterinary drug use in poultry farms and determination of antimicrobial drug residues in commercial eggs and slaughtered chicken in Kaduna State, Nigeria. Food Control 15 (2), 99–105.

Kampire, E., Kiremire, B.T., Nyanzi, S.A., Kishimba, M., 2011. Organochlorine pesticide in fresh and pasteurized cow's milk from Kampala markets. Chemosphere 84 (7), 923–927.

Keith, J.N., Nicholls, J., Reed, A., Kafer, K., 2011. The prevalence of self-reported lactose intolerance and the consumption of dairy foods among African American adults are less than expected. J. Natl. Med. Assoc. 103 (1), 36–45.

Keyvan, E., Yurdakul, O., Demirtas, A., Yalcin, H., Bilgen, N., 2020. Identification of methicillin-resistant *Staphylococcus aureus* in bulk tank milk. Food Sci. Technol. 40 (1), 150–156.

Khaniki, G.R.J., 2007. Chemical contaminants in milk and public health concerns: a review. Int. J. Dairy Sci. 2 (2), 104–115.

Kiessling, K.H., Pettersson, H., Sandholm, K., Olsen, M., 1984. Metabolism of aflatoxin, ochratoxin, zearalenone, and three trichothecenes by intact rumen fluid, rumen protozoa, and rumen bacteria. Appl. Environ. Microbiol. 47 (5), 1070–1073.

Korlath, J.A., Osterholm, M.T., Judy, L.A., Forfang, J.C., Robinson, R.A., 1985. A point-source outbreak of campylobacteriosis associated with consumption of raw milk. Int. J. Infect. Dis. 152 (3), 592–596.

Kotinagu, K., Krishnaiah, N., 2015. Organochlorine and organophosphorus pesticide residues in fodder and milk samples along Musi river belt, India. Vet. World 8 (4), 545.

Kumar, B.V., Vijayendra, S.V.N., Reddy, O.V.S., 2015. Trends in dairy and non-dairy probiotic products-a review. J. Food Sci. Technol. 52 (10), 6112–6124.

Kussendrager, K.D., Van Hooijdonk, A.C.M., 2000. Lactoperoxidase: physico-chemical properties, occurrence, mechanism of action and applications. Br. J. Nutr. 84 (1), 19–25.

Latorre, A.A., Van Kessel, J.S., Karns, J.S., Zurakowski, M.J., Pradhan, A.K., Boor, K.J., Schukken, Y.H., et al., 2010. Biofilm in milking equipment on a dairy farm as a potential source of bulk tank milk contamination with Listeria monocytogenes. J. Dairy Sci. 93 (6), 2792–2802.

Latorre, G., Besa, P., Parodi, C.G., Ferrer, V., Azocar, L., Quirola, M., Chianale, J., 2014. Prevalence of lactose intolerance in Chile: a double-blind placebo study. Digestion 90 (1), 18–26.

Laurent, M.R., Bravenboer, N., Van Schoor, N.M., Bouillon, R., Pettifor, J.M., Lips, P., 2018. Rickets and osteomalacia. In: Hoboken, N.J. (Ed.), Primer on the Metabolic Bone Diseases and Disorders of Mineral Metabolism. Wiley, pp. 684–694.

Le Loir, Y., Baron, F., Gautier, M., 2003. *Staphylococcus aureus* and food poisoning. Genet. Mol. Res. 2 (1), 63–76.

Leite, C.Q.F., Anno, I.S., Leite, S.R., Roxo, E., Morlock, G.P., Cooksey, R.C., 2003. Isolation and identification of mycobacteria from livestock specimens and milk obtained in Brazil. Mem. Inst. Oswaldo Cruz 98 (3), 319–323.

Leong, K.N., Chow, T.S., Wong, P.S., Hamzah, S.H., Ahmad, N., Ch'ng, C.C., 2015. Outbreak of human brucellosis from consumption of raw goats' milk in Penang, Malaysia. Am. J. Trop. Med. Hyg. 93 (3), 539–541.

Lopez-Exposito, I., Recio, I., 2008. Protective effect of milk peptides: antibacterial and antitumor properties. In: Park, Y.W. (Ed.), Bioactive Components of Milk. Springer, New York, pp. 271–294.

Loss, G., Apprich, S., Waser, M., Kneifel, W., Genuneit, J., Büchele, G., van Neerven, R.J., 2011. The protective effect of farm milk consumption on childhood asthma and atopy: the GABRIELA study. J. Allergy Clin. Immunol. 128 (4), 766–773.

Lucey, J.A., 2015. Raw milk consumption: risks and benefits. Nutr. Today 50 (4), 189–193.

MacDonald, L.E., Brett, J., Kelton, D., Majowicz, S.E., Snedeker, K., Sargeant, J.M., 2011. A systematic review and meta-analysis of the effects of pasteurization on milk vitamins, and evidence for raw milk consumption and other health-related outcomes. J. Food Prot. 74 (11), 1814–1832.

MacGibbon, A.K.H., Taylor, M.W., 2006. Composition and structure of bovine milk lipids. In: Fox, P.F., McSweeney, P.L.H. (Eds.), Advanced Dairy Chemistry Volume 2 Lipids. Springer, Boston, MA, pp. 1–42.

Madsen, C., 1997. Prevalence of food allergy/intolerance in Europe. Environ. Toxicol. Pharmacol. 4 (1–2), 163–167.

Mansour, A.S., Wagih, G.E.S., Morgan, S.D., Elhariri, M., El-Shabrawy, M.A., Abuelnaga, A.S., Elgabry, E.A., 2017. Detection of *Staphylococcus aureus* enterotoxigenic strains in bovine raw milk by reversed passive latex agglutination and multiplex polymerase chain reaction. Vet. World 10 (8), 843–847.

Markovinović, L., Ličina Kosanović, M.L., Tešić, V., Vojvodić, D., Vladušić Lucić, I., Kniewald, T., Vukas, T., Kutleša, M.,

Krajinović, L.C., 2016. An outbreak of tick-borne encephalitis associated with raw goat milk and cheese consumption, Croatia, 2015. Infection 44 (5), 661–665.

Matraszek-Zuchowska, I., Wozniak, B., Posyniak, A., 2017. Determination of hormones residues in milk by gas chromatography-mass spectrometry. Food Anal. Methods 10 (3), 727–739.

McCarthy, R.J., Ross, R.P., Fitzgerald, G.F., Stanton, C., 2015. The immunological consequences of pasteurisation: comparison of the response of human intestinally-derived cells to raw versus pasteurised milk. Int. Dairy J. 40, 67–72.

McFarland, L.V., 2015. From yaks to yogurt: the history, development, and current use of probiotics. Clin. Infect. Dis. 60, 85–90.

McLauchlin, J., Mitchell, R., Smerdon, W.J., Jewell, K., 2014. *Listeria monocytogenes* and listeriosis: a review of hazard characterisation for use in microbiological risk assessment of foods. Int. J. Food Microbiol. 92 (1), 15–33.

Melgar, M.J., Santaeufemia, M., Garcia, M.A., 2010. Organophosphorus pesticide residues in raw milk and infant formulas from Spanish northwest. J. Environ. Sci. Health B 45 (7), 595–600.

Melini, F., Melini, V., Luziatelli, F., Ruzzi, M., 2017. Raw and heat-treated milk: from public health risks to nutritional quality. Beverages 3 (4), 54.

Miller, D.M., Wilson, D.M., 1994. Veterinary diseases related to aflatoxins. In: Eaton, D., Groopman, J. (Eds.), The Toxicology of Aflatoxins. Academic Press, pp. 347–364.

Müller, B., Dürr, S., Alonso, S., Hattendorf, J., Laisse, C.J., Parsons, S.D., Zinsstag, J., 2013. Zoonotic *Mycobacterium bovis*–induced tuberculosis in humans. Emerg. Infect. Dis. 19 (6), 899–908.

Mummah, S., Oelrich, B., Hope, J., Vu, Q., Gardner, C.D., 2014. Effect of raw milk on lactose intolerance: a randomized controlled pilot study. Ann. Fam. Med. 12 (2), 134–141.

Nag, S.K., 2010. Contaminants in milk: routes of contamination, analytical techniques and methods of control. In: Griffiths, M.W. (Ed.), Improving the Safety and Quality of Milk Volume 1: Milk Production and Processing. Woodhead Publishing, Cambridge, pp. 146–178.

Ng, J.S.Y., Eastwood, K., Walker, B., Durrheim, D.N., Massey, P.D., Porigneaux, P., Kemp, R., McKinnon, B., Laurie, K., Miller, D., Bramley, E., Ryan, U., 2012. Evidence of Cryptosporidium transmission between cattle and humans in northern New South Wales. Exp. Parasitol. 130 (4), 437–441.

Ning, P., Guo, M., Guo, K., Xu, L., Ren, M., Cheng, Y., Zhang, Y., 2013. Identification and effect decomposition of risk factors for *Brucella* contamination of raw whole milk in China. PLoS One 8 (7), e68230.

O'Callaghan, T.F., Sugrue, I., Hill, C., Ross, R.P., Stanton, C., 2019. Nutritional aspects of raw milk: a beneficial or hazardous food choice. In: Nero, L.A., De Carvalho, A.F. (Eds.), Raw Milk: Balance Between Hazards and Benefits. Academic Press, pp. 127–148.

O'Mahony, M., Fanning, S., Whyte, P., 2009. The safety of raw liquid milk. In: Tamime, A.Y. (Ed.), Milk Processing and Quality Management. Blackwell Publishing, pp. 139–167.

Obaidat, M.M., Stringer, A.P., 2019. Prevalence, molecular characterization, and antimicrobial resistance profiles of *Listeria monocytogenes, Salmonella enterica,* and *Escherichia coli* O157:H7 on dairy cattle farms in Jordan. J. Dairy Sci. 102 (10), 8710–8720.

Ohlsson, J.A., Johansson, M., Hansson, H., Abrahamson, A., Byberg, L., Smedman, A., Lundh, Å., 2017. Lactose, glucose and galactose content in milk, fermented milk and lactose-free milk products. Int. Dairy J. 73, 151–154.

Oliver, S.P., Jayarao, B.M., Almeida, R.A., 2005. Foodborne pathogens in milk and the dairy farm environment: food safety and public health implications. Foodborne Pathog. Dis. 2 (2), 115–129.

Oliver, S.P., Boor, K.J., Murphy, S.C., Murinda, S.E., 2009. Food safety hazards associated with consumption of raw milk. Foodborne Pathog. Dis. 6 (7), 793–806.

Painter, J.A., Hoekstra, R.M., Ayers, T., Tauxe, R.V., Braden, C.R., Angulo, F.J., Griffin, P.M., 2013. Attribution of foodborne illnesses, hospitalizations, and deaths to food commodities by using outbreak data, United States, 1998–2008. Emerg. Infect. Dis. 19, 407–415.

Park, Y.W., Marnet, P.G., Yart, L., Haenlein, G.F., 2013. Mammary secretion and lactation. In: Park, Y.W., Haenlein, G.F.W. (Eds.), Milk and Dairy Products in Human Nutrition: Production, Composition and Health. Wiley, Blackwell, pp. 31–45.

Pärn, T., Hallanvuo, S., Salmenlinna, S., Pihlajasaari, A., Heikkinen, S., Telkki-Nykänen, H., Ollgren, J., Huusko, S., Rimhanen-Finne, R., 2015. Outbreak of Yersinia pseudotuberculosis O:1 infection associated with raw milk consumption, Finland, spring 2014. Euro Surveill. 20 (40), 30033.

Parodi, P.W., 2004. Milk fat in human nutrition. Aust. J. Dairy Technol. 59, 3–59.

Peh, H.Y., Tan, W.D., Liao, W., Wong, W.F., 2016. Vitamin E therapy beyond cancer: tocopherol versus tocotrienol. Pharmacol. Ther. 162, 152–169.

Perin, L.M., Pereira, J.G., Bersot, L.S., Nero, L.A., 2019. The microbiology of raw milk. In: Nero, L.A., De Carvalho, A.F. (Eds.), Raw Milk: Balance Between Hazards and Benefits. Academic Press, pp. 45–64.

Perkin, M.R., Strachan, D.P., 2006. Which aspects of the farming lifestyle explain the inverse association with childhood allergy? J. Allergy Clin. Immunol. 117 (6), 1374–1381.

Pihlanto, A., Korhonen, H., 2003. Bioactive peptides and proteins. Adv. Food Nutr. Res. 47, 175–276.

Platts-Mills, J.A., Kosek, M., 2014. Update on the burden of *Campylobacter* in developing countries. Curr. Opin. Infect. Dis. 27 (5), 444.

Porcellato, D., Aspholm, M., Skeie, S.B., Monshaugen, M., Brendehaug, J., Mellegård, H., 2018. Microbial diversity of consumption milk during processing and storage. Int. J. Food Microbiol. 266, 21–30.

Prado, F.C., Parada, J.L., Pandey, A., Soccol, C.R., 2008. Trends in non-dairy probiotic beverages. Food Res. Int. 41 (2), 111–123.

Prandini, A., Tansini, G., Sigolo, S., Filippi, L., Laporta, M., Piva, G., 2009. On the occurrence of aflatoxin M1 in milk and dairy products. Food Chem. Toxicol. 47 (5), 984–991.

Quigley, L., O'Sullivan, O., Beresford, T.P., Ross, R.P., Fitzgerald, G.F., Cotter, P.D., 2011. Molecular approaches to analysing the microbial composition of raw milk and raw milk cheese. Int. J. Food Microbiol. 150 (2–3), 81–94.

Quigley, L., O'Sullivan, O., Stanton, C., Beresford, T.P., Ross, R.P., Fitzgerald, G.F., Cotter, P.D., 2013. The complex microbiota of raw milk. FEMS Microbiol. Rev. 37 (5), 664–698.

Rafiq, S., Huma, N., Pasha, I., Sameen, A., Mukhtar, O., Khan, M.I., 2016. Chemical composition, nitrogen fractions and amino acids profile of milk from different animal species. Asian Australas. J. Anim. Sci. 29 (7), 1022–1028.

Rahimi, E., Ameri, M., Momtaz, H., 2010. Prevalence and antimicrobial resistance of *Listeria* species isolated from milk and dairy products in Iran. Food Control 21 (11), 1448–1452.

Ramos, J.M., Bernal, E., Esguevillas, T., Lopez-Garcia, P., Gaztambide, M.S., Gutierrez, F., 2008. Non-imported brucellosis outbreak from unpasteurized raw milk in Moroccan immigrants in Spain. Epidemiol. Infect. 136 (11), 1552–1555.

Rana, M.S., Lee, S.Y., Kang, H.J., Hur, S.J., 2019. Reducing veterinary drug residues in animal products: a review. Food Sci. Anim. Resour. 39 (5), 687–703.

Riedler, J., Braun-Fahrländer, C., Eder, W., Schreuer, M., Waser, M., Maisch, S., Carr, D., Schierl, R., Nowak, D., von Mutius, E., 2001. The ALEX Study Team: exposure to farming in early life and development of asthma and allergy: a cross-sectional survey. Lancet 358 (9288), 1129–1133.

Ringot, D., Chango, A., Schneider, Y.J., Larondelle, Y., 2006. Toxicokinetics and toxicodynamics of ochratoxin A, an update. Chem. Biol. Interact. 159 (1), 18–46.

Rios, E.A., Santos, J., García-Meniño, I., Flament-Simon, S.C., Blanco, J., García-López, M.L., Rodríguez-Calleja, J.M., 2019. Characterisation, antimicrobial resistance and diversity of atypical EPEC and STEC isolated from cow's milk, cheese and dairy cattle farm environments. LWT-Food Sci. Technol. 108, 319–325.

Rosenthal, M., Pedersen, R., Leibsle, S., Hill, V., Carter, K., Roellig, D.M., 2015. Cryptosporidiosis associated with consumption of unpasteurized goat milk—Idaho, 2014. MMWR 64 (7), 194–195.

Saad, N.M., Hussein, A.A., Ewida, R.M., 2018. Occurrence of *Toxoplasma gondii* in raw goat, sheep, and camel milk in Upper Egypt. Vet. World 11 (9), 1262–1265.

Salomskiene, J., Macioniene, I., Zvirdauskiene, R., Jonkuviene, D., 2013. Impact of the residues of detergents and disinfectants used in dairy farms on the results of inhibitor tests for raw milk. Adv. Biosci. Biotechnol. 4, 266–272.

Sanchez de la Campa, A.M., De la Rosa, J.D., Sánchez-Rodas, D., Oliveira, V., Alastuey, A., Querol, X., Ariza, J.G., 2008. Arsenic speciation study of PM2. 5 in an urban area near a copper smelter. Atmos. Environ. 42 (26), 6487–6495.

Sanchez, L., Calvo, M., Brock, J.H., 1992. Biological role of lactoferrin. Arch. Dis. Child. 67 (5), 657–661.

Sant'Ana, A., Rosenthal, A., Massaguer, P.R., 2009. Heat resistance and the effects of continuous pasteurization on the inactivation of Byssochlamys fulva ascospores in clarified apple juice. J. Appl. Microbiol. 107 (1), 197–209.

Sarowska, J., Choroszy-Król, I., Regulska-Ilow, B., Frej-Madrzak, M., Jama-Kmiecik, A., 2013. The therapeutic effect of probiotic bacteria on gastrointestinal diseases. Adv. Clin. Exp. Med. 22 (5), 759–766.

Sawant, A.A., Sordillo, L.M., Jayarao, B.M., 2005. A survey on antibiotic usage in dairy herds in Pennsylvania. J. Dairy Sci. 88 (8), 2991–2999.

Singhal, P., Kaushik, G., Hussain, C.M., Chel, A., 2020. Food safety issues associated with milk: a review. In: Grumezescu, A.M., Holban, A.M. (Eds.), Safety Issues in Beverage Production. Academic Press, pp. 400–401.

Sozańska, B., Pearce, N., Dudek, K., Cullinan, P., 2013. Consumption of unpasteurized milk and its effects on atopy and asthma in children and adult inhabitants in rural Poland. Allergy 68 (5), 644–650.

Stelwagen, K., Carpenter, E., Haigh, B., Hodgkinson, A., Wheeler, T.T., 2009. Immune components of bovine colostrum and milk. J. Anim. Sci. 87 (Suppl. 13), 3–9.

Stolker, A.A.M., Rutgers, P., Oosterink, E., Lasaroms, J.J.P., Peters, R.J.B., Van Rhijn, J.A., Nielen, M.W.F., 2008. Comprehensive screening and quantification of veterinary drugs in milk using UPLC–ToF-MS. Anal. Bioanal. Chem. 391 (6), 2309–2322.

Sugrue, I., Tobin, C., Ross, R.P., Stanton, C., Hill, C., 2019. Foodborne pathogens and zoonotic diseases. In: Nero, L.A., De Carvalho, A.F. (Eds.), Raw Milk: Balance Between Hazards and Benefits. Academic Press, pp. 259–272.

Swagerty Jr., D.L., Walling, A., Klein, R.M., 2002. Lactose intolerance. Am. Fam. Physician 65 (9), 1845–1850.

Tamime, A.Y., Saarela, M.A.K.S., Sondergaard, A.K., Mistry, V.V., Shah, N.P., 2005. Production and maintenance of viability of probiotic microorganisms in dairy products. Probiotic Dairy Prod. 1, 39–63.

Tan, X.T., Li, Z.M., Deng, L.G., Zhao, S.C., Wang, M.L., 2016. Analysis of 13 kinds of steroid hormones in raw milk using modified QuEChERS method combined with UPLC-QTOF-MS. J. Integr. Agric. 15 (9), 2163–2174.

Tate, P.L., Bibb, R., Larcom, L.L., 2011. Milk stimulates growth of prostate cancer cells in culture. Nutr. Cancer 63 (8), 1361–1366.

Tormo, H., Lekhal, D.A.H., Roques, C., 2015. Phenotypic and genotypic characterization of lactic acid bacteria isolated from raw goat milk and effect of farming practices on the dominant species of lactic acid bacteria. Int. J. Food Microbiol. 210, 9–15.

Turkoglu, C., Keyvan, E., 2019. Determination of aflatoxin M_1 and ochratoxin A in raw, pasteurized and UHT milk in Turkey. Acta Sci. Vet. 47, 1626.

Van Asselt, E.D., van der Fels-Klerx, H.J., HJP, M., Van Bokhorst-van de Veen, H., Groot, M.N., 2017. Overview of food safety hazards in the European dairy supply chain. Compr. Rev. Food Sci. Food Saf. 16 (1), 59–75.

Van, T.T.H., Yidana, Z., Smooker, P.M., Coloe, P.J., 2020. Antibiotic use in food animals in the world with focus on Africa: pluses and minuses. J. Glob. Antimicrob. Resist. 20, 170–177.

Veldman, A., Meijs, J.A.C., Borggreve, G.J., Heeres-Van der Tol, J.J., 1992. Carry-over of aflatoxin from cows' food to milk. J. Anim. Sci. 55 (2), 163–168.

Verraes, C., Claeys, W., Cardoen, S., Daube, G., De Zutter, L., Imberechts, H., Dierick, K., Herman, L., 2014. A review of the microbiological hazards of raw milk from animal species other than cows. Int. Dairy J. 39 (1), 121–130.

Walstra, P., Wouters, J.T.M., Geurts, T.J. (Eds.), 2006. Dairy Science and Technology. Vol. 2. CRC Press, New York, pp. 18–108.

Wang, H., Ren, L., Yu, X., Hu, J., Chen, Y., He, G., Jiang, Q., 2017. Antibiotic residues in meat, milk and aquatic products in Shanghai and human exposure assessment. Food Control 80, 217–225.

Wang, B., Timilsena, Y.P., Blanch, E., Adhikari, B., 2019. Lactoferrin: structure, function, denaturation and digestion. Crit. Rev. Food Sci. Nutr. 59 (4), 580–596.

Waser, M., Michels, K.B., Bieli, C., Flöistrup, H., Pershagen, G., Von Mutius, E., Ege, M., Riedler, J., Schram-Bijkerk, D., Brunekreef, B., Van Hage, M., Lauener, R., Braun-Fahrländer, C., the PARSIFAL Study team, 2007. Inverse association of farm milk consumption with asthma and allergy in rural and suburban populations across Europe. Clin. Exp. Allergy 37 (5), 661–670.

Wedholm, A., Hallén, E., Bach Larsen, L., Lindmark-Månsson, H., Hans Karlsson, A., Allmere, T., 2006. Comparison of milk protein composition in a Swedish and a Danish dairy herd using reversed phase HPLC. Acta Agric. Scand. Sect. A Anim. Sci. 56 (1), 8–15.

Wesley, I.V., Wells, S.J., Harmon, K.M., Green, A., Schroeder-Tucker, L., Glover, M., Siddique, I., 2000. Fecal shedding of Campylobacter and Arcobacter spp. in dairy cattle. Appl. Environ. Microbiol. 66 (5), 1994–2000.

World Health Organisation (WHO), 2015. Estimates of the Global Burden of Foodborne Diseases. Foodborne Diseases Burden Epidemiology Reference Group 2007–2015. https://www.who.int/foodsafety/areas_work/foodborne-diseases/ferg/en/. (Accessed 15 April 2020).

World Health Organization (WHO), 2007. Health risks of heavy metals from long-range transboundary air pollution. In: Joint WHO/Convention Task Force on the Health Aspects of Air Pollution. Germany.

Xu, H., Huang, W., Hou, Q., Kwok, L.Y., Sun, Z., Ma, H., Zhang, H., 2017. The effects of probiotics administration on the milk production, milk components and fecal bacteria microbiota of dairy cows. Sci. Bull. 62 (11), 767–774.

Yang, Y., He, M., Cui, H., Bian, L., Wang, Z., 2000. The prevalence of lactase deficiency and lactose intolerance in Chinese children of different ages. Chin. Med. J. (Engl.) 113 (12), 1129–1132.

Yiannikouris, A., Jouany, J.P., 2002. Mycotoxins in feeds and their fate in animals: a review. Anim. Res. 51 (2), 81–99.

CHAPTER 7

Nonbovine Milk Products as Probiotic and Prebiotic Food

LUCIA FIGLIOLA • ANTONELLA SANTILLO • MARIA GIOVANNA CILIBERTI • MARIANGELA CAROPRESE • MARZIA ALBENZIO
Department of Agriculture, Food, Natural Science and Engineering, University of Foggia, Foggia, Italy

NONBOVINE MILK ALTERNATIVES AS FUNCTIONAL FOOD

The relationship between nutrition and human health has gained great attention as scientific research on biotechnology, animal production, nutrigenomics, and food science has added fundamental information on the potential bioactivity of foods that were traditionally consumed only for their nutritional properties (Turkmen et al., 2019). Hasler (2002) defined functional foods as whole, fortified, enriched, or enhanced foods or food compounds that exert positive health benefits on the human body. The main functional foods in the market are carotenoids, dietary fibers, fatty acids, minerals, prebiotics, probiotics, symbiotics, vitamins, and minerals (Turkmen et al., 2019).

Milk is the first food consumed by newborns in each mammalian species and it represents a food with a high nutritional value; it is generally consumed by humans throughout their lives in the form of liquid milk or dairy products. World milk production is mainly led by cow milk, representing about 81% of total production, followed by buffalo milk (15%), goat milk (2%), ewe milk (1.30%), and camel milk (0.4%). The other animal species with minor relevance contribute only about 0.3% (FAOSTAT, 2018).

Although milk from minor species is used less than that from cows, there is a growing interest in goat, ewe, buffalo, camel, mare, and donkey milks for their bioactive components with functional immunological, anticarcinogenic, antimicrobial, and neurological properties (Chiofalo et al., 2004) that define these as functional foods (Turkmen et al., 2019). However, the gross composition of milk is well characterized for the principal and minor dairy species, as shown in Table 1.

Milk is a suitable carrier for probiotics and prebiotics. Indeed, fermented dairy products mainly made using cow milk such as yogurt, fermented sour milk, and cheese remain the products at the forefront of probiotic food development at present (Ranadheera et al., 2019) and account for more than 40% of the functional food market (Turkmen et al., 2019). In this scenario, dairy products from nonbovine milk have been proposed as potentially useful probiotic carrier foods for human nutrition rather than cow milk due to their peculiar therapeutic and functional characteristics (Turkmen et al., 2019).

The fermentation process in milk can happen spontaneously or by the addition of specific starter cultures (Jatmiko et al., 2019). Natural fermentation was used as a traditional method to preserve nutritional characteristics due to the lowering of pH performed by the raw milk microflora; later, the use of selected bacterial cultures led to standardized production processes and the ameliorated quality of the dairy products. Nowadays, improved industrial production has permitted the diversification of fermented milk products thanks to the addition of probiotics, prebiotics, and additive agents to enhance the organoleptic properties of fresh fruits or even different milk types. Milk and dairy products are the most suitable matrices as probiotic carriers because they are characterized by an environment with the high nutritive value required for wild-type bacterial growth, especially lactic acid bacteria (LAB) (Jatmiko et al., 2019). Every type of milk, each with its own typical nutritional composition, could represent the best vehicle for specific probiotics and prebiotics. Probiotic cultures also yield a complex outfit of proteolytic and lipolytic enzymes able to rule the production of molecules such as essential amino acids, bioactive peptides, polyunsaturated fatty acids, and conjugated linoleic acids (CLA) (Albenzio et al., 2013a) during fermentation or ripening.

Yogurt is defined as fermented milk obtained by lactic acid fermentation due to the presence of *Lactobacillus delbrueckii* ssp. *bulgaricus* and *Streptococcus salivarius* ssp. *thermophilus* in milk. Yogurt is considered

Probiotics and Prebiotics in Foods. https://doi.org/10.1016/B978-0-12-819662-5.00007-0

TABLE 1
Gross Milk Composition of Principal Mammal Species.

Milking	COMPOSITION (%)								
Species	Dry Matter	Protein	Casein	Whey Protein	Fat	Lactose	Ash	Energy (kJ/kg)	References
Bovine	10–13	2.54–4.19	2.46–2.80	0.55–0.70	3.23–6.40	4.40–5.60	0.7–0.8	2709–2843	Gantner et al. (2015), Claeys et al. (2014), Barłowska et al. (2011)
Ewe	18–20	3.35–6.60	4.20–4.60	1.02–1.10	3.35–9.00	3.35–6.60	0.8–1.0	4038–4439	Gantner et al. (2015), Claeys et al. (2014), Barłowska et al. (2011)
Goat	12–16	2.38–5.20	2.33–4.63	0.37–0.70	2.38–8.16	2.38–5.00	0.7–0.9	2802–2900	Gantner et al. (2015), Barłowska et al. (2011), Claeys et al. (2014), Albenzio et al. (2016)
Buffalo	16–17	2.70–6.29	3.20–4.00	0.38–0.60	3.44–11.00	3.20–6.29	0.8–0.9	4244–4779	Gantner et al. (2015), Barłowska et al. (2011), Claeys et al. (2014)
Donkey	8–12	1.40–2.00	0.64–1.03	0.49–0.80	0.28–1.82	5.87–7.40	0.3–0.5	1607–1803	Gantner et al. (2015), Barłowska et al. (2011), Claeys et al. (2014)
Mare	9–12	1.30–2.00	0.94–1.36	0.74–0.91	0.3–7.2	5.60–7.20	0.3–0.5	1936–2050	Gantner et al. (2015), Barłowska et al. (2011), Claeys et al. (2014)
Camel	10–11	2.06–5.23	2.21–2.60	0.59–0.81	2.35–6.67	2.77–5.85	0.7–0.9	4100–3286	Gantner et al. (2015), Korlepara et al. (2017), Barłowska et al. (2011), Claeys et al. (2014)

a functional food. In fact, yogurt contains properties or components that can benefit human health (Ibeagha-Awemu et al., 2009). Typical yogurt cultures such as *S. thermophilus* and *L. delbrueckii* ssp. *bulgaricus* are claimed to offer some health benefits, but they are not natural inhabitants of the intestine. Currently, cultures for health benefits are being introduced in the yogurt industry. *Lactobacillus acidophilus*, *Lactobacillus casei*, and *Bifidobacterium* have been used significantly in various countries. New fermented products containing *L. acidophilus*, *Bifidobacterium* spp., *L. casei Shirota*, *Lactobacillus gasseri*, *Lactobacillus rhamnosus* GG, and *Lactobacillus reuteri* have been developed in Europe and Asia. However, *L. acidophilus*, *L. casei*, and *Bifidobacterium* spp. are the most commonly used as probiotics (Ibeagha-Awemu et al., 2009).

As reported by Serafeimidou et al. (2013), Greek yogurt made from ewe milk has between 0.405 and 1.250 g CLA/100 g fat. Moreover, the cis-9 and trans-11 CLA accounts for approximately 78% of the polyunsaturated fatty acids and 2% of the total fatty acids in ewe milk yogurt (Balthazar et al., 2016). Balthazar et al. (2016) showed that the most predominant fatty acids in ewe milk yogurt are oleic acid (C18:1n9), palmitic acid (C16:0), and myristic acid (C14:0), respectively. Human diets high in oleic acid are reported to decrease the level of LDL cholesterol, whereas HDL cholesterol levels are not significantly affected (Molkentin, 2000; Balthazar et al., 2017a).

Among dairy products, cheese represents a good alternative for the delivery of probiotic bacteria into the human gut, and it has been the subject of various marketing and research studies in recent years (Gomes da Cruz et al., 2009). Cheeses show a great potential to transport probiotic microorganisms due to their naturally low acidity, high buffering capacity, nutrient density, and low-oxygen content that protect the probiotic bacteria during storage and gastrointestinal transit (Albenzio et al., 2013a; Vasconcelos et al., 2019). In the development of probiotic ovine cheese, the sensory evaluation by consumers is a crucial and essential step that rules product innovation (Santillo and Albenzio, 2015). Overall, the commercial success of any functional food, especially that containing probiotic strains, ultimately depends on the taste, appearance, price, and health claims that appeal to consumers (Ranadheera et al., 2018).

EWE MILK AND DAIRY PRODUCTS

Ewe milk is sporadically consumed in liquid form and mainly used for dairy production. So, milk quality is relevant for the production of high-quality products with high yields per liter of milk. When compared with other mammal milks, ewe milk shows a high nutritional value due to its higher concentrations of proteins, fats, vitamins, and minerals (Balthazar et al., 2017b), although it does not have as high a production yield as cow and goat milk due to productive seasonality (Albenzio et al., 2016).

The chemical composition of ewe milk, and in particular the lipid content, varies according to genetic factors (species and breed), the lactation stage, parity, season, environmental temperature, lactation efficiency, animal age and nutrition, and diseases of the udder (Claeys et al., 2014). Ewe milk has high concentrations of fat globules that are smaller than cow milk; these globules average a diameter of about 3.6 μm and 4.0 μm, respectively (Gantner et al., 2015). Agglutinin is absent in ewe milk, providing better digestibility compared to cow milk (Park et al., 2007). Ewe milk is rich in protein (almost twice as much protein as goat and cow milks) and in particular in casein (Balthazar et al., 2017a). Ewe milk casein micelles have higher mineralization degrees and are less hydrated and heat stable over cow milk (Raynal-Ljutovac et al., 2007). As reported by Masoodi and Shafi (2010), the α_{s1} and α_{s2} protein sequences of ewes diverge markedly from the corresponding proteins in cow milk; this suggests that ewe milk can lower allergic sensitivity. A strong casein genotype is significantly associated with higher fat, protein, and lactose contents and a higher cheese yield. In contrast, recessive genes provide lower α_{S1}-casein content and a softer curd (Albenzio et al., 2016).

Regarding the lipidic fraction, triacylglycerols in ewe milk have a higher percentage of medium-chain fatty acids (C26–C36) than cow milk and a lower proportion of long-chain fatty acids (C46–C54). Ewe milk fat has a higher content of saturated fatty acids (65–75 g/100 g of total fatty acids), very similar to cow milk; however, ewe milk contains more saturated fatty acids such as caproic (C6:0), caprylic (C8:0), and capric (C10:0) acids (Balthazar et al., 2017c). It also has low concentrations of butyric acid, which contributes to the in vitro inhibition of human cancer cell lines, although caproic, caprylic, and capric acids could reduce body weight and fat (Rasmussen et al., 2010; Foglietta et al., 2014). The greater digestibility of capric acid, due to the triacylglycerol position, permits the direct transfer of this fatty acid from the intestine to blood circulation as well as a rapid energy supply, which is especially useful for subjects with malnutrition or fat malabsorption syndrome (Recio et al., 2009; Balthazar et al., 2017c). Moreover, it may contribute to lower total circulating cholesterol, especially the low density lipoprotein (LDL) fraction (Raynal-Ljutovac et al., 2008). Among ruminants, ewe milk fat contains the highest levels of conjugated linoleic acid and a large amount of its physiological precursor of vaccenic acid (VA) (Balthazar et al., 2017c). It is well known that among polyunsaturated fatty acids, conjugated linoleic acids have some beneficial functional activity as well as anticarcinogenic and lipolytic activities (Luna et al., 2005; Park and Pariza, 2007; Gómez-Cortés et al., 2008). Regarding mineral elements, ewe milk is rich in calcium and phosphorus, fundamental elements for growth and bone maintenance that are essential for neonates (Al-Wabel, 2008). Calcium bound to casein is more available during the milk digestion process (Gueguen and Pointillart, 2000) and its bioavailability in ewe milk is strongly related to high casein levels. For this reason, the bioavailability of these minerals makes ewe milk a valuable source of these elements. Ewe milk is high in riboflavin compared to cow, goat, and buffalo milks and can also be considered a source of vitamin C (Wijesinha-Bettoni and Burlingame, 2013). In ewe milk, Minervini et al. (2003) found ACE inhibitory peptides derived from α_{S1}-casein fractions and α_{S2}-casein fractions produced by the microorganism *Lactobacillus helveticus* PR4. Gómez-Ruiz et al. (2008) showed antioxidant peptides derived from the hydrolysis of k-casein in ewe milk by enzymes such as pepsin, trypsin, and chymotrypsin. Ewe milk has a peculiar flavor, namely sheepy, derived from a specific fatty acid composition; it has a sweet aroma and a creamy texture due to dispersed small fat globules (Park et al., 2007). Texture and compositional features of sheep milk allow to obtain dairy products with denser consistence, that can give effective protection for probiotic bacteria during the gastrointestinal digestion and commercial storage (Balthazar et al., 2017c).

Principal probiotic and prebiotic dairy products made using ewe milk, detailed in the recent literature, are reported in Table 2.

Pasta filata cheese is produced by a thermizing and texturing process that occurs when the curd is dipped

TABLE 2
Dairy Products From Ewe Milk Characterized by the Addition of Probiotics and/or Prebiotics.

Products	Time of Ripening/ Storage	Probiotics	Prebiotics and Technologies	Viability of Probiotics in Trials	References
Scamorza cheese	15 days ripened	*Bifidobacterium longum* 46, *Bifidobacterium lactis* BB-12, *Lactobacillus acidophilus*	None	*L. acidophilus* 7.55 log cfu/g of cheese *B. longum* and *B. lactis* 9.9 log cfu/g of cheese	Albenzio et al. (2013a,b, 2015)
Ovine cheese	45 days ripened	Lamb rennet paste containing *Lactobacillus acidophilus*, and a mix of *Bifidobacterium lactis* and *Bifidobacterium longum*	None	*L. acidophilus* 7 log cfu/g of cheese and *B. longum* and *B. lactis* 7 log cfu/g of cheese	Santillo and Albenzio (2015)
Pecorino cheese	60 days ripened	Lamb rennet paste containing *Lactobacillus acidophilus*, and a mix of *Bifidobacterium lactis* and *Bifidobacterium longum*	None	*L. acidophilus* 8 log cfu/g of cheese and *B. longum* and *B. lactis* 9 log cfu/g of cheese	Santillo and Albenzio (2008), Santillo et al. (2009)
Pecorino cheese	30 days ripened	Lamb rennet paste containing *Lactobacillus acidophilus*, and a mix of *Bifidobacterium lactis* and *Bifidobacterium longum*	None	*L. acidophilus* 8 log cfu/g of cheese and *B. longum* and *B. lactis* 9 log cfu/g of cheese	Albenzio et al. (2010)
Pecorino cheese	1, 15, 30, 60, and 120 days ripened	Lamb rennet paste containing *Lactobacillus acidophilus*, and a mix of *Bifidobacterium lactis* and *Bifidobacterium longum*	Encapsulation	*L. acidophilus* 7.52 log cfu/g and *B. longum* and *B. lactis* 6.84 log cfu/g of cheese	Santillo et al. (2012)

TABLE 2
Dairy Products From Ewe Milk Characterized by the Addition of Probiotics and/or Prebiotics—cont'd

Products	Time of Ripening/ Storage	Probiotics	Prebiotics and Technologies	Viability of Probiotics in Trials	References
Canestrato Pugliesa hard cheese	56 days ripened	*Bifidobacterium bifidum* Bb02 and/or *Bifidobacterium. longum* Bb46	None	*Bifidobacterium bifidum* Bb02 > 6 log cfu/g	Corbo et al. (2001)
Argentinean cheese	15 days ripened	*Lactobacillus acidophilus* (LA-5), *Bifidobacterium lactis* (BB-12)	None	6 log cfu/g	Perotti et al. (2014)
Feta	1, 7, 15, 30, and 60 days ripened	*Propionibacterium freudenreichii* subsp. *shermanii*	None	9 log cfu/g	Angelopoulou et al. (2017)
Ice cream	1 day of storage	None	Inulin, fructo-oligosaccharide, galacto-oligosaccharide, short-chain fructo-oligosaccharide, resistant starch, corn dietary oligosaccharide, and polydextrose	–	Balthazar et al. (2017a,b)
Ice cream	150 days of storage	*Lactobacillus casei* 01	Inulin	7 log cfu/mL	Balthazar et al. (2018)
Ice cream and fermented beverage	1, 14, and 28 days of storage	*Lactobacillus casei* 431	None	12 log cfu/g	Nadelman et al. (2017)
Greek ewe milk yogurt	1, 2, 5, 10, 19, and 26 days of storage	*L. paracasei* subsp. *paracasei* DC412	None	–	Papadimitriou et al. (2007)
Yogurt (mix to cow milk)	28 days of storage	*Streptococcus thermophilus*, *Lactobacillus delbrueckii* spp. *bulgaricus*, and *Lactobacillus acidophilus*-5	None	*Streptococcus thermophilus* 12 log cfu/g, *Lactobacillus delbrueckii* spp. *bulgaricus* 8 log cfu/g, and *Lactobacillus acidophilus*-5.9 log cfu/g	Vianna et al. (2017)
Yogurt		None	Inulin	–	Balthazar et al. (2015)
Yogurt	0, 8, 16, 24, 32, and 40 days of storage	None	Inulin	–	Kiss et al. (2014)
Milk-juice beverage	0, 15, and 30 days of storage	*Lactobacillus plantarum*	Inulin	*Lactobacillus plantarum* over 9 log cfu/mL	Balthazar et al. (2019)

in hot water and worked mechanically to a plastic consistency, followed by molding into a different shape and size. Pasta filata cheeses encompass a wide range of cheeses such as mozzarella (fresh cheese), scamorza (short-ripened cheese), provolone, and caciocavallo (medium- and long-ripened cheese). These originated primarily in the northern Mediterranean region (Italy, Greece, the Balkans, and Turkey), and are generally obtained using cow milk, buffalo milk, or both. In Italy, the only referenced pasta filata cheese from raw ewe milk is Vastedda della Valle del Belice, recognized as Protected Designation of Origin by the European Union no. 283/28.10.2010 (Albenzio et al., 2013b). Short-ripened pasta filata cheese and traditional pecorino-type cheese (Albenzio et al., 2013a,b, 2015; Santillo and Albenzio, 2015) were produced using milk from the autochthonous ovine breed Gentile di Puglia and probiotic bacteria as a strategy to promote dairy ewe enterprises by product innovation. The survival of probiotic bacteria during the heat stress of pasta filata cheesemaking represents the main issue for the success of such product innovation. The authors proposed the incorporation of *L. acidophilus* and a mix of *Bifidobacterium longum* and *Bifidobacterium lactis* directly into lamb rennet paste as an approach that doesn't modify the traditional step in pasta filata cheese production. As concerns the sensorial attributes studied using a consumer panel, the innovative scamorza cheese resulted in higher structural uniformity, lower friability, and higher creaminess and graininess than the control cheese due to the presence of high levels of probiotic cells, in particular bifidobacteria, throughout the 2 weeks of ripening. Furthermore, the peptide profile of the pasta filata ewe cheese ripened for 15 days highlighted specific peptides derived from the presence and activity of probiotic bacteria. In particular, the presence of bioactive sequences previously reported in long-ripened cheeses demonstrated the ability of probiotics to control and enhance the proteolytic process and generate peptides in the cheese matrix that could be delivered upon cheese consumption. Two fragments derived from β-CN and αS1-CN were identified by Albenzio et al. (2015) with structural homology with previously described bioactive peptides in ovine milk and dairy products and in particular, the residue of peptides GPFPI and ENLLRF, which have been reported to have ACE-inhibitory properties. Two other fragments were identified: the first was derived from β-galactosidase associated with *B. longum* activity and the second was derived only from *B. longum* and *B. lactis* activity consisting of three peptide fragments of different origin, One fragment was derived from κ-CN hydrolysis, the second was of α_{S1}-CN origin, and the third coeluting fragment was identified as a sequence of endonuclease ascribed to *B. longum*. Some authors reported the presence of ACE-inhibitory peptides derived from some types of ewe milk cheese and Greek ewe milk yogurt (Papadimitriou et al., 2007; Meira et al., 2012; Politis and Theodorou, 2016). Rizzello et al. (2005) observed antibacterial peptides in Italian cheese from ewe milk.

Pecorino cheese is a typical product of the Mediterranean region, and such cheese has been characterized in relation to several factors associated with the initial milk quality, the presence of raw or selected microflora, the production protocol, and the ripening process.

Corbo et al. (2001) proposed that the survival of probiotic cultures is dependent on the strain used. Indeed, in Canestrato Pugliese ovine cheese, *Bifidobacterium bifidum* Bb02 leads to a concentration of up to 6 log cfu/g after 56 days of ripening at 12°C, whereas *B. longum* Bb46 survived for 35 days and was no longer detectable at the end of the ripening time.

Perotti et al. (2014) evaluated the survival of a mix of probiotic microorganisms (*L. acidophilus* La-5 and *Bifidobacterium animalis* ssp. *lactis* Bb12) in a semihard ovine cheese. Probiotic strains in cheese reached levels over 10^6 cfu/g; probiotic bacteria in cheese did not modify gross composition, nitrogen fractions, lipolysis, fatty acid profiles including conjugated linoleic acids (CLA) and volatile profile. Those results highlighting that semihard ovine cheese could be an adequate matrix to deliver dietary probiotics.

Different results were obtained in Pecorino cheese by Santillo and Albenzio (2015). The authors found that the addition of a mix of *B. longum* and *B. lactis* strains in pecorino cheese sustained greater proteolysis, leading to more complex soluble peptide and FAA profiles. The *L. acidophilus* strain was mainly involved in the lipolytic process and was able to significantly increase vaccenic and oleic acids, CLA, and short chain free fatty acids. The sensorial profile of the probiotic ovine cheese after 45 day of ripening was also influenced, showing lower humidity and gumminess and higher ratings for salty and pungent attributes compared to the control cheese (Santillo and Albenzio, 2015).

The ability of *L. acidophilus* to enrich the cheese matrix with a great amount of free fatty acids (FFA) and CLA was further confirmed in a study of probiotic Pecorino foggiano cheese (Santillo and Albenzio, 2008; Santillo et al., 2009). *Bifidobacteria* added in cheese was able to produce high levels of linoleic acid as well as to induce greater proteolysis that was associated with lower cheese hardness. A study highlighted that probiotic bacteria may not survive in high numbers when freely incorporated into dairy products, and encapsulation was proposed as a method to increase the survival and delivery of probiotics (Sultana et al., 2000; Anal et al., 2006). Santillo et al. (2012) proposed the addition of encapsulated probiotics into rennet paste and investigated the effect of this method on the biochemical features of Pecorino cheese manufactured from Gentile di Puglia ewe milk at different maturation times. The probiotic-containing alginate beads underwent a progressive disaggregation process, leading to the liberation of live microorganisms. Therefore, encapsulation may sustain the production of long-ripened cheese, maintaining high levels of live cells throughout the maturing process. After 60 days of ripening, the beads disappeared, and the probiotic cells remained at a level of about 7 log 10 cfu/g of cheese up to 120 days of ripening.

Feta cheese is the most known Greek PDO cheese. It is a white-brined cheese produced from ewe milk or a mixture of ewe and goat milk (the latter up to 30%) and it has a salty and slightly pungent flavor. Angelopoulou et al. (2017) produced Feta cheese using *P. freudenreichii* subsp. *shermanii*, a probiotic able to survive technological stresses imposed throughout Feta cheese manufacturing and ripening. The final product resulted white, soft, acid, contained viable cells at a level of 6.8–9 log cfu/g, and was characterized by flavor profile similar to the Feta cheese made without probiotic cultures. Furthermore, the tested probiotic was shown to survive the gastrointestinal tract and to reach the intestine alive and able to exert its probiotic potential.

Inulin is a storage carbohydrate of plant origin containing beta (2,1)-linked fructosyl residues, typically terminating with a glucose molecule. The physical effect of its addition on different low-fat dairy systems has been studied by several authors, in particular for yogurt. Dietary fibers added to products can be selectively utilized by host microorganisms, promoting health benefits and wellness to individuals (Gibson et al., 2017). The addition of inulin fiber was tested in different products made from small ruminant milk. Balthazar et al. (2015) evaluated the adjunct of different concentrations (2 or 6 g/100 g) of inulin in smoothie yogurt made from ewe milk, highlighting that inulin did not interfere with the sensory acceptance of the products while also improving the flavor acceptance. The effect of the addition of inulin (0–6 g/100 g) was also studied on the physicochemical and microbiological characteristics of ewe milk yogurts during refrigerated storage (4 °C for 28 days) (Balthazar et al., 2016). Inulin addition was confirmed not to interfere in the cell counts of the starter culture, as previously reported by Kiss et al. (2014). Balthazar et al. (2017a, 2017b) assessed the effects of different prebiotic dietary oligosaccharides replaced in ewe milk fat on the nutritional and rheological parameters as well as the physical and thermal properties and microstructural characteristics of nonfat ewe milk ice cream. The effects of the addition of *L. casei* 01 and inulin to ewe milk ice cream were also studied. The product proved to be a suitable food matrix for *L. casei* 01 growth and adhesion to intestinal Caco-2 epithelial cells in vitro, ensuring a therapeutic effect to the host during 150 days of frozen storage (−18 °C). Furthermore, the addition of inulin improved the potential nutritional benefits as the functional ewe milk ice creams (probiotic and symbiotic) presented good antihypertensive (ACE-inhibitory activity) and antioxidant properties, improving the bioactivity of the product (Balthazar et al., 2018).

Papadimitriou et al. (2007) analyzed Greek ewe milk yogurts produced with a normal yogurt culture (*L. bulgaricus*, *S. thermophilus*) mixed with *Lactobacillus paracasei* subsp. *paracasei* DC412. They showed that probiotic strains did not affect the physicochemical properties and proteolysis patterns, but revealed the presence of a peptide exerting both antihypertensive and opiate-like activity. Vianna et al. (2017) proposed a probiotic yogurt containing an equal proportion of cow and ewe milks as an advisable technological alternative for probiotic dairy industries.

Balthazar et al. (2019) developed novel formulations for functional fermented beverages manufactured with semiskimmed ewe milk; strawberry pulp (*Fragaria × ananassa* Duch.); prebiotic ingredients such as inulin, potato starch, or both; and probiotic *Lactobacillus plantarum* CECT 8328. The in vitro simulated digestion of the product during storage showed the ability of *L. plantarum* to survive the digestive process in a freshly prepared beverage and in a beverage stored up to 30 days. Also, the lactic acid bacteria were able to improve the antioxidant, antihypertensive, and antidiabetic activities of the beverages, especially when combined with inulin fiber.

GOAT MILK AND DAIRY PRODUCTS

A great number of people in the world use goat milk, either consuming it directly or in a variety of dairy products (Pulina et al., 2018). Goat milk in comparison to cow milk differs in alkalinity, buffering capacity, and digestibility while also having nutritional properties and lower allergenicity, especially in sensitized children (Ibeagha-Awemu et al., 2009; Albenzio and Santillo, 2011). This has led to increased interest for certain beneficial characteristics that make it potentially useful in human nutrition as a functional food (Slačanac et al., 2010). Ranadheera et al. (2019) reviewed differences in physicochemical, technological, textural, and sensory characteristics between goat and cow milk. Fat globules in goat milk are characterized by a smaller mean size (it varies from 2.5 to 3 μm) compared to cow mill (3–5 μm). Globule size combined with a lack of agglutinin, which facilitates fat globule coalescence upon cooling, determines a poorer creaming ability than cow milk (Ranadheera et al., 2019). Fatty acid composition shows a higher content of free fatty acids and short chain fatty acids, in particular of caprylic acid and capric acid, that contribute to the "goaty" flavor. Medium chain triglycerides and long chain fatty acids with greater ω-6 and ω-3 fatty acids, EPA, and DHA than cow milk are considered beneficial to human physiology (Ranadheera et al., 2019). Goat milk has variable amounts of β-casein and αs-casein fractions due to the high genetic polymorphism at the breed and individual level (Albenzio et al., 2015). The low αs-casein fraction expressed in milk impairs the coagulating ability of goat milk, thus influencing the quality of dairy products (Ranadheera et al., 2019) and the cheese typology that can be produced using such milk. Compared with cow milk, goat milk contains greater amounts of six essential amino acids: threonine, lysine, isoleucine, cysteine, tyrosine, and valine (Rani et al., 2017). Although the lactose content is reported to be slightly lower in goat milk (4.16 g/100 mL) than in cow milk (4.76 g/100 mL) (Ranadheera et al., 2019), goat milk is much richer in lactose-derived oligosaccharides (lactulose, lactitol, lactobionic acid, and galactooligosaccharides), which are beneficial to humans due to their prebiotic and anti-infective properties (Turkmen, 2017; Ranadheera et al., 2019). Because goats can transform all dietary carotene to vitamin A, goat milk is typically whiter than other milks and contains greater level of vitamin A; it is also rich in niacin, thiamine, riboflavin, and pantothenate (Turkmen, 2017).

It has been reported that about 50% of the human population that is intolerant to cow milk may be able to tolerate goat milk (Slačanac et al., 2010). Due to the variability of type, level, and structural properties of the main proteins, goat milk has been reported as more favorable in the case of cow milk allergy. Moreover, it undergoes a faster in vitro-simulated human gastrointestinal digestion process. Goat milk is advantageous in carrying probiotics throughout the whole shelf life of the product in adequate amounts required for health benefits due to the milk's favorable pH, buffering capacity, and nutrient content (Ranadheera et al., 2019). Yogurt, fermented milk, cheese, and ice cream are the most commonly used goat dairy vehicles for probiotic cultures (Ranadheera et al., 2018). Different typologies of dairy products obtained from goat milk and the probiotic and/or prebiotics they contain are presented in Table 3.

Research studies have focused on the effects of goat probiotic products on an in vivo model of human disease. Abeijòn Mukdsi et al. (2013) evaluated the effect of the intake of goat milk cheese manufactured with *Lactobacillus fermentum* CRL1446 on intestinal feruloyl esterase activity and oxidative status in Swiss albino mice. They confirmed that goat cheese elaborated with *L. fermentum* increased the bioavailability of antioxidant ferulic acid in the gut, and thus enhanced the consumer's oxidative status. Rodrigues et al. (2018) evaluated the effect of *L. rhamnosus* EM1107, isolated (EM1107) or added to goat cheese, on acetic acid-induced colitis in rats, suggesting that the probiotic strain alone or added to goat cheese exerts a preventive effect against acetic acid-induced damage and may be beneficial in inflammatory bowel diseases.

Radulović et al. (2017) proved that spray-dried cells of the probiotic bacteria *L. plantarum 564* could be successfully used in soft goat cheeses, as they maintained a high viable cell level during 8 weeks of storage without any adverse effect on the cheese chemical composition, pH values, or sensory quality. Also, *L. mucosae* CNPC007, used in Coalho goat cheese production, showed good sensory evaluation, technological aptitude, and a high viability for more than 28 days of refrigerated storage (de Moraes et al., 2018). Analogous to what has been previously reported in ewe milk cheese, probiotic strains were found to contribute to proteolysis and lipolysis during goat cheese aging. *Lactococcus lactis* subsp. *lactis* and *L. lactis* subsp. *cremoris* (R704), *L. acidophilus* (LA-5), *L. paracasei* subsp. *paracasei* (*L. casei* 01), and *B. animalis* subsp. *lactis* (BB 12) influenced proteolysis in Coalho cheese, resulting in a higher content of soluble proteins and a greater release of amino acids compared to the control cheese. *B. lactis* in particular showed the highest proteolytic activity (Bezerra et al., 2016). In a further study focused on the volatile profile of goat Coalho cheese added with the same probiotic lactic bacteria (in isolated or combined form), the authors concluded that the use of different cultures of probiotic lactic bacteria and the length of storage affected the volatile profile (Bezerra et al., 2017).

It is generally assumed that the viability of probiotic organisms in dairy products should be in the range of 10^6–10^8 cfu/g or 10^8–10^{10} cfu/day throughout the storage period, and bacteria should remain viable

TABLE 3
Dairy Products From Goat Milk Characterized by Probiotic and/or Prebiotic Addition.

Products	Probiotics	Prebiotics and Method of Addition	References
Yogurt, kefir, smearcase	None	Fish oil, inulin	Kiss et al. (2019)
Yogurt	*Bifidobacterium animalis* subsp. *lactis* BB-12 (10^6–10^7 cfu/mL)	Alginate encapsulation, inulin	Prasanna and Charalampopoulos (2019)
Coalho cheese (28 day ripened)	*Lactobacillus mucosae*	None	de Moraes et al. (2018)
Coalho cheese (28 day ripened)	*Lactococcus lactis* subsp. *lactis* and *L. lactis* subsp. *cremoris* (R704); *Lactobacillus acidophilus* (LA-5); *Lactobacillus paracasei* subsp. *paracasei* (*L. casei* 01); *Bifidobacterium animalis* subsp. *lactis* (BB-12) (6.5–7 log cfu/g)	None	Bezerra et al. (2016, 2017)
Fresh cheese	None	Inulin	Salvatore et al. (2014)
Beverage	*Bifidobacterium lactis* (6–8 log cfu/mL)	Inulin and oligofructose	da Silveira et al. (2015)
Boursin cheese (35 day ripened)	*Bifidobacterium animalis* subsp. *lactis*; *Lactobacillus rhamnosus* (6–8 log cfu/mL)	None	Martins et al. (2018)
Cheese 30 and 60 days of ripening	*Lactobacillus fermentum* CRL1446; 8–9 log cfu/g	None	Abeijòn Mukdsi et al. (2013)
Cheese	*Lactobacillus rhamnosus* EM1107 10^8 and 10^9 cfu/g	None	Rodrigues et al. (2018)
Soft goat cheese 8 week ripened	*Lactobacillus plantarum* 564 10^7–10^9 cfu/g	Spray-dried	Radulović et al. (2017)
Chocolate-flavored Ice cream	*Lactobacillus acidophilus* La-05, *Bifidobacterium animalis* subsp. *lactis* BB-12, *Propionibacterium jensenii* 10^7–10^8 cfu/g	None	Ranadheera et al. (2013)

until gastrointestinal digestion to guarantee significant health effects to the host (da Silveira et al., 2015). Martins et al. (2018) evaluated the suitability of two strains of probiotic bacteria (*B. animalis* subsp. *lactis* and *L. rhamnosus*) incorporated individually into Boursin-type goat cheese and their in vitro resistance in the cheese through the gastrointestinal tract. Both probiotic cultures were stable throughout 35 days of storage at 4°C at levels higher than 7 log cfu/g; however, after simulated passage through the gastrointestinal tract, *B. animalis* showed a higher resistance than *L. rhamnosus*. Apart from the probiotic culture added, consumer acceptance of the goat cheese did not change compared to the control cheese.

Strains of *L. acidophilus* LA-5, *B. animalis* subsp. *lactis* BB-12, and *Propionibacterium jensenii* 702 were tested in ice cream made from goat's milk. Adequate cell viability was also confirmed in this fresh product with counts ranging from 10^7 to 10^8 cfu/g for up to 52 weeks at −20°C whether it was packaged in polyethylene, polypropylene, or glass, ensuring a good organoleptic quality to the product (Ranadheera et al., 2013). Furthermore, chocolate goat dairy beverages containing probiotic *B. lactis* were produced and analyzed in a study by da Silveira et al. (2015). The authors evaluated the effects of goat cheese whey and prebiotics (inulin and oligofructose) on the physicochemical parameters and sensory features of the beverages. Goat cheese whey was found to sustain probiotic viability with counts between 6 and 8 log cfu/mL, and the addition of prebiotic inulin and oligofructose improved the viscosity and sensory features of the goat milk beverage.

Salvatore et al. (2014) observed that the replacement of milk fat with inulin in a fresh cheese gave differences in terms of the rheological attributes of the cheese due to the effect of inulin on the protein/fat network formation and on the aggregation of inulin during cheesemaking. Alginate goat milk mixed with inulin was used to encapsulate probiotic strains to improve their viability in goat milk yogurt; in particular, encapsulated *B. animalis* subsp. *lactis* BB-12 survived at a high count during refrigerated storage and under simulated gastrointestinal conditions (Prasanna and Charalampopoulos, 2019).

Inulin can be associated with other molecules that exert positive effects on human health. Recently, a study conducted by Kiss et al. (2019) evaluated the effect on the fatty acid composition of yogurt, kefir, and smearcase made from goat and ewe milk added with inulin and fish oil (rich in omega-3). Fortification of the products with inulin resulted in the enhanced survival of lactic acid bacteria and facilitated the production of unsaturated fatty acids, providing an excellent opportunity to increase the functionality of fermented milk products. The addition of fish oil contributed to the enhancement of the extent of lipolysis during storage.

BUFFALO MILK AND DAIRY PRODUCTS

Buffalo milk is characterized by higher amounts of fat, protein, lactose, minerals, and vitamins than cow milk and its gross composition varies according to breed, geographical location, and feeding condition (Han et al., 2012). Milk fat represents the major constituent of buffalo milk and its content is affected mainly by animal physiology, lactation stage, season, feed, breed, time, and milking sequence (El-Salam and El-Shibiny, 2011). The average fat globule size in buffalo milk (4.07 μm) is higher than that found in milk from the most commonly used dairy species. Buffalo milk fat contains higher amounts of the total saturated fatty acids C16:0 and C12:0 and lower monounsaturated fatty acids than cow milk fat (Talpur, 2007); the CLA content is comparable in buffalo and cow milk (Han et al., 2012). Compared to cow milk, buffalo milk contains higher α_{s2}-CN and κ-casein; the great content of κ-CN is related to the rapidity of the enzymatic phase of rennet coagulation and the requirement of a smaller quantity of chymosin in cheese manufacture (El-Salam and El-Shibiny, 2011). Buffalo milk has slightly higher concentrations of β-Lg than cow milk. This major whey protein is lacking in human milk, and different peptides derived from the proteolytic digestion of β-Lg have been shown to have inhibitory activity against ACE in cow milk studies. As in human milk, lysozymes in buffalo milk possess greater positive charges than those from egg-white lysozymes, and they are a more efficient antimicrobial agent. Buffalo milk has been found to contain more minerals and a higher viscosity than cow milk. The pH of buffalo milk has values slightly higher than cow milk and tends to decrease more slowly during acidification due to the higher buffering capacity of buffalo milk, resulting from the high casein and inorganic phosphate contents (El-Salam and El-Shibiny, 2011).

Due to its high nutrient content, water and pH raw buffalo milk allows the growth of several complex microbial communities. It has been reported that such milk represents excellent source of potential probiotics (Kalhoro et al., 2019; Rizqiati et al., 2015). Thus, buffalo milk offers excellent prospects for the development of fermented dairy foods and various types of yogurt, resulting in creamy textures and rich flavor profiles (Han et al., 2012; Murtaza et al., 2017). In Table 4, different types of dairy products obtained from buffalo milk that contain probiotics and/or prebiotics are presented.

In Southern European countries, buffalo milk is mainly used to produce pasta filata cheese. In particular in Italy, Mozzarella di Bufala Campana has been recognized with Protected Denomination of Origin (PDO) status under European Union regulations. Therefore, cheesemaking practices must comply with the rules established by the National Standard, including milk origin and the geographical area within which the regulations permit it to be transformed into PDO cheese (Addeo et al., 2007). Among whey cheeses, ricotta is the most popular and traditional variety of unripened cheese obtained by thermal coagulation of the whey resulting from the cheesemaking process. Ricotta cheese may provide several benefits compared to other foods as a delivery vehicle for probiotic microorganisms due to its intrinsic characteristics such as high moisture and pH, low salt concentration, and low oxygen level. All these combined features may offer additional protection to probiotic organisms during storage and gastric transit. Indeed, probiotic *L. acidophilus* La-05, incorporated as a pellet in the ricotta from buffalo milk, maintained a concentration of 7.8 ± 0.2 log cfu/g in the final product without affecting its compositional, physicochemical, textural and sensory quality, and acceptability (Sameer et al., 2020).

Studies have been conducted to produce and characterize novel varieties of cheeses from water buffalo milk, such as Camembert-, Gouda-, and Stracchino-like cheeses, but without appreciable success (Addeo et al., 2007). However, some studies reported the production of cheddar-type cheese from buffalo milk

TABLE 4
Dairy Products From Buffalo Milk Characterized by Probiotic and/or Prebiotic Addition.

Products	Milk Type	Probiotics	Prebiotics and Technologies	Viability of Probiotics in Trials	References
Ricotta cheese	Buffalo	*Lactobacillus acidophilus* La-05	None	7.8 ± 0.2 log cfu/g	Sameer et al. (2020)
Yogurt	Buffalo	*Lactobacillus acidophilus*; *Bifidobacterium bifidum*	Lactulose, oligofructose, and inulin	7.0 log cfu/mL	Ehsani et al. (2016)
Yogurt	Buffalo	*Lactobacillus acidophilus*; *Bifidobacterium* sp.; *Streptococcus thermophilus*	Manna from dry yeast *Saccharomyces cerevisiae*	*Lactobacillus acidophilus* 6.8–7.85 log cfu/g; *Bifidobacterium* sp. 5.8–7.15 log cfu/g *Streptococcus thermophilus* 6.74–7.22 Log cfu/g	Al-Manhel and Niamah (2017)
Cheddar cheese (180 day ripened)	Buffalo	*Lactobacillus acidophilus* LA-5; *Bifidobacterium bifidum* Bb-11; *Bifidobacterium longum* BB536	None	> 7.0 log10 cfu/g	Murtaza et al. (2017)

(Murtaza et al., 2012, 2014). In particular, the impact of ripening on the survival of starters and adjunct probiotic cultures (*L. acidophilus* LA-5, *B. bifidum* Bb-11, and *B. longum* BB536) was evaluated in such cheese. Irrespective to the temperature (at 4–6 °C or 12–14 °C) applied for the ripening process, cheddar from buffalo milk was found to be a suitable carrier of probiotic bacteria (probiotic cultures were > 7.0 log10 cfu/g) throughout 180 days of ripening.

Buffalo milk yogurt is considered a creamy delicacy free from artificial thickeners due to a higher dry matter content and lower cholesterol compared to cows, which impacts its textural properties. Probiotic *L. acidophilus* and *B. bifidum* together with prebiotic lactulose, oligofructose, and inulin have been tested in yogurt. Changes in probiotic counts and sensorial properties in yogurt samples after 21 days of storage at 4 °C showed that prebiotics effectively improved the physicochemical properties and enhanced probiotic bacteria survival (7 log cfu/mL throughout the study period). However, probiotic yogurt scored a higher rate for preference than symbiotic yogurt by the sensory panel because it was reported to have the typical yogurt flavor, which better met consumer expectations (Ehsani et al., 2016). The positive interaction between prebiotic and probiotic cultures was confirmed in buffalo milk yogurt by Al-Manhel and Niamah (2017). The authors found that mannan extract from dry yeast *Saccharomyces cerevisiae* cell increases the viability of probiotic bacteria (*L. acidophilus*, *Bifidobacterium* spp., and *S. thermophilus*) during 28 days of cold storage.

MINOR DAIRY SPECIES: CAMEL, DONKEY, AND MARE

Principal dairy products obtained from camel, mare, and donkey milk and containing probiotics and/or prebiotics are presented in Table 5.

Camel and Dromedary Milk and Dairy Product

Camel milk has an important role in human nutrition, especially in hot regions and arid countries such as Saudi Arabia where it is used fresh or fermented. Also, in countries such as India, Russia, and the Sudan, camel milk is recognized as a treatment for a series of diseases such as dropsy, jaundice, tuberculosis, asthma, and leishmaniasis or kala-azar. Camel milk contains all the essential nutrients found in cow milk (Al haj and Al Kanhal, 2010) and recently was also reported to exert therapeutic properties such as anticarcinogenic (Magjeed, 2005), antidiabetic (Agrawal et al., 2007), and antihypertensive (Quan et al., 2008). It is also useful in infant nutrition as an alternative to cow milk in case of allergy (El-Agamy et al., 2009) (Al haj and Al Kanhal, 2010). Although some authors indicate that camel milk is only suitable for drinking

TABLE 5
Dairy Products From Camel, Mare, and Donkey Milk Characterized by Probiotic and/or Prebiotic Addition.

Products	Milk Type	Probiotics	Prebiotics and Technologies	Viability of Probiotics in Trials	References
Fermented milk	Camel	*Enterococcus faecium*; *Streptococcus macedonicus*	none	7.2–7.6 log cfu/mL	El Hatmi et al. (2018)
Fermented milk	Camel	*Lactococcus lactis* KX881782	none	> 8.5 log cfu/mL	Ayyash et al. (2018)
Yogurt	Camel	*Streptococcus thermophilus* DSM 20617, *Lactobacillus delbrueckii* ssp. *bulgaricus* DSM 20080, *Bifidobacterium infantis* DSM 20088; *Bifidobacterium angulatum* DSM 20098	none	*Bifidobacteria* 8.2 log cfu/mL	Al-Saleh et al. (2011)
Butter	Camel	*Lactobacillus plantarum*	None	n.d.	Maurad and Meriem (2008)
Akawi cheese (21 day storage)	Bovine with probiotic from camel milk	*Lactobacillus plantarum* from camel milk	None	> 7.0 log cfu/g	Al-Dhaheri et al. (2017)
Fermented milk	Mare	*Lactobacillus rhamnosus* strains, FSMM15, FSMM22, and FSMM26	None	n.d.	Aryantini et al. (2017)
Fermented milk	Mare	*Lactobacillus acidophilus*, *Bifidobacterium longum*, *Lactobacillus casei*	None	1.4 10^9 cfu/g	Murti et al. (2016)
Beverage	Donkey	*Lactobacillus acidophilus* DSM 20079, *Lactobacillus bulgaricus* DSM 20081, *Lactobacillus paraplantarum* DSM 10667, *Lactobacillus plantarum* DSM 20174, *Lactobacillus pentosus* DSM 20314, and *Lactobacillus rhamnosus* DSM 20711	None	>9.0 log cfu/mL	Nazzaro et al. (2010)
Yogurt	Donkey	*Lactobacillus acidophilus*, *Lactobacillus casei*	None	*Lactobacillus acidophilus*, 6.75 log cfu/mL *Lactobacillus casei* 6.56 log cfu/mL	Perna et al. (2015)
Beverage	Donkey	*Streptococcus thermophilus* and *Lactobacillus plantarum*	None	7.0 log cfu/mL	Turchia et al. (2017)
Kefir	Donkey	None	Rosemary oil and sulla honey	–	Perna et al. (2019)

(Yagil et al., 1984), different dairy products have been produced from dromedary/camel milk including soft cheese, fermented milk, yogurt, ice cream, and butter (Al haj and Al Kanhal, 2010). Another interesting characteristic of this milk emerged when Abushelaibi et al. (2017) isolated and identified 9 LAB from camel milk with potential probiotic characteristics. Among these, *L. lactis* KX881782 was found to be the most promising probiotic. In light of these findings Ayyash et al. (2018) investigated the anticancer, antihypertensive, antidiabetic, and antioxidant in vitro activities of camel milk fermented with extracted autochthonous probiotics, then compared these with cow milk fermented with the same probiotics. Interestingly, the health-promoting benefits, particularly the antioxidant, ACE-inhibition, and antiproliferative activities, in fermented camel milk were markedly higher than those in fermented cow milk. The greater proteolysis observed in camel milk than in bovine milk fermented by *Lc.K782* may be attributed to the major susceptibility of camel milk caseins to proteolytic enzymes yielded by the probiotic strain; this could be associated with the higher accumulation of bioactive peptides. Indeed, proteolytic activity promotes the formation of small peptides that could confer health-promoting benefits to the fermented product. Further strains isolated from spontaneous fermented dromedary milk, *Enterococcus faecium* and *Streptococcus macedonicus*, were tested in fermented dromedary milk. The sensory features of the dairy product from dromedary milk scored lower overall acceptability than the same product made using cow milk, with the most penalizing scores associated with taste and consistency (El Hatmi et al., 2018).

Akawi cheese is a well-known Middle Eastern cheese that is extensively consumed in North Africa, the Middle East, and states in the Gulf Cooperation Council (Ayyash et al., 2012). Among LAB isolated and identified from camel milk, *L. plantarum* KX881772 and *L. plantarum* KX881779 were found to have probiotic characteristics and EPS (exopolysaccharide) production capabilities. They were used by Al-Dhaheri et al. (2017) to produce probiotic low-fat akawi cheese made from cow milk.

Maurad and Meriem (2008), tested the probiotic potential different strains of *L. plantarum* isolated from traditional butter made from camel milk. Four strains (*L. plantarum* SH5, SH12, SH24, and SH32) were sufficiently resistant to pH 2.0 for 2–6 h incubation periods, indicating they are able to survive to the gastric phase of digestion. In particular, the *L. plantarum* SH12 and SH24 strains showed rapid acidification activity, good proteolytic activity, and a high survival percentage after freeze drying. Furthermore, when the same probiotic strains were used as starter cultures in the production of novel probiotic dairy products from camel milk (such as fermented milk, butter, or cheese), the consumer panel highlighted the favorable flavor and sensory qualities.

The role of lactose is essential in the definition of textural properties of ice cream due to the involvement of disaccharide in lowering the freezing point and enhancing the sandiness of the ice cream mixture. Indeed, beyond flavor, the consumer acceptance of ice cream depends largely on the perceived textural quality. Accordingly, in frozen yogurt, a portion of lactose ferments to lactic acid, which can alter its physical properties. Al-Saleh et al. (2011) compared the physicochemical properties of probiotic frozen yogurt made from camel and cow milks enhanced with *L. delbrueckii* ssp. *bulgaricus* and *S. thermophilus* bacteria. They concluded that camel milk is suitable for the production of probiotic frozen yogurt.

Donkey Milk and Dairy Products

Donkey milk composition is more comparable to human milk than milk from other dairy species, in particular for lactose, total protein, and whey protein contents. The protein fraction of donkey milk is rich in whey proteins, and thus it gives the special property of forming a soft curd during digestion in the infant gut, similar to human milk. The softness of the curd is also due to the lower ratio of soluble calcium. This condition may explain why, in some parts of the world, donkey milk is used as a human milk substitute for infants (Aspri et al., 2016). The lipid fraction of donkey milk is comparable to that of human and mare milk and is characterized by high levels of essential fatty acids and low saturated fatty acids. Donkey milk is also rich in polyunsaturated fatty acids with a high percentage mainly of linoleic acid (Aspri et al., 2016). In a study on the availability of fatty acids of milk from different species for intestinal absorption, Santillo et al. (2018) highlighted that after in vitro digestion, the digestate from donkey milk showed a major concentration of linoleic acid compared to cow, goat, ewe, human, and formula milk.

Donkey milk has recently gained attention due to its nutraceutical properties. Its low casein content does not allow cheesemaking, so the production of fermented milk and yogurt represents an alternative strategy to increase the donkey milk shelf life (Turchia et al., 2017). Strains of *S. thermophilus* and *L. plantarum* were used for the production of a novel donkey milk-fermented beverage; the probiotic viability resulted in higher than 7.00 log cfu/mL until the end of the storage period. Moreover, *L. plantarum* showed a promoting activity toward *S. thermophilus* replication and viability (Turchia et al., 2017). Nazzaro et al. (2010) exploited the potential specific health benefits offered by donkey milk as well as its recognized pleasant taste by a new type of fermented milk drink containing probiotic strains of *Lactobacillus* spp., namely *L. acidophilus*, *L. bulgaricus*, *L. paraplantarum*, *L. plantarum*, *L. pentosus*, and *L. rhamnosus*. From the results of this study, donkey milk emerged as an excellent base ingredient for many probiotic and therapeutic food preparations. *L. acidophilus* and *L. casei* were investigated for the production of a donkey milk probiotic yogurt. The authors reported the increased antioxidant activity of probiotic yogurt as well as high scores for acceptability of the novel fermented beverages rated by a consumer panel (Perna et al., 2015). Moving from these encouraging results, Perna et al. (2019) produced donkey kefir fortified with sulla honey and rosemary essential oil and evaluated the phenolic content and antioxidant activity during refrigerated storage. Both added polyphenols influenced the antioxidant activity of donkey kefir and the formation of polyphenol-protein complexes able to influence the availability of proteic bioactive components. Sensory analysis demonstrated that the donkey kefir is well accepted by consumers; in particular, kefir fortified with sulla honey showed the highest acceptability.

Mare Milk and Dairy Products

The increasing interest in mare milk is due to its greater similarity, like donkey milk, in gross composition to human milk rather than ruminant milk. In particular, the whole protein system, both whey protein and nonprotein nitrogen fraction concentrations, of mare milk is similar to human milk. The richness and pattern of the whey protein of mare milk make it more favorable than cow milk for human nourishment and more suitable than cow milk as nourishment for infants (Malacarne et al., 2002). The fat content of mare milk is significantly lower in comparison with human and cow milk. Also, lipids are dispersed as emulsified globules organized in very little globules of about 2–3 µm compared to humans (about 4 µm) and cows (3–5 µm). Mare milk fat contains fewer triglycerides, but it is richer in free fatty acids and phospholipids; the fatty acid profile mainly is constituted by medium chain fatty acids (Malacarne et al., 2002; Pieszka et al., 2016). Mare milk is characterized by a low content of cholesterol and a high content of unsaturated fatty acids, especially linoleic (precursor of prostaglandin E, which influences the prevention of intestinal ulcers in humans) and α-linoleic, which is essential in the human diet (Malacarne et al., 2002). The high content of lactoferrin and lysozyme of mare milk with antibacterial, antiviral, and fungicidal activities makes this milk interesting for medical use. Mare milk constituents have a modulatory effect on inflammatory processes through their influence on the chemotaxis process and lowering of the respiratory burst; hence, the milk may be helpful when curing illnesses with recurring inflammation (Ellinger et al., 2002; Pieszka et al., 2016).

Mare milk, besides being the most important nutritional resource for foals, is also one of the most important basic foodstuffs for the human population in central Asia, where a lactic-alcoholic beverage called Koumiss is traditionally produced through its fermentation. Indeed, due to its low content of casein and fat, mare milk is not appropriate for cheese and butter production (Pieszka et al., 2016), and it is generally consumed as a fermented beverage.

Murti et al. (2016) developed fermented mare milk by using different combinations of the following cultures of probiotic bacteria: *L. acidophilus*, *B. longum*, and *L. casei*. Apart from the different combinations tested, probiotic bacteria were able to grow in mare milk, influencing the biochemical profile but showing low acceptability by entrained panelists.

Aryantini et al. (2017) investigated the safety and probiotic characteristics of three *L. rhamnosus* strains, FSMM15, FSMM22, and FSMM26, as potential probiotics from fermented mare milk, concluding a favorable aspect of *L. rhamnosus* FSMM15 as a probiotic strain.

CONCLUSION

In this chapter, the aspects related to the milk composition of nonbovine species such as ewe, goat, buffalo, donkey, mare, dromedary, and camel milk have been reviewed, with particular regard to the nutritional and health content of the derived dairy products. The milk composition is specific for each treated species and impacts the potential for dairy production, which includes fermented drinks, fresh and short-aged products, and long-ripened products. The addition of probiotics and prebiotics in the production of traditional or innovative dairy products must overcome the gastric barrier unharmed and finally play a beneficial role in the host intestine. Research is also oriented toward selecting effective probiotics naturally present in nonbovine milk that could be used at the industrial level. To sustain innovative dairy products from nonbovine milk, it is also important to preserve the peculiarity of sensorial characteristics of typical productions and to maximize the acceptability of innovative products. Although milk and dairy products from small ruminants have been more widely studied and investigated for their health value, their role as alternatives to bovine milk, and for their peculiar acidic and proteic composition, there is still a need to reveal the potential in the probiotic dairy production of minor species such as the buffalo, donkey, mare, dromedary, and camel.

It has to be considered that most of the reported species are characterized by nonintensive rearing and productive context, although derived dairy production sustains a large part of the global economy and has gained increasing importance in recent years. The production of these innovative foods can be a flywheel to sustain production of milk from minor species and therefore to improve the economy of marginal areas.

REFERENCES

Abeijòn Mukdsi, M.C., Haro, C., Gonzàlez, S.N., Medina, R.B., 2013. Functional goat milk cheese with feruloyl esterase activity. J. Funct. Foods 5, 801–809.

Abushelaibi, A., Al-Mahadin, S., El-Tarabily, K., Shah, N.P., Ayyash, M., 2017. Characterization of potential probiotic lactic acid bacteria isolated from camel milk. Lebensm. Wiss. Technol. 79, 316–325.

Addeo, F., Alloisio, V., Chianese, L., Alloisio, V., 2007. Tradition and innovation in the water buffalo dairy products. Ital. J. Anim. Sci. 6, 51–57.

Agrawal, R.P., Budania, S., Sharma, P., Gupta, R., Kochar, D.K., 2007. Zero prevalence of diabetes in camel milk consuming Raica community of North-west Rajasthan, India. Diabetes Res. Clin. Pract. 76, 290–296.

Al haj, O.A., Al Kanhal, H.A., 2010. Compositional, technological and nutritional aspects of dromedary camel milk. Int. Dairy J. 20, 811–821.

Albenzio, M., Santillo, A., 2011. Biochemical characteristics of ewe and goat milk: effect on the quality of dairy products. Small Rumin. Res. 101, 33–40.

Albenzio, M., Santillo, A., Caroprese, M., Marino, R., Trani, A., Faccia, M., 2010. Biochemical patterns in ovine cheese: influence of probiotic strains. J. Dairy Sci. 93, 3487–3496.

Albenzio, M., Santillo, A., Caroprese, M., Braghieri, A., Sevi, A., Napolitano, F., 2013a. Composition and sensory profiling of probiotic Scamorza ewe milk cheese. J. Dairy Sci. 96, 2792–2800.

Albenzio, M., Santillo, A., Caroprese, M., Ruggieri, D., Napolitano, F., Sevi, A., 2013b. Physicochemical properties of Scamorza ewe milk cheese manufactured with different probiotic cultures. J. Dairy Sci. 96, 2781–2791.

Albenzio, M., Santillo, A., Marino, R., Della Malva, A., Caroprese, M., Sevi, A., 2015. Identification of peptides in functional Scamorza ovine milk cheese. J. Dairy Sci. 98, 8428–8432.

Albenzio, M., Santillo, A., Avondo, M., Nudda, A., Chessa, S., Pirisi, A., Banni, S., 2016. Nutritional properties of small ruminant food products and their role on human health. Small Rumin. Res. 135, 3–12.

Al-Dhaheri, A.S., Al-Hemeiri, R., Kizhakkayil, J., Al-Nabulsi, A., Abushelaibi, A., Shah, N.P., Ayyash, M., 2017. Health-promoting benefits of low-fat akawi cheese made by exopolysaccharide-producing probiotic *Lactobacillus plantarum* isolated from camel milk. J. Dairy Sci. 100, 7771–7779.

Al-Manhel, A.J., Niamah, A.K., 2017. Mannan extract from *Saccharomyces cerevisiae* used as prebiotic in bioyogurt production from buffalo milk. Int. Food Res. J. 24, 2259–2264.

Al-Saleh, A.A., Ali, A., Metwalli, M., Ismail, E.A., 2011. Physicochemical properties of probiotic frozen yoghurt made from camel milk. Int. J. Dairy Technol. 64, 557–562.

Al-Wabel, N.A., 2008. Mineral contents of milk of cattle, camels, goats, and sheep in the central region of Saudi Arabia. Asian J. Biochem. 3, 373–375.

Anal, A.K., Stevens, W.F., Remuñán-López, C., 2006. Ionotropiccross-linked chitosan microspheres for release of ampicillin. Int. J. Pharm. 312, 166–173.

Angelopoulou, A., Alexandraki, V., Georgalaki, M., Anastasiou, R., Manolopoulou, E., Tsakalidou, E., Papadimitriou, K., 2017. Production of probiotic Feta cheese using *Propionibacterium freudenreichii subsp. shermanii* as adjunct. Int. Dairy J. 66, 135–139.

Aryantini, N.P.T., Yamasaki, E., Kurazono, H., Sujaya, I.N., Urashima, T., Fukuda, K., 2017. In vitro safety assessments and antimicrobial activities of *Lactobacillus rhamnosus* strains isolated from a fermented mare's milk. Anim. Sci. J. 88, 517–525.

Aspri, M., Economou, N., Papademas, P., 2016. Donkey milk: an overview on functionality, technology and future prospects. Food Rev. Int. 33 (3), 316–333.

Ayyash, M., Al-Dhaheri, A.S., Al Mahadin, S., Kizhakkayil, J., Abushelaibi, A., 2018. In vitro investigation of anticancer, antihypertensive, antidiabetic, and antioxidant activities of camel milk fermented with camel milk probiotic: a comparative study with fermented bovine milk. J. Dairy Sci. 101, 900–911.

Ayyash, M.M., Sherkat, F., Shah, N.P., 2012. The effect of NaCl substitution with KCl on Akawi cheese: chemical composition, proteolysis, angiotensin-converting enzyme-inhibitory activity, probiotic survival, texture profile, and sensory properties. J. Dairy Sci. 95, 4747–4759. https://doi.org/10.3168/jds.2011-4940.

Balthazar, C.F., Gaze, L.V., Silva, H.L.A., Pereira, C.S., Franco, R.M., Conte-Junior, C.A., Freitas, M.Q., Silva, A.C.O., 2015. Sensory evaluation of ovine milk yoghurt with inulin addition. Int. J. Dairy Technol. 68, 281–290.

Balthazar, C.F., Conte-Junior, C.A., Moraes, J., Costa, M.P., Raices, R.S.L., Franco, R.M., Cruz, A.G., Silva, A.C.O., 2016. Physicochemical evaluation of sheep milk yogurts containing different levels of inulin. J. Dairy Sci. 99, 4160–4168.

Balthazar, C.F., Silva, H.L.A., Vieira, A.H., Neto, R.P.C., Cappato, L.P., Coimbra, P.T., Moraes, J., Andrade, M.M., Calado, V.M.A., Granato, D., Freitas, M.Q., Tavares, M.I.B., Raices, R.S.L., Silva, M.C., Cruz, A.G., 2017a. Assessing the effects of different prebiotic dietary oligosaccharides in sheep milk ice cream. Food Res. Int. 91, 38–46.

Balthazar, C.F., Silva, H.L.A., Cavalcanti, R.N., Esmerino, E.A., Cappato, L.P., Abud, Y.K.D., Moraes, J., Andrade, M.M., Freitas, M.Q., Sant'Anna, C., Raices, R.S.L., Silva, M.C., Cruz, A.G., 2017b. Prebiotics addition in sheep milk ice cream: a rheological, microstructural and sensory study. J. Funct. Foods 35, 564–573.

Balthazar, C.F., Pimentel, T.C., Ferrão, L.L., Almada, C.N., Santillo, A., Albenzio, M., Mollakhalili, N., Mortazavian, A.M., Nascimento, J.S., Silva, M.C., Freitas, M.Q., Sant'Ana, A.S., Granato, D., Cruz, A.G., 2017c. Sheep milk: physicochemical characteristics and relevance for functional food development. Compr. Rev. Food Sci. Food Saf. 16, 247–262.

Balthazar, C.F., Hugo, L.A., Silva, E.A., Esmerino, R.S., Rocha, J., Moraes, M.A.V., Carmo, L., Azevedo, I., Camps, Y.K.D., Abud, C., Sant'Anna, R.M., Franco, M.Q., Freitas, M.C., Silva, R.S.L., Raices, G.B., Escher, D., Granato, C.S., Ranadheera, F., Nazarro, A., Cruz, G., 2018. The addition of inulin and *Lactobacillus casei 01* in sheep milk ice cream. Food Chem. 246, 464–472.

Balthazar, C.F., Santillo, A., Guimarães, J.T., Capozzi, V., Russo, P., Caroprese, M., Marino, R., Esmerino, E.A., Renata, S.L., Raices, M.C., Silva, H.L.A., Silva, M.Q., Freitas, D., Granato, A.G., Cruz, Albenzio, M., 2019. Novel milk–juice beverage with fermented sheep milk and strawberry (Fragaria × ananassa): nutritional and functional characterization. J. Dairy Sci. 102, 10724–10736.

Barłowska, J., Szwajkowska, M., Litwinczuk, Z., Krol, J., 2011. Nutritional value and technological suitability of milk from various animal species used for dairy production. Compr. Rev. Food Sci. Food Saf. 10, 291–302.

Bezerra, T.K.A., de Araujo, A.R.R., do Nascimento, E.S., de Matos Paz, J.E., Gadelha, C.A., Gadelha, T.S., Bertoldo Pacheco, M.T., de Cássia, R., do Egypto Queiroga, R., de Oliveira, M.E.G., Madruga, M.S., 2016. Proteolysis in goat "coalho" cheese supplemented with probiotic lactic acid bacteria. Food Chem. 196, 359–366.

Bezerra, T.K.A., de Oliveira Arcanjo, N.M., de Araújo, A.R.R., de Queiroz, A.L.M., de Oliveira, M.E.G., Gomes, A.M.P., Madruga, M.S., 2017. Volatile profile in goat coalho cheese supplemented with probiotic lactic acid bacteria. LWT Food Sci. Technol. 76, 209–215.

Chiofalo, B., Liotta, L., Zumbo, A., Chifalo, V., 2004. Administration of olive cake from ewe feeding effect on milk yield and composition. Small Rumin. Res. 55, 169–176.

Claeys, W.L., Verraes, C., Cardoen, S., De Block, J., Huyghebaert, A., Reas, K., Dewettinck, K., Herman, L., 2014. Consumption of raw or heated milk from different species: an evaluation of the nutritional and potential health benefit: a review. Food Control 42, 188–201.

Corbo, M.R., Albenzio, M., De Angelis, M., Sevi, A., Gobbetti, M., 2001. Microbiological and biochemical properties of Canestrato Pugliese hard cheese supplemented with *Bifidobacteria*. J. Dairy Sci. 84, 551–561.

da Silveira, E.O., et al., 2015. The effects of inulin combined with oligofructose and goat cheese whey on the physicochemical properties and sensory acceptance of a probiotic chocolate goat dairy beverage. LWT Food Sci. Technol. 62 (Part 2), 445–451.

de Moraes, G.M.D., dos Santos, K.M.O., de Barcelos, S.C., Lopes, S.A., do Egito, A.S., 2018. Potentially probiotic goat cheese produced with autochthonous adjunct culture of *Lactobacillus mucosae*: microbiological, physicochemical and sensory attributes. LWT Food Sci. Technol. 94, 57–63.

Ehsani, A., Banihabib, E.K., Hashemi, M., Saravani, M., Yarahmadi, E., 2016. Evaluation of various properties of symbiotic yoghurt of buffalo milk. J. Food Process. Preserv. 40, 1466–1473.

El Hatmi, H., Jrad, Z., Oussaief, O., Nasri, W., Sbissi, I., Khorchani, T., Canabady-Rochelle, L.L.S., 2018. Fermentation of dromedary camel (*Camelus dromedarius*) milk by *Enterococcus faecium*, *Streptococcus macedonicus* as a potential alternative of fermented cow milk. Int. Dairy J. 34, 252–256.

El-Agamy, E.I., Nawar, M., Shamsia, S.M., Awad, S., Haenlein, G.F.W., 2009. Are camel milk proteins convenient to the nutrition of cow milk allergic children. Small Rumin. Res. 82, 1–6.

Ellinger, S., Linscheid, K.P., Jahnecke, S., Goerlich, R., Enbergs, H., 2002. The effect of mare's milk consumption on functional elements of phagocytosis of human neutrophil granulocytes from healthy volunteers. Food Agric. Immunol. 14 (3), 191–200.

El-Salam, M.H.A., El-Shibiny, S., 2011. A comprehensive review on the composition and properties of buffalo milk. Dairy Sci. Technol. 91, 663–699.

FAOSTAT, 2018. Dairy Production and Products. Retrieved from www.fao.org/agriculture/dairygateway/produccionlechera/animalslecheros/pequenosrumiantes/es/#.VDJ-ImeSySo.

Foglietta, F., Serpe, L., Canaparo, R., Vivenza, N., Riccio, G., Imbalzano, E., Gasco, P., Zara, G.P., 2014. Modulation of butyrate anticancer activity by solid lipid nanoparticle delivery: an in vitro investigation on human breast cancer and leukemia cell lines. J. Pharm. Pharm. Sci. 17, 231–247.

Gantner, V., Mijic, P., Baban, M., Skrtic, Z., Turalija, A., 2015. The overall and fat composition of milk of various species: a review. Mljekarstvo 65, 223–231.

Gibson, G.R., Hutkins, R., Sanders, M.E., Prescott, S.L., Reimer, R.A., Salminen, S.J., Scott, K., Stanton, C., Swanson, K.S., Cani, P.D., Verbeke, K., Reid, G., 2017. The International Scientific Association for Probiotics and Prebiotics (ISAPP) consensus statement on the definition and scope of prebiotics. Nat. Rev. Gastroenterol. Hepatol. 14, 491–502.

Gomes da Cruz, A., Buriti, F.C.A., Batista de Souza, C.H., Fonseca Faria, J.A., Isay, Saad, S.M., 2009. Probiotic cheese: health benefits, technological and stability aspects. Trends Food Sci. Technol. 20, 344–354.

Gómez-Cortés, P., Hervás, G., Mantecón, A.R., Juárez, M., de la Fuente, M.A., Frutos, P., 2008. Milk production, CLA content, and in vitro ruminal fermentation in response to high levels of soybean oil in dairy ewe diet. J. Dairy Sci. 91, 1560–1569.

Gómez-Ruiz, J.A., López-Expósito, I., Pihlanto, A., Ramos, M., Recio, I., 2008. Antioxidant activity of ovine casein hydrolysates: identification of active peptides by HPLC-MS/MS. Eur. Food Res. Technol. 227, 1061–1067.

Gueguen, L., Pointillart, A., 2000. The bioavailability of dietary calcium. J. Am. Coll. Nutr. 19, 119S–136S.

Han, X., Lee, F.L., Zhang, L., Guo, M.R., 2012. Chemical composition of water buffalo milk and its low-fat symbiotic yogurt development. Funct. Foods Health Dis. 2, 86–106.

Hasler, C.M., 2002. Functional foods: benefits, concerns and challenges a position paper from the American Council on Science and Health. J. Nutr. 132, 3772–3781.

Ibeagha-Awemu, M., Liu, J.-R., Zhao, X., 2009. Bioactive components in yogurt products eveline. In: Park, Y.W. (Ed.), Bioactive Components in Milk and Dairy Products, pp. 235–250 (Chapter 9).

Jatmiko, Y.D., Howarth, G.S., Barton, M.D., 2019. Naturally fermented milk and its therapeutic potential in the treatment of inflammatory intestinal disorders. AIP Conf. Proc. 2019, 060009. https://doi.org/10.1063/1.5061918.

Kalhoro, M.S., Visessanguan, W., Nguyen, L.T., Anal, A.K., 2019. Probiotic potential of *Lactobacillus paraplantarum* BT-11 isolated from raw buffalo (*Bubalus bubalis*) milk and characterization of bacteriocin-like inhibitory substance produced. J. Food Process. Preserv. 43, 14015.

Kiss, A., Naár, Z., Daróczi, L., Kukovics, S., Kukovics, F., Mirmazloum, I., Némedi, E., 2019. Changes of free fatty acid composition and number of lactic acid bacteria in three functional goat and sheep milk products fortified with inulin or fish oil. J. Food 17, 51–59. https://doi.org/10.1080/19476337.2018.1551936.

Kiss, A., Naár, Z., Daróczi, L., Némedi, E., Kukovics, S., 2014. Development of functional sheep and goat milk products and the investigation of functionality in various packaging materials. J. Hyg. Eng. Des. 7, 174–179.

Korlepara, R., Gopal, A., Kalla, M., Manthani, V., Keerthi, S., 2017. Camel milk an white gold of dessert—a review. Int. Arch. Appl. Sci. Technol. 8, 74–83.

Luna, P., Fontecha, J., Juárez, M., de al Fuente, M.A., 2005. Changes in the milk and cheese fat composition of ewes fed commercial supplements containing linseed with special reference to the CLA content and isomer composition. Lipids 40, 445–454.

Magjeed, N.A., 2005. Corrective effect of milk camel on some cancer biomarkers in blood of rats intoxicated with aflatoxin B1. J. Saudi Chem. Soc. 9, 253–263.

Malacarne, M., Martuzzi, F., Summer, A., Mariani, P., 2002. Protein and fat composition of mare's milk: some nutritional remarks with reference to human and cow's milk. Int. Dairy J. 12, 869–877.

Martins, A., Deliza, I.B.A., dos Santos, R., Walter, K.M.O., Martins, E.H.M., Rosenthal, J.M., 2018. Viability of probiotics in goat cheese during storage and under simulated gastrointestinal conditions. Food Bioprocess Technol. 11, 853–863.

Masoodi, T.A., Shafi, G., 2010. Analysis of casein alpha S1 & S2 proteins from different mammalian species. Bioinformation 4, 430–435.

Maurad, K., Meriem, K.H., 2008. Probiotic characteristics of *Lactobacillus plantarum* strains from traditional butter made from camel milk in arid regions (Sahara) of Algeria. Grasas Aceites 59, 218–224.

Meira, S.M.M., Daroit, D.J., Helfer, V.E., Corre^a, A.P.F., Segalin, J., Carro, S., Brandelli, A., 2012. Bioactive peptides in water-soluble extracts of ovine cheeses from Southern Brazil and Uruguay. Food Res. Int. 48, 322–329.

Minervini, F., Algaron, F., Rizzello, C.G., Fox, P.F., Monnet, V., Gobbetti, M., 2003. Angiotensin I-converting-enzyme-inhibitory and antibacterial peptides from *Lactobacillus helveticus* PR4 proteinase-hydrolyzed caseins of milk from 6 species. Appl. Environ. Microbiol. 69, 5297–5305.

Molkentin, J., 2000. Occurrence and biochemical characteristics of natural bioactive substances in bovine milk lipids. Br. J. Nutr. 84, 47–53.

Murtaza, M.A., Rehman, S.U., Anjum, F.M., Huma, N., Tarar, O.M., Mueen-ud-Din, G., 2012. Organic acids contents of buffalo milk Cheddar cheese as influenced by accelerated ripening and sodium salt. J. Food Biochem. 36, 99–106.

Murtaza, M.A., Huma, N., Sameen, A., Murtaza, M.S., Mahmood, S., Mueenud-Din, G., Meraj, A., 2014. Texture, flavor and sensory quality of buffalo milk Cheddar cheese as influenced by reducing sodium salt. J. Dairy Sci. 97, 6700–6707.

Murtaza, M.A., Huma, N., Shabbir, M.A., Murtaza, M.S., Rehman, M.A.U., 2017. Survival of micro-organisms and organic acid profile of probiotic Cheddar cheese from buffalo milk during accelerated ripening. Int. J. Dairy Technol. 70, 562–571.

Murti, T.W., Robiyati, E., JundiL, H.L., Ramadhani, F., Rustamadji, B., Suranindyah, Y.Y., 2016. Development of fermented mare's milk using mixed probiotic cultures. Media Peternakan vol. 39, 9–13.

Nadelman, P., Frazão, J.V., Vieira, T.I., Balthazar, C.F., Andrade, M.M., Alexandria, A.K., Cruz, A.G., Fonseca-Gon çalves, A., Maia, L.C., 2017. The performance of probiotic fermented sheep milk and ice cream sheep milk in inhibiting enamel mineral loss. Food Res. Int. 97, 184–190.

Nazzaro, F., Fratianni, F., Orlando, P., Coppola, R., 2010. The use of probiotic strains in the production of a donkey milk-based functional beverage. Int. J. Probiotics Prebiotics 5, 91–96.

Papadimitriou, C.G., Vafopoulou-Mastrojiannaki, A., Silva, S.V., Gomes, A.M., Malcata, F.X., Alichanidis, E., 2007. Identification of peptides in traditional and probiotic sheep milk yoghurt with angiotensin I-converting enzyme (ACE)-inhibitory activity. Food Chem. 105, 647–656.

Park, Y.W., Pariza, M.W., 2007. Mechanisms of body fat modulation by conjugated linoleic acid (CLA). Food Res. Int. 40, 311–323.

Park, Y.W., Juárez, M., Ramos, M., Haenlein, G.F.W., 2007. Physico-chemical characteristics of goat and sheep milk. Small Rumin. Res. 68, 88–113.

Perna, A., Intaglietta, I., Simonetti, A., Gambacorta, E., 2015. Donkey milk for manufacture of novel functional fermented beverages. J. Food Sci. 80, S1352–S1359.

Perna, A., Simonetti, A., Gambacorta, E., 2019. Phenolic content and antioxidant activity of donkey milk kefir fortified with sulla honey and rosemary essential oil during refrigerated storage. Int. J. Dairy Technol. 72, 74–81.

Perotti, M.C., Wolf, I.V., Addis, M., Comunian, R., Paba, A., Meinardi, C.A., 2014. Incorporation of probiotic bacteria (*Lactobacillus acidophilus* and *Bifidobacterium* spp.) in Argentinean ovine cheese. Dairy Sci. Technol. 94, 255–267.

Pieszka, M., Łuszczyński, J., Zamachowska, M., Augustyn, R., Długosz, B., Hędrzak, M., 2016. Is mare milk an appropriate food for people?—a review. Ann. Anim. Sci. 16, 33–51.

Politis, I., Theodorou, G., 2016. Angiotensin I-converting (ACE)-inhibitory and antiinflammatory properties of commercially available Greek yoghurt made from bovine or ovine milk: a comparative study. Int. Dairy J. 58, 46–49.

Prasanna, P.H.P., Charalampopoulos, D., 2019. Encapsulation in an alginate–goats' milk–inulin matrix improves survival of probiotic *Bifidobacterium* in simulated gastrointestinal conditions and goats' milk yoghurt. Int. J. Dairy Technol. 72, 132–141.

Pulina, G., Milán, M.J., Lavín, M.P., Theodoridis, A., Morin, E., Capote, J., Thomas, D.L., Francesconi, A.H.D., Caja, G., 2018. Invited review: current production trends, farm structures, and economics of the dairy sheep and goat sectors. J. Dairy Sci. 101, 6715–6729.

Quan, S., Tsuda, H., Miyamoto, T., 2008. Angiotensin I-converting enzyme inhibitory peptides in skim milk fermented with *Lactobacillus helveticus* 130B4 from camel milk in Inner Mongolia, China. J. Sci. Food Agric. 88, 2688–2692.

Radulović, Z., Cinović, J.M., Mirković, N., Mirković, M., Paunović, D.S., Ivanović, M., Seratlić, S., 2017. Survival of spray-dried and free-cells of potential probiotic *Lactobacillus plantarum* 564 in soft goat cheese. Anim. Sci. J. 88, 1849–1854.

Ranadheera, C.S., Evans, C.A., Adams, M.C., Baines, S.K., 2013. Production of probiotic ice cream from goat's milk and effect of packaging materials on product quality. Small Rumin. Res. 112, 174–180.

Ranadheera, C.S., Naumovski, N., Ajlouni, S., 2018. Non-bovine milk products as emerging probiotic carriers: recent developments and innovations. Curr. Opin. Food Sci. 22, 109–114.

Ranadheera, C.S., Evans, C.A., Baines, S.K., Balthazar, C.F., Cruz, A.G., Esmerino, E.A., Freitas, M.Q., Pimentel, T.C., Wittwer, A.E., Naumovski, N., Graça, J.S., Sant'Ana, A.S., Ajlouni, S., Vasiljevic, T., 2019. Probiotics in goat milk products: delivery capacity and ability to improve sensory attributes. Compr. Rev. Food Sci. Food Saf. 18, 867–882.

Rani, S., Pooja, K., Pal, G.K., 2017. Exploration of potential angiotensin converting enzyme inhibitory peptides generated from enzymatic hydrolysis of goat milk proteins. Biocatal. Agric. Biotechnol. 11, 83–88.

Rasmussen, N., Andersen, J.H., Jespersen, H., Mouritsen, O.G., Ditzel, H.J., 2010. Effect of free fatty acids and lysolipids on cellular uptake of doxorubicin in human breast cancer cell lines. Anti-Cancer Drugs 21, 674–677.

Raynal-Ljutovac, K., Park, Y.W., Gaucheron, F., Bouhallab, S., 2007. Heat stability and enzymatic modifications of goat and sheep milk. Small Ruminant Res. 68, 07–20.

Raynal-Ljutovac, K., Lagriffoul, G., Paccard, P., Guillet, I., Chilliard, Y., 2008. Composition of goat and sheep milk products: an update. Small Rumin. Res. 79, 57–72.

Recio, I., de la Fuente, A., Juárez, M., Ramos, M., 2009. Bioactive compounds in sheep milk. In: Park, Y.W. (Ed.), Bioactive Compounds in Milk and Dairy Products. Wiley-Blackwell, Iowa, pp. 83–104 (Chapter 4).

Rizqiati, H., Sumantri, C., Noorc, R.R., Damayanthi, E., Rianti, E.I., 2015. Characteristics of indigenous probiotic from river buffalo milk in North Sumatera Indonesia. Int. J. Sci. Basic Appl. Res. 22, 113–120.

Rizzello, C.G., Losito, I., Gobbetti, M., Carbonara, T., De Bari, M.D., Zambonin, P.G., 2005. Antibacterial activities of peptides from the water-soluble extracts of Italian cheese varieties. J. Dairy Sci. 88, 2348–2360.

Rodrigues, R., Guerra, G., Soaresc, J., Santos, K., Rolim, F., Assis, P., Araújo, D., de Araújo Júnior, R.F., Garcia, V.B., de Araújo, A.A., Queiroga, R., 2018. *Lactobacillus rhamnosus* EM1107 in goat milk matrix modulates intestinal inflammation involving NF-κB p65 and SOCs-1 in an acid-induced colitis model. J. Funct. Food 50, 78–92.

Salvatore, E., Pes, M., Mazzarello, V., Pirisi, A., 2014. Replacement of fat with long-chain inulin in a fresh cheese made from caprine milk. Int. Dairy J. 34, 1–5.

Sameer, B., Ganguly, S., Khetra, Y., Sabikhi, L., 2020. Development and characterization of probiotic buffalo milk ricotta cheese. LWT Food Sci. Technol. 121, 108944. https://doi.org/10.1016/j.lwt.2019.108944.

Santillo, A., Albenzio, M., 2008. Influence of lamb rennet paste containing probiotic on proteolysis and rheological properties of pecorino cheese. J. Dairy Sci. 91, 1733–1742.

Santillo, A., Albenzio, M., 2015. Sensory profile and consumers' liking of functional ovine cheese. Foods 4, 665–677.

Santillo, A., Albenzio, M., Quinto, M., Caroprese, M., Marino, R., Sevi, A., 2009. Probiotic in lamb rennet paste enhances rennet lipolytic activity and conjugated linoleic acid and linoleic acid content in Pecorino cheese. J. Dairy Sci. 92, 1330–1337.

Santillo, A., Albenzio, M., Bevilacqua, A., Corbo, M.R., Sevi, A., 2012. Encapsulation of probiotic bacteria in lamb rennet paste: effects on the quality of Pecorino cheese. J. Dairy Sci. 95, 3489–3500.

Santillo, A., Figliola, L., Ciliberti, M.G., Caroprese, M., Marino, R., Albenzio, M., 2018. Focusing on fatty acid profile in milk from different species after in vitro digestion. J. Dairy Res. 85, 257–262.

Serafeimidou, A., Zlatanos, S., Kritikos, G., Tourianis, A., 2013. Change of fatty acid profile, including conjugated linoleic acid (CLA) content, during refrigerated storage of yogurt made of cow and sheep milk. J. Food Compos. Anal. 31, 24–30.

Slačanac, V., Božanić, R., Hardi, J., Szabó, J.R., Lučan, M., Krstanović, V., 2010. Nutritional and therapeutic value of fermented caprine milk. Int. J. Dairy Technol. 63, 171–189.

Sultana, K., Godward, G., Reynolds, N., Arumugaswamy, R., Peiris, P., Kailasapathy, K., 2000. Encapsulation of probiotic bacteria with alginate-starch and evaluation of survival in simulated gastro-intestinal conditions and in yoghurt. Int. J. Food Microbiol. 62, 47–55.

Talpur, F.N., 2007. Fatty Acid Composition of Ruminant Milk, Meat and Dairy Products of Livestock in Sindh, Pakistan (Ph.D. thesis). University of Sindh, Jamshoro.

Turchia, B., Pedonese, F., Torracca, B., Fratini, F., Mancini, S., Galiero, A., Montalbano, B., Cerri, D., Nuvoloni, R., 2017. *Lactobacillus plantarum* and *Streptococcus thermophilus* as starter cultures for a donkey milk fermented beverage. Int. J. Food Microbiol. 256, 54–61.

Turkmen, N., 2017. The nutritional value and health benefits of goat milk components. In: Watson, R.R., Collier, R.J., Preedy, V.R. (Eds.), Nutrients in Dairy and Their Implications on Health and Disease. Academic Press, Cambridge, MA, pp. 441–449.

Turkmen, N., Akal, C., Özer, B., 2019. Probiotic dairy-based beverages: a review. J. Funct. Foods 53, 62–75.

Vasconcelos, F.M., Silva, H.L., Poso, S.M., Barroso, M.V., Lanzetti, M., Rocha, R.S., et al., 2019. Probiotic Prato cheese attenuates cigarette smoke-induced injuries in mice. Food Res. Int. 123, 697–703.

Vianna, F.S., Canto, A.C.V.C.S., da Costa-Lima, B.R.C., Salima, A.P.A.A., Costa, M.P., Balthazar, C.F., Oliveira, B.R., Rachid, R.P., Franco, R.M., Conte-Junior, C.A., Silva, A.C.O., 2017. Development of new probiotic yoghurt with a mixture of cow and sheep milk: effects on physicochemical, textural and sensory analysis. Small Ruminant Res. 149, 154–162.

Wijesinha-Bettoni, R., Burlingame, B., 2013. Milk and dairy product composition. In: Muehlhoff, E., Bennett, A., McMahon, D. (Eds.), Milk and Dairy Products in Human Nutrition. Food and Agriculture Organization of the United Nations, Rome.

Yagil, R., Saran, A., Etzion, Z., 1984. Camels milk: for drinking only? Comp. Biochem. Physiol. 78, 263–266.

CHAPTER 8

Fermented Milk: The Most Famous Probiotic, Prebiotic, and Synbiotic Food Carrier

NUR IMANINA ABDULLAH THAIDI[a,b] • LEONARDO RIOS-SOLIS[c] • MURNI HALIM[a,b]
[a]Department of Bioprocess Technology, Faculty of Biotechnology and Biomolecular Sciences, Universiti Putra Malaysia, Serdang, Selangor, Malaysia, [b]Bioprocessing and Biomanufacturing Research Center, Faculty of Biotechnology and Biomolecular Sciences, Universiti Putra Malaysia, Serdang, Selangor, Malaysia, [c]School of Engineering, Institute for Bioengineering, The University of Edinburgh, Edinburgh, United Kingdom

FERMENTED MILK

In recent years, the consumers' interest into the nutritious effects of foods they are consuming has greatly increased. The attention given has led to increasing demands on functional foods such as fermented milk-based products which are famous for probiotic, prebiotic, and synbiotic carriers (Sohail et al., 2013). The new concept of functional foods has led to the varieties in the production of foods that deliver not only essential nutrition but can also warrant good health and longevity (Fazilah et al., 2018).

According to International Dairy Federation, fermented milks could be briefly defined as follows: "Fermented milks are prepared from milk and/or milk products (i.e., any one or combinations of whole, partially or fully skimmed, concentrated or powdered milk, buttermilk powder, concentrated or powder whey, milk protein (such as whey proteins, whey concentrates, soluble milk proteins, edible casein and caseinates), cream, butter or milkfat—all of which have been manufactured from raw materials that have been at least pasteurized) by the action of specific microorganisms, which results in a reduction of the pH and coagulation" (IDF, 1992). Meanwhile, Codex Alimentarius defines fermented milk as "A milk product obtained by the fermentation of milk, which may have been manufactured from products obtained from milk with or without compositional modification, by the action of suitable microorganisms and resulting in reduction of pH with or without coagulation (isoelectric precipitation). These starter microorganisms shall be viable, active, and abundant in the product to the date of minimum durability. If the product is heat-treated after fermentation, the requirement for viable microorganisms does not apply" (Codex Alimentarius, 2020). Around the world, the traditional and industrial fermented milk products are known for at least 400 different names based on their methods of fermentation and processing (Surono and Hosono, 2011). The traditional fermented milk products are occasionally made by spontaneous fermentation in which the microbial identity is not known, and its inoculum is taken from the previous fermentation. In contrast, the industrially fermented milk products are developed based on scientific principles whereby the microbial cultures are identified and well characterized, and production processes are optimized. In general, the fermented milk products consumed in different countries may be broadly classified into three categories (Cho et al., 2018):

1. Sour and very high sour types, e.g., yogurt, curd;
2. Acid-cum alcohol in addition to lactic acid, e.g., kefir, kumiss;
3. Moderately sour type with pleasant aroma, e.g., culture milk.

Among these, dairy products such as yogurt and kefir are the most popular and widely used as the probiotic carriers. In particular, innovative processing of functional food products includes the addition of probiotics itself or with the incorporation of prebiotic, which is termed as synbiotic, and other bioactive components to improve the nutritional food value, rheological and physiochemical characteristics, and sensory profile of the food product (Fazilah et al., 2018).

Probiotics and Prebiotics in Foods. https://doi.org/10.1016/B978-0-12-819662-5.00012-4

Yogurt

Yogurt or yogurt is a long time known-appreciated dairy food product available in various textures (i.e., liquid, set, smooth), fat contents (luxury, low-fat, virtually fat-free), and flavors (natural, fruit, cereal) (Fazilah et al., 2018). It is traditionally made from the spontaneous or induced lactic acid fermentation of milk. Basically, yogurt can be classified into two groups, which are standard culture yogurt and bioyogurt or also known as functional or probiotic yogurt (Pandey et al., 2017).

Nowadays, the demand in yogurt consumption has tremendously increased due to the public's realization on the adverse health benefits by the consumption of probiotics as suggested by Metchnikoff in the 1990s (Mackowiak, 2013). Hence, wide variety types of yogurt have been formulated to satisfy consumers' need. Yogurt has generally been considered as an excellent source of high-quality protein, calcium, potassium, and vitamin B (Shah, 2007). Yogurt eating has been associated with a number of health benefits, including improvement of lactose metabolism, antimutagenic properties, anticarcinogenic properties, management of hypertension, antidiarrheal properties, immune system stimulation, and improvement in inflammatory bowel disease (Shakerian et al., 2015). The beneficial health effects of yogurt have been partly linked to the proteolysis products produced during fermentation and storage.

Standard yogurt is typically manufactured from the conventional starter culture strains, *Lactobacillus delbrueckii* ssp. *bulgaricus* and *Streptococcus thermophilus* (Mackowiak, 2013). Meanwhile, bioyogurt, functional or probiotic yogurt, is supplemented with probiotic strains such as *Bifidobacterium* and *Lactobacillus acidophilus* that are claimed to have numerous health benefits and should remain live at adequate numbers upon consumption (Fazilah et al., 2018). Standard yogurt is classified as regular, low-fat, and low-calorie yogurt, while functional yogurt is often classified for yogurt enriched with antioxidants or fiber. In comparison, standard yogurt and functional yogurt can be differentiated by their sensory qualities. Mainly, the functional yogurts or probiotic yogurts show a significantly less uncommon syneresis and higher smoothness, and are more homogenous than the conventional or standard yogurts (Majchrzak et al., 2010).

The conventional processing for manufacturing of yogurt involves several steps: starting from the initial treatment of milk (an optional step for using a high quality of raw milk (i.e., grade A or grade B milk as defined under the US Pasteurized Milk Ordinance, Food and Drug Administration (FDA) (Murphy et al., 2016), standardization of milk (supplementation with dairy ingredients to increase protein concentration), homogenization of the fortified milk, heat treatment, fermentation process, cooling and ending with the packaging of the final yogurt product) (Sfakianakis and Tzia, 2014) (Fig. 1).

Besides cow's milk, there are also different types of milk that have been used to produce various types of yogurt, such as camel's milk, goat's milk, ewe's milk, soy milk, coconut and tigernut milk, oat-based milk, and corn milk (Fazilah et al., 2018). These types of yogurts are developed to satisfy the market demand for consumer's allergenicity and the desire for vegetarian alternatives (Farnworth et al., 2007). Generally, organoleptic, rheological, textural, and microstructural properties of yogurt depend on several factors such as fermentation process, types of milk, starter cultures and probiotic strains, packaging as well as storage conditions (Fazilah et al., 2018). The expected rheology of yogurt is pseudoplastic (shear-thinning) behavior without initial resistance to deformation as the shear rate increased and the apparent viscosity decreased (Levinson et al., 2016).

Kefir

Kefir is a fermented milk drink with an acidic taste and creamy consistency produced by bacterial fermentation of kefir grains. The term "kefir" is derived from the word kef, which means "pleasant taste" in Turkish. Kefir grains are the natural starter for kefir and are recovered after the fermentation process. The grains contain a mixture of microorganisms (yeasts, acetic acid bacteria, and lactic acid bacteria), which coexist and interact to produce a unique fermented dairy product (Prado et al., 2015).

Kefir is a popular ethnic dairy product that is continuously subjected to process development mainly focusing on different additives of flavors, types of milk (i.e., cow, goat, camel, buffalo, or mare), fibers, grains, and many more in satisfying the demand of consumers (Farag et al., 2020). Kefir's chemical composition depends not only on the starter-kefir grains but also on its geographical origin, temperature, and time-related conditions of fermentation, and especially on the type and volume of the milk used (Prado et al., 2015; Zhong et al., 2016). The characteristic smell and flavor of kefir are due to the volatile and nonvolatile compounds generated upon fermentation via lipolysis, glycolysis, and proteolysis (Karaçalı et al., 2018). The physicochemical properties of kefir include an acidic pH of 4.6, alcohol of 0.5%–2%, acidic taste, and yeasty flavor. Additionally, carbon dioxide produced by the yeast flora contributes to its sharp acid and yeasty flavor.

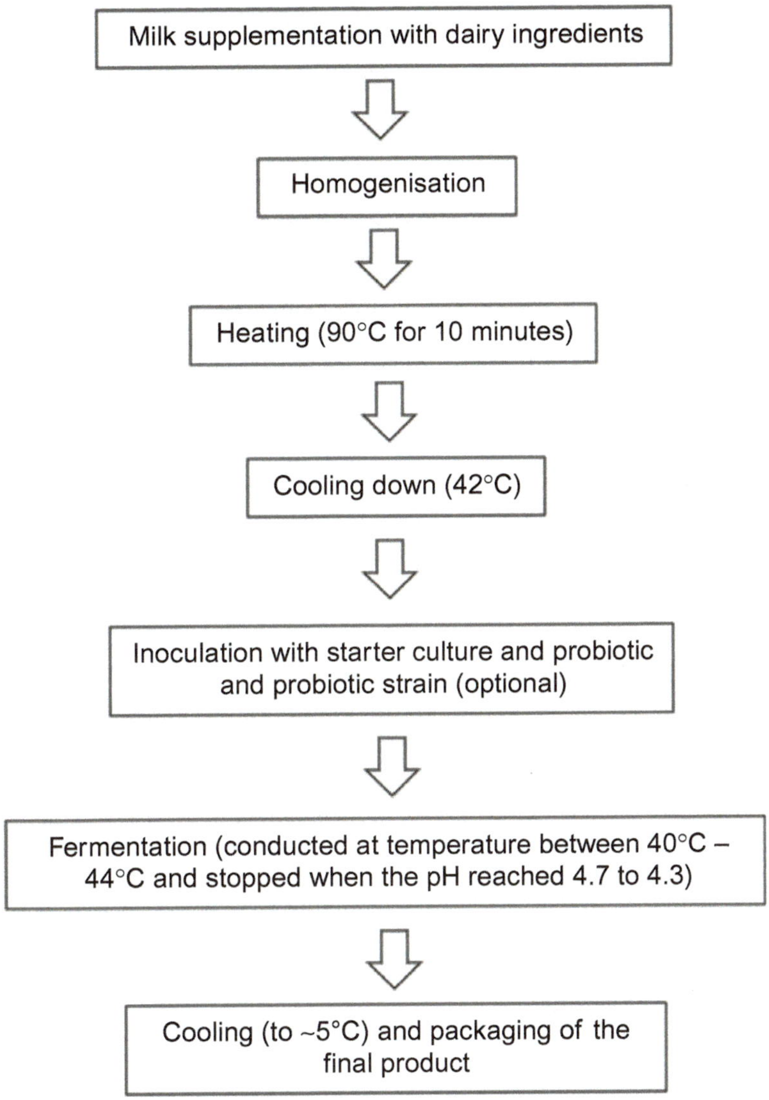

FIG. 1 Probiotic yogurt manufacturing processing steps.

Several schemes can be employed for kefir production, all sharing the same underlying principle. Kefir is first prepared by mixing two types of milk, such as mare, goat, or sheep milk (Cais-Sokolińska, 2016), or by adding additives such as native inulin, to improve its beneficial effect and final texture (Glibowski and Kowalska, 2012). An alternative way of producing kefir is to utilize nondairy substrates such as fruits and molasses to produce sugary kefir, which has unique sensory properties, such as a refreshing flavor due to the presence of ethanol, a fruity aroma due to esters as well as a body and texture attributed to its glycerol content (Fiorda et al., 2017).

Although there is an ideal relationship between the grains and the fermentation substrate (1:30 to 1:50 (w/v) in the case of animal milk), in practice, the measures are made empirically (Farnworth, 2005). Fermentation typically occurs at temperatures ranging from 8°C to 25°C, in a partially closed container, at a variable time from 10 to 40h (Fig. 2). However, the most common incubation time is 24h (Rosa et al., 2017). After fermentation, the grains are separated from the fermented milk by filtration using a sieve (Lopitz-Otsoa et al., 2006). When milk is used as a substrate, the kefir is similar to yogurt. The higher the fat

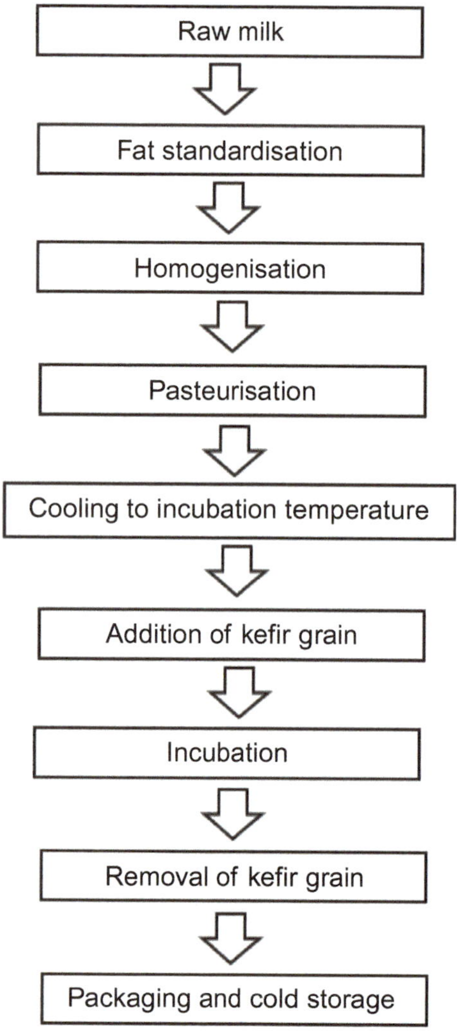

FIG. 2 Kefir manufacturing process steps.

content in the milk, the thicker and creamier the kefir (Rosa et al., 2017). Kefir grains may increase in size by up to 2% of the original to form a new biomass, which allows continuous production, since the grains can be further added to a fermentation substrate (Farnworth, 2005). Pure starter and lyophilized culture can be used, eliminating the step of recovering the kefir grains. Kefir can be consumed immediately after grain separation or can be refrigerated for later consumption. During the cooling step, alcoholic fermentation leads to the accumulation of CO_2, ethanol, and vitamin B complex (Rosa et al., 2017). This maturation step reduces the lactose content, making the product desirable for consumption by individuals with lactose intolerance and diabetes (Farnworth, 2005).

In general, kefir production faces more than one challenge, owing to the unique and diverse microflora of kefir grain, milk type, incubation time, and storage conditions. Furthermore, the sensory and physicochemical properties and the quality of kefir products hindered the mass production of kefir on an industrial scale (Fiorda et al., 2017). Such limitations may be due to the microbial diversity and interactions that influence the final product quality. Hence, additional studies are essential to improve and standardize production at an industrial level (de Oliveira Leite et al., 2013). Due to kefir's short shelf life and costly storage and packaging, the trend towards having dry kefir in powder form appears pertinent. Both spray drying and freeze drying are usually used to produce powdered kefir (Teijeiro et al., 2018).

FERMENTED MILK PRODUCTS AS PROBIOTIC FOOD CARRIERS

The term "probiotic" was derived from two Greek words which mean "for life" (Martín et al., 2015). The Food and Agriculture Organization (FAO) and the World Health Organization (WHO) of the United Nations described probiotic as a live microorganism that confers health benefits to the host by improving the microbial balance when administrated in an adequate amount (FAO/WHO, 2002). The first probiotic was introduced in 1907 by a Russian scientist and Nobel laureate Élie Metchnikoff, who found that a modification of gut microbiota is possible with the replacement of harmful bacteria with useful bacteria (de Simone, 2019). In recent years, new terms have been added to probiotic terminology to comprehensively reflect all the beneficial aspects of probiotics. Besides terms such as "paraprobiotics" (dead or inactivated cells of probiotics), "postbiotics" (healthful metabolites or probiotics), and psychobiotic (probiotics when ingested at adequate amount, exert beneficial effects on mental health), Zendeboodi et al. (2020) have just proposed several other new terminologies to explicitly cover all healthful effects of probiotics when present in different forms such as "true probiotic" that refers to viable and active microbial cells, "pseudoprobiotic" that implies viable but inactive cell in the forms of vegetative or spore, and "ghost probiotic" that entails nonviable cell in the forms of intact and ruptured. Probiotics could be an excellent substitute for the widespread use of antibiotics.

With the development of new food carriers, probiotics have started to be applied in many types of food products. Several important criteria for the selection of probiotics for use in food industry include the following: the bacteria must be nonpathogenic, nontoxic, acid-tolerant, and bile-tolerant to survive passage through the gastrointestinal tract; viable and present in sufficient quantity during consumption; colonize at the target site; and most importantly, they must be able to withstand the extreme conditions during manufacturing processes (McKinley, 2005; Vasiljevic and Shah, 2008).

In dairy fermentation, probiotic plays a role in assisting the preservation of milk by generating lactic acid and possibly antimicrobial compounds, and production of desirable flavor compounds (i.e., acetaldehyde, diacetyl in yogurt) and other metabolites (Fazilah et al., 2018). These properties provide a product with organoleptic properties desired by the consumers, improve the nutritional value of food and provision of special therapeutic or prophylactic properties such as cancer (Davoodi et al., 2013), and control the serum cholesterol level (Flore et al., 2016). Nevertheless, to be considered as probiotics, the strain should be viable (10^6–10^8 CFU/mL) during processing and storage as well as able to withstand in harsh gastrointestinal conditions (Burgain et al., 2011). There is no universal strain that can provide all purposed benefits, not even strains of the same species because health benefits presented by probiotic bacteria are very strain specific (Vasiljevic and Shah, 2008).

The two most common types of microbes being used as probiotics are *Lactobacillus* and bifidobacteria (Solanki et al., 2013). The core group consist of four genera: *Bifidobacterium* (*adolescentis, animalis, bifidum, breve,* and *longum*), *Lactobacillus* (*acidophilus, casei, fermentum, gasseri, johnsonii, reuteri, paracasei, plantarum, rhamnosus,* and *salivarius*), and *Pediococcus* and *Streptococcus* (de Simone, 2019; Khalid, 2011). These bacteria are also considered as the Generally Recognized as Safe (GRAS) (Oakey et al., 1995).

Based on the extensive studies conducted over the years, probiotics have been proven to confer various health benefits to the consumer. The benefits include lowering cholesterol level (Zeng et al., 2011), inhibition of undesired microorganism against food poisoning, lactic acid production for the fermentation process, and bacteriocin production for immune system enhancement (Elliott and Facklam, 1996). Probiotics have been mostly applied in dairy products, especially fermented milk products such as yogurt and kefir. Other potential therapeutic effects of fermented milk products containing different probiotic strains are summarized in Table 1.

Yogurt as Probiotic Food Carrier

Yogurt is considered the most popular vehicle for the delivery of probiotics for the consumers. The National Yogurt Association (NYA) of the United States specifies that bioyogurt products must contain 10^8 CFU/g lactic acid bacteria (LAB) at the time of manufacture to using "Live and Active Culture" logo, while the Australian Food Standards Code regulations require that the LAB used in yogurt fermentation must be present in a viable form in the final product; nonetheless, the numbers of CFU/g are not specified (Pandey et al., 2017).

In general, probiotic strains are selected based on their safety, nutritive value, and health-promoting properties besides other valuable features that may influence the shelf life, texture, and appearance of the probiotic yogurt (Casarotti et al., 2014). Furthermore, the selection criteria of probiotic strains must also consider the possible interactions among strains and dairy products as well as starter culture bacteria to optimize their performance and survival during storage. Conventional yogurt starter culture strains, *L. delbrueckii* ssp. *Bulgaricus* and *S. thermophilus*, lack the ability to survive passage through the intestinal tract (McFarland, 2015). These starter culture strains may not play a significant role as probiotics in the human gut due to their incapability of colonizing the human intestine. Therefore, the current trend is to add other probiotic strains during yogurt fermentation along with the starter culture bacteria to induce the probiotic effect (Fazilah et al., 2018).

It is also a common practice to combine probiotic strains with yogurt starter culture bacteria to reduce fermentation time (Fazilah et al., 2019; Damin et al., 2008). Nevertheless, some probiotic bacteria grow slowly in milk due to the lack of essential proteolytic activity, and their acidifying characteristic may affect the product texture (Sodini et al., 2002). In comparison with yogurt starter culture, probiotic bacteria are often having a poor acidification performance in milk (Almeida et al., 2008). The addition of probiotic culture can reduce the acid accumulation during the storage period (Kailasapathy, 2006). Furthermore, post-exopolysaccharides have been observed in yogurts supplemented with probiotic cultures compared with yogurt without probiotics (Han et al., 2016). High exopolysaccharides may provide a better texture for yogurt. The microbial exopolysaccharides may improve body and texture of fermented products and serve as emulsifying or gelling agents, thickening and stabilizing agents (Fazilah et al., 2018).

Among various lactic acid bacteria, *Bifidobacterium* and *Lactobacillus* are the commonly selected genera

TABLE 1
Potential therapeutic effects of fermented milk products.

Product	Probiotic strain	Potential therapeutic effects	Reference
Fermented milk	*Lactococcus lactis* KLDS4.0325	Alleviate folate deficiency	Jiao et al. (2020)
Yogurt fortified with *Synsepalum dulcificum*	*Streptococcus thermophilus* and *Lactobacillus delbrueckii* ssp. *Bulgaricus*	Potential inhibition of α-amylase and α glucosidase enzymes (antidiabetic properties)	Fazilah et al. (2020)
Kefir	*Lactococcus lactis* ssp. *lactis, Lactococcus lactis* ssp. *cremoris, Lactococcus lactis* ssp. *diacetylactis, Leuconostoc mesenteroides* ssp. *cremoris, Lactobacillus kefir, Kluyveromyces marxianus*, and *Saccharomyces unisporus*	Potential improvements in glycemic status, inflammation-related indicators, and blood pressure	Bellikci-Koyu et al. (2019)
Commercial yogurt	*Lactobacillus acidophilus, L. bulgaricus*, and *Streptococcus thermophilus*	Reduce the risk of depression and cardiovascular diseases by increasing serotonin	Kim et al. (2018)
Commercial fermented milk Actimel (Danone)	*Lactobacillus casei* CNCM I-1518, *Streptococcus thermophilus* and *Lactobacillus delbrueckii* ssp. *bulgaricus*	Lower incidence of rhinopharyngitis in children (3- to 6-year-old)	Andrey et al. (2016)
Kefir	*Streptococcus thermophiles; Lactobacillus casei, Lactobacillus acidophilus* and *Bifidobacterium lactis*	Decreased the fasting blood glucose and HbA1C levels in diabetic patients	Ostadrahimi et al. (2015)
Probiotic yogurt	*Lactobacillus acidophilus* LA-5 and *Bifidobacterium lactis* BB-12	Improved hepatic enzymes, serum total cholesterol, and low-density lipoprotein cholesterol levels in studied subjects and might be useful in the management of NAFLD risk factors	Nabavi et al. (2014)
Probiotic yogurt	*Lactobacillus reuteri* (NCIMB 30242)	Significant reductions in LDL cholesterol, total cholesterol, and non-HDL cholesterol in hypercholesterolemic subjects	Jones et al. (2012)
Probiotic fermented mare's milk	*Streptococcus thermophilus, Lactobacillus acidophilus*, and *Bifidobacterium bifidum*	Provide a protective effect against brain and kidney alterations produced by mercury toxicity	Abdel-Salam et al. (2010)

added in the probiotic yogurt products (Chen et al., 2017). Generally, the efficiency of added probiotic bacteria in yogurt depends on dose level and their viability must be maintained throughout storage, and the probiotic bacteria survive in the gut environment (Aryana and McGrew, 2007). Several brands of probiotic yogurt available in the market have been analyzed to have an inadequate presence of viable cells of probiotic strains such as L. *acidophilus* and bifidobacteria (Shah, 2000). *Lactobacillus gasseri* 4/13 was successfully applied as an adjunct culture to yogurt starters (*L. delbrueckii* ssp. *bulgaricus* and *S. thermophilus* in combination with a commercial Direct Vat Set (DVS) yogurt starter cultures (LBB 41–8 or LBB 5-54V or LBB 435)) producing yogurt with the well-accepted taste (Baltova and Dimitrov, 2014). The concentration of viable L. *gasseri* 4/13 remained above the critical threshold of 10^6 CFU/mL for 21-day storage period. Furthermore, the human-origin probiotic strain, *L. gasseri* 4/13, could also be considered as an attractive adjunct monoculture in the production of functional foods as the strain was demonstrated to have a high rate of adhesion to Caco-2 human epithelial cells, to have good ability in reducing cholesterol, and to be capable of inducing the production of interferon gamma. Damin et al. (2008) reported that the combination of probiotic bacterium *Bifidobacterium animalis* ssp. *lactis* BL 04 with *S. thermophilus* produces rheological characteristics similar to yogurt, and hence, it was suitable to be used in the production of probiotic fermented milk. *L. rhamnosus* GR-1 and RC-14 are other probiotic strains with the ability to be delivered in a yogurt form with good survival rate, resulting in palatable taste and texture (Hekmat and Reid, 2006).

Upon consumption, the probiotic bacteria must be resistant to low pH, bile acids, and digestive enzymes to remain viable during their passage through the gastrointestinal tract (Halim et al., 2017). Encapsulation is an excellent way to protect the probiotics against severe environmental factors and create a microenvironment for the probiotics to remain viable during processing, storage, and passage through the gastrointestinal tract. Spray drying, extrusion, emulsion, and encapsulation in polymer system are among the commonly reported techniques. An assessment study conducted by Krasaekoopt and Tandhanskul (2008) found that the consumer acceptances for plain and fruit yogurt containing probiotic beads were as high as 82.3% and 94.9%, respectively. Probiotic cells can also be incorporated in yogurt via immobilization in natural supports, including fruits and grains. For instance, yogurt supplemented with immobilized L. *casei* on fresh apple pieces, wheat grains, or dried raisins showed an improved cell viability (7 log CFU/g) after 60 days of storage at 4°C than that obtained in yogurt with free probiotic cells (Bosnea et al., 2017). Other improved cell survival of encapsulated probiotics while incorporated in yogurt is tabulated in Table 2.

Kefir as Probiotic Food Carrier

Kefir is a fermented product involving coevolution of a microbial consortium in the presence of bacteria and yeasts in a symbiotic matrix. Nowadays, there is an increasing commercial interest in using kefir as an effective food matrix for the delivery of probiotic microorganisms. The microorganisms present in kefir possess probiotic potential that is nutritionally beneficial and proven to inhibit a number of pathogen and spoilage microorganisms (Paucean and Carmen, 2008). The microbiota in kefir can produce antagonistic substances, such as organic acids and bacteriocins (Silva et al., 2009), and interfere with the adherence of pathogenic bacteria in the intestinal mucosa (Xie et al., 2012), potentially contributing to the improvement of gut health. Furthermore, numerous bacterial species isolated from kefir demonstrated high resistance to the low pH and bile salts in the gastrointestinal tract, and can adhere to the intestinal mucus (Golowczyc et al., 2008).

The identification of microbiota present in kefir and its grains is important since it is directly related to the quality of the product (Garbers et al., 2004). Nowadays, the understanding of microbial ecology of foods has dramatically changed. The use of a combined approach using culture-dependent and culture-independent methods, such as functional genomics, transcriptomics, proteomics, and metabolomics, is encouraged to understand the behavior of microorganisms in foods (Ercolini, 2013). In particular, different methodologies have been applied to study the microbiota of kefir; however, the classical approach of culturing microorganisms in nutrient media (universal and selective) and identifying isolated cultures is still being performed (Chen et al., 2009). As tabulated in Table 3, various lactic acid bacteria and yeast strains have been isolated and identified in kefir and kefir grains from different countries.

Kefir is proposed as one of the factors associated with the long life expectancy of the people of Caucasus, owing to its many health benefits mainly attributed to kefir's protein, vitamin, lipid, mineral, amino acid, and microelement compositions (Farag et al., 2020). Moreover, the fermentation process enriches the contents of vitamins B_1, B_{12}, K, folic acid, calcium, and amino acids, adding to kefir's health benefits. The vitamin composition of kefir is affected by the type of milk and microbiological flora used in its production. *Propionibacterium peterssoni* and *Propionibacterium pituitosum* produced vitamin B_{12}, whereas *Freudenreichii* ssp. *Propionibacterium shermanii* supported more vitamin B_6 production (Sarkar, 2007). Kefir is rich in amino acids serine, threonine, alanine, lysine, valine, isoleucine, methionine, phenylalanine, and tryptophan, which play a significant role in the central nervous system. The branched-chain amino acids that are also found in kefir improve the cognitive recovery of patients with severe traumatic brain injury (Bifari and Nisoli, 2017). Kefir also contains partially digested proteins (e.g., caseins) that aid in its digestion and absorption by the body (Simova et al., 2006). Macroelements enriched in kefir include calcium, magnesium, potassium, and sodium that aid in the utilization of carbohydrates, fats, and proteins for cell growth, maintenance, and energy (Bakircioglu et al., 2018). Kefir also contains microelements, including iron, zinc, and copper, which are of special value in cellular metabolism and blood production. Kefir possesses a myriad of bioactive properties, including anticancer, antimicrobial, anti-inflammatory, hypocholesterolemia, wound healing, antioxidant, and gastrointestinal (Farag et al., 2020).

The regular consumption of probiotics could improve blood glucose levels (Gomes et al., 2014). This effect has been mainly attributed to the probiotic ability to positively modulate the composition of the intestinal microbiota and hence reduce intestinal

TABLE 2
Size and cell viability of encapsulated probiotic beads in yogurts.

Probiotic strains	Encapsulation technique	Size of bead	Probiotic cell viability/survival	Reference
L. acidophilus ATTC-4356	Extrusion technique using sodium alginate and carrageenan	Mean sizes of sodium alginate and carrageenan beads are 714 μm and 726 μm, respectively	The numbers of viable encapsulated cells decreased by just 1.17 log CFU/mL (and 1.5 log CFU/mL) (carrageenan beads), while in case of nonencapsulated cells decreased by 3.85 log CFU/mL during 21 days of storage	Afzaal et al. (2019)
Saccharomyces boulardii and *Enterococcus faecium*	Emulsion and internal gelation	Microbeads exhibited a spherical and smooth shape, sizes between 300 and 500 μm	The number of viable encapsulated cells decreased by just 1.76 log CFU/g, while in case of nonencapsulated cells decreased by 4.82 log CFU/g	Qi et al. (2019)
Lactococcus lactis Gh1	Microencapsulation with gum arabic and *Synsepalum dulcificum* via spray drying	–	Higher probiotic cell viability (~ 10^7 CFU/mL) when incorporated into yogurt compared with non-microencapsulated cells (10^5 CFU/mL)	Fazilah et al. (2019)
W. paramesenteroides (CP023501.1), *L. paraplantarum* (AB362736.1), *E. faecalis* (HQ802261.1), and L. *plantarum* (MF369875.1)	Extrusion technique using arrowroot starch, maltodextrin, and whey protein	Microsphere size between 318 and 681.34 μm	Cell survival rates of maltodextrin-coated L. *paraplantarum*, *E. faecalis*, *L. plantarum*, *W. paramesenteroides* were 68.2%, 73.63%, 77.34%, and 73.73%, respectively, while for arrowroot starch-coated *L. paraplantarum*, *E. faecalis*, *L. plantarum*, *W. paramesenteroides* were 69.45%, 79.24%, 77.85%, and 72.77%, respectively, during storage at 4°C for 90 days.	Samedi and Charles (2019)
Lactobacillus rhamnosus	Extrusion technique using sodium alginate and carrageenan	Round in shape with a diameter of 0.33 mm and 0.5–1 mm for sodium alginate and carrageenan beads, respectively	Yogurts containing alginate-microencapsulated probiotic retained culture values of 6–7 log CFU/g up to 10 days of storage at 4°C, while yogurts containing carrageenan beads and free cells only managed to retain the same values up to 5 days of cold storage	Kumar and Kumar (2016)

permeability, oxidative stress, and inflammation. Oral administration of milk and milk kefir in mice inoculated with sarcoma 180 ascites tumor resulted in the inhibition of 64.8% and 70.9% of tumor growth, respectively, compared with the administration of unfermented milk (Liu et al., 2002). Furthermore, kefir was able to induce cell lysis by apoptosis and increase levels of IgA in the intestinal mucosa of animals after 30 days of consumption. These data suggest that kefir is a promising probiotic in cancer prevention. In addition, there are some evidences which indicate that probiotic bacteria or their fermented products play an important role in controlling blood pressure. For instance, Quirós et al. (2005) found that kefir can inhibit the activity of angiotensin-converting enzyme (ACE) through the action of bioactive peptides generated from casein during the milk fermentation process. In the meantime, Maeda et al. (2004) stated that the antihypertensive activity observed in their study was due to the ability of kefiran to inhibit ACE activity.

TABLE 3
Microorganisms found in kefir and kefir grains from different countries.

Microorganism	Country	Reference
Lactobacillus kefiri, *Lactobacillus kefiranofaciens*, *Leuconostoc mesenteroides*, *Lactococcus lactis*, *Lactococcus lactis* spp. *cremoris*, *Gluconobacter frateurii*, *Acetobacter orientalis*, *Acetobacter lovaniensis*, *Kluyveromyces marxianus*, *Naumovozyma* sp., *Kazachstania kefir*	Belgium	Korsak et al. (2015)
Lactobacillus kefiranofaciens, *Dekkera anomala*, *Streptococcus thermophilus*, *Lactococcus lactis*, *Acetobacter* sp., *Lactobacillus lactis*, *Enterococcus* sp., *Bacillus* sp., *Acetobacter fabarum*, *Acetobacter lovaniensis*, *Acetobacter orientalis*	Italy	Garofalo et al. (2015)
Saccharomyces boulardii, *Kazachstania unispora*, *Saccharomyces cerevisiae*, *Kodamaea ohmeri*	Malaysia	Azhar and Munaim (2019)
Lactobacillus kefiranofaciens, *Saccharomyces cerevisiae*, *Saccharomyces unisporus*, *Issatchenkia occidentalis*, *Kluyveromyces marxianus*	Taiwan	Diosma et al. (2013)
Lactobacillus kefiranofaciens spp., *Kefirgranum*, *Lactobacillus parakefiri*, *Lactobacillus kefiri*, *Kluyveromyces marxianus*, *Kazachstania exigua*, *Rhodosporidium kratochvilovae*	Slovenia	Vardjan et al. (2013)
Lactobacillus helveticus, *Lactobacillus kefir*, *Lactobacillus parakefir*	Russia	Mainville et al. (2006)

FERMENTED MILK PRODUCTS AS PREBIOTIC FOOD CARRIERS

Prebiotic was initially defined as "a non-digestible food ingredient that beneficially affects the host by selectively stimulating the growth and/or activity of one or a limited number of bacteria in the colon, and thus improves host health" (Gibson and Roberfroid, 1995). The definition was updated in 2016 at the meeting of the International Scientific Association of Probiotics and Prebiotics to "a substrate that is selectively utilized by host microorganisms conferring health benefit" (ISAPP, 2020). Most of prebiotics are from fiber sources, but not all fibers can be classified as prebiotics. Basically, there are three requirements needed for a substance to be considered as prebiotic:

- Resists gastric acidity, hydrolysis by mammalian enzymes, and absorption in the upper gastrointestinal tract.
- Is fermented by the intestinal microflora.
- Selectively stimulates the growth and/or activity of intestinal bacteria potentially associated with health and well-being.

Prebiotics can be categorized into three types: saccharide derivatives, proteins or peptides, and lipids (Ozcan et al., 2016). Oligosaccharides are the food ingredients that have attracted a lot of interest as prebiotics (Van Loo et al., 1995). Glycosidic bonds allow resistance to digestive enzymes but can be metabolized by bacteria. Oligosaccharides occur naturally in commodities such as artichokes, onions, chicory, garlic, leeks, and to a lesser extent cereals. Meanwhile, the most extensively studied nondigestible oligosaccharide (NDO) prebiotics include inulin (such as Jerusalem artichoke, jicama, and chicory root), fructooligosaccharides, galactooligosaccharides, and lactulose. Prebiotics are commercially extracted and concentrated from fruits and vegetables through the hydrolysis of polysaccharides from dietary fibers or starch, or through enzymatic generation.

Nowadays, prebiotics are increasingly formulated as food ingredients especially in fermented milk products primarily to enhance the efficacy of probiotics, promote the growth of autochthonous beneficial bacteria, and improve the nutritionally quality of the products. For example, pectin-derived poly- and oligosaccharides isolated from apple pomace exhibited a prebiotic effect that may serve as a carrier of phytochemicals and a source of nutrients that stimulate the numbers and/or activities of probiotic bacteria and simultaneously reduce the adherence of various enteric pathogens (Wilkowska et al., 2019; Islamova et al., 2017). The health potentials of prebiotics include increased urinary hippurate concentrations (Dewulf et al., 2013), allergy-protective effects (Osborn and Sinn, 2013), improved gut barrier functions, and less inflammation (Cani et al., 2009). Prebiotics also play a role in defense mechanism of the immune system (Frei et al., 2015). Moreover, prebiotics are also shown to increase the absorption of calcium and magnesium, influence blood glucose levels, and improve plasma lipids (Csutak, 2010). In terms of sensory properties, some researchers reported on better flavor and texture profile that leads to good consumer acceptance (Issar et al., 2017; Jovanović et al., 2020), while some indicate prebiotic-containing fermented milk products have no significant differences in comparison with control (Kailasapathy, 2006; Hekmat and Reid, 2006).

Yogurt as Prebiotic Food Carrier

Yogurt has been shown to be an effective delivery vehicle for numerous prebiotic compounds. Most studies on prebiotic fortified yogurts emphasized on physicochemical properties (Wang et al., 2019; Wang et al., 2020), optimal amount of added prebiotic and its functionality (Jovanović et al., 2020), biological activity as well as sensorial properties (Fernandes et al., 2019; Issar et al., 2017).

In terms of production, the prebiotic is typically added into the yogurt starter culture-inoculated milk prior followed by incubation under ideal conditions of the microorganisms. For instance, stirred-type yogurts fortified with apple pomace flour were prepared by mixing the ingredient (1%, 3%, 5%, and 20%) with *L. acidophilus*-, *B. bifidum*-, and *S. thermophilus*-inoculated pasteurized cow's milk and fermented at 43°C (Jovanović et al., 2020). Once the pH reached 4.6, the yogurt was stirred and stabilized by cooling at 4°C for 24h. The biochemical, biological, sensorial, and textural properties of the formulated yogurt were found to be correlated with the amount of apple pomace flour in a dose-dependent manner. Five percent (5%) of apple pomace flour resulted in the highest total phenolic compound, which showed antioxidant and cytotoxic activities against colon cancer lines, while 3% of apple pomace flour is the optimal amount that represents the best sensorial and textural properties of the yogurt produced.

Previously, the supplementation of partially hydrolyzed guar gum to act as a soluble fiber enrichment was studied while formulating a functional yogurt (Mudgil et al., 2016). In this study, guar gum was first subjected to enzymatic hydrolysis by cellulase from *Aspergillus niger* and freeze-dried to a powder form prior to application in yogurt. Guar gum was observed to have prominent effects on several characteristics of yogurt. In comparison with control yogurt, guar gum-fortified yogurt showed an increase in pH, viscosity, water-holding capacity but lower titratable acidity and was generally well accepted in terms of functional and sensory qualities. In contrast, it was reported that the addition of 2.5% guar gum or 0.5% cress seed mucilage did not affect the pH, fermentation time, and proteolysis extent of set-yogurt throughout the 15-day storage period at 5 ± 2°C (Hassan et al., 2015). Nevertheless, the presence of these potential prebiotic compounds improved the quality of set-yogurt compared with the polysaccharide-free yogurt. Likewise, the nonflavored yogurts fortified with increasing concentrations of oligofructose also showed no influence on the pH, proteolysis, and the viability of yogurt starter cultures, *Streptococcus thermophilus*, and *Lactobacillus bulgaricus* during 28 days of cold storage (Cruz et al., 2013).

The accepted potential candidates of prebiotics must first be evaluated for their sensory attributes before commercial inclusion. The important criterion for prebiotic-enriched yogurt is texture that depends on several aspects such as type and dose of added prebiotics, protein and fat contents as well as fiber properties (Jovanović et al., 2020). Some prebiotics such as inulin may potentially alter the sensory perception of yogurt products due to their capacity as fat substitute, bulking agent, sweetener, and texture modifier (Guggisberg et al., 2009). Yogurts with good organoleptic (in terms of flavor, appearance, body, and texture) acceptance were previously formulated with several polysaccharides extracted from taro corm (*Arum colocasia*), mature okra fruit (*Hibiscus esculents*), and whole plant Jew's-mallow (*Corchorus olitorius*) (Hussein et al., 2011). Also, Allgeyer et al. (2010) reported a sensory evaluation of yogurts formulated with three different prebiotics such as soluble corn fiber, polydextrose, and chicory inulin with clover honey aroma, buttermilk aroma, butter aroma, sweetness, sourness, chalky mouthfeel, and viscosity, which were identified as their significant attributes. The yogurts produced can also be claimed as a good source of fiber as each contained between 2.5g and 5g fiber/serving.

Kefir as Prebiotic Food Carrier

Reports on the beneficial properties and nutritional value of kefir have led to an increase in research, production, and consumption. During milk fermentation by kefir grains, numerous functional compounds like bioactive peptides (i.e., with antihypertensive, antioxidative, antiallergenic, antitumor, antimicrobial, anti-inflammatory, and cholesterol-lowering activities), antimicrobial compounds (i.e., organic acids, alcohols, carbon dioxides, and bacteriocins), and heteropolysaccharides (i.e., kefiran) with potential prebiotic activity are formed (Nejati et al., 2020).

Nowadays, several fruits and fruit wastes with prebiotic potential have been formulated into kefir to serve as substrate fermentation besides offering alternative variation of taste. For instance, pomegranate that has a potential prebiotic effect (Li et al., 2015) has been supplemented with kefir in forms of juice and peel extracts as reported by Lagouri et al. (2019). The pomegranate juice and peel extract were separately added into kefir after it was harvested from the fermentation process. Both pomegranate juice- and pomegranate peel extract-enriched kefir have shown an increase in antioxidant activity and total phenolic content during 28 days of storage. Furthermore, the addition of pomegranate juice and peel extract decreased the pH and viscosity but increased the flow behavior index of kefir produced. Earlier, Montanuci et al. (2012) studied the effect of inulin addition on the whole and skim milk kefir that is prepared from the inoculation of either kefir grains or commercial starter culture. Inulin was added to the whole or skim milk and underwent heat treatment in a water batch (at 90 °C for 2–3 min) and cooling to 25 °C prior inoculation with kefir grains or starter culture and incubated at 25 °C for 24 h. The addition of inulin had no effect on the chemical characteristics and the viability of lactic acid bacteria and *Leuconostoc* but significantly affected firmness, syneresis, and the survival of *Lactococcus*, yeast, and acetic acid bacteria.

FERMENTED MILK PRODUCTS AS SYNBIOTIC FOOD CARRIERS

Synbiotic is a combination of probiotics and prebiotics that affects the host beneficially by improving the survival and implantation of selected live microbial strains in the gastrointestinal tract (Mohanty et al., 2018). Utilization of prebiotics by probiotics should be a prerequisite for synbiotic selection, to maintain a good synergy between the two and to maximize their beneficial effects upon formulation in fermented milk products. Among the commonly used probiotic strains for synbiotic product formulations are lactobacilli, *Bifidobacteria* spp., *Saccharomyces boulardii*, and *Bacillus coagulans*, whereas the major prebiotics used include oligosaccharides such as fructooligosaccharide (FOS), galactooligosaccharides (GOS), xyloseoligosaccharides (XOS), inulin, and prebiotic from natural sources like yacon roots and chicory (Pandey et al., 2015).

Synbiotics have significant benefits to health such as antimicrobial, anticancer, anti-allergic, and immune-stimulating properties (Cadieux et al., 2008). The combination of probiotic bacteria with the prebiotic compound can cause the release of antibacterial substances such as bacteriocin, which can retard the growth of pathogenic bacteria (Fazilah et al., 2018). Synbiotics favor the survival of the probiotic bacteria in the gastrointestinal tract as well as the absorption of bioactive components by the body (Cho et al., 2018). A study proved that the administration of synbiotics containing 4×10^8 CFU/m *L. casei* and 400 mg of inulin can give positive influences on the human plasma antioxidant capacity and antioxidant enzyme activities (Kleniewska et al., 2016). In addition, synbiotics can improve the absorption of minerals, prevent diarrhea, and optimize the assimilation of nutrients (Cadieux et al., 2008).

To date, many clinical studies have shown that synbiotic fermented milk products can further enhance the health-promoting effect of fermented milk (Knight et al., 2009; Wallace et al., 2011). Oral administration of a fermented milk product (prepared using fresh milk, sugar, and stabilizer and fermented with starter culture strains, *L. delbrueckii* ssp. *bulgaricus* and *Streptococcus thermophilus*) enriched with probiotics, *Lactobacillus acidophilus* NCFM and *Bifidobacterium lactis* Bi-07, and prebiotic isomaltooligosaccharide showed an increase in fecal bifidobacterial and lactobacilli in healthy adults compared with control subjects (Wang et al., 2012). After intervention with the fermented milk, a similar phenomenon was observed for the intestinal microbiota of mice with a significant increase in delayed-type hypersensitivity, plaque-forming cells, and half-hemolysis value. In addition, a synbiotic fermented milk formulated with prebiotics, fructooligosaccharide and isomaltooligosaccharide and probiotic *Lactobacillus acidophilus* ATCC® 4357™ have shown an antimicrobial effect (against *Escherichia coli* and *Staphylococcus aureus*) and antibiotic potentials with the reduction in blood glucose, urea, and creatinine levels of diabetic rabbits (Shafi et al., 2018).

Yogurt as Synbiotic Food Carrier

Among the functional foods consumed worldwide, yogurt is always of interest, emphasized for their interesting probiotic and synbiotic variations. Synbiotic yogurts are commonly developed to deliver the beneficial effects of probiotic-prebiotic combination besides improved consumer's acceptability, and offer alternative to the plain yogurt.

In yogurt, prebiotic acts as a substrate for the growth of probiotic bacteria and hence provides a positive influence on probiotic bacteria multiplication. For instance, the survivability of probiotic yeast, *Saccharomyces boulardii*, in synbiotic yogurt containing inulin as prebiotic was higher compared with that in yogurt without inulin (Sarwar et al., 2019). There was also an increase in cell count of *Bifidobacterium* in yogurt incorporated with rice compared with the plain yogurt due to the prebiotic effect (Kumari et al., 2015). Meanwhile, the supplementation of total dietary fibers from apple and banana to probiotic yogurt also increased the shelf life of probiotic strains, *L. acidophilus* and *B. animalis* ssp. *lactis* (do Espírito Santo et al., 2010). The effects can be associated with the high content of pectin and fructooligosaccharides in both fruits (Emaga et al., 2008). In contrast, Allgeyer et al. (2010), indicated a 2 to 3 log decrease in the survival of probiotics (*Bifidobacterium lactis* BB-12 and *Lactobacillus acidophilus* LA-5) for all yogurts fortified with soluble corn fiber, polydextrose, and chicory inulin during 30 days of cold storage.

In terms of sensory properties, the probiotic yogurt enriched with passion fruit fiber received a good score for appearance, color and odor, but the intensity of the flavor was considered weak (Fazilah et al., 2018). Synbiotic yogurt formulated with yeast, *S. boulardii*, and 1% inulin attained higher sensory score for taste, odor, color, and appearance as compared with plain yogurt (Sarwar et al., 2019). The production of carbon dioxide and alcohol by yeast strain contributes to foamy and refreshing taste (Wang et al., 2008) while the addition of inulin may improve mouthfeel and taste of fermented dairy products (Golob et al., 2004). The combination of yeast and inulin in a synbiotic yogurt also contributes to better textural attributes in term of hardness, cohesiveness, and adhesiveness compared with yogurt containing yeast without inulin (Sarwar et al., 2019). In the meantime, the addition of passion fruit rind that is rich in pectin in yogurt containing L. *acidophilus* and *B. animalis* ssp. *lactis* exhibited a higher viscosity than the control yogurt (Espírito-Santo et al., 2013).

Kefir as Synbiotic Food Carrier

Nowadays, recent progress has been achieved towards producing synbiotic kefir drink aiming to improve its quality in terms of both physicochemical and microbiological properties. Selecting the best prebiotic is the key for promoting the growth of the probiotic of interest and develop a synbiotic effect. A study by Likotrafiti et al. (2015) reported that *Lactobacillus kefiri* B6 isolated from kefir grains showed properties associated with probiotic bacteria and was able to grow well on a range of prebiotics (fructooligosaccharides, galactooligosaccharide (GOS), and lactulose) that facilitated the development of synbiotic combinations. In a medium supplemented with prebiotic galactooligosaccharide, *L. kefiri* B6 showed its antimicrobial effect from the significant inactivation of *Listeria monocytogenes*. The addition of inulin in kefir beverages prepared using whole and skim milk affects the survival of yeasts, *Lactococcus*, and acetic acid bacteria (Montanuci et al., 2012). The cell counts of yeasts, *Leuconostoc*, and acetic acid bacteria were increased or remained unchanged while 1- to 2-log decreases were observed for lactic acid bacteria and *Lactococcus* during 28 days of refrigerated storage. In the meantime, the addition of *Dioscorea esculenta* puree that contained about 14% inculin in a synbiotic kefir formulation increased the viability of both yeast and lactic acid bacteria, and produced a thicker kefir texture with increased viscosity (Kartika et al., 2020).

Synbiotic kefir is a highly nutritious drink with the combination of both probiotics and prebiotics that further enhanced the good effects on health. Positive effects on oxidative stress, systemic inflammation, blood cell damage, and cognitive dysfunction have been reported for Alzheimer's patients after the consumption of synbiotic-supplemented milk fermented with kefir grains (2 mL/kg/daily) for 90 days (Ton et al., 2020). A continuous dietary supplement of synbiotic kefir grains is hence recommended for Alzheimer's patients as the majority of the patients exhibited a marked improvement in memory, visual-spatial/abstraction abilities as well as language functions. The synergic effects of prebiotics grape seed flour (PGSF) and probiotics (kefir-derived LAB) have been systemically demonstrated in induced obese (DIO) mice (Cho et al., 2018). The health-beneficial effects of polyphenol-rich grape seed flour were closely associated with the modulation of the intestinal microbiota in DIO mice, indicating the prebiotic role of PGSF.

SUMMARY

Fermented milk products have always been considered as the most popular carriers for probiotic and prebiotic with products such as yogurts and kefir, which are widely employed as dairy matrices for probiotics and various prebiotic ingredients. Nevertheless, the inclusion of probiotics, prebiotics, and their synbiotic combination into fermented milk products encompasses several microbiological and technological challenges. These challenges include maintaining and enhancing the viability of probiotic cells during shelf life and utilizing the appropriate technology and optimum processing conditions to manufacture quality products that meet the market demand. In optimizing the delivery of probiotics, prebiotics, and synbiotics via fermented milk foods, several aspects must be taken into consideration that include probiotic strain selection, type, and concentration of prebiotic ingredients as well as processing and storage strategies. Hence, to maintain the consumer preference on the fermented food products, maintenance of product quality via constant research and technology innovation is necessary. Furthermore, the ability to regulate the composition of the microbiota by prebiotic dietary substances and probiotic microorganisms is an exciting approach in the control and treatment of some major diseases. Considering the fast evolution of fermented milk products at either the research or commercial stages, further development should demand an accurate measure of quality, safety, and efficacy to meet consumers' expectations on quality and claimable health benefits. The confirmation of health-promoting properties and efficacy would involve a broader range of studies from in vitro experiments to in vivo and clinical studies.

REFERENCES

Abdel-Salam, A.M., Al-Dekheil, A., Babkr, A., et al., 2010. High fiber probiotic fermented mare's milk reduces the toxic effects of mercury in rats. N. Am. J. Med. Sci. 2 (12), 569–575.

Afzaal, M., Khan, A.U., Saeed, F., et al., 2019. Functional exploration of free and encapsulated probiotic bacteria in yogurt and simulated gastrointestinal conditions. Food Sci. Nutr. 7 (12), 3931–3940.

Allgeyer, L.C., Miller, M.J., Lee, S.-Y., 2010. Sensory and microbiological quality of yogurt drinks with prebiotics and probiotics. J. Dairy Sci. 93 (10), 4471–4479.

Almeida, K.E., Tamime, A.Y., Oliveira, M.N., 2008. Acidification rates of probiotic bacteria in Minas frescal cheese whey. LWT- Food Sci. Technol. 41, 311–316.

Andrey, P., Violeta, N., Juergen, S., et al., 2016. Fermented milk consumption and common infections in children attending day-care centers: a randomized trial. J. Pediatr. Gastroenterol. Nutr. 63 (5), 534–543.

Aryana, K., McGrew, P., 2007. Quality attributes of yogurt with *Lactobacillus casei* and various prebiotics. LWT- Food Sci. Technol. 40, 1808–1814.

Azhar, M.A., Munaim, M.S.A., 2019. Identification and evaluation of probiotic potential in yeast strains found in kefir drink samples from Malaysia. Int. J. Food Eng. 2019, 20180347.

Bakirciogl u, D., Topraksever, N., Yurtsever, S., et al., 2018. Investigation of macro, micro and toxic element concentrations of milk and fermented milks products by using an inductively coupled plasma optical emission spectrometer, to improve food safety in Turkey. Microchem. J. 136, 133–138.

Baltova, K., Dimitrov, Z., 2014. Probiotic and cultural characteristic of strain *Lactobacillus gasseri* 4/13 of human origin. Biotechnol. Biotechnol. Equip. 28, 1084–1088.

Bellikci-Koyu, E., Sarer-Yurekli, B.P., Akyon, Y., et al., 2019. Effects of regular kefir consumption on gut microbiota in patients with metabolic syndrome: a parallel-group, randomized, controlled study. Nutrients 11 (9), 2089.

Bifari, F., Nisoli, E., 2017. Branched-chain amino acids differently modulate catabolic and anabolic states in mammals: a pharmacological point of view. Br. J. Pharmacol. 174, 1366–1377.

Bosnea, L.A., Kopsahelis, N., Kokkali, V., et al., 2017. Production of a novel probiotic yogurt by incorporation of *L. casei* enriched fresh apple pieces, dried raisins and wheat grains. Food Bioprod. Process. 102, 62–71.

Burgain, J., Gaiani, C., Linder, M., et al., 2011. Encapsulation of probiotic living cells: from laboratory scale to industrial applications. J. Food Eng. 104, 467–483.

Cadieux, P., Wind, A., Sommer, P., et al., 2008. Evaluation of reuterin production in urogenital probiotic *Lactobacillus reuteri* RC-14. Appl. Environ. Microbiol. 74, 4645–4649.

Cais-Sokolińska, D., 2016. Rheological, texture and sensory properties of kefir from mare's milk and its mixtures with goat and sheep milk. Mljekarstvo 66, 272–281.

Cani, P.D., Possemiers, S., Van de Wiele, T., et al., 2009. Changes in gut microbiota control inflammation in obese mice through a mechanism involving GLP-2-driven improvement of gut permeability. Gut 58 (8), 1091–1103.

Casarotti, S., Monteiro, D., Moretti, M., et al., 2014. Influence of the combination of probiotic cultures during fermentation and storage of fermented milk. Food Res. Int. 59.

Chen, T.H., Wang, S.-Y., Chen, K.-N., et al., 2009. Microbiological and chemical properties of kefir manufactured by entrapped microorganisms isolated from kefir grains. J. Dairy Sci. 92, 3002–3013.

Chen, C., Zhao, S., Hao, G., et al., 2017. Role of lactic acid bacteria on the yogurt flavour: a review. Int. J. Food Prop. 20, S316–S330.

Cho, Y.-J., Lee, H.G., Seo, K.-H., et al., 2018. Antiobesity effect of prebiotic polyphenol-rich grape seed flour supplemented with probiotic kefir-derived lactic acid bacteria. J. Agric. Food Chem. 66, 12498–12511.

Codex Alimentarius, 2020. Codex Standard for Fermented Milks (Codex Stan 243-2003). http://www.fao.org/fao-who-codexalimentarius/codex-texts/list-standards/en/. (Accessed 21 September 2020).

Cruz, A.G., Cavalcanti, R.N., Guerreiro, L.M.R., et al., 2013. Developing a prebiotic yogurt: rheological, physico-chemical and microbiological aspects and adequacy of survival analysis methodology. J. Food Eng. 114 (3), 323–330.

Csutak, E., 2010. Effect of various prebiotics on LA-5 and BB-12 probiotic bacteria multiplication, and on probiotic yoghurt production. Acta Univ Sapientiae, Alimentaria 3, 35–52.

Damin, M.R., Minowa, E., Alcantara, M.R., et al., 2008. Effect of cold storage on culture viability and some rheological properties of fermented milk prepared with yogurt and probiotic bacteria. J. Texture Stud. 39, 40–55.

Davoodi, H., Esmaeili, S., Mortazavian, A.M., 2013. Effects of milk and milk products consumption on cancer: a review. Compr. Rev. Food Sci. Food Saf. 12, 249–264.

de Oliveira Leite, A.M., Miguel, M.A.L., Peixoto, R.S., et al., 2013. Microbiological, technological and therapeutic properties of kefir: a natural probiotic beverage. Braz. J. Microbiol. 44, 341–349.

de Simone, C., 2019. The unregulated probiotic market. Clin. Gastroenterol. Hepatol. 17 (5), 809–817.

Dewulf, E.M., Cani, P.D., Claus, S.P., et al., 2013. Insight into the prebiotic concept: lesson from an exploratory, double blind intervention study with inulin-type fructans in obese women. Gut 62 (8), 1112–1121.

Diosma, G., Romanin, D.E., Rey-Burusco, M.F., et al., 2013. Yeasts from kefir grains: Isolation, identification, and probiotic characterization. World J. Microbiol. Biotechnol. 30, 43–53.

do Espírito Santo, A.P., Silva, R.C., Soares, F.A., et al., 2010. Açai pulp addition improves fatty acid profile and probiotic viability in yoghurt. Int. Dairy J. 20, 415–422.

Elliott, J.A., Facklam, R.R., 1996. Antimicrobial susceptibilities of *Lactococcus lactis* and *Lactococcus garvieae* and a proposed method to discriminate between them. J. Clin. Microbiol. 34, 1296–1298.

Emaga, T.H., Robert, C., Ronkart, S.N., et al., 2008. Dietary fibre components and pectin chemical features of peels during ripening in banana and plantain varieties. Bioresour. Technol. 99, 4346–4354.

Ercolini, D., 2013. High-throughput sequencing and metagenomics: moving forward in the culture-independent analysis of food microbial ecology. Appl. Environ. Microbiol. 79 (10), 3148–3155.

Espírito-Santo, A.P., Lagazzo, A., Sousa, A.L.O.P., et al., 2013. Rheology, spontaneous whey separation, microstructure and sensorial characteristics of probiotic yoghurts enriched with passion fruit fiber. Food Res. Int. 50, 224–231.

FAO/WHO, 2002. Guidelines for the Evaluation of Probiotics in Food. Report of a Joint FAO/WHO Working Group on Drafting Guidelines for the Evaluation of Probiotics in Food; Ontario, Canada. April 30, May 1,.

Farag, M.A., Jomaa, S.A., El-Wahed, A.A., et al., 2020. The many faces of kefir fermented dairy products: quality characteristics, flavour chemistry, nutritional value, health benefits, and safety. Nutrients 12 (2), 346.

Farnworth, E., 2005. Kefir—a complex probiotic. Food Sci. Tech. Bull. Funct. Foods 2, 1–17.

Farnworth, E.R., Mainville, I., Desjardins, M.P., et al., 2007. Growth of probiotic bacteria and bifidobacteria in a soy yogurt formulation. Int. J. Food Microbiol. 116, 174–181.

Fazilah, N.F., Ariff, A.B., Khayat, M.E., et al., 2018. Influence of probiotics, prebiotics, synbiotics and bioactive phytochemicals on the formulation of functional yogurt. J. Funct. Foods 48, 387–399.

Fazilah, N.F., Hamidon, N.H., Ariff, A.B., et al., 2019. Microencapsulation of *Lactococcus lactis* Gh1 with gum arabic and *Synsepalum dulcificum* via spray drying for potential inclusion in functional yogurt. Molecules 24, 1422.

Fazilah, N.F., Ariff, A.B., Khayat, M.E., et al., 2020. Anti-diabetic properties of *Synsepalum dulcificum* and its potential inclusion in functional yogurt. IOP Conf. Ser. Mater. Sci. Eng. 716 (2020), 012004.

Fernandes, P.A., Ferreira, S.S., Bastos, R., et al., 2019. Apple pomace extract as a sustainable food ingredient. Antioxidants 8 (6), 189.

Fiorda, F.A., de Melo Pereira, G.V., Thomaz-Soccol, V., et al., 2017. Microbiological, biochemical, and functional aspects of sugary kefir fermentation—a review. Food Microbiol. 66, 86–95.

Flore, E., Ngongang, T., Tiencheu, B., et al., 2016. Effects of probiotic bacteria from yogurt on enzyme and serum cholesterol levels of experimentally induced hyperlipidemic Wistar albino rats. Am. J. Biol. Life Sci. 4 (6), 48–55.

Frei, R., Akdis, M., O'Mahony, L., 2015. Prebiotics, probiotics, synbiotics, and the immune system. Curr. Opin. Gastroenterol. 31 (2), 158.

Garbers, I., Britz, T.J., Witthuhn, R.C., 2004. PCR-based denaturing gradient gel electrophoretictypification and identification of the microbial consortium present in Kefir grains. World J. Microbiol. Biotechnol. 20, 687–693.

Garofalo, C., Osimani, A.V., Aquilanti, L., et al., 2015. Bacteria and yeast microbiota in milk kefir grains from different Italian regions. Food Microbiol. 49, 123–133.

Gibson, G.R., Roberfroid, M.B., 1995. Dietary modulation of the human colonic microbiota: introducing the concept of prebiotics. J. Nutr. 125, 1401–1412.

Glibowski, P., Kowalska, A., 2012. Rheological, texture and sensory properties of kefir with high performance and native inulin. J. Food Eng. 111, 299–304.

Golob, T., Micovic, E., Bertoncely, J., et al., 2004. Sensory acceptability of chocolate with inulin. Acta Agric. Slov. 83, 221–231.

Golowczyc, M.A., Gugliada, M.J., Hollmann, A., et al., 2008. Characterization of homofermentative lactobacilli isolated from kefir grains: potential use as probiotic. J. Dairy Res. 75, 211–217.

Gomes, A.C., Bueno, A.A., de Souza, R.G.M., et al., 2014. Gut microbiota, probiotics and diabetes. Nutr. J. 13, 60.

Guggisberg, D., Cuthbert-Seteven, J., Piccinali, P., et al., 2009. Rheological, microstructural and sensory characterization of low-fat and whole milk set yogurt as influenced by inulin addition. Int. Dairy J. 19 (2), 107–115.

Halim, M., Mohd Mustafa, N.A., Othman, M., et al., 2017. Effect of encapsulant and cryoprotectant on the viability of probiotic *Pediococcus acidilactici* ATCC 8042 during freeze-drying and exposure to high acidity, bile salts and heat. LWT- Food Sci. Technol. 81, 210–216.

Han, X., Yang, Z., Jing, X., et al., 2016. Improvement of the texture of yogurt by use of exopolysaccharide producing lactic acid bacteria. Biomed. Res. Int. 2016, 7945675.

Hassan, L.K., Haggag, H.F., ElKalyoubi, M.H., et al., 2015. Physico-chemical properties of yoghurt containing cress seed mucilage or guar gum. Ann. Agric. Sci. 60, 21–28.

Hekmat, S., Reid, G., 2006. Sensory properties of probiotic yogurt is comparable to standard yogurt. Nutr. Res. 26 (4), 163–166.

Hussein, M.M., Hassan, F.A.M., Abdel Daym, H.H., et al., 2011. Utilization of some plant polysaccharides for improving yoghurt consistency. Ann. Agric. Sci. 56, 97–103.

IDF, 1992. General Standard of Identity for Fermented Milks. International Dairy Federation, Brussels. 1992; Standard no. 163.

ISAPP, 2020. Prebiotics. https://isappscience.org/for-scientists/resources/prebiotics/. (Accessed 21 September 2020).

Islamova, Z.I., Ogai, D.K., Abramenko, O.I., et al., 2017. Comparative assessment of the prebiotic activity of some pectin polysaccharides. Pharm. Chem. J. 51, 288–291.

Issar, K., Sharma, P.C., Gupta, A., 2017. Utilization of apple pomace in the preparation of fiber-enriched acidophilus yoghurt. J. Food Process. Preserv. 41, e13098.

Jiao, W., Wang, S., Guan, J., et al., 2020. Milk fermented with *Lactococcus lactis* KLDS4.0325 alleviates folate status in deficient mice. Food Funct. 11, 4571–4581.

Jones, M.L., Martoni, C.J., Parent, M., et al., 2012. Cholesterol-lowering efficacy of a microencapsulated bile salt hydrolase-active *Lactobacillus reuteri* NCIMB 30242 yoghurt formulation in hypercholesterolaemic adults. Br. J. Nutr. 107 (10), 1505e13.

Jovanović, M., Petrović, M., Miočinović, J., et al., 2020. Bioactivity and sensory properties of probiotic yogurt fortified with apple pomace flour. Foods 9 (6), 763.

Kailasapathy, K., 2006. Survival of free and encapsulated probiotic bacteria and their effect on the sensory properties of yoghurt. LWT- Food Sci. Technol. 39, 1221–1227.

Karaçalı, R., Özdem İr, N., Çon, A.H., 2018. Aromatic and functional aspects of kefir produced using soya milk and *Bifidobacterium* species. Int. J. Dairy Technol. 71, 921–933.

Kartika, Rahayuningsih, M., Setyaningsih, D., 2020. Production and characterization of synbiotic kefir cow milk with addition of *Dioscorea esculenta* puree. IOP Conf. Ser. Earth Environ. Sci. 472, 012026.

Khalid, K., 2011. An overview of lactic acid bacteria. Int. J. Biosci. 1, 1–13.

Kim, H.-K., Kim, S.-H., Jang, C.-S., et al., 2018. The combined effects of yogurt and exercise in healthy adults: implications for biomarkers of depression and cardiovascular diseases. Food Sci. Nutr. 6 (7), 1968–1974.

Kleniewska, P., Hoffmann, A., Pniewska, E., et al., 2016. The influence of probiotic *Lactobacillus casei* in combination with prebiotic inulin on the antioxidant capacity of human plasma. Oxid. Med. Cell. Longev. 2016, 1340903.

Knight, D.J., Gardiner, D., Banks, A., et al., 2009. Effect of synbiotic therapy on the incidence of ventilator associated pneumonia in critically ill patients: a randomised, double-blind, placebo-controlled trial. Intensive Care Med. 35, 854–861.

Korsak, N., Taminiau, B., Leclercq, M., et al., 2015. Short communication: evaluation of the microbiota of kefir samples using metagenic analysis targeting the 16S and 26S ribosomal DNA fragments. J. Dairy Sci. 98, 3684–3689.

Krasaekoopt, W., Tandhanskul, A., 2008. Sensory and acceptance assessment of yogurt containing probiotic beads in Thailand. Kasetsart J. (Nat. Sci.) 42, 99–106.

Kumar, A., Kumar, D., 2016. Development of antioxidant rich fruit supplemented probiotic yogurts using free and microencapsulated *Lactobacillus rhamnosus* culture. J. Food Sci. Technol. 53 (1), 667–675.

Kumari, A.G.I.P., Ranadheera, S., Prasanna, P.H.P., et al., 2015. Development of a rice incorporated synbiotic yogurt with low retrogradation properties. Int. Food Res. J. 22, 2032–2040.

Lagouri, V., Dimitreli, G., Kouvatsi, A., 2019. Effects of Greek pomegranate extracts in the antioxidant properties and storage stability of kefir. Curr. Bioact. Compd. 15 (4), 437–441.

Levinson, Y., Ish-Shalom, S., Segal, E., et al., 2016. Bioavailability, rheology and sensory evaluation of fat-free yogurt enriched with VD3 encapsulated in re-assembled casein micelles. Food Funct. 7, 1477–1482.

Li, Z., Summanen, P.H., Komoriya, T., et al., 2015. Pomegranate ellagitannins stimulate gut bacteria in vitro: implications for prebiotic and metabolic effects. Anaerobe 34, 164–168.

Likotrafiti, E., Valavani, P., Argiriou, A., et al., 2015. In vitro evaluation of potential antimicrobial synbiotics using *Lactobacillus kefiri* isolated from kefir grains. Int. Dairy J. 45, 23–30.

Liu, J.-R., Wang, S.-Y., Lin, Y.-Y., et al., 2002. Antitumor activity of milk kefir and soy milk kefir in tumor-bearing mice. Nutr. Cancer 44, 183–187.

Lopitz-Otsoa, F., Rementeria, A., Elguezabal, N., et al., 2006. Kefir: a symbiotic yeasts-bacteria community with alleged healthy capabilities. Rev. Iberoam. Micol. 23, 67–74.

Mackowiak, P.A., 2013. Recycling metchnikoff: probiotics, the intestinal microbiome and the quest for long life. Front. Public Health 1, 52.

Maeda, H., Zhu, X., Omura, K., et al., 2004. Effects of an exopolysaccharide (kefiran) on lipids, blood pressure, blood glucose, and constipation. Biofactors 22, 197–200.

Mainville, I., Robert, N., Lee, B., et al., 2006. Polyphasic characterization of the lactic acid bacteria in kefir. Syst. Appl. Microbiol. 29, 59–68.

Majchrzak, D., Hartjes, B., Dürrschmid, K., 2010. Conventional and probiotic yogurts differ in sensory properties but not in consumers' preferences. J. Sens. Stud. 25, 431–446.

Martín, M., Lara-Villoslada, F., Martínez, M., et al., 2015. Microencapsulation of bacteria: a review of different technologies and their impact on the probiotic effects. Innov. Food Sci. Emerg. Technol. 27, 15–25.

McFarland, L.V., 2015. From yaks to yogurt: the history, development, and current use of probiotics. Clin. Infect. Dis. 60, S85–S90.

McKinley, M.C., 2005. The nutrition and health benefits of yoghurt. Int. J. Dairy Technol. 58, 1–12.

Mohanty, D., Mishra, S., Mohapatra, S., et al., 2018. Prebiotics and synbiotics: recent concepts in nutrition. Food Biosci. 26, 152–160.

Montanuci, F.D., Pimentel, T.C., Garcia, S., et al., 2012. Effect of starter culture and inulin addition on microbial viability, texture, and chemical characteristics of whole or skim milk Kefir. Ciênc. Tecnol. Aliment. Campinas 32 (4), 840–861.

Mudgil, D., Barak, S., Khatkar, B.S., 2016. Development of functional yoghurt via soluble fiber fortification utilizing enzymatically hydrolyzed guar gum. Food Biosci. 14, 28–33.

Murphy, S.C., Martin, N.H., Barbano, D.M., et al., 2016. Influence of raw milk quality on processed dairy products: how do raw milk quality test results relate to product quality and yield? J. Dairy Sci. 99, 10128–10149.

Nabavi, S., Raffaf, M., Somi, M.H., et al., 2014. Effects of probiotic yogurt consumption on metabolic factors in individuals with nonalcoholic fatty liver disease. J. Dairy Sci. 97 (12), 7386–7393.

Nejati, F., Junne, S., Neubauer, P., 2020. A big world in small grain: a review of natural milk kefir starters. Microorganisms 8 (2), 192. https://doi.org/10.3390/microorganisms8020192.

Oakey, H.J., Harty, D.W.S., Knox, K.W., 1995. Enzyme production by lactobacilli and the potential link with infective endocarditis. J. Appl. Bacteriol. 78, 142–148.

Osborn, D., Sinn, J., 2013. Prebiotics in infants for prevention of allergy. Cochrane Database Syst. Rev. 2013 (3), CD006474.

Ostadrahimi, A., Taghizadeh, A., Mobasseri, M., et al., 2015. Effect of probiotic fermented milk (kefir) on glycemic control and lipid profile in type 2 diabetic patients: a randomized double-blind placebo-controlled clinical trial. Iran. J. Public Health 44 (2), 228–237.

Ozcan, O., Ozcan, T., Yilmaz-Ersan, L., et al., 2016. The use of prebiotics of plant origin in functional milk products. Food Sci. Technol. 4 (2), 15–22.

Pandey, K.R., Naik, S.R., Vakil, B.V., 2015. Probiotics, prebiotics and synbiotics—a review. J. Food Sci. Technol. 52, 7577–7587.

Pandey, A., Du, G., Sanromán, M.Á., Soccol, C.R., et al., 2017. Current Developments in Biotechnology and Bioengineering Food and Beverages Industry. Elsevier, Amsterdam.

Paucean, A., Carmen, S., 2008. Probiotic activity of mixed cultures of kefir's lactobacilli and non-lactose fermentating yeasts. Bull. UASVM Agric. 65 (2), 329–334.

Prado, M.R., Blandón, L.M., Vandenberghe, L.P.S., et al., 2015. Milk kefir: composition, microbial cultures, biological activities, and related products. Front. Microbiol. 6, 1177.

Qi, W., Liang, X., Yun, T., et al., 2019. Growth and survival of microencapsulated probiotics prepared by emulsion and internal gelatin. J. Food Sci. Technol. 56 (3), 1398–1404.

Quirós, A., Hernández-Ledesma, B., Ramos, M., et al., 2005. Angiotensin-converting enzyme inhibitory activity of peptides derived from caprine kefir. J. Dairy Sci. 88, 3480–3487.

Rosa, D.D., Dias, M.M.S., Grześkowiak, Ł.M., et al., 2017. Milk kefir: nutritional, microbiological and health benefits. Nutr. Res. Rev. 30, 82–96.

Samedi, L., Charles, A.L., 2019. Viability of 4 probiotic bacteria microencapsulated with arrowroot starch in the simulated gastrointestinal tract (GIT) and yoghurt. Foods 8, 175.

Sarkar, S., 2007. Potential of kefir as a dietetic beverage—a review. Br. Food J. 109, 280–290.

Sarwar, A., Aziz, T., Al-Dalali, S., et al., 2019. Physicochemical and microbiological properties of synbiotic yogurt made with probiotic yeast *Saccharomyces boulardii* in combination with inulin. Foods 8, 468.

Sfakianakis, P., Tzia, C., 2014. Conventional and innovative processing of milk for yogurt manufacture, development of texture and flavor: a review. Foods 3, 176–193.

Shafi, A., Raja, H.N., Farooq, U., et al., 2018. Antimicrobial and antidiabetic potential of synbiotic fermented milk: a functional dairy product. Int. J. Dairy Technol. 72 (1), 15–22.

Shah, N.P., 2000. Effects of milk-derived bioactives: an overview. Br. J. Nutr. 84 (Suppl 1), S3–S10.

Shah, N.P., 2007. Functional cultures and health benefits. Int. Dairy J. 17, 1262–1277.

Shakerian, M., Razavi, S.H., Ziai, S.A., et al., 2015. Proteolytic and ACE-inhibitory activities of probiotic yogurt containing non-viable bacteria as affected by different levels of fat, inulin and starter culture. J. Food Sci. Technol. 52, 2428–2433.

Silva, K.R., Rodrigues, S.A., Filho, L.X., et al., 2009. Antimicrobial activity of broth fermented with kefir grains. Appl. Biochem. Biotechnol. 152, 316–325.

Simova, E., Simov, Z., Beshkova, D., et al., 2006. Amino acid profiles of lactic acid bacteria, isolated from kefir grains and kefir starter made from them. Int. J. Food Microbiol. 107, 112–123.

Sodini, I., Lucas, A., Oliveira, M.N., et al., 2002. Effect of milk base and starter culture on acidification, texture, and probiotic cell counts in fermented milk processing. J. Dairy Sci. 85, 2479–2488.

Sohail, A., Turner, M., Coombes, A., et al., 2013. The viability of *Lactobacillus rhamnosus* GG and *Lactobacillus acidophilus* NCFM following double encapsulation in alginate and maltodextrin. Food Bioproc. Tech. 6, 2763–2769.

Solanki, H.K., Pawar, D.D., Shah, D.A., et al., 2013. Development of microencapsulation delivery system for long-term preservation of probiotics as biotherapeutics agent. Biomed. Res. Int. 2013, 620719.

Surono, I.S., Hosono, A., 2011. Types and standards of identity. In: John, W.F., Patrick, F.F., Paul, L.H.M. (Eds.), Encyclopedia of Dairy Sciences. vol. 2. Academic Press Elsevier Inc, London, pp. 470–476.

Teijeiro, M., Pérez, P., De Antoni, G., et al., 2018. Suitability of kefir powder production using spray drying. Food Res. Int. 112, 169–174. https://doi.org/10.1016/j.foodres.2018.06.023.

Ton, A., Campagnaro, B., Alves, G., et al., 2020. Oxidative stress and dementia in Alzheimer's patients: effects of synbiotic supplementation. Oxid. Med. Cell. Longev. 2020, 2638703.

Van Loo, J., Coussement, P., De Leenheer, L., et al., 1995. On the presence of Inulin and Oligofructose as natural ingredients in the western diet. Crit. Rev. Food Sci. Nutr. 35, 525–552.

Vardjan, T., Mohar Lorbeg, P., Rogelj, L., et al., 2013. Characterization and stability of lactobacilli and yeast microbiota in kefir grains. J. Dairy Sci. 96, 2729–2736.

Vasiljevic, T., Shah, N.P., 2008. Probiotics-from Metchnikoff to bioactives. Int. Dairy J. 18, 714–728.

Wallace, T.C., Guarner, K., Madsen, M.D., et al., 2011. Human gut microbiota and its relationship to health and disease. Nutr. Rev. 69, 392–403.

Wang, S.Y., Chen, H.C., Liu, J.R., et al., 2008. Identification of yeasts and evaluation of their distribution in Taiwanese kefir and viili starters. J. Dairy Sci. 91, 3798–3805.

Wang, S., Zhu, H., Lu, C., et al., 2012. Fermented milk supplemented with probiotics and prebiotics can effectively alter the intestinal microbiota and immunity of host animals. J. Dairy Sci. 95 (9), 4813–4822.

Wang, X., Kristo, E., LaPointe, G., 2019. The effect of apple pomace on the texture, rheology and microstructure of set type yogurt. Food Hydrocoll. 91, 83–91.

Wang, X., Kristo, E., LaPointe, G., 2020. Adding apple pomace as a functional ingredient in stirred-type yogurt and yogurt drinks. Food Hydrocoll. 100, 105453.

Wilkowska, A., Nowak, A., Antczak-Chrobot, A., et al., 2019. Structurally different pectic oligosaccharides produced from apple pomace and their biological activity in vitro. Foods 8, 365.

Xie, N., Zhou, T., Li, B., 2012. Kefir yeasts enhance probiotic potentials of *Lactobacillus paracasei* H9: The positive effects of coaggregation between the two strains. Food Res. Int. 45, 394–401.

Zendeboodi, F., Khorshidian, N., Mortazavian, A.M., et al., 2020. Probiotic: conceptualization from a new approach. Curr. Opin. Food Sci. 32, 103–123.

Zeng, X.Q., Pan, D.D., Zhou, P.D., 2011. Functional characteristics of *Lactobacillus fermentum* F1. Curr. Microbiol. 62, 27–31.

Zhong, Z., Hou, Q., Kwok, L., et al., 2016. Bacterial microbiota compositions of naturally fermented milk are shaped by both geographic origin and sample type. J. Dairy Sci. 99, 7832–7841.

CHAPTER 9

Probiotic and Prebiotic Plant Milk Dairy Foods

SWATI SETHI[a] • RAHUL KUMAR ANURAG[b]
[a]Food Grains & Oilseeds Processing Division, ICAR-Central Institute of Post-Harvest Engineering and Technology, Ludhiana, Punjab, India, [b]Agricultural Structures and Environmental Control Division, ICAR-Central Institute of Post-Harvest Engineering and Technology, Ludhiana, Punjab, India

INTRODUCTION

Foods deliberate to enhance gut well-being are often categorized as probiotics, prebiotics, and synbiotics (Stanton et al., 2005). The probiotic market is dominated by dairy-based beverages or products; however, increased interest in creating nondairy alternatives has progressed towards the development of novel probiotic formulations to address lactose intolerance and milk protein allergy, and to manage cholesterol and saturated fatty acid levels in plasma (Kumar et al., 2010). Presently, an upsurge in demand for probiotics in nonlactic matrices based on cereals, legumes, and nuts has gained the attention of food scientists and technologists towards the development of novel nondairy probiotic milk and yogurts (Ranadheera et al., 2018). Plant-based milk/nondairy milk is in high demand and presents a great opportunity due to either the mounting problems related to milk intolerance, allergies, and malabsorption, or the deviations in diet inclinations and changing lifestyle (Fiocchi et al., 2010). These milk products differ compositionally with every type of grain used for their manufacture and to that of bovine milk; however, they practically look similar in appearance and resemble bovine milk. These milk products are emulsions that are prepared by disintegration and extraction of plant materials such as cereals, legumes, pseudocereals, seeds, and nuts, etc. (Sethi et al., 2016). Due to compositional deviations to that of bovine milk, such as high starch and low-fat content, low mineral content, these milk products require adequate processing steps (homogenization, heat treatment, enzymatic treatment, fortification, etc.) to produce alterations in the structural arrangement of various constituents to achieve required adaptations in the physical stability of the final product (Bernat et al., 2015). Plant-based beverages/milk products have been confirmed as functional beverages owing to the presence of nutritious and health-promoting components, e.g., prebiotics, polyphenols, phytoestrogens, etc. Since these plant materials are a natural source of prebiotics (soluble fibers, complex carbohydrates, etc.), the beverages or milk prepared out of these sources might serve as an efficient delivery vehicle for probiotics and might function as synbiotics. Furthermore, apart from adding functionality, fermentation with the help of probiotic culture may improve the low palatability attributes of plant materials. The application of starter cultures with identified flavor- and texture-augmenting abilities suggests an encouraging means for in situ product improvement, innovation, and expansion of plant-based beverages (Peyer et al., 2016). The ancient grains/coarse grains/minor grains are gaining a significant interest in this segment of functional beverages over staple/conventional grains due to their superior bioactive profile. The coarse grains such as sorghum, millets, spelt, amaranth, quinoa, and buckwheat are nutritionally rich grains with significant quantities of bioactive compounds (dietary fiber, resistant starch, phenolic compounds, etc.) and have created renewed interest, particularly in developed countries where these grains are commonly used as an animal feed (Coda et al., 2014). These grains are prospective sustainable delivery systems as they comprise nutrients that are easily absorbed by probiotics (Herrera-Ponce et al., 2014). They can also assist in the efficient passage of lactobacilli through the extreme conditions of the gastrointestinal (GI) tract, and can encourage the progression of single- and mixed-culture fermentations by probiotics (Charalampopoulos et al., 2002). Food grains could be used to formulate nondairy probiotic products with acceptable physicochemical quality and organoleptic properties (Salmeron et al., 2015). However, their assimilation in varied food

Probiotics and Prebiotics in Foods. https://doi.org/10.1016/B978-0-12-819662-5.00017-3

matrices (plant-based beverages from different grains) is a noteworthy challenge, primarily considering the maintenance of culture viability during production and storage.

The fermented foods sector has progressed from traditional fermented food products to newer and innovative products designed to improve gastrointestinal health and overall well-being with the accurate release of active constituents using nanotechnology and microencapsulation. Probiotic effects of food products can be either conferred by directly introducing probiotic bacteria in the colon through fermented foods or combined with prebiotics in the food matrix which will help in increasing the number of beneficial bacteria, such as bifidobacteria, in the intestinal microbiota. In this category of plant-based milk alternatives, few grains/nuts such as soy, almonds, etc. have attracted consumers in the past and now have a well-established market for the same; however, the utilization of other grains/seeds/nuts is now experiencing an upward trend. Legumes and nuts are low in starch and can be easily converted into beverages using the steam extraction process; however, cereals and pseudocereals contain a large proportion of starch, which thickens and forms a slurry upon thermal treatment above the gelatinization temperature (55–65°C). This can be prevented by enzymatic liquefaction and high-pressure homogenization. Therefore, to overcome such processing limitations, technological scientific interventions and additional processing steps are required. Also, these plant materials impart a typical characteristic flavor to the beverage, which needs to be removed or reduced to increase the consumer acceptability; fermentation and germination are the two promising technologies for the flavor improvement. These technologies help in bringing necessary changes due to the breakdown of chemical constituents and sometimes with the generation of newer flavors. Lactic acid bacteria fermentation is known to improve the sensorial and nutritional value of the food product (Leroy and De Vuyst, 2004). Plant milk alternatives can be fermented to produce nondairy milk and yogurt-like products with enhanced palatability. Lactic acid bacteria is known for its applications in cereal fermentations since ages, and many cereal constituents are recognized to maintain their growth; however, low concentrations of fermentable sugars in such grains may present a technological limitation (Zannini et al., 2012). These limitations can be overcome by employing the germination of grains before fermentation. Germination helps in increasing the concentrations of fermentable sugars and amino acids in cereals, pseudocereals, and legumes, which in turn supports the growth of probiotic strains (Charalampopoulos et al., 2002). For textural improvements and stability enhancement in plant-based beverages and yogurts, high-pressure homogenization, and the use of food additives, such as carboxymethylcellulose, calcium citrate, milk powder, and gelatin, showed encouraging results (Cheng et al., 2006). This chapter gives a comprehensive view on the prebiotics present in potential raw material for the manufacture of nondairy probiotic foods, designing, and development of plant-based synbiotic beverages, technological improvements for better commercialization of these products in the market through advanced processing techniques, consumer acceptance towards these products, and present and future trends in nondairy probiotic foods.

PREBIOTIC COMPONENTS IN WHOLE GRAINS, SEEDS, AND NUTS

Prebiotics were initially prompted as dietary means of modifying the colonic microbiota (Gibson and Roberfroid, 1995), and more precisely, they are defined as nondigestible polysaccharides and oligosaccharides, which stimulate the progression of favorable colonic lactic acid bacteria and limit the proliferation of pathogenic microorganisms such as *Salmonella* sp. or *Escherichia coli* through the antagonistic approach. In nature, a range of prebiotics occurs with different sources and chemical characteristics. Existing prebiotics are classified based on common principles by Stowell (2007). Fructooligosaccharides, galactooligosaccharides, inulin, lactulose, and polydextrose are documented as the well-established prebiotics, while isomaltooligosaccharides, xylooligosaccharides, and lactitol are classified as evolving prebiotics. These oligosaccharides work as selective promoters for the growth of beneficial gut microbes that aid in improving gut health and preventing intestinal ailments. Mannitol, maltodextrin, raffinose, lactulose, and sorbitol also expressed prebiotic properties with associated health benefits (Yeo and Liong, 2010). Wheat bran-derived arabinoxylooligosaccharides and xylooligosaccharides demonstrated their prebiotic potential through numerous applications (Xu et al., 2009). Resistant starch is another important category of prebiotics in whole grains that are linked to several health benefits. These starches are not absorbed by the small intestine of healthy individuals and are fermented by colonic microflora to produce short-chain fatty acids (SCFA) (Vaidya and Sheth, 2010). Soluble dietary fibers present in cereal grains such as β-glucan in oats are also identified to express prebiotic effects and produce SCFA via colonic fermentation (Lin et al., 2011). The positive health benefits linked to the consumption of prebiotics include maintenance of gut health, cholesterol management, prevention of colitis, inhibition of cancer, reduced risk of cardiovascular ailments, immunopotentiation, weight management, prevention of constipation, etc. Owing to positive health benefits, prebiotics find their way in several applications. The well-accepted applications of prebiotics include starter culture formulation, production of bacteriocin, cattle, poultry, fish feed, and pet food, as supplements to treat gastrointestinal diseases. However, the ever-increasing demand for prebiotics has increased inclination towards prebiotic supplementation in food/beverages, designing of novel probiotic and synbiotic food products, and extraction of prebiotics using advanced techniques.

Legumes are nutrition-packed plant material and are now attracting scientific interest towards their prebiotic potential. They are known to be a significant source of raffinose family of oligosaccharides (RFOs) also recognized as α-galactosides, which promote the growth of *bifidobacteria*. Chemically, these are water-soluble low-molecular-weight oligosaccharides made by α-(1-6)-galactosides interconnected to a glucose moiety of sucrose at the C-6 position and are broadly constituted by raffinose (α-Gal-(1 → 6)-α-Glu-(1 → 2)-β-Fru), stachyose ($[\alpha$-Gal-$(1 \rightarrow 6)]_2$-α-Glu-(1 → 2)-β-Fru), verbascose ($[\alpha$-Gal-$(1 \rightarrow 6)]_3$-α-Glu-(1 → 2)-β-Fru), and some other oligosaccharides. These prebiotics maintain viability under harsh gastric conditions (acidic) and are not digestible in the upper GI tract, which permits the stimulation of bacterial growth and activity in the intestine. Galactosidases are the enzymes accountable for the breakdown of α-D-galactosides, comprising galactose oligosaccharides, galactomannans, and galactolipids; however, the human system lacks this enzyme making them indigestible in the upper GI tract. Lactobacilli contains enzyme galactosidase which brings down the hydrolysis of α-D-galactosides during fermentation. Various probiotic strains (*L. rhamnosus* 910, *L. casei* CLbBCV1, *B. breve* S46, *B. longum* R0175, *L. acidophilus* La-2, and *L. fermentum* ATCC 14931) effectively utilized raffinose and stachyose for their growth enhancement; however, not all strains expressed α-galactosidase activity. The second important category of plant prebiotics, especially whole grains, is resistant starch (RS) and is associated with positive health benefits on the functioning of the digestive tract. These types of starches are not assimilated by the small intestine of healthy people and are fermented by colonic microflora to produce short-chain fatty acids, mainly acetate, propionate, butyrate, and gases. Resistant starch serves as food to colonic microflora and manages the blood cholesterol and glucose level, which in turn assists in controlling diabetes and cardiovascular ailments. Butyrate functions as a prime source of energy for colonocytes and reduces the jeopardy of malicious alteration in the colon (McGarr et al., 2005).

Resistant starch refers to the portion of starch and starch products that resist digestion as they pass through the gastrointestinal tract. According to different studies, resistant starch is comparatively low-molecular-weight (1.2×10^5 Da) starch, which is linearly composed of α-1, 4-D-glucan, principally acquired from the retrograded segment of the amylose (Tharanathan, 2002). RS is categorized into four different categories (RS1–4) based on physical and chemical characteristics (Nugent, 2005). These starch fractions are not digested in small intestine even after 120 min of consumption but are fermented in the colon by colonic microflora. The digestion behavior of resistant starch is similar to that of soluble dietary fibers; therefore, they are considered as a type of dietary fibers. Various factors such as the size of the starch-containing fragments, the arrangement, and conformation of whole starch granules, and the creation of retrograded crystallites due to chemical alteration generated by processing methods may make starch resistant to digestion. RS commonly occurs in many foods, such as cereals, legumes, seeds, and nuts (Topping and Clifton, 2001). The slowly digestible nature of resistant starch makes it a potential ingredient for the creation of newer functional foods with prebiotic properties and low glycemic index that can address the issues of glucose and weight management, and also improves the gastrointestinal health (Fuentes-Zaragoza et al., 2010). An average daily intake of 5–6 g is required to achieve its stated health benefits as mentioned by researchers (Kay et al., 2006). Resistant starch serves as a source of energy and carbon for the colonic microflora (400–500 bacterial species) and in turn is fermented under anaerobic conditions to yield short-chain fatty acids (SCFA). The short-chain fatty acids are composed of butyric, propionic, and acetic acids, which can alter the colonic environment by lowering the pH and hence modify the colonic microflora composition and its metabolic activities. The lowering of pH also has the potential in reducing the risk of the formation of cancerous tumors (Yao et al., 2009). Resistant starch forms a significantly higher proportion of butyric acid via colonic fermentation in comparison with dietary fibers and other indigestible carbohydrates. These butyrates are known to be the main vigor source for the colonocytes and thus own an important relation with colonic well-being (Scourboutakos, 2010). These butyrates augment the growth of beneficial probiotic microflora by serving as a source of energy/nutrition and also manipulate microbiota's gene expression at the same time. Resistant starch can be utilized in two ways for expression of their prebiotic behavior: (1) as a substrate for the growth of probiotics such as lactobacilli and bifidobacteria and (2) as a coating material/protectant for the encapsulation of probiotics (to achieve better stability) (Fuentes-Zaragoza et al., 2011).

Soybean (*Glycine* max) is an important legume crop for the vegan population of the world due to its significantly high protein content. It contains nearly 40%–45% of protein, 20% of total lipids, 5% of total minerals, and 30%–35% of carbohydrates. The total carbohydrate in soybean is estimated between 30% and 35%, out of which 10% is comprised of soluble sugars, 25% by insoluble fibers, and < 1% by starch. The soluble sugars comprise 5% sucrose, 4% stachyose, and 1% raffinose (Berk, 1992). Soybean has been extensively studied for its potential in the preparation of plant-based milk, and hence, soy milk is considered an important source of protein for sportspersons and people on diet or lactose-intolerant population. However, soybean are also a significant source of oligosaccharides, namely, stachyose, raffinose, and verbascose, which act as prebiotics, and hence, the consumption of soy milk can significantly contribute to the growth and proliferation of colonic probiotic bacteria. A substantial upsurge in the in vivo population of *bifidobacteria* due to soybean oligosaccharides was reported by Bouhnik et al. (2004). Another important legume from the prebiotic potential point of view is chickpea. Chickpea is also an important source of protein and is the most commonly consumed legume in the world (Zia-U1-Haq et al., 2007). It is one of the cheap sources of quality protein in the developing world and is also a good source of minerals and carbohydrates. Chickpea contains significantly higher levels of α-galactooligosaccharides (5%–10%). Chickpea oligosaccharides consist of ciceritol, raffinose, stachyose, and verbascose. Ciceritol, the chief glycoside and a characteristic prebiotic in chickpea, is accounted for about 50% of the total α-galactooligosaccharides followed by 35% of stachyose. The levels of oligosaccharides such as raffinose, ciceritol, stachyose, and verbascose, in chickpea, ranged between 0.46% and 0.92%, 3.00% and 5.10%, 1.60% and 3.10%, and 0.27% and 0.70%, respectively (Xiaoli et al., 2008). The variability in levels of oligosaccharides depends on genotypes and environmental conditions. Apart from α-GOS, chickpea is a good source of resistant starch. Desi variety of chickpea expressed a resistant starch in the range of 3.4%–10.3%, while Kabuli variety showed 16.43% resistant starch (de Almeida Costa et al., 2006; Bravo, 1999). Chickpea seeds also contain a significant quantity of nonstarch polysaccharides that act as prebiotics and are further classified into soluble or insoluble (Wood and Grusak, 2006). Soluble nonstarch polysaccharides constitute hemicellulose and pectic substances that are present around cellulose in amorphous matrices and work as binding agents between cells in the middle lamella. In chickpea, hemicellulose and pectic substances constitute 3.5%–9% and 1.5–4%, respectively, whereas nonsoluble polysaccharides, namely, uronic acid and arabinose, constitute 3.41% (Singh, 1984; Bravo, 1999). Fernando et al. (2010) reported the possible favorable effects due to an increase in *Bifidobacterium* spp. through chickpea diets and a decline in pathogenic and putrefactive bacteria, including *Clostridium* cluster I/II and XI. Lentil is an important source of protein and micronutrients, holds a significant proportion of prebiotic carbohydrates, and also expresses a low glycemic index. Lentil prebiotics include raffinose-family oligosaccharides, fructooligosaccharides, sugar alcohols, and resistant starch. The levels of RFOs in lentil varies between 2.5% and 7.2% (Wang et al., 2009). Lupin is another potential legume that is rich in RFOs. Lupin seed contains remarkably high levels of total α-galactosides, which ranges between 9.5% and 12.3%, out of this 7.4% is constituted by stachyose and 3.1% by verbascose. However, a wide variation was observed between different cultivars (Martinez-Villaluenga et al., 2005). Lupin seed RFOs express bifidogenic effects in probiotic fermented milk and can be used as a functional food component. The inclusion of RFOs in dairy milk augmented populations of *B. lactis* Bb-12 and *L. acidophilus* in comparison with control. Also, lupin RFOs reduced fermentation time from 12 to 10 h; these RFOs were preferably used by *B. lactis* Bb-12 and *L. acidophilus* as a carbon source (57.7%) in comparison with lactose (23.7%) at the end of fermentation (Martinez-Villaluenga et al., 2005). Pea accessions from Spain were assessed for prebiotics level; the results depicted the values for α-D-galactosides, stachyose, verbascose, raffinose, and sucrose, which ranged between 22.6 and 63.4, 10.7 and 26.7, 0.0 and 26.7, 4.1 and 10.3, and 11.6 and 25.4 g/kg DW, respectively. Out of total raffinose family oligosaccharides, raffinose, stachyose, and verbascose shared 14.9%, 38.0%, and 47.1%, respectively (Vidal-Valverde et al., 2003).

Among pseudocereals, quinoa is an important source of quality protein with a balanced amino acid profile and high lysine content (5.8 g/100 g). It also contains a significant proportion of phenolic compounds, such as α-tocopherol (Fischer et al., 2013). Quinoa is rich in dietary fibers, which function as prebiotics. Quinoa flour comprises approximately 7% of dietary fibers, where 22% of total dietary fibers are constituted by soluble fibers, mainly glucose, galacturonic acid, and arabinose subunits (Hager et al., 2012). Quinoa expressed prebiotic properties via in vitro studies and promoted the growth of probiotic bacteria with the production of SCFAs (Gullón et al., 2016). Quinoa polysaccharides displayed immune-regulating properties during animal studies (Hu et al., 2017). Similarly, amaranth, another important pseudocereal, contains 11% dietary fibers, where 22% share is occupied by soluble fibers, mainly xylose, which acts as prebiotic (Lamothe et al., 2015).

Among cereals, oats are the richest source of fermentable dietary fibers, predominantly β-glucan followed by arabinoxylan. Oat β-glucan ranged between 3.2% and 6.8% and is dependent on cultivar and environmental factors. β-Glucans, a nonstarch polysaccharide, is made up of D-glucose monomers linked by β-linkages. Oat β-glucans are a mixture of unbranched β-glucans with β-1,3- and β-1,4-glycosidic linkages; these structural variation explains the differences in functional activities and associated health benefits (Sargautiene et al., 2018). It has well demonstrated in various research investigations about the role of oat β-glucan in improving lipid and glucose metabolism, weight management, and immune regulation, which may be due to the absorption of the fermentation products of oat fibers in the gut (Chen and Raymond, 2008). The potential prebiotic effects of supplementation of different oat fractions on the colonic microflora upon digestion were studied by Abbeele et al. (2018). All oat fractions supported the growth of *Bifidobacterium* with an average increment of 0.73 log CFU; also, the fractions indicated high fermentability with the production of metabolites such as propionates, acetates, and lactates.

Almond is a rich source of proteins and fat; however, it is an important source of dietary fibers, which represents 11%–12% share. Also, almond's cell wall material comprises pectic constituents that are majorly contributed by arabinose. Finely ground whole and defatted almonds were assessed for their prebiotic potential by in vitro gastric and duodenal digestion and the subsequent effect on the composition and metabolic function of gut microflora. Finely ground whole almonds resulted in a significant increase in the populations of bifidobacteria and *Eubacterium rectale*, with a higher prebiotic index of 4.43 in comparison with 4.08 index of commercial prebiotic fructooligosaccharides after 24 h of incubation (Mandalari et al., 2008). A significant enhancement in the counts of *Bifidobacterium* spp. and *Lactobacillus* spp. was perceived in fecal specimens after supplementation with almond or almond skin. Also, it suppressed the growth of pathogen *Clostridium perfringens*. This supplementation resulted in the modification of the intestinal microbiota and resultant bacterial enzyme activities. Increase in fecal β-galactosidase activity and decreases in fecal β-glucuronidase, nitroreductase, and azoreductase activities were also observed (Liu et al., 2014).

PROBIOTIC STRAIN SELECTION

The human intestinal tract is a composite microbial network with a bacterial count of up to 10^{12} colony-forming units (CFU)/g; however, small intestine displayed a bacterial count of 10^{8} CFU/g, while in the stomach, it was found as low as 10^{2} CFU/g due to highly acidic pH. More than 400 bacterial species have been described in human feces (Gorbach, 1996). The human intestinal tract has a fair balance of favorable and pathogenic microbes; though, different conditions such as diseases or infections, antibiotic treatments, age, weakening of immune system, and dietary changes disturb this biosystem's balance. Supplementing the gut with probiotics through diet might help in restoring the balance of microflora, which in turn strengthens the immune system (Hemarajata and Versalovic, 2013). Probiotics are live microorganisms that, upon ingestion, beneficially affect their host by improving the balance of the intestinal microflora. The definition of probiotics has been modified and expanded over the years. Latest definition with new probiotic terminologies includes "paraprobiotics" (dead/inactivated cells of probiotics) and "postbiotics" (healthful metabolites of probiotics). Recent researches have presented that dead cells (intact cells or ruptured cells/cell extracts) along with the cell metabolites could render substantial impacts on human health. However, these terms are not precise and inclusive to exhibit the anticipated meanings related to the probiotic effects. Therefore, a new methodology and conceptualization in probiotic terminology for universal practice is proposed by Zendeboodi et al. (2020). The terminologies include three main categories of probiotic, namely, "true probiotic" denoting viable and active probiotic cell; "pseudo-probiotic" denoting viable and inactive cell, in the forms of vegetative or spore; and "ghost probiotic" denoting dead/nonviable cell, in the form of intact or ruptured cell. Each of these classes is further classified into two groups based on their site of action/impact, i.e., in vivo or in vitro. Probiotics might help in treating various disorders affecting the gastrointestinal tract by altering the intestinal pH and secreting antimicrobial constituents, for instance, bacteriocins, organic acids, and hydrogen peroxide. The prebiotics selectively stimulate the growth of probiotics by serving as a source of energy. The prebiotics are resistant to digestion in the small intestine and are fermented in the large intestine by saccharolytic bacteria. The energy obtained during fermentation is utilized by bifidobacteria for their growth, which ultimately leads to reduced pH in the intestine. Low pH moderates the concentrations of harmful metabolites and balances the microflora for the maintenance of a healthy gut (Verbeke et al., 2015). The most common genera of bacteria used as probiotics include *Bifidobacteria* and *Lactobacillus*. The selection of probiotic strain is an important phase and needs to be assured in terms of safety and efficacy. The strain selection is accomplished in three steps: (1) selection and characterization of strains, (2) in vitro capacity assessment of probiotics, and (3) in vivo capacity assessment of probiotics (Rubio et al., 2008). Selection and characterization of strain is the first step where the strains are screened and identified, and their growth conditions such as growth kinetics, substrates, pH and temperature, preservation, and metabolic stability are studied to check the probability of industrial-scale production. Thereafter, in the second stage, the strains are assessed for in vitro efficacy in terms of production of enzymes, vitamins, and amino acids, adherence capacity, the antagonistic effect on pathogenic microorganisms, tolerance to bile salts, production of bacteriocins, resistance to gastric juices, reduction of cholesterol levels, and immune system stimulation (Morell, 2007). In the third or last stage, the strains are assessed for in vivo efficacy where their ability is estimated based on clinical trials. Since it is difficult to get probiotic characteristics in a single strain, it is often essential to comprise characteristics of quite a few strains in a single product (Lee and Salminen, 2009). Lactic acid bacteria (LAB) is an important group of probiotic bacteria that ferment glucose and lactose to produce lactic acid, which results in a decrease in pH, ultimately leading to the destruction of pathogenic microbes. LAB can also secrete antimicrobials such as bacteriocin, which works against the growth of pathogenic microbes. Furthermore, LAB also expresses immunoregulatory characteristics (Cross et al., 2001). LAB exists as aerobic, anaerobic, and facultative anaerobes. Microbes interact with each other for their survival in three ways: (1) competition for space, (2) competition for nutrients, and (3) production of antimicrobials (Rodríguez and Cardozo, 2012). The most commonly used LAB genre in the formulation of probiotic food products includes *Lactobacillus*, *Leuconostoc*, *Streptococcus*, *Bifidobacterium*, and *Pediococcus*. Probiotic fermented beverages may play an important role in treating diarrhea and postantibiotic therapy intestinal infections. Commonly used probiotics in plant-based beverages include *L. plantarum*, *L. paracasei*, *L. acidophilus*, *L. rhamnosus*, *L. fermentum*, *L. reuteri*, *B. lactis*, and *B. bifidum*.

HEALTH BENEFITS OF PROBIOTICS AND PREBIOTICS

The relationship between prebiotics and probiotics on positive benefits on human gastrointestinal health has established through scientific facts long back. The health benefits of prebiotics and probiotics are interdependent and occur due to their synergistic association (Fig. 1). Initially, it was observed by Tissier that gut microflora from healthy breastfed infants showed bifidobacteria, which were not found in formula-fed infants suffering from diarrhea. From there, the concept evolved with the further exploration of probiotics in human well-being. The beneficial effects of prebiotics and probiotics on human health include the improvement of intestinal health, relief from symptoms of lactose intolerance, immunostimulation, competitive exclusion of pathogens, and reduced risk of many other diseases such as coronary heart diseases, cancer, etc. Probiotics are of particular interest in treating different types of diarrhea such as infectious diarrhea, antibiotic-associated diarrhea, traveler's diarrhea, *Clostridium difficile*-induced diarrhea, and inflammatory bowel diseases. Antibiotic therapy-associated diarrhea disturbs the balance of gut microflora and encourages the growth of pathogenic microbes. A meta-analysis on evaluating the probiotics for their efficacy in treating such kind of diarrhea specified that the administration of probiotic strains such as *L. rhamnosus*, *L. casei*, and the yeast *S. boulardii* results in a reduced risk of antibiotic-associated diarrhea (Martinez-Villaluenga et al., 2006). Also, it was observed that the intake of 12 g/d FOS for 30 days resulted in lesser episodes of *C. difficile*-induced diarrhea (Lewis et al., 2005). Probiotics also help in treating other infectious acute diarrhea, which arises due to pathogens such as rotavirus, which is the most common cause of diarrhea in infants. Studies have demonstrated the positive benefits of probiotic strains *L. rhamnosus* GG, *L. reuteri*, *L. casei* Shirota, and *B. animalis* BB-12 in shortening the duration of acute rotavirus diarrhea (Szajewska and Mrukowicz, 2001). Such benefits lead to the supplementation of infant formulas with probiotic strains to prevent or treat such infections. The proposed mechanisms by which probiotics treat diarrheal infections include obstructing the receptor site

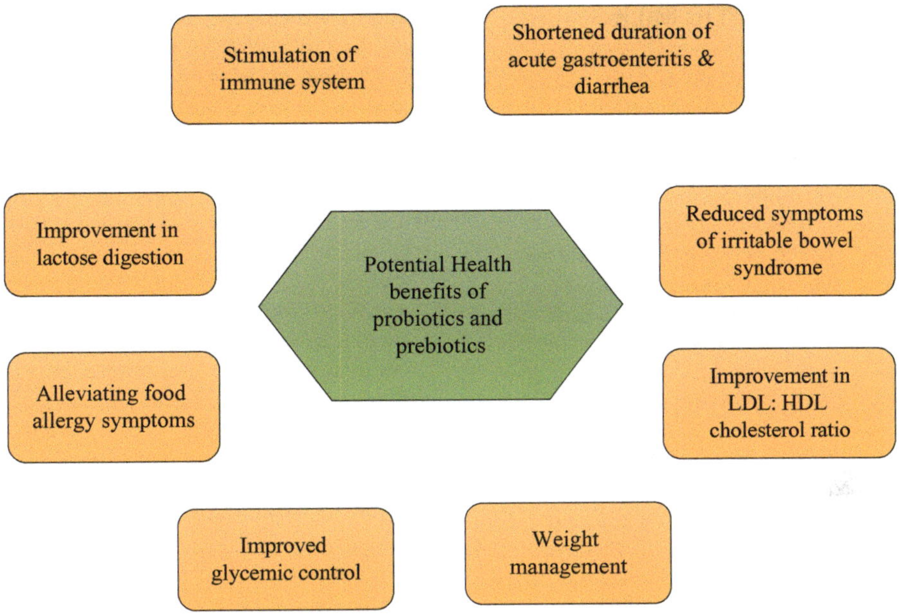

FIG. 1 Potential Health benefits of probiotics and prebiotics.

signals, which control the secretory and motility defenses, augmentation of the immunological reactions, and secretion of antimicrobials. Also, it was reported that 10 g/d ingestion of prebiotic inulin 2 weeks before and 2 weeks after travel reduced the frequency and resulted in less severity of traveler's diarrhea (Cummings et al., 2001). Administration of trans-GOS for 4 weeks improved the fecal bifidobacterial count at 3.5 g/day in patients suffering from irritable bowel syndrome (Silk et al., 2009). Lactose intolerance is another important genetically determined condition benefitted by the use of probiotics (De Vrese et al., 2001). This condition results due to the inability to hydrolyze lactose into glucose and galactose owing to the deficiency of β-galactosidase. Due to this, the undigested lactose reaches the large intestine where it gets degraded by bacterial enzymes leading to osmotic diarrhea. Probiotics also play an important protective role against allergic reactions. Probiotics improve mucosal barrier function, which aids in moderating allergic reactions particularly involving food allergy and atopic dermatitis. Breastfed infants showed less eczema severity; furthermore, high-risk infants fed on mothers supplied with *L. rhamnosus* GG prenatally showed protection against atopic eczema Isolauri et al., 2001). In addition to these effects, *Lactobacillus* and *Bifidobacterium* spp. moderate the concentrations of carcinogenetic enzymes generated by colonic microflora. The proposed mechanism of this effect includes the regulation of intestinal permeability, the balance of microflora, the secretion of antimutagenic organic acids, and improvement in immune responses (Kumar et al., 2010). Intake of probiotic strains through dairy products showed a better management of bacterial vaginosis in women by balancing the normal vaginal lactobacilli microflora (Falagas et al., 2007). Probiotics can also contribute to reducing the risk of coronary heart disease by decreasing serum cholesterol levels and also help in controlling blood pressure. Apart from contributing to gastrointestinal health by supporting the growth of probiotics, prebiotics are known to reduce the occurrence of some chronic diseases such as cardiovascular disease, type-2 diabetes, and obesity. Prebiotics such as dietary fibers are considered indispensable for the favorable digestive health. A diet containing soy isoflavones and prebiotic expresses a considerable decline in levels of LDL cholesterol and the LDL/HDL cholesterol ratio (Wong et al., 2010). Prebiotics have also proven to controlling obesity in a study, which showed reduced fecal concentrations of Bacteroidetes in obese people in comparison with lean people (Armougom et al., 2009). The health benefits of prebiotics and probiotics are surely encouraging; however, controlled human studies to establish the appropriate dosage to achieve the stated efficacy are required to be taken up by researchers in the future. Also, there is a need to set global regulatory guidelines to furnish the effectiveness, safety, and endorsement of health assertions on food labels (Kechagia et al., 2013).

PLANT-BASED PROBIOTIC MILK AND YOGURTS

Lactose intolerance, cow milk allergy, obesity, hypercholesterolemia, and gastrointestinal health are stronger drivers of nondairy probiotic food products (Valencia-Flores et al., 2013). Even though dairy products are the most commonly used vehicle for the delivery of probiotics, recent times witnessed a surge for newer probiotic foods in nondairy mediums, based on legumes, cereals, seeds, and nuts (Ranadheera et al., 2018). The consumer's interest in plant-based milk alternatives increased dramatically as these materials are an excellent source of functional ingredients, namely, prebiotics and phenolic compounds etc. Nevertheless, recent confirmations of health benefits owing to the synergetic association between plant components and probiotics promote further developments in the area of nondairy probiotic or synbiotic foods (Duda-Chodak et al., 2015; Valdés et al., 2015; Ozdal et al., 2016). Several research investigations have assessed the prospects of legumes, cereals, seeds, and nuts as a substrate for the growth of probiotic strains. These functional beverages have the potential to manage well-being through food. Furthermore, the progression in dietetic regimes, such as veganism, is also stimulating the demand for plant milk and associated products. The international dairy alternatives market is predicted to grow from USD 21.4 billion in 2020 to USD 36.7 billion by 2025, at a compound annual growth rate (CAGR) of 11.4% (https://www.marketsandmarkets.com/Market-Reports/dairy-alternative-plant-milk-beveragesmarket-677.html). Also, the global probiotic market is valued at USD 42.55 billion in 2017 and is projected to reach at USD 74.69 billion by the end of 2025 with an anticipated CAGR of 7.32% (https://www.fortunebusinessinsights.com/industry-reports/probiotics-market-100083). Soy milk dominates the universal dairy alternatives market followed by almond, rice, and coconut milk. Geographically, the Asia-Pacific region leads the alternative dairy market due to the increasing number of lactose-intolerant people and inclination towards a vegan diet; furthermore, globally two-thirds of the population is lactose-intolerant. The potential cereal-based raw materials for the preparation of nondairy milk include oat, rice, and spelt, while, in the case of legumes, they include peanut, lupin, and lentils. Furthermore, the seeds and nuts include potential choices of sesame, hemp, and almond, cashew, tiger nut, hazelnut, coconut, respectively. It has been observed by researchers that phenolic compounds and complex carbohydrates may work in synergism with probiotics (Puupponen-Pimiä et al., 2002; Selma et al., 2009), which gives an edge to plant-based probiotic product inventors. In the case of plant materials, each food medium is distinctive in composition; therefore, formulations for plant-based probiotics or synbiotics need to be optimized for each product based on the required sensory and physicochemical characteristics, prolonged shelf life, product stability, and cost economics. The successful development of plant-based probiotic products depends on explanatory parameters such as pH and ionic strength of the medium, oxygen levels, water activity, macro- and microstructure, presence of inhibitors, and the existence of competing microorganisms that have a direct impact on the survival of probiotic culture. Several researchers reported promising results with oats, maize, barley, spelt, and rice as mediums that encourage the progression of colonic lactobacillus strains (Angelov et al., 2006; Laine et al., 2003). Due to a large array of health-promoting components including prebiotics, in cereals, legumes, seeds, and nuts, they can be an important matrix for probiotic foods. Also, these are economic raw materials and present amply in every part of the world. However, the survival of probiotic strains and the ability to support the growth of probiotic strains in these matrices vary with each type of raw material and storage conditions of the developed product. In recent researches, scientists have reported that the physicochemical responses are attributable to the probiotic action of plant-based matrices. Also, it is important to assess the effects of probiotication on the final product. Various unit operations used in the preparation of plant-based probiotic products, such as milling/disintegration, thermal treatment, enzymatic hydrolysis, and fermentation, may result in the generation of newer flavor compounds. Fermentation of grain-based probiotic formulation is expected to result in major changes in flavor and aroma profile in the final product and also during storage. Since each grain has its typical flavor and aroma, fermentation may help in improving the flavor of the product in a better way. A relative assessment for prebiotic effects of barley, wheat, and barley malt extracts on the growth and sustainability of human-derived *L. plantarum* strain during refrigerated storage revealed better performance by malt extract due to the presence of raffinose and dietary fibers in the beverage matrix (Charalampopoulos and Pandiella, 2010). Few researchers have assessed the probability of oat matrix for the development of probiotic products such as milk and yogurt (Kedia et al., 2008; Angelov et al., 2006).

β-Glucan in oats expresses prebiotic properties, and that is why this grain is considerably explored for its prebiotic potential in supporting the growth of probiotic microflora. Gupta et al. (2010) reported the positive effects of oat fibers on the growth of *L. plantarum* in oat-based nondairy milk. The oat fibers reported supporting the complex nutritional requirements of probiotic strains. In another study, the oat-based synbiotic beverage prepared through the mixed-culture fermentation by *L. plantarum*, *Lactobacillus paracasei* subsp. *casei*, and *L. acidophilus* showed the survival of probiotic strains *L. plantarum* and *L. paracasei* subsp. *casei* (10^7–10^8 CFU/mL) during 10-week storage period under refrigeration conditions. Angelov et al. (2006) also prepared a synbiotic beverage from whole oats, fermented by a probiotic *L. plantarum B28* under measured conditions for 8h. Another study explored the possibilities of preparation of functional nondairy milk from oat flakes. Additionally, a cocktail of xylanase and α-amylase enzymes is used to support the growth of culture and to shorten the fermentation duration. The functional beverage fermented with *L. plantarum LP09* and containing 25% oat flakes was reported to be optimal based on quality parameters (Luana et al., 2014). Like oats, barley is also an important source of β-glucan and is a potential raw material for the preparation of nondairy, plant-based milk. Barley and malt mixed-cereal substrates were explored for their fermentation potential using a mixed probiotic culture of *L. plantarum* NCIMB 8826 and *L. acidophilus* NCIMB 8821. It was reported that the addition of malt accelerated the growth of lactic acid bacteria. A cereal-based alternate probiotic beverage to dairy-based probiotics using barley and malt, fermented by *L. plantarum* and *L. acidophilus* in a range between 7.9 and 8.5 log CFU/mL, was developed by Rathore et al. (2012). Rice- and millet-based probiotic beverage with added pumpkin and sesame seed extract, fermented with a mixed starter culture of *Streptococcus thermophilus*, *L. acidophilus*, and *Bifidobacterium* BB-12, resulted in a functional beverage with very good sensory acceptability (Hassan et al., 2012). A probiotic functional beverage prepared with a combination of grains, namely, white and red sorghum, millet, and wheat, fermented by probiotic yeasts *Pichia kudriavzevii* OG 32, showed improved sensory and physicochemical properties (Ogunremi et al., 2015). Sorghurt, a sorghum-based yogurt-like product, expressed a higher viable count of > 8 log10 CFU/mL in the product, with adequate sensory scores (Sanni et al., 2013).

Among legumes, soybean has gained noteworthy consideration from the scientists owing to its protein superiority. It is also revealed by various researchers that soy milk is an appropriate medium to support the growth of the probiotic lactic acid bacteria, especially bifidobacteria (Chou and Hou, 2000). Several investigations stated the possibilities of utilizing soybeans in the preparation of probiotic milk and yogurts (Fuchs et al., 2005). Also, other legumes are equally potential raw materials for the preparation of probiotic beverages and yogurts. A novel nondairy beverage was developed from garbanzo chickpeas by Wang et al. (2018). The fermented beverage showed an enhancement in the microbial population from about 5×10^4 to 7.4×10^6 after a 16-h fermentation period. Another important legume lupin is a rich source of raffinose oligosaccharides. Lupin RFOs were studied for their bifidogenic effects while preparing a nondairy probiotic fermented milk. Lupin RFOs were assessed for their effect on the mixed culture of *B. lactis* BB-12 and *Lactobacillus acidophilus* (1:1). The results expressed an increment in the population of *B. lactis* Bb-12 and *L. acidophilus* with a reduction in fermentation time from 12 to 10h in comparison with control samples. The researchers also reported that the lupin RFOs (57.7%) were preferred over lactose (23.7%) as a source of carbon by probiotic culture (Martinez-Villaluenga et al., 2005). In a similar research investigation, the effect of raffinose family of oligosaccharides (RFOs) as prebiotics on six bifidobacteria strains in the manufacture of fermented milk was studied by in vitro approaches. Under this study, strain growth and acidification rates were compared in milk systems with and without added RFOs (2%). Different bifidobacterial strains expressed a varied growth and acidification rate; however, the strains with greater levels of α-galactosidase activity showed an enhanced growth and acidification rate, which also proves the preference of RFOs over lactose as a carbon source by probiotic culture (Martinez-Villaluenga and Gomez, 2007). Verni et al. (2020) designed and characterized fermented beverages from lentils using seven different strains of *Lactobacillus*. *L. acidophilus* ATCC 4356, *L. fermentum* DSM 20052, and *L. paracasei* subsp. paracasei DSM 20312 showed the highest growth and acid production rate. The fermented product maintained the viability of bacterial strains at greater than 7 log CFU/mL after 28 days of storage under refrigerated conditions. The fermented beverage also showed a reduction in the phytic acid (antinutritional factor) by 30% and in the raffinose oligosaccharides content by 12%. A study estimated the survival rate of probiotic strain *Lactobacillus rhamnosus* GR-1 in nondairy probiotic beverages made out of soybean, almonds, and peanut during 28 days of storage period. *L. rhamnosus* GR-1 continued to be viable with a cell count of 10^7 CFU/mL in all types of nondairy probiotic milk over 28 days of storage under refrigerated conditions (He and Hekmat, 2015). Murevanhema and Jideani (2015) reviewed the possibilities of Bambara groundnut in the preparation of nondairy probiotic milk. Amarteifio and Moholo (1998) have reported the compositional resemblances between soy and Bambara groundnut, which in turn indicates the greater possibilities of creating a probiotic beverage from Bambara groundnut. The peanut-soy milk, inoculated with six different lactic acid bacteria (LAB), including probiotic strains and yeasts for 24 h at 37 °C followed by another 24 h at ± 4 °C, expressed significant amounts of lactic acid (3.35 g/L) production by probiotic strain *L. acidophilus* (LACA 4); however, it was also observed that co-culture resulted in better utilization of carbohydrates as an energy source in comparison with single culture (Santos et al., 2014).

Probiotic almond milk with added inulin, fermented with probiotic strains *Lactobacillus reuteri* and *Streptococcus thermophiles*, was developed by Bernat et al. (2014). Various factors including glucose, fructose, inulin, and starters concentrations were optimized to reassure the probiotic survivals throughout the storage period of 28 days under refrigeration. High probiotic count, i.e., > 10^7 CFU/mL, was obtained with the optimized formulation of 0.75 g of glucose/100 mL, 0.75 g of fructose/100 mL, 2 g/100 mL of inulin, and 6 mL/100 mL of probiotic inoculum. The authors also reported a noteworthy survival of 51% of probiotic strains to the in vitro digestion, which is related to the presence of prebiotic inulin. Cashew nut milk was assessed as a medium to carry commercial probiotic strains *B. animalis*, *L. acidophilus*, and *L. plantarum* for the development of nondairy probiotic milk. The results expressed the survival of all probiotic strains in cashew milk stored at 4 °C for 30 days. During the said storage period, *B. animalis* counts were above 10^7 CFU/mL with no significant difference in the quality of the product (Bruno et al., 2020). Chickpea and coconut beverage added with different levels of sugar (0%–7.5%) and inoculated with *L. paracasei* LBC 81 were assessed for fermentative process, chemical composition, and stability of beverages during storage of 10 days. The results expressed a prominent growth in samples with no added sugar, and viability of *L. paracasei* LBC 81 was estimated at greater than 8 log CFU/mL, indicating chickpea and coconut as a suitable matrix for delivering probiotics (Mesquita et al., 2020). Commercial hemp seed-based beverage fermented with probiotics strains *L. fermentum*, *L. plantarum*, and *B. bifidum* was formulated and evaluated for prebiotic activity by cultural and quantitative PCR methods. The findings of this research displayed a strong prebiotic activity, the ability to support the growth of probiotics, and increased levels of some bioactive compounds in hemp seed beverage (Nissen et al., 2020). In a study, coconut milk fermented with *Lactobacillus reuteri* LR 92 or DSM 17938 was optimized for the temperature of preparation and pulp/water ratio. Both the strains exhibited varied production rates and levels of organic acids. Storage at 4 °C after 30 days displayed a cell viability of 7.55 ± 0.07 log CFU/mL for *L. reuteri* LR 92 and 8.57 ± 0.09 log CFU/mL for DSM 17938 in coconut milk (Mauro and Garcia, 2019). When desiccated coconut and nondehulled seeds of hemp were used to prepare substitute probiotic milk using probiotic strain *L. casei* subsp. Rhamnosus, the prepared milk expressed a sufficient viability meeting the therapeutic criterion after 21 days of refrigerated storage (Szparaga et al., 2019). The cited literature confirms the potential of plant-based materials in the development of novel prebiotic, probiotic, and synbiotic functional food products. However, further research is required to upscale laboratory-based findings to a large scale for better commercialization of such products. The plant-based prebiotic and probiotic beverages are summarized in Table 1.

DAIRY-BASED SYNBIOTICS WITH PLANT CONSTITUENTS AS PREBIOTICS

Functional dairy-based beverages supplemented with plant constituents can boost the immune system and be used as an important component of supportive therapy. Prebiotics and probiotics can be used alone in the creation of newer products with the stated health benefits; however, considerable research attention is fixated on the collective use of probiotics and prebiotics, in the creation of synbiotic products with synergistic health properties. Milk is a complex assortment of biologically active constituents, i.e., immunoglobulin, enzymes, antimicrobial peptides, oligosaccharides, fatty acids, saccharides, hormones, cytokines, and growth factors, which makes it a physiologically functional food (Bharti et al., 2012). The dairy industry has also evolved from traditional products to innovative products with health-promoting components in response to consumers' alertness and insight into functional foods. The conception of newer functional dairy-based products such as yogurt, cheese, and fermented drinks could work as an alternative to enhance health-promoting microbial flora in the intestine. Milk products lack prebiotics but are excellent carriers for probiotics; however, plant foods are a natural source of prebiotics; therefore, combining these two can result in novel synbiotic products with functional effectiveness of both probiotics and prebiotics. Synbiotic food products are of particular importance as probiotics can abide by harsh conditions of food matrix such as oxygen, low pH,

TABLE 1
Plant-Based Prebiotic and Probiotic Beverages.

Type of Product	Prebiotic Component	Probiotic Strain	Reference
Barley, wheat, and barley malt extracts	Raffinose and dietary fibers	*L. plantarum*	Charalampopoulos and Pandiella (2010)
Oat-based nondairy milk	Oat fibers	*L. plantarum*	Gupta et al. (2010)
Barley and malt	Barley fibers	L. *plantarum* and *L. acidophilus*	Rathore et al. (2012)
Chickpea-based nondairy beverage	Ciceritol		Wang et al. (2018)
Lupin-based non-dairy probiotic milk	Raffinose oligosaccharides	*Bifidobacterium lactis* Bb-12 and *Lactobacillus acidophilus*	Martinez-Villaluenga et al. (2005)
Fermented beverage from lentils	Raffinose oligosaccharides	*L. acidophilus* ATCC 4356, *L. fermentum* DSM 20052, and *L. paracasei* subsp. paracasei DSM 20312	Verni et al. (2020)
Nondairy probiotic beverages made out of soybean, almonds, and peanut	Raffinose oligosaccharides, dietary fibers	*Lactobacillus rhamnosus* GR-1	He and Hekmat (2015)
Probiotic almond milk	Inulin, arabinose	*Lactobacillus reuteri* and *Streptococcus thermophiles*	Bernat et al. (2014)
Cashew nut milk	–	*Bifidobacterium animalis*, *Lactobacillus acidophilus*, and *Lactobacillus plantarum*	Bruno et al. (2020)
Chickpea and coconut beverage	Ciceritol	*L. paracasei* LBC 81	Mesquita et al. (2020)
Hemp seed-based beverage	Presence of different terpenes inhibits the growth of enteropathogens, and high levels of acetate, propionate, and butyrate produced during fermentation contributed to the prebiotic effect	*Lactobacillus fermentum*, *L. plantarum*, and *Bifidobacterium bifidum*	Nissen et al. (2020)
Coconut milk	Naturally occurring sucrose in the matrix (c. 4.4 g/L)	*Lactobacillus reuteri* LR 92 or DSM 17938	Mauro and Garcia (2019)

and unfavorable temperature and can endure in the digestive system in the presence of prebiotics (Ozcan et al., 2016). Presently, plant-based components are supplemented in dairy products to promote the growth of probiotics and improve the nutritional profile of probiotic products by the addition of plant bioactive. Consequently, the inclusion of prebiotics into dairy-based probiotic products would possibly create a product with enhanced mucosal surfaces of host and production of antimicrobial compounds to impede the growth of pathogens, in addition to fermentation and biopreservation of the food (Oliveira et al., 2011).

The addition of lupin seed reduced fermentation time and increased probiotic bacterial count in fermented dairy milk (Martinez-Villaluenga et al., 2006). Supplementation of oat bran improved the survival rate of *L. casei* (LC-1) in a dairy fruit beverage (Guergoletto et al., 2010). The addition of quinoa flour to fermented milk increased the nutritive value and improved the appeal of the product to consumers (Casarotti et al., 2014). Supplementation of plant-based prebiotics also improved the quality of dairy-based yogurts. The addition of triticale bran in yogurt not only supported the growth of probiotic strains but also improved the antioxidant profile of the product (Agil and Hosseinian, 2012). The addition of various pulse-based ingredients (pea protein, chickpea flour, lentil flour, pea fiber, soy protein concentrate, and soy flour) in dairy yogurts improved the growth and metabolic activity of probiotic and yogurt starter culture (Zare et al., 2012). Hoppert et al. (2013) developed reduced sugar yogurt through supplementation with a cereal grain mixture (wheat, rye, oat, barley, and millet); thus, the developed product displayed an improved acceptance of fiber-enriched yogurt. Chickpea, a rich source of protein, fiber, and prebiotics, was explored for its prebiotic effects in dairy-based probiotic yogurt at a level ranging from 1% to 5%. The incorporation of chickpea flour to the yogurt mix stimulated the growth of bacteria, reduced the incubation time, and sustained higher probiotic counts throughout the storage period of 21 days under refrigeration. The results demonstrated that the enrichment of yogurt with 1% and 2% chickpea flour not only supported the growth of probiotic bacteria but also improved the protein content and viscosity of the yogurt (Hussein et al., 2020). Similarly, skim milk supplemented with lentil flour in a range of 1%–3%, fermented with the yogurt culture, and stored under refrigeration conditions enhanced acid production during fermentation (Zare et al., 2011). A yogurt-like product from hazelnut slurry added with skimmed milk powder was developed by Ilyasoglu et al. (2015). A high population of starter cultures > 10^7 CFU/g of *S. thermophilus* and 10^4 cfu/g of *L. bulgaricus* was estimated after 28 days of storage under refrigeration. Quinoa flour, an important source of prebiotics, was assessed in levels ranging from 1% to 3%, for acidification kinetics; deviations in postacidification viable counts of *Streptococcus thermophiles*, *B. animalis* subsp. *lactis* BB-12, and *L. acidophilus* LA-5; the resistance of probiotics to simulated gastrointestinal conditions; and in vitro adhesion of probiotics to intestinal cells during 28 days of refrigerated storage in fermented dairy milk (Casarotti et al., 2014). A set-type synbiotic yogurt was developed using cow milk and quinoa milk in the ratio of 7:3 with 7% sugar and 1% inulin using probiotics strains of *Bifidobacterium* BB-12TM and *L. acidophilus* LA-5TM (Zhang et al., 2018). The addition of 3% amaranth flour resulted in yogurt with 20% reduced syneresis and a total viable count of 4.8×10^4 CFU/mL *Lactobacillus bulgaricus* after 21 days of refrigerated storage (Dimitrova and Kartalska, 2015). Supplementation of bovine whole milk with buckwheat flour and oat bran, inoculated with *L. rhamnosus* IMC 501, *L. paracasei* IMC 502 alone, and in a combination of 1:1, showed a significant faster lowering of the pH and enhanced stability of probiotic strains during 28 days of storage at 4°C. Human feeding trials to validate the product claim showed about 10^9 viable cells in fecal samples (Coman et al., 2013). Fermented ice cream prepared with *Lactobacillus acidophilus* (La-05) or *B. bifidum* (Bb-12) using cow, soy, or coconut milk as well as the combination of cow milk or coconut milk was assessed for suitability of fermentation medium. The composite milk ice cream (75% soy milk and 25% coconut milk) augmented the growth rate of LA-05 and Bb-12 by 1.55 and 1.07 log10 CFU/g, respectively (Aboulfazli et al., 2016). The plant milk, dairy-based prebiotic and probiotic beverages are summarized in Table 2.

NONTHERMAL TECHNOLOGIES FOR PROCESSING AND PRESERVATION OF PLANT-BASED DAIRY ANALOGS

Plant-based probiotic beverages are usually preserved by thermal processing, particularly ultra-high-temperature (UHT) treatment before being packaged, to guarantee suitably extended shelf life. However, conventional thermal processing can result in undesirable reactions affecting the stability, texture, flavor, and appearance of the final product. Also, thermal processing promotes the degradation of heat-sensitive nutrients and bioactive compounds. In an attempt to minimize these limitations, nonthermal food processing technologies such as high hydrostatic pressure, ultrasound, and pulsed electric fields have been explored. The techno-scientific assessment of these advanced nonthermal processing technologies expressed encouraging outcomes in plant-based beverages segment by providing enhanced shelf life, preserved nutritional constituents, and bioactive components while maintaining the textural stability of the product. Although these technologies are underexplored in the case of plant-based milk alternatives but have been investigated significantly on dairy milk, they can serve as potential preservation techniques in this segment also.

High hydrostatic pressure processing employs high pressures in the range of 100–1000 MPa for microbial and enzyme inactivation. It can be combined with mild heat to achieve synergistic effects. This technique lets most foods to be preserved with an insignificant influence on product characteristics like taste, texture, or nutritional properties. Manassero et al. (2016) employ high-pressure treatment on calcium-fortified soy milk to enhance protein solubility and inactivation of trypsin inhibitors and lipoxygenase. The authors optimized the levels of pressure, temperature, and calcium to attain a product with maximum stability. Soy milk treated with

TABLE 2
Plant Milk, Dairy-Based Prebiotic and Probiotic Beverages.

Type of Product	Prebiotic Component	Positive Effects on Product	Reference
Fermented dairy milk	Lupin seed	Reduced fermentation time and increased probiotic bacterial count	Martinez-Villaluenga et al. (2006)
Dairy fruit beverage	Oat bran	Improved survival rate of *Lactobacillus casei* (LC-1)	Guergoletto et al. (2010)
Fermented milk	Quinoa flour	Increased nutritive value as well as improved product appeal to the consumers	Casarotti et al. (2014)
Yogurt	Triticale bran	Supported the growth of probiotic strains and improved antioxidant profile of the product	Agil and Hosseinian, 2012
Yogurt	Chickpea flour	Stimulated the growth of bacteria, reduced the incubation time, and sustained higher probiotic counts throughout the storage	
Skim milk yogurt	Lentil flour	Enhanced acid production during fermentation	Zare et al. (2011)
Yogurt-like product	Hazelnut slurry	High population of starter cultures > 10^7 cfu/g of *S. thermophilus* and 10^4 cfu/g of *L. bulgaricus*	Ilyasoglu et al. (2015)
Yogurt	Amaranth flour	Reduced syneresis	Dimitrova and Kartalska, 2015
Fermented milk	Buckwheat flour and oat bran	Significant faster lowering of the pH and enhanced stability of probiotic strains	Coman et al. (2013)

high pressure at 400–600 MPa at 75°C resulted in the total inhibition of aerobic bacteria for 28 days (Smith et al., 2009). In another attempt, high-pressure processing was employed to preserve the soy smoothies without affecting the sensory and nutritional properties of the product (Andrés et al., 2016). Soy smoothies treated with high pressures of 450–650 MPa for 3 min at 20°C showed a total reduction in microorganisms and maintained a microbial stability until the storage period of 45 days under refrigeration. No significant difference in terms of sensorial quality was observed in soy smoothies treated at 450 MPa. Pulsed electric field is another interesting nonthermal processing technology where the intended food is placed between two electrodes and is exposed to pulses varied from few milliseconds/even in microseconds at high voltages of 1–80 kV/cm, to achieve microbial destruction. Like HPP, this technique also maintains the natural characteristics of the original product. A range of voltage is used depending on the principal requirement. In the plant-based beverage segment, the application of high-intensity pulsed electric field was deliberated on tiger nut milk by Selma et al. (2009); the results displayed an increase in the lag phase of *Enterobacter aerogenes* inoculated to tiger milk. Pulsed electric field in the range of 20–40 kV/cm and pulses with a duration of 0–547 μs showed the inactivation of *Escherichia coli* and *Staphylococcus aureus* in soy milk; however, the rate of inactivation was proportional to the increasing intensity and treatment exposure. Xiang et al. (2011) also evaluated the effect of PEF treatment on rheological and color properties of soy-based milk. The soy milk was treated with PEF of 18–22 kV/cm; the number of pulses varied from 25 to 100 at 26°C. The results demonstrated an increment in apparent viscosity from 6.62 to 7.46×10^{-3} Pa s when an electric field intensity increased from 18 to 22 kV/cm and pulses from 0 to 100. The observed effect could be linked to the development of transient networks from the denatured molecules.

Ultrasonication is another promising nonthermal processing technology but is poorly utilized in the segment of plant-based beverages. Homogenization of plant-based milk alternatives is important to achieve product stability, and hence, ultrasound finds its importance in such applications. Iswarin and Permadi (2012) studied the effect of ultrasonic waves on the droplet diameter of coconut-based milk. The ultrasonic waves with varying power levels (2.5–7.0 W) and exposure durations (5–25 min) were applied to achieve homogenization. The results showed that particle size was significantly affected by power levels of ultrasound.

High-pressure or ultra-high-pressure homogenization is another advanced promising technology to achieve high product stability and microbial safety in plant-based milk alternatives while maintaining the quality of the product. This technique employs the use of high pressures ranging from 200 to 600 MPa and temperatures ranging from 30°C to 85°C to attain a reduced particle size with uniform emulsions. A significant reduction of particle size in soy milk was observed after UHPH treatment at 300 MPa. UHPH-treated soy-based milk expressed better stability than ultra-high temperature-treated processed samples (Cruz et al., 2007). HPH-treated yogurts also expressed improved firmness due to early onset of gelation and decreased aggregation rate and gel network density upon UHPH treatment (Cruz et al., 2009). Complete destruction of bacteria, spores, and *Bacillus cereus* growth was observed by UHPH at a combination of 200 MPa and 75°C and 300 MPa and 55–75°C in almond milk (Valencia-Flores et al., 2013).

FLAVOR AND TEXTURE MODULATIONS IN NONDAIRY PROBIOTIC BEVERAGES

The most important step used in the preparation of probiotic nondairy beverages is fermentation, which is in itself an approach to complement functionality and improves the organoleptic characteristics of plant materials. The appropriate selection of starter cultures with identified flavor- and texture-improving characteristics offers an encouraging means for in situ product improvement, novelty, and expansion of cereal-based beverages. Fermentation of cereal and legume-based beverages, by LAB, has contributed positive results in terms of improving protein digestibility (Taylor and Taylor, 2002), the bioavailability of minerals and other micronutrients (Greffeuille et al., 2011), shelf life of a product (Gupta et al., 2010), and most important organoleptic attributes (Peyer et al., 2015). The occurrence of favorable carbohydrates, amino acids, peptides, nucleotides, vitamins, macroelements, and fatty acids in cereals, legumes, seeds, and nuts has presented them as an excellent vehicle for LAB fermentation (Blandino et al., 2003). However, raw cereals and legumes are not considered organoleptically pleasant due to their bitter, flat, and green taste or odor. Legume-based nondairy milk alternatives such as soybean and peanut milk tend to smell beany and earthy, which limits its consumption by mainstream customers who are regular consumers of dairy milk. Oxidation of plant lipids releases volatile constituents such as *n*-hexanal and *n*-hexanol, which are responsible for such off-flavors. Additionally, phenolic compounds, namely, tannins, saponins, glucosinolates, terpenes, and flavonoids, present in outer layers of cereals and legumes contribute to bitterness and astringency in taste, based on their molecular weights, and hence result in lower palatability and poor acceptance (Heini€o et al., 2011). Furthermore, bioactives such as isoflavonoids also impart unpleasant aftertaste (Matsuura et al., 1989). In the past, the researches on the creation of novel plant-based milk alternatives were focused on the use of plant constituents as a source of energy for the growth of probiotics and the survival of probiotic strains during storage; however, in recent years, researchers have shifted their inclination towards the flavor and textural alterations triggered by the measured inoculation of LAB starter cultures in grain-based beverages (Salmeron et al., 2014; Prado et al., 2008). To promise product safety and quality (uniform sensory characteristics and functionality) of such nondairy, plant-based probiotic beverages, investigators have attempted to replace the usual "backslop" cultures with defined cultures of single or mixed LAB (Peyer et al., 2016). The mainstream work on novel cereal-based fermented beverages used *L. plantarum* as a robust starter culture under matrix conditions of low pH (Charalampopoulos et al., 2002), which gives this strain an edge over other indigenous microflora of the grains. This strain also delivered a range of flavor components with respect to the substrate. The common flavoring compounds resulted from this strain-specific fermentation are diacetyl, acetoin, and acetaldehyde; the relativeness of these flavoring compounds to that of dairy products adds a pleasant organoleptic profile in the nondairy probiotic products (Salmeron et al., 2015). The cereal substrates, namely, oat, wheat, barley, and barley malt, fermented by *L. plantarum* NCIMB 8826 displayed different microbial flavor compounds at variable concentrations, demonstrating a complex interactive behavior between bacterial culture and substrate components (Salmeron et al., 2009). The predominant volatile compounds detected in four cereal matrices, viz., oat, wheat, barley, and malt, during fermentation were oleic acid, linoleic acid, acetic acid, and 5-hydroxymethylfurfural, respectively (Salmeron et al., 2014, 2015). Carbohydrates, amino acids, organic acids, and fatty acids present in the food matrix or released as intermediate compounds during fermentation by LAB can be routed into different metabolic pathways for specific flavor generation (Gänzle et al., 2007). Pyruvate, an important intermediate compound, releases different flavor and taste active constituents through different metabolic pathways (Liu, 2003). The generated flavor compounds are of both nonvolatile and volatile nature. The nonvolatile components include sugars and some carboxylic acids that are responsible for the sweet and sour taste of the probiotic product, while the volatile fractions comprise carboxylic acids, alcohols, aldehydes, ketones, and esters (Lasekan and Lasekan, 2012). The malt-based probiotic beverage fermented with *L. plantarum* exhibited greater sensorial acceptance in terms of flavor profile due to the highest concentration of acetaldehyde in the final product (Salmeron et al., 2015). The formation or generation of flavor compounds by LAB is one aspect, and maintaining the flavor or preventing the flavor changes during storage is an important area of research in probiotic products; however, only very few researches have assessed this side in the development of nondairy probiotic products. The flavor of nondairy probiotic beverages or yogurts might see changes over time because of the accumulation of staling compounds or the degradation of pleasant-flavoring compounds (Peyer et al., 2016). Commercial and experimental barley malt-based beverage fermented with a *Lactobacillus* spp., assessed for flavor modulation under simulated conditions (storage under accelerated temperature and shaking), expressed aging in terms of the concentration changes of eight indicator compounds (3-methyl butanal, 2-furfural, benzaldehyde, phenyl-acetaldehyde, hexadienal, heptanal, methional, and b-damascenone) associated with thermal and/or oxidative stress (Krahl et al., 2009).

Apart from fermentation, various mechanical and thermal preprocessing operations such as dehulling, sprouting/germination, roasting, soaking, blanching, and cooking have shown positive effects in enhancing the organoleptic acceptability of plant-based probiotic products. These preprocessing treatments have shown a flavor improvement by reducing or masking the off-flavors and also expressed improve mouthfeel and color (Dakwa et al., 2005). These treatments also result in the reduction of antinutritional factors and hence improve the nutritional profile of developed products. Germination is one of the important food processing techniques, which increases the bioavailability of nutrients by releasing the bound nutrients that are present in the food matrix. It also improves the bioavailability of starch and proteins, which are important sources of energy for the growth of microorganisms. Germinated grains exhibit a better nutritional profile than their raw counterparts and are an important source of bioavailable vitamins, minerals, amino acids, phytochemicals, and easily digestible starch and proteins. Generalized method for the preparation of probiotic beverage using germination as pretreatment is illustrated in Fig. 2.

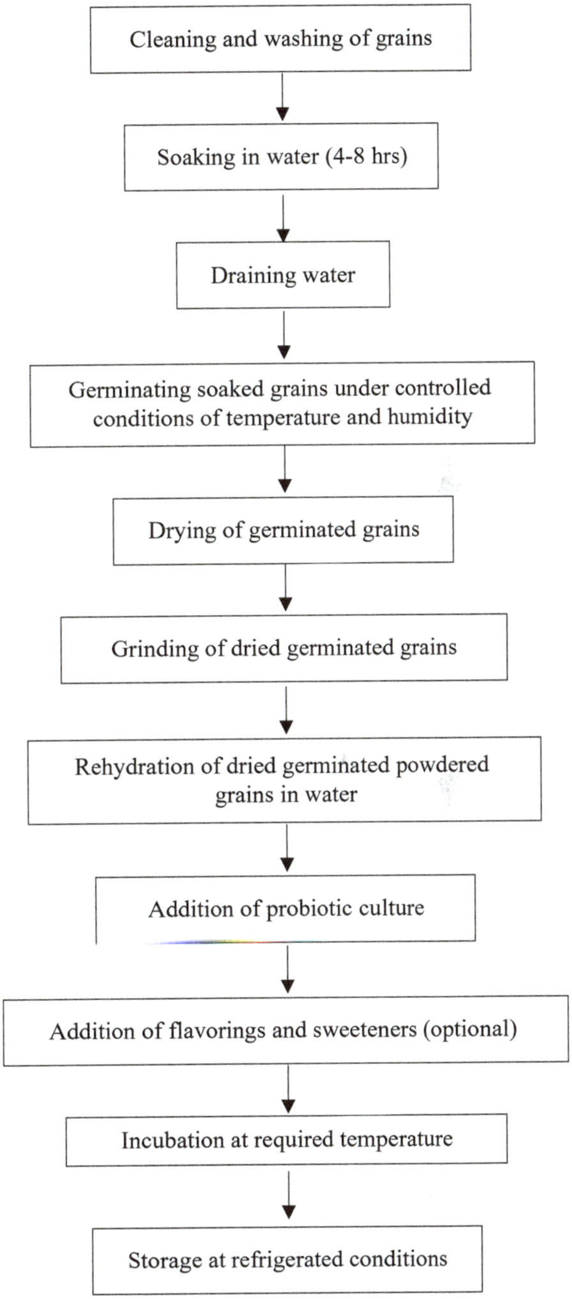

FIG. 2 Generalized flowchart for preparation of probiotic beverage using germination as pretreatment.

The addition of malted grains or hydrolytic enzymes such as amylases, glucanases, and peptidases improved the bioavailability of nutrients in cereal matrices (Luana et al., 2014; Tangyu et al., 2019). Better growth of LAB in liquid barley malt along with higher levels of fermentable sugars (glucose, fructose, maltose, and sucrose) and free amino acids was observed in comparison with raw barley and wheat (Charalampopoulos et al., 2002). Another similar study reported that LAB exhibits an exponential growth phase at a faster rate in the germinated oat matrix than in raw oats (Herrera-Ponce et al., 2014). Probiotic beverages (soy, almond, and coconut) added with germinated barley, finger millet, and moth bean and inoculated with *Lactobacillus acidophilus* expressed the improved overall acceptability and functional properties of beverage in comparison with beverages prepared using ungerminated grains in the similar formulation. Coconut milk-based probiotic drink showed the maximum acceptance based on sensory scores (Chavana et al., 2018).

The preparation of plant-based beverages involves the disintegration of plant constituents with water. The presence of particles of varied sizes depends on the type of grain and size reduction technique; however, the textural integrity and stability of the final product are governed by the water-soluble components (Sethi et al., 2016). A chalky or sandy consistency and a thin mouthfeel owing to the presence of large insoluble particles adversely affect the buyer's readiness for purchasing plant-based beverages (Peyer et al., 2016). Physical instability and phase separation in beverages and syneresis in yogurt is an important concern in the development of novel plant-based probiotic beverages and yogurts. Processes such as fermentation, germination, enzymatic pretreatments, and roasting alter the solubilization of plant constituents in a better way and hence may play a crucial role in improving the sensorial, nutritional, bioavailability, and techno-functional properties of the end product. These processes enhance the solubility of the proteins and bioactive constituents (phenolic compounds) in the extraction medium and can result in an aqueous extract with improved protein extraction and enhanced nutritional and functional values (Buenoa et al., 2020). Some lactic acid bacterial species can excrete high-molecular-weight polysaccharides, also known as exopolysaccharides (EPS), which can increase the viscosity of the aqueous medium. These EPS are homo- or heteropolysaccharides made via polymerization of repeating glucose or fructose subunits. In situ production of EPS is of specific concern in the preparation of fermented cereal-based milk alternatives. EPS can be an effective natural additive to maintain the stability of plant-based beverages and yogurts (Badel et al., 2011). The production rate of exopolysaccharides can be controlled by altering the level of fermentable sugars in the medium. Mridula and Sharma (2014) developed nondairy probiotic drinks using sprouted wheat, barley, pearl millet, and green gram distinctly with oat, stabilizer, and sugar, soy milk, and distilled water fermented with *L. acidophilus* NCDC14. A 100% oat-based yogurt-like product was developed using EPS-producing strain *Pediococcus damnosus* in addition to yogurt culture (Mårtensson et al., 2002). No significant difference in terms of sensorial parameters was observed in nondairy products and its dairy equivalent as control. In another study, an oat-based medium showed an improved viscosity due to the release of relatively lower molecular mass EPS by *L. delbrueckii* subsp. bulgaricus NCFB 2772, in the presence of glucose as a supplementary carbon source (Grobben et al., 1997). Furthermore, lower temperatures and longer incubation time in addition to the appropriate carbon/nitrogen ratio result in substantial amounts of EPS. In this way, EPS can improve physical stability and can reduce the chances of phase separation in nondairy grain-based probiotic products. Germination also plays an important role in improving the texture of plant-based probiotic beverages. Conversion of starch and protein into simpler sugars and amino acids enhances the solubility of these macromolecules, which in turn provides better sensorial attributes in terms of texture and stability of plant-based probiotic beverages. Cereal grains particularly are rich in starch and soluble fibers such as β-glucan, which during the preparation of plant-based beverage upon the application of thermal treatment, converts into a thick mass owing to gelatinization of starch and water absorption by soluble fibers (Lorri and Svanberg, 2009). To encounter such technological issues, LAB cultures with enzymatic activity have been used for the partial degradation of such biopolymers (Onyango et al., 2004). Amylolytic LABs (*L. manihotivorans, L. fermentum, L. amylovorus, L. amylophilus, L. plantarum*, and *L. amylolyticus*) have the capability of degrading polysaccharides and can be employed as a culture medium during the manufacture of plant-based beverages (Reddy et al., 2008). Humblot et al. (2014) observed the gene expression and enzymatic action of various starch-degrading enzymes produced by *L. plantarum* A6 throughout the fermentation of a pearl millet-based gruel. Alternatively, cereal malt can serve as an economic source of amylolytic enzymes in reducing the viscosity of beverage formulations; however, the variability in their amylolytic activity and longer production duration limit its applications in probiotic beverage formulations. It was also reported that high levels of glucose, maltose, or sucrose in the food matrix can inhibit the synthesis of amylases by amylolytic LABs. To overcome this drawback, co-culture fermentation has been used by some researchers, where they found quicker starch-hydrolyzing rates and reduced viscosity compared with amylolytic LAB alone (Mukisa et al., 2012). In co-culture applications, amylolytic LAB in combination with a fast-fermenting LAB accelerates the depletion of free fermentable sugars. In co-culture, faster acidification rates and high lactic acid yields could ensure a better safety and stability of the products.

CONSUMER ACCEPTABILITY OF PLANT-BASED MILK ALTERNATIVES

Although the plant-based milk alternatives witnessed an increasing demand from consumers due to changing lifestyles and inclination towards vegan foods, the reluctance of the typical consumers to attempt unacquainted foods that are professed as unattractive appears to be a strong limiting factor. Advanced scientific and technological interventions have led to the improvement in flavor profile and textural stability of the plant-based milk; however, this segment carries a stigma because of less appealing products in the past (Wansink et al., 2005). Taste acceptance is the key attribute in buying criteria of novel/unfamiliar foods; also, an increased willingness was observed in case of information about a good and/or familiar taste (Magnusson et al., 2001). A study conducted by Dupont on US consumers showed the taste (26%) as the prime criterion for buying preference followed by protein content (19%), thickness and creaminess (14.8%), calories (13.8%), sugar content (13.3%), and health and wellness claims (11.9%). The USA witnessed plant-based milk sales exceeded $1.8 billion in 2018 with a 9% rise from 2017 (according to Nielsen data for the 52 weeks ending, August 11, 2018). Also, it was reported that more than one in three U.S. households purchase plant-based milk. Nondairy yogurt segment witnessed a 55% growth in 2018 and was valued at $162 million. The drivers towards the sale of nondairy yogurts are the same as that in the case of plant-based milk, which includes perceived health benefits and avoidance of dairy products. Since the plant-based yogurt segment is still evolving, prospects exist for innovation to accomplish consumer desires for improved taste and texture, along with perceived health claims. Probable health benefits are important criteria in deciding the buyer preference and willingness to try newer food products (Luckow et al., 2006; Kihlberg et al., 2005). Studies have shown that frequent acquaintance has been shown to augment the likeness in the cases of plant-based probiotic beverages (Stein et al., 2003). In the case of rice milk, a five-day exposure increased the pleasant perception towards taste and overall acceptability (Russell and Delahunty, 2004). Over 3 weeks' consumption of oat milk by male subjects, who initially gave lower scores, expressed higher scores (Onning et al., 1998). In the creation of novel functional products, it is essential to comprehend the sensory influences of nonconventional or newer constituents on the color, aroma, flavor, and texture and to regulate how these constituents impact the acceptability and consumer's inclination (Granato et al., 2010).

ADVANCED TECHNIQUES FOR CONTROLLED DELIVERY OF PROBIOTICS IN PLANT-BASED MILK

The food industry is continuously working on technological expansions to encounter the present limitations and changing consumer preferences. Also, food science and technology have evolved from serving basic nutritional needs or formulating nutritious foods towards designing functional foods/health foods with health-promoting characteristics. The technological advances are directed to target or address health-related problems associated with a particular group of people. In this context, the improvement in gastrointestinal health through functional foods such as probiotics, prebiotics, and synbiotics holds a promising area of research for food scientists and nutritionists. Since its inception, the focus of this segment was based on or revolves around dairy foods; however, in the last few years due to technological advancements, the focus has shifted towards grain-based probiotic and synbiotic products. Improved delivery or controlled release of probiotics through advanced food processing techniques helps in designing and development of novel plant-based probiotic and synbiotic products such as nondairy milk and yogurts (Fig. 3). The techniques commonly used for controlled delivery of probiotics and for maintaining the viability of culture throughout processing and storage include the addition of cell protectants, microencapsulation, and nanotechnology. These techniques might help in formulating probiotic and synbiotic nondairy products, through an amalgamation of microcapsules or nanoparticles with exclusive and precise bioactivity. These technological interventions have unwrapped an innovative prospect in this sector of functional foods, shaped to progress the overall well-being of consumers. Nanotechnology is the technique for the development of nanosized food particles that have found numerous applications in the design and development of novel functional foods and food ingredients. The theory of engineered nanomaterials has been defined by the European Commission

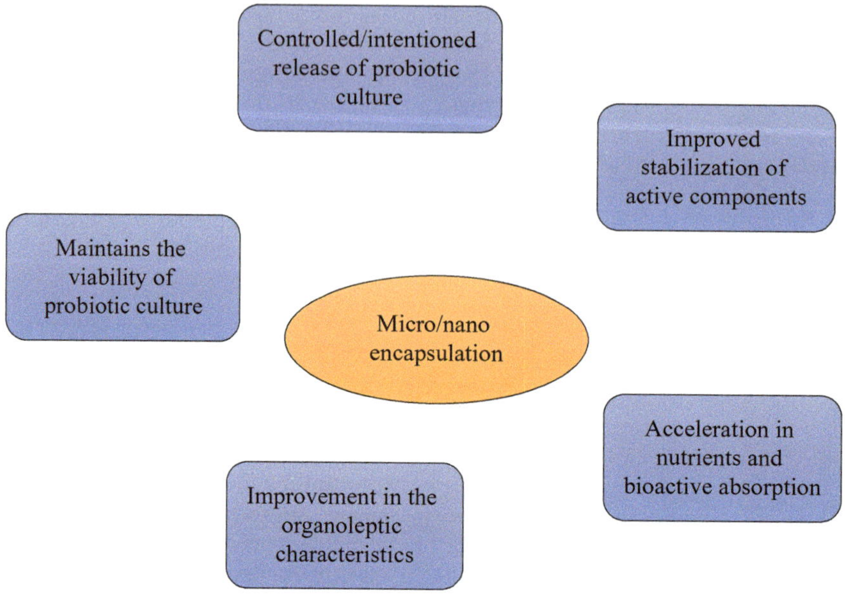

FIG. 3 Benefits of micro/nanoencapsulation in preparation of plant-based probiotic and synbiotic beverages.

as "any purposefully manufactured material, comprising particles in an unbound form or as an aggregate or as an agglomerate, where 50% or more of the particles in the number size distribution are in the size range of 1–100 nm" European_Union, 2011). Some applications of nanotechnology in the food industry include controlled delivery approaches for bioactive, antimicrobials, food additives, and enzymes (Trujillo et al., 2016). Apart from maintaining the viability and controlled/intentioned release of probiotic culture, this technique might help in the improvement of the organoleptic characteristics, acceleration in nutrients and bioactive absorption, and improved stabilization of active components in plant-based nondairy food products (Ranjan et al., 2014). The designing of GRAS nanoparticles expresses a pronounced perspective for their inclusion into novel grain-based beverages or nondairy milk. However, extensive in vivo or clinical studies are essential to measuring the consequence of their passage throughout the GI tract until they reach the anticipated target location. Presently, there are not enough in vivo studies that assess the contact of nanoparticles to the human GI tract; a significant cause to reflect is the passage time employed by the nanoparticles after ingestion. The travel exposure of designed nanoparticles to the gut pH, bile salts, and time needs to be closely monitored to control the bioactivity of these particles. The commonly used nanomaterials are based on biopolymers, proteins, and carbohydrates. Plant-based milk alternatives are colloidal dispersions that are generally defined as "dispersions that consist of at least two phases: one or more dispersed or internal phases, and a continuous or external phase known as a dispersion medium." These dispersions consist of one or more constituents that have at least one element in the range of 1–10 nm towards the lower side and a small number of elements in the range of micrometers at the upper side (Bowman et al., 2006). There is an excessive possibility to apply colloidal dispersions in the form of nanoemulsions to deliver nanoparticles encapsulated with water-insoluble fragments that were not suitable to be used previously owing to their poor solubility. Consequently, a myriad collection of functional plant-based or nondairy probiotic and synbiotic food products can be designed with nanoparticles of specific interest without adversely affecting the physicochemical, rheological, and sensorial characteristics of the original product.

Similarly, microencapsulation, a technique for enclosing solids, liquids, or gaseous constituents in small, closed capsules that can deliver their elements in specified quantities under defined conditions, can be employed. A microcapsule comprises a core (solid or liquid), with a diameter ranging from few microns to 1 mm, which in turn is surrounded by a semipermeable outer membrane. Active constituents within the capsule remain protected from harsh environmental conditions such as acidity, oxygen, and gastric conditions with the help of coating. Apart from maintaining the viability of culture, these capsules ensure the controlled release of active constituents in the human system. Frequently used material for the encapsulation of probiotic culture consists of polysaccharides of different origins. It includes alginate and κ-carrageenan (from seaweeds), gum arabic, starch and its derivatives (from plant various sources), xanthan and gellan (from bacteria), and gelatin or milk proteins (from animal sources) (Rokka and Rantamäki, 2010). The microcapsules of probiotics are usually manufactured by three important techniques, namely, emulsion, extrusion, and spray drying. The size of microcapsules prepared by emulsion, extrusion, and spray drying techniques varies from 20 μm to 2 mm, 150 μm to 4 mm, and 5 to 80 μm, respectively (Muthukumarasamy et al., 2006; Picot and Lacroix, 2003). A 50 μm of average starch granule, which works as the adhesion base for *bifidobacteria*, is reported by Mattila-Sandholm et al. (2002). Low-temperature techniques such as spray chilling and freeze drying are widely used in food and pharmaceutical industries for the encapsulation of probiotics. These techniques give a better viability to the probiotic culture and also maintain the stability of microcapsules during transportation and storage. The microcapsules are prepared by first allowing probiotic culture to grow under optimal conditions which are then used as suspension after centrifugation or further dried with suitable drying technology (preferably by freeze drying) for long-term convenient usage by the food industry. For better protection and stability of probiotic culture, matrix encapsulation is also studied by researchers and industries. In this technique, a combination of encapsulation material is used for the preparation of microcapsules. Encapsulation by MicroMAX technology used proteins, lipids, and carbohydrates for the protection of probiotics during spray drying and storage, and also during the passage over the stomach. Coating capsules with additional layer improve their stability in low pH as well as with exposure to oxygen. Promising coating constituents comprise alginate, chitosan, starch, gum, poly-L-lysine, and gelatin. The survival of probiotic bacteria during processing, storage, and in stomach gastric environments is greatly reliant on the type of strain selected, the material used for encapsulation, and the food matrix environment. Characteristic survival rates in the range of 70%–85% were observed with spray drying and freeze drying processes. The addition of oxygen scrubber and desiccant has been observed to improve the survival rate in probiotics (Lian et al., 2003). Till date, the probiotic market is captured by dairy-based products; its applications in novel nondairy product development is very challenging due to the compositional variations in the food matrix. Designing novel grain-based products, where food constituents support probiotic survival in the GI tract in amalgamation with micro- or nanoencapsulation techniques, may demonstrate to be effective. Micro- and nanoencapsulation techniques are, however, not fully established at the large scale; therefore, scaling up of these techniques requires additional scientific interventions. Successful scaling up of encouraging laboratory-scale results will benefit food industries in further refining and commercialization of these products. A complete scientific methodology is required in further integrating micro- and nanoencapsulation with emerging nonthermal food processing technologies that can help in maintaining the viability of probiotics culture. Advanced processing and packaging techniques such as high-pressure processing, pulsed electric field, active and intelligent packaging systems may help in the creation of grain-based novel probiotic or synbiotic products with high sensory acceptability. This requires sequentially integrated optimization of process parameters specific to each product. Also, the development of new strains with increased stress resistance through gene technology will perform a chief role in successful future applications of probiotics and synbiotics.

CONCLUSION

The market for plant-based probiotic products is continuously expanding. However, the maintenance of probiotic viability, prevention of generation of off-odors and flavors, unbalanced nutrition relative to dairy-based probiotic products, and textural stability of beverage limit their consumption and pose technological challenges in the development of newer plant-based probiotic products. The use of mixed-culture fermentation for better sensory profile culture protection through micro- and nanoencapsulation, and nonthermal food processing technologies for dispersion stability throughout the storage, carries immense potential for enhancing the nutritional and sensory values of such products. Effects of fermentation on plant bioactive compounds such as phenolics need to be studied for establishing positive health benefits of such constituents. Complete understanding of plant matrix fermentation and advanced realistic formulations with functional claims and improvements in existing products is the need of the hour. With increasingly diverse markets, the fermentation of bioactive-rich plant materials, especially legumes, cereals, seeds, and nuts, will become an important trend in the functional beverage market. Critical selection of probiotic strains in a combination of prebiotics that promotes their growth in plant-based food matrices with predictable synergistic relations would be extremely important in the direction of developing efficient fermentation processes and quality products.

REFERENCES

Abbeele, P.V., Kamil, A., Fleige, L., Chung, Y., Chavez, P., Marzorati, M., 2018. Different oat ingredients stimulate specific microbial metabolites in the gut microbiome of three human individuals in-vitro. ACS Omega 3, 12446–12456.

Aboulfazli, F., Shori, A.B., Baba, A.S., 2016. Effects of the replacement of cow milk with vegetable milk on probiotics and nutritional profile of fermented ice cream. LWT- Food Sci. Technol. 70, 261–270.

Agil, R., Hosseinian, F., 2012. Dual functionality of triticale as a novel dietary source of prebiotics with antioxidant activity in fermented dairy products. Plant Foods Hum. Nutr. 67 (1), 88–93.

Amarteifio, J.O., Moholo, D., 1998. The chemical composition of four legumes consumed in Botswana. J. Food Compos. Anal. 11, 329–333.

Andrés, V., Mateo-Vivaracho, L., Guillamón, E., Villanueva, M.J., Tenorio, M.D., 2016. High hydrostatic pressure treatment and storage of soy-smoothies: colour, bioactive compounds and antioxidant capacity. LWT-Food Sci. Technol. 69, 123–130.

Angelov, A., Gotcheva, V., Kuncheva, R., Hristozova, T., 2006. Development of a new oat-based probiotic drink. Int. J. Food Microbiol. 112, 75–80.

Armougom, F., Henry, M., Vialettes, B., Raccah, D., Raoult, D., 2009. Monitoring bacterial community of human gut microbiota reveals an increase in Lactobacillus in obese patients and Methanogens in anorexic patients. PLoS ONE 4, e7125.

Badel, S., Bernardi, T., Michaud, P., 2011. New perspectives for *Lactobacilli* exopolysaccharides. Biotechnol. Adv. 29 (1), 54–66.

Berk, Z., 1992. Technology of production of edible flours and protein products from soybeans. FAO Agric. Serv. Bull. 97, 178.

Bernat, N., Cháfer, M., Chiralt, A., González Martínez, C., 2014. Development of a non-dairy probiotic fermented product based on almond milk and inulin. Food Sci. Technol. Int. 21 (6), 440–453.

Bernat, N., Chafera, M., Chiralt, A., Gonzalez-Martinez, C., 2015. Probiotic fermented almond "milk" as an alternative to cow-milk yoghurt. Int. J. Food Stud. 4, 201–211.

Bharti, S.K., Sharma, N.K., Murari, K., Kumar, A., 2012. Functional aspects of dairy foods in human health: an overview. Crit. Rev. Pharm. Sci. 2 (1), 35–42.

Blandino, A., Al-Aseeri, M.E., Pandiella, S.S., Cantero, D., Webb, C., 2003. Cereal based fermented foods and beverages. Food Res. Int. 36 (6), 527–543.

Bouhnik, Y., Raskine, L., Simoneau, G., Vicaut, E., Neut, C., Flouriéb, B.F., Bornet, F., 2004. The capacity of nondigestible carbohydrates to stimulate fecal bifidobacteria in healthy humans: a double-blind, randomized, placebo-controlled, parallel-group, dose-response relation study. Am. J. Clin. Nutr. 6, 1658–1664.

Bowman, B.J., Ofner, C.M., Schott, H., 2006. Colloidal dispersions. In: Troy, D.B. (Ed.), Remington: The Science and Practice of Pharmacy. Lippincott Williams & Wilkins, Baltimore, MD, USA, p. 293.

Bravo, L., 1999. Effect of processing on the non-starch polysaccharides and in-vitro digestibility of legumes. Food Sci. Technol. Int. 5, 415–423.

Bruno, L.M., Lima, J.R., Wurlitzer, N.J., Rodrigues, T.C., 2020. Non-dairy cashew nut milk as a matrix to deliver probiotic bacteria. Food Sci. Technol. 40, 604–607.

Buenoa, D.B., Júnior, S.I.S., Chiarotto, A.B.S., Cardoso, T.M., Neto João, A., Reis Guilherme, C.L., Glória, M.B.A., Tavano, O.L., 2020. The germination of soybeans increases the water-soluble components and could generate innovations in soy-based foods. LWT- Food Sci. Technol. 117, 108599.

Casarotti, S.N., Carneiro, B.M., Penna, A.L.B., 2014. Evaluation of the effect of supplementing fermented milk with quinoa flour on probiotic activity. J. Dairy Sci. 97 (10), 6027–6035.

Charalampopoulos, D., Pandiella, S.S., 2010. Survival of human derived *Lactobacillus plantarum* in fermented cereal extracts during refrigerated storage. LWT- Food Sci. Technol. 43 (3), 431–435.

Charalampopoulos, D., Pandiella, S.S., Webb, C., 2002. Growth studies of potentially probiotic lactic acid bacteria in cereal-based substrates. J. Appl. Microbiol. 92, 851–859.

Chavana, M., Gat, Y., Harmalkar, M., Waghmare, R., 2018. Development of non-dairy fermented probiotic drink based on germinated and ungerminated cereals and legume. LWT- Food Sci. Technol. 91, 339–344.

Chen, J., Raymond, K., 2008. Beta-glucans in the treatment of diabetes and associated cardiovascular risks. Vasc. Health Risk Manag. 4, 1265–1272.

Cheng, Y., Thompson, L., Brittin, H., 2006. Sogurt, a yogurt-like soybean product: development and properties. J. Food Sci. 55, 1178–1179.

Chou, C.C., Hou, J.W., 2000. Growth of *bifidobacteria* in soymilk and their survival in the fermented soymilk drink during storage. Int. J. Food Microbiol. 56 (2–3), 113–121.

Coda, R., Di Cagno, R., Gobbetti, M., Rizzello, C.G., 2014. Sourdough lactic acid bacteria: exploration of non-wheat cereal-based fermentation. Food Microbiol. 37, 51–58.

Coman, M.M., Verdenelli, M.C., Cecchini, C., Silvi, S., Vasile, A., Bahrim, G.E., Orpianesi, C., Cresci, A., 2013. Effect of buckwheat flour and oat bran on growth and cell viability of the probiotic strains *Lactobacillus rhamnosus* IMC 501®, *Lactobacillus paracasei* IMC 502® and their combination SYNBIO®, in synbiotic fermented milk. Int. J. Food Microbiol. 167 (2), 261–268.

Cross, M.L., Stevenson, L.M., Gill, H.S., 2001. anti-allergy properties of fermented foods: an important inmunoregulatory mechanism of lactic acid bacteria? Int. Immunopharmacol. 1, 891–901.

Cruz, N., Capellas, M., Hernández, M., Trujillo, A.J., Guamis, B., Ferragut, V., 2007. Ultra high pressure homogenization of soymilk: microbiological, physicochemical and microstructural characteristics. Food Res. Int. 40 (6), 725–732.

Cruz, N.S., Capellas, M., Jaramillo, D.P., Trujillo, A.J., Guamis, B., Ferragut, V., 2009. Soymilk treated by ultra-high-pressure homogenization: acid coagulation properties and characteristics of a soy-yogurt product. Food Hydrocoll 23 (2), 490–496.

Cummings, J.H., Christie, S., Cole, T.J., 2001. A study of fructo oligosaccharides in the prevention of travellers' diarrhoea. Aliment. Pharmacol. Ther. 15, 1139–1145.

Dakwa, S., Sakyi-Dawson, E., Diako, C., Annan, N.T., Amoa-Awua, W.K., 2005. Effect of boiling and roasting on the fermentation of soybeans into dawadawa (soy-dawadawa). Int. J. Food Microbiol. 104 (1), 69–82.

de Almeida Costa, G.E., da Silva, Q.M.K., Reis, S.M.P.M., de Oliveira, A.C., 2006. Chemical composition, dietary fiber and resistant starch contents of raw and cooked pea, common bean, chickpea and lentil legumes. Food Chem. 94, 327–330.

De Vrese, M., Stegelmann, A., Richter, B., Fenselau, S., Laue, C., Schrezenmeir, J., 2001. Probiotics-compensation for lactase insufficiency. Am. J. Clin. Nutr. 73 (2), 421S–429S.

Dimitrova, K., Kartalska, Y., 2015. Technological characteristics of yogurt supplemented with 3% ground amaranth seeds during refrigerated storage. Appl. Sci. Rep. 12 (1), 27–32.

Duda-Chodak, A., Tarko, T., Satora, P., et al., 2015. Interaction of dietary compounds, especially polyphenols, with the intestinal microbiota: a review. Eur. J. Nutr. 54 (3), 325–334.

European_Union, 2011. Commission recommendation of 18 October 2011 on the definition of nanomaterial. Off. J. Eur. Union 275, 38–40.

Falagas, M.E., Betsi, G.I., Athanasiou, S., 2007. Probiotics for the treatment of women with bacterial vaginosis. Clin. Microbiol. Infect. 13 (7), 657–664.

Fernando, W.M.U., Hill, J.E., Zello, G.A., Tyler, R.T., Dahl, W.J., Van Kessel, A.G., 2010. Diets supplemented with chickpea or its main oligosaccharide component raffinose modify faecal microbial composition in healthy adults. Benefic. Microbes 1 (2), 197–207.

Fiocchi, A., Brozek, J., Schuenemann, H., Bahna, S.L., von Berg, A., Beyer, K., Bozzola, M., Bradsher, B., Compalati, E., Ebisawa, M., Guzman, M.A., Li, H., Heine, R.G., Keith, P., Lack, G., Landi, M., Martelli, A., Rancé, F., Sampson, H., Stein, A., Terracciano, L., Vieths, S., 2010. World allergy organization (WAO) diagnosis and rationale for action against cow's milk allergy (dracma) guidelines. Pediatr. Allergy Immunol. 21 (21), 1–125.

Fischer, S., Wilckens, R., Jara, J., Aranda, M., 2013. Variation in antioxidant capacity of quinoa (*Chenopodium quinoa* Will) subjected to drought stress. Ind. Crop Prod. 46, 341–349.

Fuchs, R.H.B., Borsato, D., Bona, E., Hauly, M.C.O., 2005. Soy yogurt supplemented with oligofructose and inulin. Food Sci. Technol. (Campinas) 1 (25), 175–181.

Fuentes-Zaragoza, E., Riquelme-Navarrete, M.J., Sánchez-Zapata, E., Pérez-Álvarez, J.A., 2010. Resistant starch as functional ingredient: a review. Food Res. Int. 43, 931–942.

Fuentes-Zaragoza, E., Sanchez-Zapata, Elena, S., Esther, S.E., Navarro, C., Fernandez-Lopez, J., Perez-Alvarez, J.A., 2011. Resistant starch as prebiotic: a review. Starch/Starke 63, 406–415.

Gänzle, M.G., Vermeulen, N., Vogel, R.F., 2007. Carbohydrate, peptide and lipid metabolism of lactic acid bacteria in sourdough. Food Microbiol. 24 (2), 128–138.

Gibson, G.R., Roberfroid, M.B., 1995. Dietary modulation of the human colonic microbiota: introducing the concept of prebiotics. J. Nutr. 125, 1401–1412.

Gorbach, S.L., 1996. Microbiology of the gastrointestinal tract medical microbiology. In: Medical Microbiology, fourth ed. University of Texas Medical Branch at Galveston, Galveston (TX). Chapter 95.

Granato, D.B.G.F., Nazzaro, F., Cruz, A.G., Faria, J.A.F., 2010. Functional foods and nondairy probiotic food development: trends, concepts, and products. Compr. Rev. Food Sci. Food Saf. 9 (3), 292–302.

Greffeuille, V., Polycarpe Kayode, A.P., Icard-Vernire, C., Gnimadi, M., Rochette, I., Mouquet-Rivier, C., 2011. Changes in iron, zinc and chelating agents during traditional African processing of maize: effect of iron contamination on bioaccessibility. Food Chem. 126 (4), 1800–1807.

Grobben, G.J., Van Casteren, W.H.M., Schols, H.A., Oosterveld, A., Sala, G., Smith, M.R., Sikkema, J., de Bont, J.A.M., 1997. Analysis of the exopolysaccharides produced by *Lactobacillus delbrueckii* subsp. bulgaricus NCFB 2772 grown in continuous culture on glucose and fructose. Appl. Microbiol. Biotechnol. 48 (4), 516–521.

Guergoletto, K.B., Magnani, M., San-Martin, J., Andrade, J.C.G.T., Garcia, S., 2010. Survival of *Lactobacillus casei* (LC-1) adhered to prebiotic vegetal fibers. Innov. Food Sci. Emerg. Technol. 11 (2), 415–421.

Gullón, B., Gullón, P., Tavaria, F.K., Yáñez, R., 2016. Assessment of the prebiotic effect of quinoa and amaranth in the human intestinal ecosystem. Food Funct. 7 (9), 3782–3788.

Gupta, S., Cox, S., Abu-Ghannam, N., 2010. Process optimization for the development of a functional beverage based on lactic acid fermentation of oats. Biochem. Eng. J. 52 (2–3), 199–204.

Hager, A.S., Wolter, A., Jacob, F., Zannini, E., Arendt, E.K., 2012. Nutritional properties and ultra-structure of commercial gluten free flours from different botanical sources compared to wheat flours. J. Cereal Sci. 56 (2), 239–247.

Hassan, A.A., Aly, M.M.A., El-Hadidie, S.T., 2012. Production of cereal-based probiotic beverages. World Appl. Sci. J. 19 (10), 1367–1380.

He, S., Hekmat, S., 2015. Sensory evaluation of non-dairy probiotic beverages. J. Food Res. 4 (1), 186–192.

Heini€o, R.L., Kaukovirta-Norja, A., Poutanen, K., 2011. Flavor in processing new oat foods. Cereal Foods World 56 (1), 21–26.

Hemarajata, P., Versalovic, J., 2013. Effects of probiotics on gut microbiota: mechanisms of intestinal immunomodulation and neuromodulation. Therap. Adv. Gastroenterol. 6 (1), 39–51.

Herrera-Ponce, A., Nevarez-Morillon, G., Ortega-Rıvas, E., Perez-Vega, S., Salmeron, I., 2014. Fermentation adaptability of three probiotic *Lactobacillus* strains to oat, germinated oat and malted oat substrates. Lett. Appl. Microbiol. 59, 449–456.

Hoppert, K., Zahn, S., Janecke, L., Mai, R., Hoffmann, S., Rohm, H., 2013. Consumer acceptance of regular and reduced-sugar yogurt enriched with different types of dietary fiber. Int. Dairy J. 28 (1), 1–7.

Hu, Y., Zhang, J., Zou, L., Fu, C., Li, P., Zhao, G., 2017. Chemical characterization, antioxidant, immune-regulating and anticancer activities of a novel bioactive polysaccharide from *Chenopodium quinoa* seeds. Int. J. Biol. Macromol. 99, 622–629.

Humblot, C., Turpin, W., Chevalier, F., Picq, C., Rochette, I., Guyot, J.P., 2014. Determination of expression and activity of genes involved in starch metabolism in *Lactobacillus plantarum* A6 during fermentation of a cereal-based gruel. Int. J. Food Microbiol. 185, 103–111.

Hussein, H., Awad, S., El-Sayed, I., Ibrahim, A., 2020. Impact of chickpea as prebiotic, antioxidant and thickener agent of stirred bio-yoghurt. Ann. Agric. Sci. 65, 49–58.

Ilyasoglu, H., Yılmaz, F., Burnaz, N.A., Baltac, C., 2015. Preliminary assessment of a yoghurt-like product manufactured from hazelnut slurry: Study using response surface methodology. LWT- Food Sci. Technol. 62, 497–505.

Isolauri, E., Sütas, Y., Kankaanpää, P., Arvilommi, H., Salminen, S., 2001. Probiotics: effects on immunity. Am. J. Clin. Nutr. 73 (2), 444S–450S.

Iswarin, S.J., Permadi, B., 2012. Coconut milk's fat breaking by means of ultrasound. Int. J. Basic Appl. Sci. 12, 1–5.

Kay, M., Behall, K.M., Daniel, J., Scholfield, D.J., Hallfrisch, J.G., Liljeberg-Elmstah, H.G.M., 2006. Consumption of both resistant starch and β-glucan improves postprandial plasma glucose and insulin in women. Diabetes Care 29, 976–981.

Kechagia, M., Basoulis, D., Konstantopoulou, S., Dimitriadi, D., Gyftopoulou, K., Skarmoutsou, N., Fakiri, E.M., 2013. Health benefits of probiotics: a review. Int. Sch. Res. Notices, 481651.

Kedia, G., Vázquez, J.A., Pandiella, S.S., 2008. Fermentability of whole oat flour, PeriTec flour and bran by *Lactobacillus plantarum*. J. Food Eng. 89, 246–249.

Kihlberg, I., Johansson, L., Langsrud, O., Risvik, E., 2005. Effects of information on liking of bread. Food Qual. Prefer. 16, 25–35.

Krahl, M., Zarnkow, M., Stürmer, F., Becker, T., 2009. Flavor stability of alternative malt-based beverages. Master Brewers Assoc. Am. Tech. Q., 1–7.

Kumar, M., Kumar, A., Nagpal, R., 2010. Cancer-preventing attributes of probiotics: an update. Int. J. Food Sci. Nutr. 61 (5), 473–496.

Laine, R., Salminen, S., Benno, Y., Arthur, C.O., 2003. Performance of *Bifidobacteria* in oat-based media. Int. J. Food Microbiol. 83, 105–109.

Lamothe, L.M., Srichuwong, S., Reuhs, B.L., Hamaker, B.R., 2015. Quinoa (*Chenopodium quinoa* W.) and amaranth (*Amaranthus caudatus* L.) provide dietary fibers high in pectic substances and xyloglucans. Food Chem. 167, 490–496.

Lasekan, O., Lasekan, A., 2012. Effect of processing and flavor fine-tuning techniques on the volatile flavor constituents of pseudocereals and some minor cereals. J. Food Agric. Environ. 10, 73–79.

Lee, Y., Salminen, S., 2009. Handbook of Probiotics and Prebiotics, second ed. John Wiley & Sons Inc., New Jersey.

Leroy, F., De Vuyst, L., 2004. Lactic acid bacteria as functional starter cultures for the food fermentation industry. Food Sci. Technol. 15, 67–78.

Lewis, S., Burmeister, S., Brazier, J., 2005. Effect of the prebiotic oligofructose on relapse of Clostridium difficile-associated diarrhea: a randomized, controlled study. Clin. Gastroenterol. Hepatol. 3, 442–448.

Lian, W., Hsiao, H., Chou, C., 2003. Survival of *bifidobacteria* after spray-drying. Int. J. Food Microbiol. 86, 293–301.

Lin, B., Gong, J., Wang, Q., Cui, S., Yu, H., Huang, B., 2011. In vitro assessment of the effects of dietary fibers on microbial fermentation and communities from large intestinal digesta of pigs. Food Hydrocoll. 25, 180–188.

Liu, S., 2003. Practical implications of lactate and pyruvate metabolism by lactic acid bacteria in food and beverage fermentations. Int. J. Food Microbiol. 83 (2), 115–131.

Liu, Z., Lin, X., Huang, G., Zhang, W., Rao, P., Ni, L., 2014. Prebiotic effects of almonds and almond skins on intestinal microbiota in healthy adult humans. Anaerobe 26 (1–6).

Lorri, W., Svanberg, U., 2009. Lactic acid-fermented cereal gruels: viscosity and flour concentration. Int. J. Food Sci. Nutr. 44 (3), 207–213.

Luana, N., Coda, R., Curiel, J.A., Kaisa, P., Marco, G., Rizzello, C.G., 2014. Manufacture and characterization of a yogurt-like beverage made with oat flakes fermented by selected lactic acid bacteria. Int. J. Food Microbiol. 185, 17–26.

Luckow, T., Sheehan, V., Fitzgerald, G., Delahunty, C., 2006. Exposure, health information and flavor-masking strategies for improving the sensory quality of probiotic juice. Appetite 47, 315–323.

Magnusson, M.K., Arvola, A., Hursti, U.K.K., Aberg, L., Sj€oden, P.O., 2001. Attitudes towards organic foods among Swedish consumers. Br. Food J. 103, 209–227.

Manassero, C.A., Vaudagna, S.R., Sancho, A.M., Añón, M.C., Speroni, F., 2016. Combined high hydrostatic pressure and thermal treatments fully inactivate trypsin inhibitors and lipoxygenase and improve protein solubility and physical stability of calcium-added soymilk. Innov. Food Sci. Emerg. Technol. 35, 86–95.

Mandalari, G., Nueno-Palop, C., Bisignano, G., Wickham, M.S.J., Narbad, A., 2008. Potential prebiotic properties of almond (*Amygdalus communis* L.) seeds. Appl. Environ. Microbiol. 74, 4264–4270.

Mårtensson, O., Oste, R., Holst, O., 2002. Texture promoting capacity and EPS formation by lactic acid bacteria in three different oat-based non-dairy media. Eur. Food Res. Technol. 214, 232–236.

Martinez-Villaluenga, C., Gomez, R., 2007. Characterization of *bifidobacteria* as starters in fermented milk containing raffinose family of oligosaccharides from lupin as prebiotic. Int. Dairy J. 17 (2), 116–122.

Martinez-Villaluenga, C., Frias, J., Vidal-Valverde, C., Gomez, R., 2005. Raffinose family of oligosaccharides from lupin seeds as prebiotics: application in dairy products. J. Food Prot. 68 (6), 1246–1252.

Martinez-Villaluenga, C., Frias, J., Gomez, R., Vidal-Valverde, C., 2006. Influence of addition of raffinose family oligosaccharides on probiotic survival in fermented milk during refrigerated storage. Int. Dairy J. 16 (7), 768–774.

Matsuura, M., Obata, A., Fukushima, D., 1989. Objectionable flavor of soy milk developed during the soaking of soybeans and its control. J. Food Sci. 54 (3), 602–605.

Mattila-Sandholm, T., Myllärinen, P., Crittenden, R., Mogensen, G., Fonden, R., Saarela, M., 2002. Technological challenges for future probiotic foods. Int. Dairy J. 12 (2/3), 173–182.

Mauro, C.S.I., Garcia, S., 2019. Coconut milk beverage fermented by *Lactobacillus reuteri*: optimization process and stability during refrigerated storage. J. Food Sci. Technol. 56 (2), 854–864.

McGarr, S.E., Ridlon, J.M., Hylemon, P.B.D., 2005. anaerobic bacterial metabolism, and colon cancer: a review of the literature. J. Clin. Gastroenterol. 39, 98–109.

Mesquita, M.C., Leandro, E.S., de Alencar, E.R., Botelho, R.B.A., 2020. Fermentation of chickpea (*Cicer arietinum* L.) and coconut (*Coccus nucifera* L.) beverages by *Lactobacillus paracasei subsp paracasei* LBC 81: The influence of sugar content on growth and stability during storage. LWT- Food Sci. Technol. 132, 109834.

Morell, M., 2007. In vitro assessment of probiotic bacteria: from survival to functionality. Int. Dairy J. 17, 1278–1283.

Mridula, D., Sharma, M., 2014. Development of non-dairy probiotic drink utilizing sprouted cereals, legume and soymilk. LWT-Food Sci. Technol. 62, 482–487.

Mukisa, I.M., Byaruhanga, Y.B., Muyanja, C.M.B.K., Aijuka, M., Schüller, R.B., Sahlstrom, S., Langsrud, T., Narvhus, J.A., 2012. Influence of co fermentation by amylolytic *Lactobacillus plantarum* and *Lactococcus lactis* strains on the fermentation process and rheology of sorghum porridge. Appl. Environ. Microbiol. 78 (15), 5220–5228.

Murevanhema, Y.Y., Jideani, V.A., 2015. Production and characterization of milk produced from bambara groundnut (*Vigna subterranea*) varieties. J. Food Process. Preserv. 39 (6), 1485–1489.

Muthukumarasamy, P., Allan, W.P., Holley, R.A., 2006. Stability of *Lactobacillus reuteri* in different types of microcapsules. J. Food Sci. 71, M20–M24.

Nissen, L., Carlo, E., Gianotti, A., 2020. Prebiotic potential of hemp blended drinks fermented by probiotics. Food Res. Int. 131, 109029.

Nugent, A.P., 2005. Health properties of resistant starch. Br. Nutr. Foundat. Nutr. Bull. 30, 27–54.

Ogunremi, O.R., Agrawal, R., Sanni, A.I., 2015. Development of cereal-based functional food using cereal-mix substrate fermented with probiotic strain—*Pichia kudriavzevii* OG 32. Food Sci. Nutr. 3 (6), 486–494.

Oliveira, R., Perego, P., Oliveira, M., Converti, A., 2011. Effect of inulin as a prebiotic to improve growth and counts of a probiotic cocktail in fermented skim milk. LWT- Food Sci. Technol. 44 (2), 520–523.

Onning, G., Akesson, B., Oste, R., Lundquist, I., 1998. Effects of consumption of oat milk, soya milk, or cow's milk on plasma lipids and antioxidative capacity in healthy subjects. Ann. Nutr. Metab. 42 (4), 211–220.

Onyango, C., Bley, T., Raddatz, H., Henle, T., 2004. Flavor compounds in backslop fermented uji (an East African sour porridge). Eur. Food Res. Technol. 218, 579–583.

Ozcan, O., Ozcan, T., Yilmaz-Ersan, L., Akpinar-Bayizit, A., Delikanli, B., 2016. The use of prebiotics of plant origin in functional milk products. Food Sci. Technol. 4 (2), 15–22.

Ozdal, T., Sela, D.A., Xiao, J., Boyacioglu, D., Chen, F., Capanoglu, E., 2016. The reciprocal interactions between polyphenols and gut microbiota and effects on bioaccessibility. Forum Nutr. 8 (2), 78–83.

Peyer, L.C., Zannini, E., Jacob, F., Arendt, E.K., 2015. Growth study, metabolite development and organoleptic profile of a malt-based substrate fermented by lactic acid bacteria. Am. Soc. Brew. Chem. J. 73 (4), 303–313.

Peyer, L.C., Zannini, E., Arendt, E.K., 2016. Lactic acid bacteria as sensory biomodulators for fermented cereal based beverages. Trends Food Sci. Technol. 54, 17–25.

Picot, A., Lacroix, C., 2003. Optimization of dynamic loop mixer operating conditions for production of O/W emulsion for cell microencapsulation. Le Lait 83 (3), 237–250. INRA Editions.

Prado, F.C., Parada, J.L., Pandey, A., Soccol, C.R., 2008. Trends in non-dairy probiotic beverages. Food Res. Int. 41 (2), 111–123.

Puupponen-Pimiä, R., Aura, A.M., Oksman-Caldentey, K.M., Myllärinen, P., Saarela, M., Mattila-Sandholm, T., Poutanen, K., 2002. Development of functional ingredients for gut health. Trends Food Sci. Technol. 13 (1), 3–11.

Ranadheera, C.S., Naumovski, N., Ajlouni, S., 2018. Non-bovine milk products as emerging probiotics carriers: recent developments and innovations. Curr. Opin. Food Sci. 22, 109–114.

Ranjan, S., Dasgupta, N., Chakraborty, A.R., Melvin, S.S., Ramalingam, C., Shanker, R., Kumar, A., 2014. Nanoscience and nanotechnologies in food industries: opportunities and research trends. J. Nanopart. Res. 16, 2464.

Rathore, S., Salmerón, I., Pandiella, S.S., 2012. Production of potentially probiotic beverages using single and mixed cereal substrates fermented with lactic acid bacteria cultures. Food Microbiol. 30 (1), 239–244.

Reddy, G., Altaf, M., Naveena, B.J., Venkateshwar, M., Kumar, E.V., 2008. Amylolytic bacterial lactic acid fermentation—a review. Biotechnol. Adv. 26 (1), 22–34.

Rodríguez, A.C.A., Cardozo, J.H.M., 2012. Biotechnological Aspects in the Selection of the Probiotic Capacity of Strains., https://doi.org/10.5772/50050.

Rokka, S., Rantamäki, P., 2010. Protecting probiotic bacteria by microencapsulation: challenges for industrial applications. Eur. Food Res. Technol. 231, 1–12.

Rubio, A., Hernandez, C., Aguirre, A., Poutou, R., 2008. In vitro preliminary identification of probiotic properties of *S. cerevisiae* strains. Rev. MVZ Córdoba 13 (1), 1157–1169.

Russell, K., Delahunty, C., 2004. The effect of viscosity and volume on pleasantness and satiating power of rice milk. Food Qual. Prefer. 15, 743–750.

Salmeron, I., Fuciños, P., Charalampopoulos, D., Pandiella, S.S., 2009. Volatile compounds produced by the probiotic strain *Lactobacillus plantarum* NCIMB 8826 in cereal-based substrates. Food Chem. 117 (2), 265–271.

Salmeron, I., Thomas, K., Pandiella, S.S., 2014. Effect of substrate composition and inoculum on the fermentation kinetics and flavor compound profiles of potentially non-dairy probiotic formulations. LWT-Food Sci. Technol. 55 (1), 240–247.

Salmeron, I., Thomas, K., Pandiella, S.S., 2015. Effect of potentially probiotic lactic acid bacteria on the physicochemical composition and acceptance of fermented cereal beverages. J. Funct. Foods 15, 106–115.

Sanni, A., Franz, C., Schillinger, U., Huch, M., Guigas, C., Holzapfel, W., 2013. Characterization and technological properties of lactic acid bacteria in the production of Sorghurt, a cereal-based product. Food Biotechnol. 27, 178–198.

Santos, C.C.A., Libeck, B.S., Schwanm, R.F., 2014. Co-culture fermentation of peanut-soy milk for the development of a novel functional beverage. Int. J. Food Microbiol. 186, 32–41.

Sargautiene, V., Nakurte, I., Nikolajeva, V., 2018. Broad prebiotic potential of non-starch polysaccharides from oats (*Avena sativa* L.): an in vitro study. Pol. J. Microbiol. 67 (3), 307–313.

Scourboutakos, M., 2010. 1 + 1 = 3. Synbiotics: combining the power of pre- and probiotics. J. Food Sci. Educ. 9, 36–37.

Selma, M.V., Espín, J.C., Tomás-Barberán, F.A., 2009. Interaction between phenolics and gut microbiota: role in human health. J. Agric. Food Chem. 57 (15), 6485–6501.

Sethi, S., Tyagi, S.K., Anurag, R.K., 2016. Plant-based milk alternatives an emerging segment of functional beverages: a review. J. Food Sci. Technol. 53 (9), 3408–3423.

Silk, D.B., Davis, A., Vulevic, J., Tzortzis, G., Gibson, G.R., 2009. Clinical trial: the effects of a trans-galactooligosaccharide prebiotic on faecal microbiota and symptoms in irritable bowel syndrome. Aliment. Pharmacol. Ther. 29, 508–518.

Singh, U., 1984. Dietary fiber and its constituents in desi and kabuli chickpea (*Cicer arietinum* L.) cultivars. Nutr. Rep. Int. 29, 419–426.

Smith, K., Mendonca, A., Jung, S., 2009. Impact of high-pressure processing on microbial shelf-life and protein stability of refrigerated soymilk. Food Microbiol. 26 (8), 794–800.

Stanton, C., Ross, R.P., Fitzgerald, G.F., Van Sinderen, D., 2005. Fermented functional foods based on probiotics and their biogenic metabolites. Curr. Opin. Biotechnol. 16, 198–203.

Stein, L.J., Nagai, H., Nakagawa, M., Beauchamp, G.K., 2003. Effects of repeated exposure and health-related information on hedonic evaluation and acceptance of a bitter beverage. Appetite 40, 119–129.

Stowell, J., 2007. Calorie control and weight management. In: Mitchell, H. (Ed.), Sweeteners and Sugar Alternatives in Food Technology. Blackwell Publishing Ltd., https://doi.org/10.1002/978047099.6003.ch4.

Szajewska, H., Mrukowicz, J.Z., 2001. Probiotics in the treatment and prevention of acute infectious diarrhea in infants and children: a systematic review of published randomized, double blind, placebo-controlled trials. J. Pediatr. Gastroenterol. Nutr. 33 (4), S17–S25.

Szparaga, A., Tabor, S., Kocira, S., Czerwinska, E., Kubon, M., Płóciennik, B., Findura, P., 2019. Survivability of probiotic bacteria in model systems of non-fermented and fermented coconut and hemp milks. Sustainability 11, 6093.

Tangyu, M., Muller, J., Bolten Christoph, J., Wittmann, C., 2019. Fermentation of plant-based milk alternatives for improved flavor and nutritional value. Appl. Microbiol. Biotechnol. 103, 9263–9275.

Taylor, J., Taylor, J.R.N., 2002. Alleviation of the adverse effect of cooking on sorghum protein digestibility through fermentation in traditional African porridges. Int. J. Food Sci. Technol. 37 (2), 129–137.

Tharanathan, R.N., 2002. Food-derived carbohydrates: structural complexity and functional diversity. Crit. Rev. Biotechnol. 22 (1), 65–84.

Topping, D.L., Clifton, P.M., 2001. Short-chain fatty acids and human colonic function: roles of resistant starch and non-starch polysaccharides. Physiol. Rev. 81, 1031–1064.

Trujillo, L.E., Avalos, R., Granda, S., Guerra, L.S., Pais-Chanfrau, J.M., 2016. Nanotechnology applications for food and bioprocessing industries. Biol. Med. (Aligarh) 8, 289–294.

Vaidya, R.H., Sheth, M.K., 2010. Processing and storage of Indian cereal and cereal products alters its resistant starch content. J. Food Sci. Technol. 48, 622–627.

Valdés, L., Cuervo, A., Salazar, N., Ruas-Madiedo, P., Gueimonde, M., González, S., 2015. The relationship between phenolic compounds from diet and microbiota: impact on human health. Food Funct. 6, 2424–2439.

Valencia-Flores, D.C., Herna'ndez-Herrero, M., Guamis, B., Ferragut, V., 2013. Comparing the effects of ultra-high-pressure homogenization and conventional thermal treatments on microbiological, physical, and chemical quality of almond beverages. J. Food Sci. 78 (2), E199–E205.

Verbeke, K.A., Boobis, A.R., Chiodini, A., Edwards, C.A., Franck, A., Kleerebezem, M., Nauta, A., Raes, J., Van

Tol, E.A.F., Tuohy, K.M., On Behalf of the ILSI Europe Prebiotics Task Force Expert Group 'Microbial Metabolism and Fermentation, 2015. Towards microbial fermentation metabolites as markers for health benefits of prebiotics. Nutr. Res. Rev. 28, 42–66.

Verni, M., Demarinis, C., Giuseppe, R.C., Baruzzi, F., 2020. Design and characterization of a novel fermented beverage from lentil grains. Foods 9 (7), 893.

Vidal-Valverde, C., Frias, J., Hernández, A.M.A., Pedro, J.S.I., Rodríguez, C., Blazquez, I., Vicente, G., 2003. Assessment of nutritional compounds and anti-nutritional factors in pea (*Pisum sativum*) seeds. J. Sci. Food Agric. 83 (4), 298–306.

Wang, N., Hatcher, D.W., Toews, R., Gawalko, E.J., 2009. Influence of cooking and dehulling on nutritional composition of several varieties of lentils (*Lens culinaris*). LWT-Food Sci. Technol. 42 (4), 842–848.

Wang, S., Chelikani, V., Serventi, L., 2018. Evaluation of chickpea as alternative to soy in plant-based beverages, fresh and fermented. LWT-Food Sci. Technol. 97, 570–572.

Wansink, B., Sonka, S., Goldsmith, P., Chiriboga, J., Eren, N., 2005. Increasing the acceptance of soy-based foods. J. Int. Food Agribus. Market. 17, 35–55.

Wong, J.M.W., Kendall, C.W.C., de Souza, R., Emam, A., Marchie, A., Vidgen Ed, H.C., Jenkins, D.J.A., 2010. The effect on the blood lipid profile of soy foods combined with a prebiotic: a randomized controlled trial. Metabolism 59 (9), 1331–1340.

Wood, J.A., Grusak, M.A., 2006. Nutritional value of chickpea. In: Yadav, S.S., Redden, R.J., Chen, W., Sharma, B. (Eds.), Chickpea Breeding and Management. Cromwell Press, Trowbridge.

Xiang, B.Y., Simpson, M.V., Ngadi, M.O., Simpson, B.K., 2011. Effect of pulsed electric field on the rheological and colour properties of soy milk. Int. J. Food. Sci. Nutr. 62 (8), 787–793.

Xiaoli, X., Liyi, Y., Shuang, H., Wei, L., Yi, S., Hao, M., Jusong, Z., Xiaoxiong, Z., 2008. Determination of oligosaccharide contents in 19 cultivars of chickpea (*Cicer arietinum* L.) seeds by high performance liquid chromatography. Food Chem. 111, 215–219.

Xu, B., Wang, Y., Li, J., Lin, Q., 2009. Effect of prebiotic xylooligosaccharides on growth performances and digestive enzyme activities of allogynogenetic crucian carp (*Carassius auratus* gibelio). Fish Physiol. Biochem. 35, 351–357.

Yao, N., Paez, A.V., White, P.J., 2009. Structure and function of starch and resistant starch from corn with different doses of mutant amylose-extender and floury-1 alleles. J. Agric. Food Chem. 57, 2040–2048.

Yeo, S.K., Liong, M.T., 2010. Effect of prebiotics on viability and growth characteristics of probiotics in soymilk. J. Sci. Food Agric. 90, 267–275.

Zannini, E., Pontonio, E., Waters, D.M., Arendt, E.K., 2012. Applications of microbial fermentations for production of gluten-free products and perspectives. Appl. Microbiol. Biotechnol. 93 (2), 473–485.

Zare, F., Boye, J.L., Orsat, V., Champagne, C., Simpson, B.K., 2011. Microbial, physical and sensory properties of yogurt supplemented with lentil flour. Food Res. Int. 44 (8), 2482–2488.

Zare, F., Champagne, C., Simpson, B.K., Orsat, V., Boye, J.L., 2012. Effect of the addition of pulse ingredients to milk on acid production by probiotic and yogurt starter cultures. LWT-Food Sci. Technol. 45 (2), 155–160.

Zendeboodi, F., Khorshidian, N., Mortazavian, A.M., da Cruz, A.G., 2020. Probiotic: conceptualization from a new approach. Curr. Opin. Food Sci. 32, 103–123.

Zhang, X.F., Liu, C.L., Wang, J.Q., Wang, C.N., 2018. Optimization of the process for developing a symbiotic set-type quinoa yogurt using response surface methodology. Int. J. Probiot. Prebiot. 13 (4), 151–160.

Zia-U1-Haq, M., Iqbal, S., Ahmad, S., Imran, M., Niaz, A., Bhanger, M.I., 2007. Nutritional and compositional study of Desi chickpea (*Cicer arietinum* L.) cultivars grown in Punjab, Pakistan. Food Chem. 105, 1357–1363.

FURTHER READING

Agarry, O.O., Nkama, I., Akoma, O., 2010. Production of Kunun-zaki (a Nigerian fermented cereal beverage) using starter culture. Int. Res. J. Microbiol. 1 (2), 18–25.

McFarland, L.V., 2007. Meta-analysis of probiotics for the prevention of traveler's diarrhea. Travel Med. Infect. Dis. 5, 97–105.

Vijaya Kumar, B., Vijayendra, S.V.N., Reddy, O.V.S., 2015. Trends in dairy and non-dairy probiotic products—a review. J. Food Sci. Technol. 52, 6112–6124.

Williams, E.A., Coxhead, J.M., Mathers, J.C., 2003. Anti-cancer effects of butyrate: use of micro-array technology to investigate mechanisms. Proc. Nutr. Soc. 62, 107–115.

CHAPTER 10

Beer With Probiotics and Prebiotics

MEI ZHI ALCINE CHAN[a] • MINGZHAN TOH[a] • SHAO-QUAN LIU[a,b]
[a]Department of Food Science & Technology, National University of Singapore, Singapore, Singapore,
[b]National University of Singapore (Suzhou) Research Institute, Suzhou, China

INTRODUCTION

Introduction to Beers and the Brewing Process

Beer is one of the world's oldest fermented beverages, with origins dating as far back to Neolithic agrarian societies. As time passed, the brewing process changed, most notably during the eighteenth and nineteenth centuries, which saw the progressive industrialization of breweries owing to equipment and technological advancements. The implementation of scientific principles during the industrial revolution shaped the finely controlled brewing process that is ubiquitously employed today, enabling reductions in costs, consistencies in product quality, and widespread beer production and consumption (Meussdoerffer, 2009). Beer is brewed with four key ingredients, namely water, barley, hops, and yeast. Despite the simple list of ingredients, factors such as water composition, barley and hop varieties, yeast strain, and fermentation temperature give rise to numerous permutations of beer styles, resulting in variations in the beer manufacturing stages. Nevertheless, the same fundamental unit operations can be applied across the different beer styles, which are shown in Fig. 1.

In the first stage, malting of barley is carried out, where cereal grains are steeped in water to initiate germination, thereby activating endogenous enzymes. Germination is terminated by kilning, a drying process that reduces moisture, in addition to imparting color and flavor to the malt via the Maillard reaction (Bamforth, 2017).

In the second stage, the malted barley is milled, mashed, and lautered to obtain wort. During mashing, milled grains are soaked in warm water for enzymatic hydrolysis of complex carbohydrates and proteins into simpler metabolizable substrates. Enzymes involved in mashing were activated from the previous malting step. To create optimal conditions for each enzyme, the temperature of the mash is increased in a step-wise manner. During lautering, the enzymes are denatured by raising the temperature to 78°C (mash-out), and the liquid extract that is separated from the mash is collected. More commonly referred to as the wort, the liquid extract is recirculated to remove debris, with the grain bed itself functioning as a filter. The grain bed is sprayed with additional water to extract residual sugars (sparging), followed by transferring the wort to a boiling tank. Wort boiling is traditionally conducted with hops, to isomerize hop α-acids into iso-α-acids (e.g., isohumulone, isocohumulone, and isoadhumulone) for flavoring and antimicrobial purposes. Boiling also precipitates proteins and polyphenols, which are removed during wort clarification (Wunderlich and Back, 2009; Kyselová and Brányik, 2015; Bamforth, 2017).

The wort is then cooled, before being pitched with yeasts in the third stage, to allow for primary fermentation (5–8 days), enabling sugars to be biotransformed into ethanol, carbon dioxide, and other aroma active metabolites (e.g., higher alcohols, esters, aldehydes). After which, "green" beers with a green apple aroma are left to mature (7–30 days), in the interest of flavor development, reducing levels of undesirable flavor compounds, and carbon dioxide fixation. Finally, downstream processing steps are taken to ensure colloidal, microbial, and flavor stability. For example, finished beers are usually filtered to remove residual yeast, pasteurized to ensure microbial stability, and stabilized to prevent haze turbidity (Wunderlich and Back, 2009; Kyselová and Brányik, 2015; Bamforth, 2017).

Introduction to Craft and Functional Beers

The characteristics of a finished beer are innately influenced by the ingredients used, and conditions applied in each step of the brewing process (Bamforth, 2017). By changing just one, or a few parameters, limitless possibilities of beer styles can be conceived, other than the traditional styles of lager and ale. This concept has been tapped on by craft brewers, who have experimented with novel ingredients (e.g., new varieties of hops, grains, fruits, herbs, spices, and yeasts), and modified traditional beer processes (e.g., unpasteurized beers, maturation in wooden barrels) to provide

Probiotics and Prebiotics in Foods. https://doi.org/10.1016/B978-0-12-819662-5.00004-5

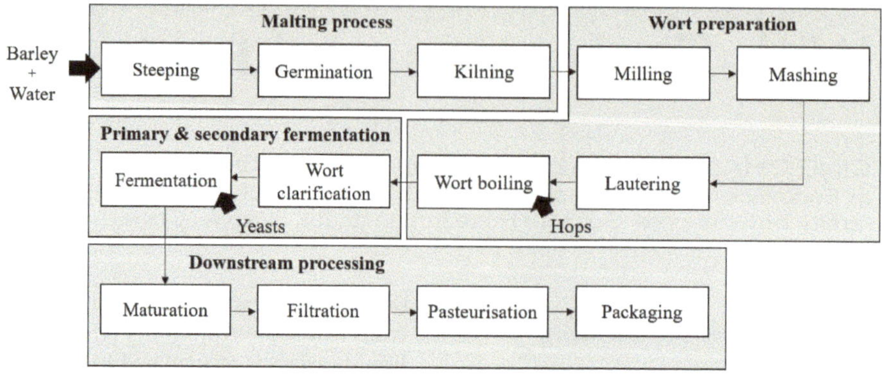

FIG. 1 Common unit operations during the beer manufacturing process.

consumers with a wide range of choices in terms of beer styles and flavor variations (Asano et al., 2007; Donadini and Porretta, 2017; Garavaglia and Swinnen, 2018). For example, beers brewed with the Bacuri fruit, which are native to the Amazonian rainforest, are available in Brazil, while Japan has its own beers containing wasabi and green tea (Euromonitor International, 2019). The appearance of these exotic beer styles that have no precedent has been fueled by increasingly affluent consumers who demand unique and premium products, valuing experiences and events over mere physical consumption. The defining features of craft brewers (e.g., limited size, story, exclusivity, autonomy, and flavor), and their strong emphasis on innovation and creativity, make craft beers well positioned to fulfill these needs (Donadini and Porretta, 2017; Euromonitor International, 2019; Garavaglia and Swinnen, 2018). As a result, the craft beer movement has witnessed significant volume growth for many years, with considerable growth potential in emerging markets such as China and Vietnam, boosted by urbanization and the emerging middle class (Euromonitor International, 2019).

In mature beer markets such as the United States and Western Europe, growth is stabilizing amid increasing competition. In a bid to continuously innovate in saturated markets, brewers have looked toward wider issues and consumer trends outside of the beer industry as new sources of inspiration (Euromonitor International, 2019). In particular, a rise in consumer health awareness has shaped the development and directions of the craft movement, with the launch of beers marketed as gluten-free, low-alcohol, low-calorie, low-purine bases, and even beers that aid in moderating sugar absorption and reducing abdominal fat (Beverage Daily, 2019; Mintel Blog, 2015; Rošul et al., 2019). On a research level, a number of studies have also developed functional beers (e.g., xanthohumol beer, estrogenic beer, and kefir beer) in response to global health and wellness trends (Yeo and Liu, 2014; Rošul et al., 2019). As breweries recognize the potential in appealing to consumers seeking "healthier" alcohol options, tremendous opportunities exist within the niched category of functional beers.

In this niched category, beers fortified with well-established and certified prebiotics and/or probiotics are a novelty. Several scientific papers evaluating the feasibilities of probiotic beer production have only been published in recent years, but none have been commercialized at the time of writing. Furthermore, the concept of prebiotic beer production remains unexplored. The inclusion of prebiotics and/or probiotics in beers is a commercially attractive feature compared to regular beers, as they are effective tools for human health promotion and possible disease prevention. In light of the popularity of functional beers, it is expected that such beers will have a direct and significant impact on the brewing industry, opening up the benefits of probiotic and prebiotic consumption to a wider audience. This chapter will focus on recent developments in using beer as a medium of probiotic and prebiotic delivery, challenges and commercial considerations, and how the brewing process can be modified to accommodate these difficulties.

PROBIOTIC BEERS

It was only recently that concrete steps were taken to assess the feasibility and practicality of probiotic beers, with numerous scientific publications on the topic appearing from the year 2018 onward. These publications, which are summarized in Table 1, have explored beer production based on different probiotic microorganisms: lactic acid bacteria (LAB; mainly *Lactobacillus* spp.), yeasts (*Saccharomyces cerevisiae* var. *boulardii*), and

TABLE 1
List of Publications That Have Reported on Probiotic Beers.

Probiotic Microorganisms	Variables/Brewing Procedure	Outcome	Reference Type
LACTIC ACID BACTERIA			
L. acidophilus LA-5 and *B. lactis* BB-12 inoculated into freshly made low-alcohol or nonalcoholic beers that were fermented by *S. cerevisiae* 70,424 and *S. rouxii* 2531, respectively	Wort fermented with either yeast strain to produce low-alcohol (2.5%) and nonalcoholic beers (< 0.5%), which was then pasteurized and inoculated (~ 9 Log CFU/mL) with either probiotic LAB strain. Probiotic viability monitored over 20 days at 5°C	Both probiotic strains saw a minimum reduction of 3 Log CFU/mL in cell counts after storage, with greater losses observed in low-alcohol compared to nonalcoholic beers	Peer-reviewed article (Sohrabvandi et al., 2010)
L. paracasei L26 in coculture with *S. cerevisiae* S-04, with single cultures as control	Unhopped wort cofermented with both yeast and probiotic LAB, followed by addition of isomerized hop extract (~ 27 IBUs) and monitoring of probiotic survival during cold and ambient storage	During mixed fermentation of unhopped wort, *L. paracasei* L26 demonstrated satisfactory growth to 8.8 Log CFU/mL, with significant production of lactic acid. Yeast performance was not compromised, with the majority of sugars attenuated, and significant amounts of ethanol and esters produced During storage, refrigeration in combination with live yeast significantly enhanced the survival of *L. paracasei* L26 in hopped wort	Patent and peer-reviewed article (Liu and Chan, 2018; Chan et al., 2019)
Free and encapsulated *L. rhamnosus* GG in alginate or alginate silica monocarriers	Addition of free and encapsulated *L. rhamnosus* GG (in alginate or alginate silica monocarrier) in beer (5% *w/v* ethanol), followed by storage at 4°C for 1 week	Encapsulation strategy limited the release of bacteria over time, with alginate silica beads more efficient than alginate beads alone The former carrier was also more effective in retaining probiotic viability in the filtrate and in the carrier matrix after 1 week of storage	Peer-reviewed article (Haffner and Pasc, 2018)
Yeasts			
Single-cultured *S. boulardii* T1 in comparison with *S. cerevisiae* Safbrew T58	Primary fermentation of wort with either yeast strain, followed by maturation, and in-bottle refermentation	Both yeast strains yielded cell counts above 5 Log CFU/mL after the process; pH was similar (~ 4.3), but a lower degree of attenuation was observed for *S. boulardii*	Patent (Marín Iniesta et al., 2016)

Continued

TABLE 1
List of Publications That Have Reported on Probiotic Beers—cont'd

Probiotic Microorganisms	Variables/Brewing Procedure	Outcome	Reference Type
Single-cultured *S. boulardii*-17	Optimization of growth parameters in yeast extract-peptone-dextrose (YPD) broth. YPD spiked with an alcohol level of 5°GL was used to verify optimal conditions in a 5-L bioreactor, as well as to assess *S. boulardii* tolerance to in vitro gastrointestinal transit	Growth was maximized with 0°GL alcohol content, pH 5.92, at 25.2°C. During fermentation, reductions in glucose, dissolved oxygen, and pH were observed, along with acetic acid and glycerol production. A 1.1 Log CFU/mL reduction was observed after simulated GI transit (final biomass: 6.9 Log CFU/mL)	Peer-reviewed article (de Paula et al., 2019)
Single-cultured *S. boulardii* (strain not specified; isolated from food supplement Biopron Forte, Walmark)	Growth characteristics evaluated in synthetic medium (containing wort sugars, ethanol, and iso-α-acids) RSM applied to optimize fermentation conditions for maximal production of alcohols and esters Stability of *S. boulardii* assessed in incomplete fermented wort (4% *w*/v, 3.945 g/L ethanol, 2°C, initial inoculum 6.6 Log CFU/mL)	Highest specific growth rate on glucose compared to maltose and maltotriose. Significant decline in specific growth rate observed at ethanol level of 5% ABV, or 50 IBUs Temperature and pitching rate influenced the formation of alcohols and esters In incomplete fermented wort, complete utilization of glucose, and 91% of maltose was utilized, with concomitant increases in ethanol after 4 weeks. From 4 to 11 weeks, gradual decline in maltotriose observed, with no further increases in ethanol content. Dextrins were not metabolized. HPP was applied to finished beers to inactivate *S. boulardii* and prevent further ethanol production	Peer-reviewed article (Senkarcinova et al., 2019)
Single-cultured *S. boulardii* CECT 1474 in comparison with *S. cerevisiae* Safale S-04	Wort fermented with either yeast strain, with monitoring of physicochemical parameters (yeast biomass, °Plato, pH, DPPH antioxidant assay, total phenolic content) Yeast growth was also monitored in bioreactor cultures	*S. boulardii* demonstrated greater viability compared to *S. cerevisiae* after 36 days of refrigerated storage. Beers fermented with the probiotic yeast exhibited a greater degree of acidification, lowered ethanol content, and enhanced antioxidant capacity, but sensorial attributes were similar between both yeast strains In batch bioreactor cultures at 24°C, *S. boulardii* exhibited faster growth and larger cell volumes compared to *S. cerevisiae*	Peer-reviewed article (Mulero-Cerezo et al., 2019)

TABLE 1
List of Publications That Have Reported on Probiotic Beers—cont'd

Probiotic Microorganisms	Variables/Brewing Procedure	Outcome	Reference Type
S. boulardii (strain not specified, isolated from Codex) in single and mixed cultures with five *S. cerevisiae* strains	Wort fermented with single and mixed cultures of *S. boulardii* and five different strains of *S. cerevisiae* isolated from ales and sourdough	After fermentation, *S. boulardii* was dominant in almost all (except for 1) mixed fermentations over *S. cerevisiae*, presenting cell counts between 6.9 and 7.8 Log CFU/mL. As a single culture, *S. boulardii* displayed lower attenuation levels and higher levels of volatile acidity, which was lowered in mixed cultures. In mixed cultures, inclusion of *S. boulardii* did not negatively impact aroma, and an increase in antioxidant capacity and total phenolic content was observed	Peer-reviewed article (Capece et al., 2018a)
Bacillus spp. spores			
Bacillus coagulans MTCC 5856 spores	Assessment of viabilities during storage after: 1. Addition of spores during different stages of the brewing process (malting, hop addition, secondary fermentation, carbonation, packing, and pasteurization) 2. Addition of spores during packing and pasteurization, followed by carbonation in mild (3.35%), regular (3.5%), strong (4.8%) lager beers, and in whiskey (42.8%)	1. A greater degree of survival observed when spores are added later into the beer manufacturing stages 2. Spores survived better in whiskey than in lager beers	Patent (Majeed et al., 2019)
B. subtilis 168 or *B. natto* CICC 24640	Addition of *Bacillus* spores into beer spiked with zearalenone, followed by incubation at 30°C for 1 h	*B. subtilis* 168 or *B. natto* CICC 24640 degraded 65% and 73% of zearalenone in beers, respectively`	Peer-reviewed article (Ju et al., 2019)

spore-forming bacteria (mainly vegetative and spore forms of *Bacillus* spp.). Regardless of the type used, it is imperative that they remain active and viable throughout the beer manufacturing and storage processes, with sufficient levels present at the time of consumption. The recommended minimum efficacious dosage is 9 Log colony-forming units (CFUs) per serving of product (Hill et al., 2014). Ensuring the viability, state, and the quantities of probiotics will enhance the acceptance by regulatory authorities of a product holding a probiotic label. Unfortunately, this is technologically challenging to adhere to, as intrinsic and extrinsic hurdles prevent probiotic growth and survival in beers. For example, conventional downstream processing steps such as pasteurization and filtration cannot be practiced, as high temperatures will kill probiotics, while filtration will remove them. In the case for craft beers, which are typically unpasteurized and unfiltered to maintain distinct freshness and cloudiness, this may not be a major issue (Asano et al., 2007; Bamforth, 2017; Rodhouse and Carbonero, 2019). However, other hurdles are not easily solvable, often requiring modifications to existing brewing practices to maintain probiotic viability. These challenges and required modifications, which are specific to the type of probiotics, are listed in Table 2, along with some of their merits.

Beers With Probiotic Lactic Acid Bacteria

A combination of intrinsic antimicrobial hurdles such as acidity (pH 3.9–4.4), ethanol (up to 10% *v*/v, typically 3%–5% *v*/v), lack of nutrients, elevated carbon dioxide levels (0.5% *w*/w), and hop iso-α-acids (17–55 ppm) can impair growth and survival of probiotic LAB in beers (Vriesekoop et al., 2012). Among these inhibitory factors, emphasis must be placed on hop iso-α-acids, as they exert the greatest physiological stress on gram-positive probiotic LAB. As mobile carrier ionophores, hop iso-α-acids are capable of transmembrane proton transport at various pH values, dissipating the pH gradient and reducing the proton motive force (pmf) across the cytoplasmic membrane (Simpson, 1993; Behr and Vogel, 2009; Suzuki, 2011). The antibacterial effects of iso-α-acids could also be extended to its role in a manganese- binding-dependent transmembrane redox reaction, thereby exerting oxidative stress on gram-positive LAB under acidic conditions (Behr and Vogel, 2010;

TABLE 2
Advantages, Disadvantages, and Some Strategies to Produce Probiotic and Prebiotic Beers.

	Advantages	Disadvantages	Possible Modifications to Existing Brewing Practices
Probiotics			
Lactic acid bacteria	• If used as starter cultures, lactic and/or acetic acid produced. Wort souring fits closely into recent revival of traditional sour styles • Greater variety of probiotic *Lactobacillus* species, translating to greater choices • Possible improvement in nutritive content due to bioactive metabolites (postbiotics) produced by LAB during fermentation.	• More susceptible to antimicrobial hurdles in beers, especially hop iso-α-acids, making it difficult to maintain probiotic viability • Causes souring when used as a starter culture. Implications on types of beers that can be produced (e.g., limited to sour beer styles). Nevertheless, could be overcome by microencapsulation • Possible production of biogenic amines not assessed • Probiotic LAB used for kettle souring may coaggregate with brewing yeast. Consequently, the removal of trubs after primary fermentation will inadvertently result in the removal of the LAB • Heterofermentative LAB may generate CO_2, leading to overpressurization and packaging explosion • Presence of probiotic LAB will increase the turbidity of the beer, which may not be visually appealing	• Removal of hops or use of hop derivatives (e.g., hop oils) to enable the growth of *Lactobacillus* probiotics • Microencapsulated probiotic LAB added to the screw cap of beers to prevent souring • Beer souring after primary fermentation instead of kettle souring to prevent the removal of probiotic LAB with trubs

TABLE 2
Advantages, Disadvantages, and Some Strategies to Produce Probiotic and Prebiotic Beers—cont'd

	Advantages	Disadvantages	Possible Modifications to Existing Brewing Practices
Yeasts	• Less susceptible to antimicrobial hurdles in beer. Easier to maintain viability • Lower attenuation rates compared to conventional brewing yeasts may be beneficial in the production of low-alcohol beers	• Lowered attenuation rates • Possible flavor issues (e.g., increased production of volatile acids, phenols, and esters) • Viability during storage not assessed • Secondary fermentation of residual sugars may cause overpressurization of the packaging • Probiotic yeast sediments could impart turbidity and a yeasty aroma to the beer	• May need to coculture with conventional beer yeasts to compensate for brewing performance and possible flavor issues • Ensuring complete attenuation prior to packaging, or use of nonmetabolizable sugars to prevent overpressurization
Bacillus spp. spores	• Spore forms are less susceptible to antimicrobial hurdles in beer. Thermal treatment (e.g., pasteurization, wort boiling) can be applied • Vegetative forms can potentially enhance nutritive (e.g., postbiotics) and safety value of beers	• Production of undesirable flavor compounds by vegetative bacilli cells • Possible cross-contamination in breweries due to resistance of spores to environmental factors • Possible production of biogenic amines not assessed	• Vegetative forms may also be susceptible to hops. Similar modifications as nonspore-forming probiotic LAB may be required • Dedicated production lines can reduce the occurrence of cross-contamination • Use of harsher cleaning regimen to sanitize production lines and prevent cross-contamination
Probiotics in general	• Potentially added health benefits derived from live probiotics compared to regular beers	• Probiotic viabilities have to be maintained throughout processing and storage. Modifications to existing brewing practices (e.g., manufacturing, packaging design, distribution conditions) may be required • Quality and technological characteristics in beer have to be taken into account (i) Flavor (ii) Sedimentation caused by excessive probiotic biomass in finished beers (iii) Others, e.g., foam, clarity	• Pasteurization and filtration cannot be practiced

Continued

TABLE 2
Advantages, Disadvantages, and Some Strategies to Produce Probiotic and Prebiotic Beers—cont'd

	Advantages	Disadvantages	Possible Modifications to Existing Brewing Practices
(Candidate) prebiotics			
Arabinoxylo-oligosaccharides (AXOS) and β-glucan	• Potentially added health benefits derived from prebiotics compared to regular beers • Nonviable entities. Viability, safety, and flavor concerns associated with live microorganisms are not applicable	• Sensorial parameters (e.g., solubility, viscosity, mouthfeel) and brewing performance (e.g., filtration rates, turbidity) may be affected, especially at the recommended daily dosage of 3g per day • Structure-function relationship not apparent, due to significant structure heterogeneity	• Adjuncts rich in arabinoxylan, AXOS, β-glucan • Addition of enzymes during mashing (e.g., endoxylanases, *endo*-β-glucanases) • Optimal conditions during mashing and malting steps to activate enzymes • Exogenous addition of prebiotic extracts into finished beers

Schurr et al., 2015). Both of these mechanisms disrupt the transmembrane potential and pmf, hampering nutrient uptake and culminating in cell death. Furthermore, when gram-positive probiotic bacteria (e.g., *Lactobacillus*, *Bifidobacterium*) are used as starter cultures in hopped beers, acidification of the medium due to the production of lactic and/or acetic acids must be closely monitored. Production of these organic acids will cause a fall in pH values, enabling a greater proportion of the iso-α-acids to exist in the undissociated form (pK_a values of 3.0–3.5; Ting and Ryder, 2017), thereby augmenting its ionophoric properties and accelerating oxidative stress-induced reactions.

Indeed, the difficulties of culturing probiotic LAB strains in beers were first demonstrated by Sohrabvandi et al. (2010). The authors observed that *Lactobacillus acidophilus* LA-5 and *Bifidobacterium lactis* BB-12 lost their viabilities by at least three Log cycles in low-alcohol and (2.5%) and nonalcohol (< 0.5%) beers over a 20-day period at 5°C. Greater losses were observed in low-alcohol beers compared to the nonalcoholic counterpart. Probiotic LAB susceptibility to iso-α-acids was also confirmed by Chan et al. (2019). In hopped wort (~ 27 International Bitterness Units (IBUs)), six probiotic strains, namely *L. acidophilus* NCFM, *L. helveticus* (previously *L. acidophilus*) L10, *L. bulgaricus* Lb-64, *L. paracasei* L26, *L. paracasei* Lpc-37, and *L. rhamnosus* HN001, were unable to grow, and live bacterial cells were no longer detectable within 7 days. The authors attributed probiotic cell death to the presence of hop iso-α-acids, as satisfactory growth of at least 1.7 Log cycles could be achieved when the LAB were propagated in unhopped wort.

To produce beers with high cell counts of live probiotics, it is essential that yeasts are used as starter cultures for the primary purpose of producing ethanol and carbon dioxide. Probiotic LAB are incapable of fulfilling this primary function and thus need to be cultured with yeasts if they are to be used as starter cultures in beers. This presents us with an opportunity to grow probiotic LAB in hopped beers: Yeasts may be able to confer probiotic LAB viability-enhancing effects, due to strong mechanistic and ecological evidence in matrices such as in acidic buffers, fermented milks, and durian pulp (Parrella et al., 2012; Karaolis et al., 2013; Suharja et al., 2014; Lim et al., 2015; Yeo et al., 2016; Ponomarova et al., 2017; Lu et al., 2018). To test this hypothesis, Chan et al. (2019) first cocultured the probiotic LAB, *L. paracasei* L26, with an ale yeast, *S. cerevisiae* S-04, in unhopped wort. After fermentation, the unhopped beer was well-attenuated, and the probiotic LAB exhibited satisfactory growth to 8.8 Log CFU/mL, along with lactic acid production. During storage of the unfiltered and unpasteurized beers, the addition of isomerized hop extract (~ 27 IBUs) accelerated probiotic death, which, interestingly, was delayed by *S. cerevisiae* S-04.

Under refrigeration, cell counts of cocultured *L. paracasei* L26 fell below 7 Log CFU/mL after 22 days of storage, compared to just 6 days for the single-cultured probiotic. Therefore, the unconventional approach of first using unhopped wort for primary fermentation to enable the growth of probiotic LAB, followed by the addition of isomerized hop extract thereafter, might be a viable option toward developing probiotic beers that have a practical and reasonable shelf life. These findings are also published in a patent (Liu and Chan, 2018).

Besides their therapeutic benefits, it is befitting to use probiotic LAB as starter cultures in the production of sour beers, due to their abilities to produce lactic acid and acidify the wort. This is particularly advantageous given the resurgence in traditional sour beer styles (e.g., Belgian lambics, Flanders red ale) that fit into recent craft beer trends (Donadini et al., 2016; Euromonitor International, 2019). Nonetheless, souring by probiotic LAB limits the styles of beer that can be produced, and it is of commercial and scientific interest to develop nonsour probiotic LAB beers. To this end, microencapsulation may be a possible solution to maintain probiotic LAB viability, while minimizing acid production. Haffner and Pasc (2018) identified alginate silica beads as an effective microencapsulation material to protect *L. rhamnosus* GG in beers (5% *w*/v ethanol). Compared to free cells and those encapsulated in alginate beads, alginate silica beads enabled slow release of probiotic cells while achieving the greatest survival rates during 1 week of storage at 4°C. The authors also suggested a unique approach to incorporate probiotic LAB to beers: the attachment of freeze-dried microencapsulated material to the screw cap of the beverage packaging, allowing instantaneous release just prior to consumption.

The suggestion provided by Haffner and Pasc (2018) exemplifies how conventional brewing steps can be modified to develop probiotic LAB-based beers, especially against the backdrop of the craft beer movement where beers do not need to adhere strictly to conventional or traditional processing steps. It is conceivable that other strategies can be made to develop probiotic LAB beers, which include reducing or totally omitting hops to avoid their antimicrobial action, different inoculation timings of probiotic LAB (e.g., after primary or secondary fermentation), and different yeast-probiotic LAB combinations. The latter is pertinent not only to produce probiotic LAB beers with different flavors arising from different yeast metabolites, but also to potentially lengthen the shelf life and enhance probiotic LAB viabilities in hopped beers, since the effects of yeast on probiotic LAB viabilities are strain dependent (Liu and Tsao, 2009).

Nevertheless, safety aspects must not be overlooked especially since biogenic amines are commonly cited as potent toxicological agents in beers, eliciting harmful effects such as headaches, heart palpitations, migraine, vomiting, nausea, and diarrhea. Biogenic amines are most commonly produced by bacteria, via substrate-specific decarboxylase enzymes. Their production is usually emphasized in alcoholic beverages, since monoamine oxidase, the enzyme responsible for the detoxification of biogenic amines in bacteria, is suppressed by ethanol (Gardini et al., 2016). Moreover, strain-dependent production of biogenic amines has been observed in *Bifidobacterium* and *L. rhamnosus* (Deepika Priyadarshani and Rakshit, 2011; Lorencová et al., 2012, 2014).

Another point to consider is the coaggregation between brewing yeasts and probiotic LAB during kettle souring, mediated by the binding of mannoproteins on yeast cell walls to cell surface proteins of LAB (Katakura et al., 2010; Hirayama et al., 2012). Consequently, probiotic bacteria will be dragged down to the fermentation tank together with flocculating yeasts to form trubs, and the conventional practice of removing trubs after primary fermentation may inadvertently lead to the removal of the LAB. In this case, probiotic LAB may be pitched after trub removal to ensure sufficient numbers of probiotic LAB are present to meet the recommended efficacious dosage of 9 Log CFUs per serving of product (Hill et al., 2014). Other points to note are whether turbidities caused by probiotic LAB are acceptable to consumers, as well as probable carbon dioxide accumulation and overpressurization of packaging vessels if carbon dioxide-producing heterofermentative LAB are to be used as probiotic starter cultures.

Beers With the Probiotic Yeast, *Saccharomyces boulardii*

Compared to gram-positive probiotic bacteria, yeasts are often less susceptible to stresses encountered in beers, making it an easier task to sustain their viabilities compared to LAB. Most importantly, yeasts are able to tolerate and prevent the cytosolic accumulation of hop iso-α-acids via entrapment on the cell wall, vacuolar sequestration, and active export across the membrane, thus making them resistant to hops at levels commonly encountered in beers (Hazelwood et al., 2010). Despite superior tolerance to antimicrobial hurdles in beer, the main challenge in developing beers based on probiotic yeasts is that their technological traits and fermentative performance have to be comparable to, or at the very least, compatible with conventional brewing yeasts.

This is to guarantee that fermentative performance is maintained, and the quality of finished beers is upheld.

The conventional beer yeasts, *Saccharomyces cerevisiae* (ale yeast; top fermenters) and *Saccharomyces pastorianus* (lager yeast; bottom fermenters), have been domesticated over many years through industry-specific selection, chosen for their fermentation efficiency, stress tolerance, sugar utilization, and flavor production (Gallone et al., 2016). In contrast, the probiotic yeast, *Saccharomyces cerevisiae* var. *boulardii* (*S. boulardii*), has origins dating back to 1920, where it was first isolated from lychee and mangosteen (McFarland, 2017). *S. boulardii* is the only yeast to be utilized widely as a probiotic, being placed at the forefront of many probiotic yeast-based food formulations. Several strains of the probiotic yeast exist; among them, the strain *S. boulardii* CNCM-I745 is the most researched, being supported by numerous clinical studies to treat acute infectious diarrhea, prevent antibiotic-associated diarrhea, and reduce side effects of *H. pylori* treatment (Dinleyici et al., 2014; Palma et al., 2015; McFarland, 2017).

Since *S. boulardii* is an undomesticated brewing strain, the probiotic yeast may not possess the optimal traits necessary for brewing. Nonconventional brewing yeasts generally have a lower attenuation performance, lowered ethanol production, and tend to produce significant levels of phenolic off-flavors (e.g., clove-like aromas) derived from the decarboxylation of ferulic acid to 4-vinyl guaiacol (Capece et al., 2018b; Cubillos et al., 2019). Furthermore, comparative genomics have revealed that although *S. cerevisiae* and *S. boulardii* are members of the same species, both species are distinct physiologically and metabolically, displaying different fermentative behaviors, growth profiles, and resistance to environmental stresses. For example, *S. boulardii* is more resistant to acidic pH conditions and bile salts, has a higher optimum growth temperature of 37°C, and demonstrates pseudohyphal activities under nitrogen limitations (Fietto et al., 2004; Edwards-Ingram et al., 2007). Due to differences in physiological properties, it is crucial to characterize basic growth properties and fermentative performance of *S. boulardii* in beers and compare them to conventional brewing yeasts. Technological traits such as attenuation rates, flocculation, ethanol tolerance, production of flavor compounds, and cell viabilities will be helpful in determining whether *S. boulardii* is suitable for beer fermentations. Studies listed in Table 1 have also applied response surface methodology (RSM) to optimize fermentation conditions favorable for growth, as well as formation of odor-active compounds (e.g., esters and higher alcohols) to ensure that flavor and quality are not compromised.

Marín Iniesta et al. (2016) published a patent in Spanish disclosing the production of beers using *S. boulardii*, compared to *S. cerevisiae* Safbrew T58. Worts were inoculated with single starter cultures of either yeast strain for primary fermentation, followed by maturation, and in-bottle fermentation. In finished beers, both yeasts yielded cell counts greater than 5 Log CFU/mL, with similar pH values observed (~ pH 4.3). However, a lower degree of attenuation was observed for *S. boulardii* (°Brix ~ 7.25) compared to *S. cerevisiae* (°Brix ~ 6). It is important to note that although methods employed in this patent may not be fully described and are not subjected to peer review, this was the earliest description of how probiotic beers based on *S. boulardii* could be produced, thus paving the way for others to apply more rigorous scientific methods (e.g., RSM) to assess the suitability of *S. boulardii* in beer brewing applications.

Recognizing the importance of probiotic viability for efficacy, de Paula et al. (2019) applied RSM to optimize the growth of *S. boulardii*-17 (Floratil) in yeast extract-peptone-dextrose (YPD) broth. The optimal conditions to achieve maximal yeast growth were identified: 0°GL (Gay-Lussac) of alcohol, pH 5.92, at 25.2°C, implying that alcohol content negatively affected cellular concentration. When these optimal conditions were applied in a 5-L bioreactor (spiked with an alcohol level of 5°GL), *S. boulardii* grew from 5.5 to 7.6 Log CFU/mL, exhibiting satisfactory resistance to simulated gastrointestinal conditions (1.1 Log CFU/mL reduction).

The basic growth capabilities of *S. boulardii* under conditions that better resemble beer environments were evaluated by Senkarcinova et al. (2019). The authors first monitored specific growth rates in a synthetic medium containing varying amounts of wort sugars (glucose, maltose, maltotriose), ethanol, and iso-α-acids. The highest specific growth rate was observed on glucose, while ethanol and iso-α-acid at levels of 5% alcohol by volume (ABV) and 50 IBUs, respectively, negatively affected growth. In addition, RSM identified the optimal conditions for flavor formation: 9%–12% wort extract, at 14°C, pitching rate of 7.1 Log CFU/mL. When tested in a real wort system, *S. boulardii* was capable of producing ethanol at 2°C, which was deemed undesirable since the authors aimed to produce low-alcohol beers. Therefore, high-pressure processing (HPP) was applied, resulting in microbial inactivation (5.4 Log CFU/mL reduction in biomass), and a complete cessation of fermentation, without adverse impacts on taste and composition. Although this study demonstrated the capabilities of *S. boulardii* to grow in a legitimate wort medium, it should be pointed out that the final product contained inactive probiotic cells. Emerging evidence suggests that nonviable probiotics or "paraprobiotics" are capable of eliciting biological effects (e.g., immunomodulation, pathogen inhibition; de Almada et al., 2016). However, as paraprobiotics are not considered as probiotics (Hill et al., 2014), beers that contain inactive *S. boulardii* cannot be identified as probiotic beers.

The fermentative performance of *S. boulardii* CECT 1474 in an actual wort system was confirmed by Mulero-Cerezo et al. (2019), who compared it to the ale yeast, *S. cerevisiae* S-04. Beers fermented with *S. boulardii* yielded significantly higher pH values, lowered alcohol levels, and enhanced antioxidant capacity (as measured by 2,2-diphenyl-1-picrylhydrazyl; DPPH assay) compared to beers fermented by *S. cerevisiae* S-04. Similar sensory attributes were observed between the two beers, as assessed by a panel of 10 trained experts, rated for appearance, aroma, flavor, and bitterness. After storage for 36 days at 4 °C, greater viability was demonstrated by *S. boulardii* (6.9 Log CFU/mL) compared to *S. cerevisiae* S-04 (5.0 Log CFU/mL). In a 20-L batch bioreactor maintained at 24 °C, the probiotic yeast exhibited faster growth and larger cell volumes compared to the brewing yeast. Based on these findings, it can be reasoned that *S. boulardii* possesses suitable technological traits that make it suitable to be used as a single starter for the production of probiotic beers.

These capabilities of *S. boulardii* as the principal wort fermenter were aligned with findings presented by Capece et al. (2018a). The authors conducted single and mixed wort fermentations with the probiotic yeast, and five other nonconventional *S. cerevisiae* strains that were isolated from ales and sourdough. Compared to the *S. cerevisiae* strains, the probiotic yeast produced beers with lower attenuation levels, together with high levels of volatile acidity, the latter observed to be less pronounced in mixed cultures. It was interesting to note that although the addition of *S. boulardii* in mixed starter cultures did not negatively impact beer aroma, it was dominant in almost all of the mixed fermentations (except for one), presenting high viable cell counts between 6.9 and 7.8 Log CFU/mL. Mixed-cultured beers also presented greater antioxidant capacities (DPPH) and total phenolic content (except for 1 mixed culture fermentation) in comparison with beers from single starter fermentations.

Taken together, the abovementioned publications have affirmed that *S. boulardii* is a promising yeast for craft beer production, with satisfactory growth and fermentation characteristics, whether in single or mixed cultures. Although complete sugar utilization by single-cultured *S. boulardii* can be achieved, lowered ethanol yields were commonly reported compared to conventional brewing yeasts, which may not necessarily have negative connotations if low-alcohol beers are desired. Nevertheless, more studies are required to validate its survival during storage, especially since the definition of probiotics dictates that microorganisms should be alive (Hill et al., 2014). In particular, efforts should be made to monitor the survival of *S. boulardii* under distribution conditions that are common in the brewing industry. Current distribution networks in the brewing industry often lack adequate temperature management systems, with ambient, elevated temperatures and/or temperature fluctuations being commonly encountered (Jaskula-Goiris et al., 2019). Brewers and/or researchers have to examine if the viability of *S. boulardii* can be preserved under these circumstances, as breweries lacking cold chain distribution infrastructure are less amenable to a probiotic beer with a short ambient life span.

Another consideration for including live yeasts in finished beers is the likelihood of refermentation, particularly in products with residual sugars. The production of carbon dioxide within beer cans and bottles could lead to overpressurization and pose a safety risk due to exploding vessels. Hence, it is imperative that beers containing probiotic *S. boulardii* are fully attenuated or contain nonmetabolizable sugars such as isomaltulose if sweetness is desired. Lastly, the possibility of flocculation by live probiotic yeasts should be studied in future works, as excessive yeast sedimentation may impart turbidity and yeasty aromas to the finished product.

Beers With Probiotic *Bacillus* spp. Spores

Probiotic strains of *Bacillus* spp. (e.g., *B. coagulans*, *B. subtilis*, *B. clausii*, *B. licheniformis*) have an overall excellent health-promoting record, demonstrating strain-specific benefits that are not limited to preventing and curing diarrhea, gingivitis, *Helicobacter pylori* infections and maintaining intestinal homeostasis. Having been approved by regulatory agencies as probiotics for human use, they are sold globally as general health-promoting pharmaceutical formulations (Elshaghabee et al., 2017; Lee et al., 2019). Besides demonstrating therapeutic efficacies in clinical trials, probiotic strains of *Bacillus* spp. have received considerable attention due to their remarkable resistance to environmental stresses, making them suitable for health-promoting food formulations that require heat treatment (Elshaghabee et al., 2017).

The resistance of *Bacillus* spores to heat presents significant advantages in a brewing system. For example, restrictions surrounding the sensitivity of vegetative probiotics to heat (e.g., *S. boulardii* or LAB) can be overcome, as heating steps in the beer brewing process (e.g., malting, wort boiling, pasteurization) can now be applied when probiotic beers are manufactured based on *Bacillus* spores. This concept has been leveraged upon in a patent disclosed by Majeed et al. (2019). When *Bacillus coagulans* MTCC 5856 spores were added in different stages of the brewing process, high recoveries in viabilities were achieved regardless of the point of addition, but a higher proportion survived when it was added later into the beer manufacturing process. For example, when spores were added during the malting step, ~ 60% of viability (~ 5 Log CFU/mL) was retained after 6 months of storage, in comparison with ~ 90% (~ 8 Log CFU/mL) when spores were added during packing and pasteurization. In addition, a greater degree of survival was observed in whiskey (42.8% ABV), compared to mild (3.35% ABV), regular (3.5% ABV), and strong lager beers (4.8% ABV), suggesting greater susceptibilities to iso-α-acids in beers, as opposed to significantly greater ethanol levels in whiskey. Nevertheless, despite greater viability losses in lager beers, satisfactory cell counts were still obtained (at least 60%, 6 Log CFU/mL), even after storage at 40°C for 6 months, with no effects on the specific gravity, pH, and alcohol content.

Although hop resistance mechanisms have been demonstrated by beer spoilage spore-forming members of *Bacillus* spp. (e.g., *B. cereus*, *B. licheniformis*, *B. thuringiensis*; Haakensen and Ziola, 2008; Munford et al., 2017), it is unclear if the same mechanisms are shared by probiotic strains of the same genus. It can be postulated that strong viabilities exhibited by *B. coagulans* MTCC 5856 may be attributed to its spore structure. Consisting of a core of condensed and inactive chromosomes, the spore is surrounded by additional layers of peptidoglycan-rich cortex and one or more layers of proteinaceous material. These layers, which form the spore coat, are highly impermeable to hydrophobic and hydrophilic molecules (Cutting, 2011; Bernardeau et al., 2017). It is reasonable to speculate that any cellular influx of hop iso-α-acids is restricted by the extremely low permeability of the inner membrane, thereby conferring hop resistance to *B. coagulans* MTCC 5856.

Besides adding dried spores directly into beers, an alternative method to formulate *Bacillus*-based probiotic beers is through cultivating the vegetative form as a starter culture. Although this makes the probiotic more susceptible to environmental perturbations (especially hops), *Bacillus* spp.-induced fermentations may potentially enhance the nutritive value and safety of beers. For example, *Bacillus* spp. are able to produce vitamins (e.g., riboflavin, cobalamin, inositol), carotenoids, bacteriocin, bacteriocin-like inhibitory substances, and enzymes (e.g., lipase, protease, carbohydrase, mycotoxin-degrading enzymes; Bernardeau et al., 2017; Elshaghabee et al., 2017). These bioactive secondary metabolites (so-called postbiotics) are exclusively produced by vegetative cells, generally during the late exponential growth phase, and their production cannot be accomplished by metabolically dormant spores (Bernardeau et al., 2017). Existing literature on the use of vegetative forms of the probiotic *Bacillus* spp. to ferment beers is scarce, but Ju et al. (2019) demonstrated that *B. subtilis* 168 and *B. natto* CICC 24640 could degrade 65% and 73% of zearalenone (*Fusarium* mycotoxin) present in beers, respectively. However, a drawback of *Bacillus*-induced food fermentations is the risk of off-flavors, particularly via proteolysis that generates pungent branched-chain fatty acids (BCFAs) and ammonia (Amoa-Awua et al., 2014; Takemura et al., 2000; Hong et al., 2017).

Despite the versatility of using probiotic *Bacillus* spp., either as an ingredient in the spore form or as a starter culture in the vegetative form, some resistance may be encountered when adopting them for wider use in the industry. *Bacillus* spp. spores are remarkably resistant to environmental conditions, and uneasiness remains with regard to possible cross-contamination to other production lines, since they may not be fully eradicated with routine cleaning. Furthermore, given that several beer spoilage bacteria are *Bacillaceae* (e.g., *B. cereus*, *B. licheniformis*; Haakensen and Ziola, 2008; Wang et al., 2017), spoilage concerns raised by brewers are understandable. For successful applications, probiotic strains of *Bacillus* spp. need to be sufficiently characterized phenotypically and genotypically, to differentiate them from beer spoilage *Bacillaceae*. In addition, it is imperative that they are handled in a controlled manner (e.g., dedicated production lines or use of harsher cleaning regimen) to avoid cross-contamination. Lastly, several *Bacillus* strains are similarly able to produce biogenic amines (Lee et al., 2019), which will need to be evaluated to safeguard consumer safety.

PREBIOTIC BEERS

Unlike probiotics, prebiotics in beer are nonviable entities and, as a result, are more shelf stable. Scientific publications that have formulated beers with established prebiotics (e.g., fructo-oligosaccharides, galacto-oligosaccharides) are unavailable. There are, however, few studies (listed in Table 3) that have revealed

TABLE 3
Publications That Have Reported on (Candidate) Prebiotic Beers.

Candidate Prebiotics	Outcome	Reference Type
Arabinoxylo-oligosaccharides (AXOS)		
Use of wheat bran as an adjunct (25% of total grist), together with xylanase	Significant increases in AXOS, ferulic acid, and soluble protein, along with improvements in body, smoothness, and warming notes. Significant decreases in acetaldehyde, bitterness, drying, and grainy characters	Peer-reviewed article (Veraverbeke et al., 2011)
1. Use of rye bran, together with xylanase 2. Addition of AXOS extracted from wheat into finished light beers at dosage levels of 2 g/L or 10 g/L 3. AXOS extracted from side stream industrial wheat processing, and addition during wort boiling	1. A 3.7-fold increase in AXOS in beers (3.5 g/L), with no changes in final ethanol content 2. A dosage level of 10 g/L significantly improved mouthfeel of light beers 3. AXOS-enriched beer (2.8 g/L) had 2.5-fold increase in AXOS, without compromising on alcohol content and original extract. AXOS-enriched beers also had significant decreases in bitterness sensation and sourness, while the quality of bitterness and mouthfeel significantly improved. Of a panel of 13, 12 preferred the AXOS-enriched beer to the control	Patent (Aerts et al., 2008)
β-glucan		
High β-glucan Shonkin barley, in comparison with low to moderate β-glucan barley. Each barley cultivar was malted and brewed separately. Rice hulls were used as a filter bed during lautering	High β-glucan Shonkin barley elevated the levels of β-glucan in worts and beers, regardless of germination time. Rice hulls may be used as a substitute for hull-less barley malt to maintain filtration efficiencies	Peer-reviewed article (Vis and Lorenz, 1998)
Partial replacement of barley malt with unmalted oats (0%–40%)	Increasing oat replacement levels enhanced the levels of β-glucan and viscosity, but adversely affected total soluble nitrogen and free amino nitrogen in the wort, with significant decreases in foam stability. A replacement level of 10% oats did not affect the pH of mash/wort and lautering performance. Beers made with 30% and 40% oat had acceptable aroma and purity of taste	Peer-reviewed article (Schnitzenbaumer et al., 2012)

Note: The aims of the abovementioned articles were not specifically to develop prebiotic beers. These articles are listed as they serve as examples as to how beers can be enriched with prebiotics, or candidate prebiotics.

enhanced levels of candidate prebiotics in beers, more specifically, in the form of arabinoxylo-oligosaccharides (AXOS) and β-glucans via in situ production. A greater number of researchers have imbued beers with polyphenols via exogenous supplementation with fruits or polyphenolics-rich ingredients. Nonetheless, it is worth noting that the intent of these researchers was not to specifically develop prebiotic beers. Rather, they aimed to create novel beers and flavors, or enhance the functional properties of beers through increasing dietary fiber content, polyphenol content, and/or antioxidant capacities. We highlight that these studies can potentially produce prebiotic (or candidate prebiotic) beers if proper requirements are met, for example, ensuring sufficient levels of (candidate) prebiotics are present, in addition to their structural characterization and demonstration of prebiotic efficacies in beers.

Beers Enriched With the Candidate Prebiotics: Arabinoxylan-Oligosaccharides and β-Glucans

Beer itself is a rich source of candidate prebiotics, which is advantageous, since in situ production makes it easier to comply with all regulatory requirements, particularly in countries with the German "purity law." Of the 1.3–3.8 g/L of dietary fiber present in beers, nonstarch polysaccharides (NSPs) predominate, with arabinoxylans (AXs) comprising 20%, while β-glucans comprise 75% of the walls surrounding the starchy endosperm of the barley grain (Bamforth and Kanauchi, 2001; Díaz-Rubio and Saura-Calixto, 2009). Conventional brewing practices seek to reduce endogenous levels of NSPs during the malting and mashing steps, as undegraded residues have been implicated in overly viscous worts, impairing filtration rates and extraction yields and generating haze and turbidity in beers (Gupta et al., 2010). Nevertheless, AX, its hydrolysis product AXOS, and β-glucans are increasingly being recognized for their nutritional properties, with strong evidence supporting their role as prebiotics (Scott et al., 2019). Therefore, enhancing endogenous AXOS and β-glucan levels via in situ production may be an effective approach in creating (candidate) prebiotic beers.

AX comprises of a backbone of β-D-xylopyranosyl residues linked by β-(1–4) glycosidic bonds, with α-L-arabinofuranosyl units being mostly attached as side chains via β-(1–2) or β-(1–3) linkages, sometimes substituted with acetic acid or ferulic acid (Fox, 2009). During malting, several enzymes, namely endo-1,4-β-xylanase (randomly cleaves internal bonds of the xylan backbone), β-xylopyranosidase (releases xylose from the nonreducing ends of the xylan backbone), α-L-arabinofuranosidase (removes arabinosyl side chains), and esterases (removes acetic acid/ferulic acid), hydrolyze AX into a mixture of AXOS and xylo-oligosaccharides (Ferreira, 2009; Fox, 2009). Depending on the beer style, the yield of AXOS ranges from 0.49 to 1.90 g/L, with an average degree of arabinose substitution ranging from 0.49 to 0.66, and an average degree of polymerization varying between 19 and 61 (Courtin et al., 2009). Dietary AX largely reaches the colon intact, where they serve as substrates for colonic bacteria to generate AXOS. The latter has been demonstrated in clinical trials to exert bifidogenic or prebiotic effects, produce short-chain fatty acids (SCFAs), normalize stool frequency, and even attenuate glycemic responses (Grootaert et al., 2007; Cloetens et al., 2010; Broekaert et al., 2011; Damen et al., 2012; Boll et al., 2016; Rivière et al., 2018).

In favor of enhancing AXOS in beers via in situ production, adjuncts such as wheat and rye that are rich in AX or AXOS (e.g., wheat, rye), coupled with the inclusion of enzymes (e.g., *endo*-1,4-β-xylanase), can be employed. As an example, Veraverbeke et al. (2011) showed that by replacing 25% of the total grist with wheat bran, followed by the addition of xylanase during mashing, significant improvements in AXOS levels can be achieved. Beers brewed with wheat bran also showed significant increases in body, smoothness, and warming notes, while significant decreases in acetaldehyde, bitterness, drying, and grainy attributes were observed. To further boost AXOS levels in finished beers, mashing and malting conditions can be carefully controlled. For instance, Li et al. (2005) observed that the degree of AX solubilization and hydrolysis can be optimized by varying the mashing time, mash thickness, grist coarseness, and degree of agitation.

The strategy of employing AX- or AXOS-rich adjuncts together with endoxylanases to boost AXOS levels in beers was similarly validated in a patent disclosed by Aerts et al. (2008). Replacing 10% of the barley malt with rye bran, together with endoxylanases, led to a 3.7-fold increase in AXOS (final 3.5 g/L), without compromising on ethanol production. Besides the enhancement of AXOS contents in beers in situ, the inventors also demonstrated that by supplementing light beers with 10 g/L wheat bran AXOS extracts, significant improvements in mouthfeel properties, without adverse effects on viscosity, can be achieved. Using a separate AXOS extract derived from a wheat pentosan concentrate, enriched beers (2.78 g/L AXOS) were rated more highly by panelists for bitterness and mouthfeel qualities compared to control beers (1.09 g/L AXOS).

The major component of the endosperm of barley cell walls is β-glucans, consisting of β-D-glucopyranosyl residues linked with a mixture of β-1,3 (~ 30%) and β-1,4 glycosidic (~ 70%) bonds to form linear homopolysaccharides (Ferreira, 2009). Levels of β-glucans in barley are among the highest in cereals, consisting of 10.89–19.7 g/100 g dry weight, which are hydrolyzed during malting to produce mainly tri- and tetra-saccharides via *endo*-β-glucanases (Ferreira, 2009; Bader Ul Ain et al., 2018). In finished beers, total β-linked glucan ranges from 0.34 to 6.0 mg/mL, with a greater proportion of low-molecular-weight β-glucan than their high-molecular-weight counterparts (Bamforth and Gambill, 2007; Kanauchi et al., 2011). Theoretically, β-linked oligomers may be further degraded to glucose by *exo*-β-glucanase, but it appears that the enzyme has limited action possibly due to poor affinities (Kanauchi and Bamforth, 2008). Barley β-glucan oligosaccharides are considered as candidate prebiotics (Scott et al., 2019) due to in vitro evidence supporting their ability to pass through the gastrointestinal tract intact, where they act as substrates for the gut microbiota, while positively stimulating several intestines associated with probiotic *Lactobacillus* strains (Lam and Cheung, 2013; Arena et al., 2014; Kanyer et al., 2017; Bader Ul Ain et al., 2018).

Similar to AXOS-enriched beers, levels of soluble β-glucan can be enhanced by using adjuncts rich in β-glucans, such as raw barley, unmodified malt, and oat. To illustrate, Vis and Lorenz (1998) successfully produced beers enriched with β-glucan by using Shonkin barley (a variety rich in β-glucans), together with rice hulls as a filter bed during lautering to maintain filtration efficacies. In another study, Schnitzenbaumer et al. (2012) partially replaced barley malt with unmalted oats (0%–40% of total grist) and noted greater levels of β-glucans in beers with increasing levels of oat adjunct. However, adverse effects on lautering time and foam stability were observed at replacement levels of 20% or more. The use of unmalted oat also negatively affected levels of total soluble nitrogen, free amino nitrogen, and extract level in worts. Nevertheless, at higher replacement levels of 30% and 40%, sensorial qualities were improved, eliciting acceptable aroma and purity of taste.

Other approaches in producing β-glucan-enriched beers include the use of specialized enzymes and barley varieties. For example, commercial β-glucanases possessing substantial endo-β-glucanase activity, but lacking exo-β-glucanase and β-glucosidase activities may improve levels of β-glucan oligosaccharides, while limiting their hydrolysis to glucose. Alternatively, a barley variety that lacks the latter two enzymes can also be used to produce β-glucan-rich beers (Bamforth and Gambill, 2007).

Collectively, these studies demonstrate that several approaches can be taken to produce (candidate) prebiotic beers: the selection of adjuncts rich in AX, AXOS, or β-glucan, addition of the corresponding enzymes during mashing, optimal conditions to suit enzymatic activity during mashing and malting, and finally, exogenous addition of prebiotic extracts into finished beers (Table 2). These strategies will be useful in raising the levels of NSPs in beers, especially since a daily prebiotic oral dosage of 3 g is recommended to elicit an effect on gut microbiota (Gibson et al., 2017), and beers only contain 0.49–1.90 g/L of AXOS and 0.34–3.79 g/L of total β-linked oligosaccharides (Bamforth and Gambill, 2007; Courtin et al., 2009).

However, as a result of increasing NSP levels, changes in sensorial qualities may occur. For instance, Krebs et al. (2019) demonstrated that spiking β-glucans at levels commonly found in beers into nonalcoholic beers positively enhanced palate fullness, while evoking viscous and thick mouthfeel. Nonetheless, it appears that any sensorial changes arising from elevated NSP levels are still within consumer acceptance, as elucidated by studies listed in Table 3. Likewise, alterations to the brewing process can be made to overcome any filtration issues related to the enhanced viscosities of NSP-enriched beers, such as the suggestion provided by Vis and Lorenz (1998) to use rice hulls as a filter bed.

Another consideration when developing prebiotic beers is the significant heterogeneity within the structures of AXOS and β-glucans, owing to differences in molecular weight, degree of polymerization, and substitution. Such differences arise due to variations in cereal types, barley varieties, and extraction methods and can translate to variations in physical properties relating to viscosities and solubilities. More importantly, structural heterogeneity often elicits varying biological activities, such as enzymatic breakdown efficiency, and bulk fermentability to beneficial SCFA (Zhu et al., 2016; Bader Ul Ain et al., 2018; Lam et al., 2019). There is a paucity of studies that have investigated if a structure-function relationship exists in human host, and it is unclear if different structures of NSPs are able to exert the same prebiotic effects on the gut microbiota. One study concluded that among five β-1,3-1,4-glucans with different molecular weights, the 28 kDa β-glucan held the greatest potential of being a prebiotic, due to favorable effects on the *Firmicutes*/*Bacteroides* ratio, along with SCFA and ammonia production after in vitro fermentation by infant fecal microbiome (Lam et al., 2019). Therefore, it would be useful in the future to characterize the structure of NSPs present in beers and determine if the same prebiotic effects can be achieved as an extract or when consumed in beers. This will substantiate any health claims and/or marketing messages.

Beers Fortified With (Candidate) Prebiotics

Opportunities to create beers rich in (candidate) prebiotics are already present, and those wishing to create prebiotic beers can look toward existing practices and ingredients from a fresh perspective, specifically from a (candidate) prebiotics angle. For instance, common starchy adjuncts such as corn, potato, and tapioca represent rich sources of resistant starch (Scott et al., 2019), while fruit beers made with persimmon (Cho et al., 2018), cherry, raspberry, peach, apricot, grape, plum, orange, apple (Nardini and Garaguso, 2020), quince (Zapata et al., 2019), cornelian cherry juice (Kawa-Rygielska et al., 2019), and grape must (Veljovic et al., 2010) are abundant in the candidate prebiotics, pectin, and polyphenols (Scott et al., 2019). This does not mean that existing beers that are rich in polyphenols or fruit pectin automatically qualify as prebiotic beers. Similar to AXOS and β-glucans, it is essential that efforts are made to identify and characterize the type of (candidate) prebiotics, in addition to quantifying levels present in finished beers. Even if sufficient amounts of (candidate) prebiotics are present in finished beers, sensorial qualities must not be compromised. For example, depending on the structure, pectin confers gelling and stabilizing effects (Naqash et al., 2017), while beer polyphenols are often implicated in haze formation, protein complexation, bitterness, and astringency in beers (Aron and Shellhammer, 2010). It is also worth noting that excessive intake of prebiotics may also lead to flatulence (Gibson et al., 2017). Brewers therefore need to balance between added therapeutic benefits derived from prebiotics by ensuring sufficient amounts are present, without compromising on brewing performance and organoleptic properties.

FUTURE TRENDS AND CONSIDERATIONS

A possible way forward is through developing beers with a synbiotic combination of probiotics and prebiotics, which is a complete novelty. Such a combination may potentially improve probiotic survival not only in the gastrointestinal tract, but also in beers, thereby strengthening additional health benefits that may be conferred to the host. As an example, β-glucans derived from barley and oat have been shown in numerous studies to improve probiotic stress tolerance in yogurts during storage (Vasiljevic et al., 2007; Rosburg et al., 2010; Lazaridou et al., 2014; Kurtuldu and Ozcan, 2018). In a study by Arena et al. (2014), pasta enriched with barley β-glucans significantly improved the growth rate of unstressed probiotics, as well as their survival after in vitro digestive tract conditions, while enhancing probiotic-enterocyte interactions. Although the same probiotic protective and enterocyte adhesion effects were not observed in a chemically defined medium fortified with oat and barley β-glucan, the fortified medium stimulated probiotic-induced anti-inflammatory effects by modulating transcription of several immune-related genes (Arena et al., 2016). Besides enhancing probiotic viabilities and efficacies, it is conceivable that synbiotic combinations may aid in in situ production of AXOS and β-glucans, thereby providing an alternative method to manufacture prebiotic beers. For example, several *Bacillus* strains are capable of producing xylanases (Grootaert et al., 2007; Elshaghabee et al., 2017), and while production has not been reported in *Bifidobacterium*, few strains of the LAB have been identified to produce α-L-arabinofuranosidase and β-xylosidase (Mendis and Simsek, 2014).

In view of the brewing industry's eagerness to counterpoise the adverse health aspects of alcoholic consumption, it is tempting to label such beers with health-promoting messages. Nonetheless, it is paramount that health claims abide by local regulations, which can vary between different countries. The US Alcohol and Tobacco Tax and Trade Bureau permits health claims for alcoholic beverages, but such claims must not be deemed misleading and should be adequately justified with scientific or medical evidence. In addition, it is necessary to disclose the health risks associated with moderate and heavy alcohol consumption (Code of Federal Regulations, 2020). Regulations are stricter in the European Union, as health claims for beverages containing more than 1.2% *v*/v ethanol are prohibited (Regulation (EC), 2006). To safeguard consumers' interests and health, these measures taken by regulatory authorities are understandable, given that alcohol is a leading risk factor for premature death and disability, with a recent comprehensive systematic analysis recommending absolute alcohol abstinence to minimize health loss across health outcomes (Griswold et al., 2018). Therefore, even if marketing messages and health claims abide by local laws and legislations, the onus is on beer manufacturers to promote responsible and moderate drinking. In this context, probiotic and/or prebiotic alcohol-free beers may be more appealing, as a means to avoid contradictions with the harmful effects of alcohol, in addition to benefiting from less stringent legal regulations and alcohol tax relief.

CONCLUSION

Rising trends in health and wellness have compelled the craft beer movement to redirect its attention to functional beers. Within this category, probiotic and prebiotic beers represent a novel research domain, mainly driven by researchers who have resonated with the potential roles that probiotic and prebiotic beers can play in providing additional therapeutic benefits compared to regular beers. Within the domain of probiotic beers, probiotic microorganisms such as yeasts (*S. boulardii*), LAB (*Lactobacillus*) and sporulating bacteria (vegetative and spore forms of *Bacillus* spp.) have been explored as starter cultures, or as nonfermentative ingredients in the form of spores or microencapsulated vegetative cells. Depending on the probiotic type used, each has its own unique set of merits and challenges, but studies published thus far have affirmed that these challenges can be overcome by modifying conventional brewing practices and conditions. For prebiotic beers, we highlight that existing brewing ingredients are potential sources of candidate prebiotics which brewers can leverage to produce prebiotic beers. Arabinoxylan-oligosaccharides and β-glucans from cereal grains, as well as pectin and polyphenols from fruit adjuncts, are just some examples of (candidate) prebiotic sources that can be derived from current brewing ingredients. Against the backdrop of growing consumer recognition and acceptance of probiotics and prebiotics, it is expected that more innovative products will appear, for example, beers containing synbiotic combinations of probiotics and prebiotics. Nevertheless, any health claims or marketing messages must adhere to local laws and regulations, and beer manufacturers have the obligation to promote responsible and moderate drinking. Overall, probiotic and prebiotic beers are a refreshing concept amid a saturated beer market, but it remains to be seen if laboratory-scale prototypes are translatable to a large-scale brewery production.

REFERENCES

Aerts, G., Broekaert, W., Courtin, C., Delcour, J., 2008. Arabinoxylo-Oligosaccharides in Beer. World Intellectual Property Organization patent WO2008098320A1,.

Amoa-Awua, W.K., Awusi, B., Owusu, M., Appiah, V., Ofori, H., Thorsen, L., Jespersen, L., 2014. Reducing the atypical odour of dawadawa: effect of modification of fermentation conditions and post-fermentation treatment on the development of the atypical odour of dawadawa. Food Control 42, 335–342.

Arena, M.P., Caggianiello, G., Fiocco, D., Russo, P., Torelli, M., Spano, G., Capozzi, V., 2014. Barley β-glucans-containing food enhances probiotic performances of beneficial bacteria. Int. J. Mol. Sci. 15 (2), 3025–3039.

Arena, M.P., Russo, P., Capozzi, V., Rascón, A., Felis, G.E., Spano, G., Fiocco, D., 2016. Combinations of cereal β-glucans and probiotics can enhance the anti-inflammatory activity on host cells by a synergistic effect. J. Funct. Foods 23, 12–23.

Aron, P.M., Shellhammer, T.H., 2010. A discussion of polyphenols in beer physical and flavor stability. J. Inst. Brew. 116 (4), 369–380.

Asano, S., Suzuki, K., Iijima, K., Motoyama, Y., Kuriyama, H., Kitagawa, Y., 2007. Effects of morphological changes in beer-spoilage lactic acid bacteria on membrane filtration in breweries. J. Biosci. Bioeng. 104 (4), 334–338.

Bader Ul Ain, H., Saeed, F., Ahmad, N., Imran, A., Niaz, B., Afzaal, M., Imran, M., Tufail, T., Javed, A., 2018. Functional and health-endorsing properties of wheat and barley cell wall's non-starch polysaccharides. Int. J. Food Prop. 21 (1), 1463–1480.

Bamforth, C.W., 2017. Progress in brewing science and beer production. Annu. Rev. Chem. Biomol. Eng. 8 (1), 161–176.

Bamforth, C.W., Gambill, S.C., 2007. Fiber and putative prebiotics in beer. J. Am. Soc. Brew. Chem. 65 (2), 67–69.

Bamforth, C.W., Kanauchi, M., 2001. A simple model for the cell wall of the starchy endosperm in barley. J. Inst. Brew. 107 (4), 235–240.

Behr, J., Vogel, R.F., 2009. Mechanisms of hop inhibition: hop ionophores. J. Agric. Food Chem. 57 (14), 6074–6081.

Behr, J., Vogel, R.F., 2010. Mechanisms of hop inhibition include the transmembrane redox reaction. Appl. Environ. Microbiol. 76 (1), 142–149.

Bernardeau, M., Lehtinen, M.J., Forssten, S.D., Nurminen, P., 2017. Importance of the gastrointestinal life cycle of *Bacillus* for probiotic functionality. J. Food Sci. Technol. 54 (8), 2570–2584.

Beverage Daily, 2019. No Beer Belly: Kirin Releases Non-Alcoholic Beer With Fat-Reducing Function (Website). https://www.beveragedaily.com/Article/2019/10/24/No-beer-belly-Kirin-releases-non-alcoholic-beer-with-fat-reducing-function. (Accessed 30 December 2019).

Boll, E.V.J., Ekström, L.M.N.K., Courtin, C.M., Delcour, J.A., Nilsson, A.C., Björck, I.M.E., Östman, E.M., 2016. Effects of wheat bran extract rich in arabinoxylan oligosaccharides and resistant starch on overnight glucose tolerance and markers of gut fermentation in healthy young adults. Eur. J. Nutr. 55 (4), 1661–1670.

Broekaert, W.F., Courtin, C.M., Verbeke, K., Van de Wiele, T., Verstraete, W., Delcour, J.A., 2011. Prebiotic and other health-related effects of cereal-derived arabinoxylans, arabinoxylan-oligosaccharides, and xylooligosaccharides. Crit. Rev. Food Sci. Nutr. 51 (2), 178–194.

Capece, A., Romaniello, R., Pietrafesa, A., Siesto, G., Pietrafesa, R., Zambuto, M., Romano, P., 2018a. Use of *Saccharomyces cerevisiae* var. *boulardii* in co-fermentations with *S. cerevisiae* for the production of craft beers with potential healthy value-added. Int. J. Food Microbiol. 284, 22–30.

Capece, A., Romaniello, R., Siesto, G., Romano, P., 2018b. Conventional and non-conventional yeasts in beer production. Ferment 4 (2), 1–11.

Chan, M.Z.A., Chua, J.Y., Toh, M., Liu, S.-Q., 2019. Survival of probiotic strain *Lactobacillus paracasei* L26 during co-fermentation with *S. cerevisiae* for the development of a novel beer beverage. Food Microbiol. 82, 541–550.

Cho, J.-H., Kim, I.-D., Dhungana, S.K., Do, H.-M., Shin, D.-H., 2018. Persimmon fruit enhanced quality characteristics and antioxidant potential of beer. Food Sci. Biotechnol. 27 (4), 1067–1073.

Cloetens, L., Broekaert, W.F., Delaedt, Y., Ollevier, F., Courtin, C.M., Delcour, J.A., Rutgeerts, P., Verbeke, K., 2010. Tolerance of arabinoxylan-oligosaccharides and their prebiotic activity in healthy subjects: a randomised, placebo-controlled cross-over study. Br. J. Nutr. 103 (5), 703–713.

Code of Federal Regulations, 2020. 27 CFR 7.54 Alcohol, Tobacco Products and Firearms. Part 7-Labelling and Advertising of Malt Beverages. Accessed 13 January 2020.

Courtin, C.M., Broekaert, W.F., Swennen, K., Aerts, G., Van Craeyveld, V., Delcour, J.A., 2009. Occurrence of arabinoxylo-oligosaccharides and arabinogalactan peptides in beer. J. Am. Soc. Brew. Chem. 67 (2), 112–117.

Cubillos, F.A., Gibson, B., Grijalva-Vallejos, N., Krogerus, K., Nikulin, J., 2019. Bioprospecting for brewers: exploiting natural diversity for naturally diverse beers. Yeast 36 (6), 383–398.

Cutting, S.M., 2011. *Bacillus* probiotics. Food Microbiol. 28 (2), 214–220.

Damen, B., Cloetens, L., Broekaert, W.F., François, I., Lescroart, O., Trogh, I., Arnaut, F., Welling, G.W., Wijffels, J., Delcour, J.A., Verbeke, K., Courtin, C.M., 2012. Consumption of breads containing in situ-produced arabinoxylan oligosaccharides alters gastrointestinal effects in healthy volunteers. J. Nutr. 142 (3), 470–477.

de Almada, C.N., Almada, C.N., Martinez, C.R., Sant'Ana, A.S., 2016. Paraprobiotics: evidences on their ability to modify biological responses, inactivation methods and perspectives on their application in foods. Trends Food Sci. Technol. 58, 96–114.

de Paula, B.P., Chávez, D.W.H., Lemos Junior, W.J.F., Guerra, A.F., Corrêa, M.F.D., Pereira, K.S., Coelho, M.A.Z., 2019. Growth parameters and survivability of *Saccharomyces boulardii* for probiotic alcoholic beverages development. Front. Microbiol. 10, 2092.

Deepika Priyadarshani, W.M., Rakshit, S.K., 2011. Screening selected strains of probiotic lactic acid bacteria for their ability to produce biogenic amines (histamine and tyramine). Int. J. Food Sci. Technol. 46 (10), 2062–2069.

Díaz-Rubio, M.E., Saura-Calixto, F., 2009. Dietary fiber complex in beer. J. Am. Soc. Brew. Chem. 67 (1), 38–43.

Dinleyici, E.C., Kara, A., Ozen, M., Vandenplas, Y., 2014. Expert opinion on biological therapy *Saccharomyces boulardii* CNCM I-745 in different clinical conditions. Expert. Opin. Biol. Ther. 14 (11), 1593–1609.

Donadini, G., Porretta, S., 2017. Uncovering patterns of consumers' interest for beer: a case study with craft beers. Food Res. Int. 91, 183–198.

Donadini, G., Fumi, M.D., Kordialik-bogacka, E., Maggi, L., Lambri, M., Sckokai, P., 2016. Consumer interest in specialty beers in three European markets. Food Res. Int. 85, 301–314.

Edwards-Ingram, L., Gitsham, P., Burton, N., Warhurst, G., Clarke, I., Hoyle, D., Oliver, S.G., Stateva, L., 2007. Genotypic and physiological characterization of *Saccharomyces boulardii*, the probiotic strain of *Saccharomyces cerevisiae*. Appl. Environ. Microbiol. 73 (8), 2458–2467.

Elshaghabee, F.M.F., Rokana, N., Gulhane, R.D., Sharma, C., Panwar, H., 2017. *Bacillus* as potential probiotics: status, concerns, and future perspectives. Front. Microbiol. 8, 1490.

Euromonitor International, 2019. What's Brewing in Craft Beer? (Report). Available from: https://www.euromonitor.com/whats-brewing-in-craft-beer-/report. (Accessed 27 December 2019).

Ferreira, I.M.P.L.V.O., 2009. 27. Beer carbohydrates. In: Preedy, V.R. (Ed.), Beer in Health Disease and Prevention. Academic Press, San Diego, pp. 291–298.

Fietto, J.L.R., Araújo, R.S., Valadão, F.N., Fietto, L.G., Brandão, R.L., Neves, M.J., Gomes, F.C.O., Nicoli, J.R., Castro, I.M., 2004. Molecular and physiological comparisons between *Saccharomyces cerevisiae* and *Saccharomyces boulardii*. Can. J. Microbiol. 50 (8), 615–621.

Fox, G.P., 2009. 29. Beer and arabinoxylan. In: Preedy, V.R. (Ed.), Beer in Health Disease and Prevention. Academic Press, San Diego, pp. 309–316.

Gallone, B., Steensels, J., Prahl, T., Soriaga, L., Saels, V., Herrera-Malaver, B., Merlevede, A., Roncoroni, M., Voordeckers, K., Miraglia, L., Teiling, C., Steffy, B., Taylor, M., Schwartz, A., Richardson, T., White, C., Baele, G., Maere, S., Verstrepen, K.J., 2016. Domestication and divergence of *Saccharomyces cerevisiae* beer yeasts. Cell 166 (6), 1397–1410.e16.

Garavaglia, C., Swinnen, J., 2018. Economics of the craft beer revolution: a comparative international perspective. In: Garavaglia, C., Swinnen, J. (Eds.), Economic Perspectives on Craft Beer: A Revolution in the Global Beer Industry. Springer International Publishing, Cham, pp. 3–51.

Gardini, F., Özogul, Y., Suzzi, G., Tabanelli, G., Özogul, F., 2016. Technological factors affecting biogenic amine content in foods: a review. Front. Microbiol. 7, 1218.

Gibson, G.R., Hutkins, R., Sanders, M.E., Prescott, S.L., Reimer, R.A., Salminen, S.J., Scott, K., Stanton, C., Swanson, K.S., Cani, P.D., Verbeke, K., Reid, G., 2017. Expert consensus document: the International Scientific Association for Probiotics and Prebiotics (ISAPP) consensus statement on the definition and scope of prebiotics. Nat. Rev. Gastroenterol. Hepatol. 14 (8), 491–502.

Griswold, M.G., Fullman, N., Hawley, C., Arian, N., Zimsen, S.R.M., Tymeson, H.D., Venkateswaran, V., Tapp, A.D., Forouzanfar, M.H., Salama, J.S., Abate, K.H., Abate, D., Abay, S.M., Abbafati, C., Abdulkader, R.S., Abebe, Z., Aboyans, V., Abrar, M.M., Acharya, P., Adetokunboh, O.O., et al., 2018. Alcohol use and burden for 195 countries and territories, 1990-2016: a systematic analysis for the global burden of disease study 2016. Lancet 392 (10152), 1015–1035.

Grootaert, C., Delcour, J.A., Courtin, C.M., Broekaert, W.F., Verstraete, W., Van de Wiele, T., 2007. Microbial metabolism and prebiotic potency of arabinoxylan oligosaccharides in the human intestine. Trends Food Sci. Technol. 18 (2), 64–71.

Gupta, M., Abu-Ghannam, N., Gallaghar, E., 2010. Barley for brewing: characteristic changes during malting, brewing and applications of its by-products. Compr. Rev. Food Sci. Food Saf. 9 (3), 318–328.

Haakensen, M., Ziola, B., 2008. Identification of novel horA-harbouring bacteria capable of spoiling beer. Can. J. Microbiol. 54 (4), 321–325.

Haffner, F.B., Pasc, A., 2018. Freeze-dried alginate-silica microparticles as carriers of probiotic bacteria in apple juice and beer. LWT- Food Sci. Technol. 91, 175–179.

Hazelwood, L.A., Walsh, M.C., Pronk, J.T., Daran, J.-M., 2010. Involvement of vacuolar sequestration and active transport in tolerance of *Saccharomyces cerevisiae* to hop iso-alpha-acids. Appl. Environ. Microbiol. 76 (1), 318–328.

Hill, C., Guarner, F., Reid, G., Gibson, G.R., Merenstein, D.J., Pot, B., Morelli, L., Canani, R.B., Flint, H.J., Salminen, S., Calder, P.C., Sanders, M.E., 2014. The International Scientific Association for Probiotics and Prebiotics consensus statement on the scope and appropriate use of the term probiotic. Nat. Rev. Gastroenterol. Hepatol. 11, 506–514.

Hirayama, S., Furukawa, S., Ogihara, H., Morinaga, Y., 2012. Yeast mannan structure necessary for co-aggregation with *Lactobacillus plantarum* ML11-11. Biochem. Biophys. Res. Commun. 419 (4), 652–655.

Hong, C., Chen, Y., Li, L., Chen, S., Wei, X., 2017. Identification of a key gene involved in branched-chain short fatty acids formation in natto by transcriptional analysis and enzymatic characterization in *Bacillus subtilis*. J. Agric. Food Chem. 65 (8), 1592–1597.

Jaskula-Goiris, B., De Causmaecker, B., De Rouck, G., Aerts, G., Paternoster, A., Braet, J., De Cooman, L., 2019. Influence of transport and storage conditions on beer quality and flavour stability. J. Inst. Brew. 125 (1), 60–68.

Ju, J., Tinyiro, S.E., Yao, W., Yu, H., Guo, Y., Qian, H., Xie, Y., 2019. The ability of *Bacillus subtilis* and *Bacillus natto* to degrade zearalenone and its application in food. J. Food Process. Preserv. 43 (10), e14122.

Kanauchi, M., Bamforth, C.W., 2008. The relevance of different enzymes for the hydrolysis of β-glucans in malting and mashing. J. Inst. Brew. 114 (3), 224–229.

Kanauchi, M., Ishikura, W., Bamforth, C.W., 2011. β-Glucans and pentosans and their degradation products in commercial beers. J. Inst. Brew. 117 (1), 120–124.

Kanyer, A.J., Bornhorst, G.M., Marco, M.L., Bamforth, C.W., 2017. Is beer a source of prebiotics? J. Inst. Brew. 123 (3), 361–365.

Karaolis, C., Botsaris, G., Pantelides, I., Tsaltas, D., 2013. Potential application of *Saccharomyces boulardii* as a probiotic in goat's yoghurt: survival and organoleptic effects. Int. J. Food Sci. Technol. 48 (7), 1445–1452.

Katakura, Y., Sano, R., Hashimoto, T., Ninomiya, K., Shioya, S., 2010. Lactic acid bacteria display on the cell surface cytosolic proteins that recognize yeast mannan. Appl. Microbiol. Biotechnol. 86 (1), 319–326.

Kawa-Rygielska, J., Adamenko, K., Kucharska, A.Z., Prorok, P., Piórecki, N., 2019. Physicochemical and antioxidative properties of cornelian cherry beer. Food Chem. 281, 147–153.

Krebs, G., Müller, M., Becker, T., Gastl, M., 2019. Characterization of the macromolecular and sensory profile of non-alcoholic beers produced with various methods. Food Res. Int. 116, 508–517.

Kurtuldu, O., Ozcan, T., 2018. Effect of β-glucan on the properties of probiotic set yoghurt with *Bifidobacterium animalis* subsp. *lactis* strain Bb-12. Int. J. Dairy Technol. 71 (S1), 157–166.

Kyselová, L., Brányik, T., 2015. 20. Quality improvement and fermentation control in beer. In: Holzapfel, W. (Ed.), Advances in Fermented Foods and Beverages. Woodhead Publishing, United Kingdom, pp. 477–500.

Lam, K.-L., Cheung, C.-K.P., 2013. Non-digestible long chain beta-glucans as novel prebiotics. Bioact. Carbohydr. Diet. Fibre 2 (1), 45–64.

Lam, K.-L., Ko, K.-C., Li, X., Ke, X., Cheng, W.-Y., Chen, T., You, L., Kwan, H.-S., Cheung, P.C.-K., 2019. *In vitro* infant faecal fermentation of low viscosity barley β-Glucan and its acid hydrolyzed derivatives: evaluation of their potential as novel prebiotics. Molecules 24 (5), 828.

Lazaridou, A., Serafeimidou, A., Biliaderis, C.G., Moschakis, T., Tzanetakis, N., 2014. Structure development and acidification kinetics in fermented milk containing oat β-glucan, a yogurt culture and a probiotic strain. Food Hydrocoll. 39, 204–214.

Lee, N.-K., Kim, W.-S., Paik, H.-D., 2019. *Bacillus* strains as human probiotics: characterization, safety, microbiome, and probiotic carrier. Food Sci. Biotechnol. 28 (5), 1297–1305.

Li, Y., Lu, J., Gu, G., 2005. Control of arabinoxylan solubilization and hydrolysis in mashing. Food Chem. 90 (1), 101–108.

Lim, P.L., Toh, M., Liu, S.Q., 2015. *Saccharomyces cerevisiae* EC-1118 enhances the survivability of probiotic *Lactobacillus rhamnosus* HN001 in an acidic environment. Appl. Microbiol. Biotechnol. 99 (16), 6803–6811.

Liu, S.-Q., Chan, M.Z.A., 2018. An Alcoholic Beverage. PCT WO 2018/192512 A1,.

Liu, S., Tsao, M., 2009. Enhancement of survival of probiotic and non-probiotic lactic acid bacteria by yeasts in fermented milk under non-refrigerated conditions. Int. J. Food Microbiol. 135 (1), 34–38.

Lorencová, E., Buňková, L., Matoulková, D., Dráb, V., Pleva, P., Kubáň, V., Buňka, F., 2012. Production of biogenic amines by lactic acid bacteria and bifidobacteria isolated from dairy products and beer. Int. J. Food Sci. Technol. 47 (10), 2086–2091.

Lorencová, E., Buňková, L., Pleva, P., Dráb, V., Kubáň, V., Buňka, F., 2014. Selected factors influencing the ability of *Bifidobacterium* to form biogenic amines. Int. J. Food Sci. Technol. 49 (5), 1302–1307.

Lu, Y., Putra, S.D., Liu, S.Q., 2018. A novel non-dairy beverage from durian pulp fermented with selected probiotics and yeast. Int. J. Food Microbiol. 267, 1–8.

Majeed, M., Nagabhushanam, K., Arumugam, S., Ali, F., Majeed, S., Beede, K., 2019. Alcoholic Beverage Composition Containing *Bacillus coagulans*. United States Patent US 20190085277,.

Marín Iniesta, F., Sánchez Rubio, M., RMy, C.R., Taboada Rodríguez, A., 2016. Beer Obtained by Fermentation With *Saccharomyces Boulardii*. Spanish Patent ES2583178A1,.

McFarland, L.V., 2017. 18. Common organisms and probiotics: *Saccharomyces boulardii*. In: Floch, M.H., Ringel, Y., Allan Walker, W. (Eds.), The Microbiota in Gastrointestinal Pathophysiology. Academic Press, Boston, pp. 145–164.

Mendis, M., Simsek, S., 2014. Arabinoxylans and human health. Food Hydrocoll. 42, 239–243.

Meussdoerffer, F.G., 2009. A comprehensive history of beer brewing. In: Eβlinger, H.M. (Ed.), Handbook of Brewing: Processes, Technology, Markets. Wiley-VCH Verlag GmbH & Co. KGaA, Weinheim, pp. 1–42.

Mintel Blog, 2015. Can Beer be Healthy? (Website). https://www.mintel.com/blog/drink-market-news/can-beer-be-healthy. (Accessed 30 December 2019).

Mulero-Cerezo, J., Briz-Redón, Á., Serrano-Aroca, Á., 2019. *Saccharomyces cerevisiae* var. *boulardii*: valuable probiotic starter for craft beer production. Appl. Sci. 9 (16).

Munford, A.R.G., Alvarenga, V.O., do Prado-Silva, L., Crucello, A., Campagnollo, F.B., Chaves, R.D., Oteiza, J.M., Sant'Ana, A.S., 2017. Sporeforming bacteria in beer: occurrence, diversity, presence of hop resistance genes and fate in alcohol-free and lager beers. Food Control 81, 126–136.

Naqash, F., Masoodi, F.A., Rather, S.A., Wani, S.M., Gani, A., 2017. Emerging concepts in the nutraceutical and functional properties of pectin—a review. Carbohydr. Polym. 168, 227–239.

Nardini, M., Garaguso, I., 2020. Characterization of bioactive compounds and antioxidant activity of fruit beers. Food Chem. 305, 125437.

Palma, M.L., Zamith-miranda, D., Martins, F.S., Bozza, F.A., Nimrichter, L., Montero-lomeli, M., Marques Jr., E.T., Douradinha, B., 2015. Probiotic *Saccharomyces cerevisiae* strains as biotherapeutic tools: is there room for improvement? Appl. Microbiol. Biotechnol. 99 (16), 6563–6570.

Parrella, A., Caterino, E., Cangiano, M., Criscuolo, E., Russo, C., Lavorgna, M., Isidori, M., 2012. Antioxidant properties of different milk fermented with lactic acid bacteria and yeast. Int. J. Food Sci. Technol. 47 (12), 2493–2502.

Ponomarova, O., Gabrielli, N., Sévin, D.C., Mülleder, M., Zirngibl, K., Bulyha, K., Andrejev, S., Kafkia, E., Typas, A., Sauer, U., Ralser, M., Patil, K.R., 2017. Yeast creates a niche for symbiotic lactic acid bacteria through nitrogen overflow. Cell Syst. 5 (4), 345–357.e6.

Regulation (EC), 2006. No 1924/2006 of the European Parliament and of the Council of 20 December 2006 on Nutrition and Health Claims Made on Foods., p. 9. Accessed 13 January 2020.

Rivière, A., Selak, M., Geirnaert, A., Van den Abbeele, P., De Vuyst, L., 2018. Complementary mechanisms for degradation of inulin-type fructans and arabinoxylan oligosaccharides among Bifidobacterial strains suggest bacterial cooperation. Appl. Environ. Microbiol. 84 (9), e02893-17.

Rodhouse, L., Carbonero, F., 2019. Overview of craft brewing specificities and potentially associated microbiota. Crit. Rev. Food Sci. Nutr. 59 (3), 462–473.

Rosburg, V., Boylston, T., White, P., 2010. Viability of bifidobacteria strains in yogurt with added oat beta-glucan and corn starch during cold storage. J. Food Sci. 75 (5), C439–C444.

Rošul, M.Đ., Mandić, A.I., Mišan, A.Č., Đerić, N.R., Pejin, J.D., 2019. Review of trends in formulation of functional beer. Food Feed Res. 46 (1), 23–35.

Schnitzenbaumer, B., Kerpes, R., Titze, J., Jacob, F., Arendt, E.K., 2012. Impact of various levels of unmalted oats (*Avena sativa* L.) on the quality and processability of mashes, worts, and beers. J. Am. Soc. Brew. Chem. 70 (3), 142–149.

Schurr, B.C., Hahne, H., Kuster, B., Behr, J., Vogel, R.F., 2015. Molecular mechanisms behind the antimicrobial activity of hop iso-α-acids in *Lactobacillus brevis*. Food Microbiol. 46, 553–563.

Scott, K.P., Grimaldi, R., Cunningham, M., Sarbini, S.R., Wijeyesekera, A., Tang, M.L.K., Lee, J.C.-Y., Yau, Y.F., Ansell, J., Theis, S., Yang, K., Menon, R., Arfsten, J., Manurung, S., Gourineni, V., Gibson, G.R., 2019. Developments in understanding and applying prebiotics in research and practice—an ISAPP conference paper. J. Appl. Microbiol. 128.

Senkarcinova, B., Graça Dias, I.A., Nespor, J., Branyik, T., 2019. Probiotic alcohol-free beer made with *Saccharomyces cerevisiae* var. *boulardii*. LWT- Food Sci. Technol. 100, 362–367.

Simpson, W.J., 1993. Ionization behaviour of hop compounds and hop-derived compounds. J. Inst. Brew. 99 (4), 317–326.

Sohrabvandi, S., Razavi, S.H., Mousavi, S.M., Mortazavian, A.M., 2010. Viability of probiotic bacteria in low alcohol-and non-alcoholic beer during refrigerated storage. Philipp. Agric. Sci. 93 (1), 24–28.

Suharja, A.A.S., Henriksson, A., Liu, S., 2014. Impact of *Saccharomyces cerevisiae* on viability of probiotic *Lactobacillus rhamnosus* in fermented milk under ambient conditions. J. Food Process. Preserv. 38 (1), 326–337.

Suzuki, K., 2011. 125th anniversary review: microbiological instability of beer caused by spoilage bacteria. J. Inst. Brew. 117 (2), 131–155.

Takemura, H., Ando, N., Tsukamoto, Y.L., 2000. Breeding of branched short-chain fatty acids non-producing natto bacteria and its application to production of natto with light smells. Nippon *Shokuhin Kagaku Kogaku Kaishi* 47 (10), 773–779.

Ting, P.L., Ryder, D.S., 2017. The bitter, twisted truth of the hop: 50 years of hop chemistry. J. Am. Soc. Brew. Chem. 75 (3), 161–180.

Vasiljevic, T., Kealy, T., Mishra, V.K., 2007. Effects of β-glucan addition to a probiotic containing yogurt. J. Food Sci. 72 (7), C405–C411.

Veljovic, M., Djordjevic, R., Leskosek-Cukalovic, I., Lakic, N., Despotovic, S., Pecic, S., Nedovic, V., 2010. The possibility of producing a special type of beer made from wort with the addition of grape must. J. Inst. Brew. 116 (4), 440–444.

Veraverbeke, W., De Cock, J., De Rouck, G., Delcour, J., Van Mellaert, H., Aerts, G., Broekaert, W., 2011. Partial substitution of barley malt by wheat bran in the grist results in lager beer with better taste profile and higher content in arabinoxylan-oligosaccharides (AXOS). BrewingScience 64 (11–12), 168–174.

Vis, R.B., Lorenz, K., 1998. Malting and brewing with a high β-glucan barley. LWT- Food Sci. Technol. 31 (1), 20–26.

Vriesekoop, F., Krahl, M., Hucker, B., Menz, G., 2012. 125th anniversary review: bacteria in brewing: the good, the bad and the ugly. J. Inst. Brew. 118 (4), 335–345.

Wang, W., Liu, Y., Sun, Z., Du, G., Li, X., 2017. Hop resistance and beer-spoilage features of foodborne *Bacillus cereus* newly isolated from filtration-sterilized draft beer. Ann. Microbiol. 67 (1), 17–23.

Wunderlich, S., Back, W., 2009. 1. Overview of manufacturing beer: ingredients, processes, and quality criteria. In: Preedy, V.R. (Ed.), Beer in Health and Disease Prevention. Academic Press, San Diego, pp. 3–16.

Yeo, H.Q., Liu, S., 2014. An overview of selected specialty beers: developments, challenges and prospects. Int. J. Food Sci. Technol. 49 (7), 1607–1618.

Yeo, A.Y.Y., Toh, M.Z., Liu, S.Q., 2016. Enhancement of bifidobacteria survival by *Williopsis saturnus* var. *saturnus* in milk. Benefic. Microbes 7 (1), 135–144.

Zapata, P.J., Martínez-Esplá, A., Gironés-Vilaplana, A., Santos-Lax, D., Noguera-Artiaga, L., Carbonell-Barrachina, Á.A., 2019. Phenolic, volatile, and sensory profiles of beer enriched by macerating quince fruits. LWT- Food Sci. Technol. 103, 139–146.

Zhu, F., Du, B., Xu, B., 2016. A critical review on production and industrial applications of beta-glucans. Food Hydrocoll. 52, 275–288.

CHAPTER 11

Cereal Bars Added With Probiotics and Prebiotics

GLENISE BIERHALZ VOSS • DÉBORA ANDREIA CAMPOS • MARIA MANUELA PINTADO
Portuguese Catholic University, CBQF—Centre for Biotechnology and Fine Chemistry—Associated Laboratory, School of Biotechnology, Porto, Portugal

INTRODUCTION

The consumption of cereal bars has grown in the past years since it has been presented as quick and easy food to consume throughout the day (Pereira et al., 2019). Cereal bars are a packed combination of dry ingredients (dried cereals) and binder components, which complement each other by conferring the correct sensory properties (Garcia et al., 2018). Cereal-based foods are mostly made from a variety of grains including wheat, maize, oats, rice, rye, barley, millet, quinoa, and sorghum, with rice, maize, and wheat comprising two-thirds of human food consumption (Day, 2013). The applied cereals and grains can be cooked and eaten whole; ground into flour to make a variety of cereal foods like bread, pasta, and noodles; or made into ready-to-eat breakfast cereals or cereal bars (McSweeney and Day, 2016). The tendency to consume healthier foods instead of sweets has led to the emergence of different types of bars, including formats using chocolate coatings or incorporation of different fruits and nuts. The development of cereals bars has taken on a variety of forms including high fiber (wholegrain), with oats, muesli, and those including other grains, legumes, and fruits to improve the nutritional composition of the food consumed (McSweeney and Day, 2016). On the other hand, the high protein bars have been increasing, including animal protein such as whey protein or egg ovalbumin, or vegetable including pea or soy protein, to improve the general nutritional content of the bar. As the consumption of cereals has expanded to any moment of the day, these products have become an excellent carrier for delivering functional ingredients to foods in the marketplace (da Silva et al., 2016). This fact was associated with the busy lifestyles and the increasing demand from consumers for meals and snacks that are quick sources of good nutrition; thus the food industry developed foods like cereal and nutrition bars that combine convenience, nutrition, and health properties (Dutcosky et al., 2006; Pereira et al., 2019). Thus, the association between cereal bars and healthy foods has become a trend in the food industry (Garcia et al., 2018; Pereira et al., 2019).

In parallel, in recent decades, the lifestyle and eating habits of the worldwide population changed, and this has led to an increase in different health problems, such as overweight, obesity, noncommunicable diseases, such as hypertension, diabetes, cardiovascular diseases, and some types of cancer (Centurión et al., 2019). In this context, the development of functional foods is an opportunity to contribute to the improvement of the quality of food and to contribute to consumer health and well-being. Considering, that many years ago, previous studies reported the need for change and the challenges faced by industrialized countries in order "to control the cost of health care; to offer to their ageing population a real opportunity to live, not only longer but also better; to provide to other and more 'busy' consumers, a choice of healthy processed or ready-to-eat foods" (Roberfroid, 1999).

The addition of molecules to such carriers might highly improve the impact on health, such as the addition of probiotics and prebiotics, both of which have high impact on human health. The incorporation of probiotics into cereal-based products is, therefore, desirable since it can combine the healthy formulation of a cereal mixture, the added value of prebiotic ingredients, and the beneficial effects of the probiotic bacteria (Henriques, 2011). However, as the probiotics are living bacteria and yeasts, the conditions of application in cereal bars might raise some questions. Encapsulation techniques have been recently applied to protect probiotic cells from production and storage conditions of cereal bars, to increase shelf-life of probiotic products, and to overcome other technological hurdles such as

Probiotics and Prebiotics in Foods. https://doi.org/10.1016/B978-0-12-819662-5.00009-4

the food production stresses (Henriques, 2011). On the other hand, the prebiotics are food ingredients, usually applied to food preparations, that can act as fermentable substrates in specific bacteria, leading to the stimulation of bacterial growth and activity. Generally, prebiotics are carbohydrates that are non-digestible by human microbiota and are delivered in the bowel, acting as positive promoting agents (Gibson and Roberfroid, 1995). The prebiotics differ from most dietary fibers that stimulate the growth of a wide variety of intestinal microorganisms and not necessarily its beneficial constituents (Gibson et al., 2017; Kerry et al., 2018). Still, according to Gibson et al. (2017), prebiotics provoke a biased metabolism toward host health promotion rather than being extensively metabolized.

Nowadays, several types of prebiotics are recognized, but the search for new prebiotics is continuous, due to the importance of prebiotics in the human gut balance and in the human general health, previously proved by Roberfroid et al. (2010) and Salminen et al. (1998). They have demonstrated that such prebiotics increase beneficial bacteria in the human gut, such as *Bifidobacterium* and *Lactobacillus* strains, inhibit pathogenic bacteria, and mainly promote the production of short-chain fatty acids and lactic acid during its metabolic degradation, thereby further contributing to their beneficial health effects (Roberfroid et al., 2010).

Therefore, the current chapter intends not only to wrap up the common sense of cereal bar production and consumption, but also to highlight the most effective natural additives that increase the value of the bar and deliver health promoters with focus on probiotics and prebiotics.

CURRENT REPERCUSSION IN THE MEDIA, THE IMPORTANCE OF HEALTH

Changing lifestyles, eating habits, and health awareness have made consumers more conscious about the role of food for a better quality of life, consequently leading to a growing search for healthy and nutritious foods with more benefits broad to health, able to prevent diseases (Panghal et al., 2018). The market of functional food ingredients has been growing worldwide due to the current trends of wellness and health and this has increased the awareness of the type and quality of foods that humans are ingesting. In 2018, the functional ingredients market was evaluated in 57.6 billion euros and was projected to reach 79.1 billion euros by 2023, at a CAGR of 6.6% (Markets and Markets, 2020).

Previous studies have been related to several impacts of probiotic, prebiotic, and symbiotic food on human health, and this has driven the growth of these foods in the market. Symbiotic products combine probiotic microorganism and prebiotic substances, helping the survival and the implantation of live microbial dietary supplements in the gut and promoting a positive interaction in vivo; besides, symbiotic might increase the stability on the final product (Martinez et al., 2015; Mondragón-bernal et al., 2017). Studies have related that the symbiotic relationship between prebiotics and probiotics has a significant influence on host health, as expected, due to the synergetic benefits occurring between probiotics and prebiotics (Kerry et al., 2018). Furthermore, previous studies have shown benefits for gut health and prevention of diseases, like cardiovascular diseases (Bedani et al., 2014; Kerry et al., 2018).

The preference of consumers for vegetarian food has stimulated the use of cereal, vegetables, legume, and fruit for the development of probiotics or symbiotic food products (Nazzaro et al., 2008; Mondragón-bernal et al., 2017). Besides, cereals and legumes are the most important plant group and show rich nutritional composition that is responsible to meet nutritional requirements worldwide (Panghal et al., 2018).

A Window Into Probiotic and Bioactive Compounds Present in Cereal Products

Previous studies have reported promising results concerning the evaluations of oats, maize, barley, wheat, and malt as matrices that support the growth of probiotic strains (Angelov et al., 2006; Salmerón et al., 2015; Ogunremi et al., 2017). As an alternative matrix for probiotic foods, the cereal products have the advantage of a large distribution rate and nutritive value since these are considered one of the most important sources of proteins, carbohydrates, vitamins, minerals, and fibers. Furthermore, the cereal may present essential fatty acids, resistant starch, phenolic compounds, and phytosterol (Rivera-Espinoza and Gallardo-Navarro, 2010; Dey, 2018). These functional compounds present in cereal grains have been related to a lower risk of chronic diseases, and these facts increased the consumption of whole cereal for people all over the world (Dey, 2018). Besides, when the cereal compounds such as oligosaccharides, resistant starch, β-glucans, and arabinoxylans are used to promote probiotic microorganisms, it is possible to obtain the combined benefits of prebiotic and probiotic effects from this combination (Lamsal and Faubion, 2009).

Health Claims

In order to promote a health effect, the oral ingestion of a probiotic will have to satisfy a predetermined threshold of viable cells, and according to previous studies, a functional product should contain between 10^6 and 10^8 colony-forming units per gram (CFU/g) of viable probiotic cells during the shelf life (Donkor et al., 2007; Martinez et al., 2015; Mondragón-bernal et al., 2017). The oral administration of probiotic food may be beneficial in a multitude of disorders in the gastrointestinal tract such as antipathogenicity, antiallergic, angiogenic, anticancer, and antiinflammatory activities extensible to other organs, proving to also have antidiabetic, antiobesity effects or even on the brain and central nervous system (Kerry et al., 2018; Brüssow, 2019).

According to Santosa et al. (2006), the main health benefit attributed to probiotic bacteria is the improvement of the intestinal-environment balance by modifying the gut's microflora; also, the specific health benefits appear to be related with the strain used. For example, some clinical trials describe the specific therapeutic effects of probiotics in cases of diarrhea, antibiotic-associated diarrhea (ADD), constipation, and irritable bowel syndrome (IBS) (D'Souza et al., 2002; Moayyedi et al., 2010). Additionally, in a previous study in which a metaanalysis of randomized clinical trials was conducted on 32 studies enrolling 1971 patients, Wang et al. (2018) showed that the different probiotic strains, such as *L. acidophilus*, *L. Lactis*, and *L. plantarum*, reduced significantly the level of total serum cholesterol. The probiotics have been associated with the cardiovascular benefits; some scientific studies report that the microbial imbalance of the gut may play a role in the pathogenesis of cardiovascular diseases (Thushara et al., 2016). Therefore, considering that hypertension also plays a major role in the development of cardiovascular diseases, some human clinical trials showed that the consumption of probiotic products may improve blood pressure (Khalesi et al., 2014). In addition, the recent studies reported that probiotics and their metabolites can be exploited to manage hypertension via mechanisms such as improving total cholesterol and LDL-cholesterol levels, reducing blood glucose levels and insulin resistance, and regulating the renin-angiotensin system (Khalesi et al., 2014; Thushara et al., 2016; Qi et al., 2020). Furthermore, Fig. 1 summarizes other health benefits conveyed by probiotic consumption.

PROBIOTIC

Probiotics are defined as live microorganisms with beneficial effects on human and animal health when given in adequate amounts (Gibson et al., 2017; Kerry et al., 2018), but the adequate dose of probiotic microorganisms depends on the strain, as well on the type of product (Martinez et al., 2015). However, during recent years, several investigators have proposed different terminologies to the probiotic, such as "paraprobiotics" (dead/inactive cells of probiotics) and "postbiotics" (healthful metabolites of probiotics). Thus, according to the new definition by Zendeboodi et al. (2020),

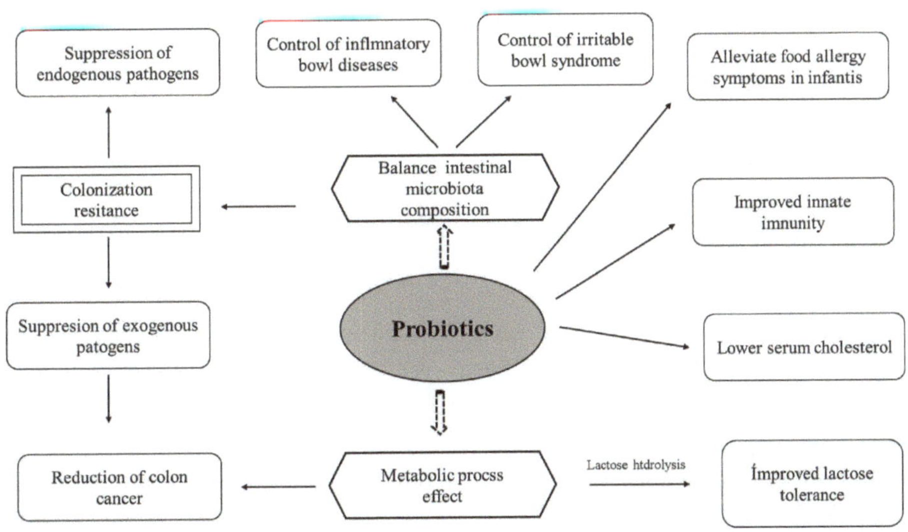

FIG. 1 Health benefits from probiotic consumption. (Adapted from Tripathi, M.K., Giri, S.K., 2014. Probiotic functional foods: survival of probiotics during processing and storage. J. Funct. Foods. 9, 225–241. https://doi.org/10.1016/j.jff.2014.04.030.)

probiotic can be defined as a viable or inviable microbial cell (vegetative or spore; intact or ruptured) that is potentially healthful to the host.

Currently, most of the consumers ingest probiotics through food products, and most of these products are categorized as functional food (Tripathi and Giri, 2014). The selection of proper probiotic strains is the first requirement for developing a probiotic food product. Evaluating the adequate amount to add in order to assure adequate viable numbers for consumption, surviving to the processing and storage conditions, comes next. But, after production and storage time, the products still have to overcome the gastrointestinal tract and digestion; thus the probiotic strains have to survive during gastric and intestinal transit and still deliver health benefits for consumers (Parvez et al., 2006; Tripathi and Giri, 2014). In that sense, FAO has stated in 2002 that any food claimed to have probiotic effect must contain at least 10^6 to 10^7 CFU/mL of viable probiotic bacteria.

Moreover, other technological and physiological characteristics of probiotic strains are important during the selection of probiotics for application in commercial food products, such as safety criteria (pathogenicity and infectivity, origin), technological (genetically stable strains, good sensory properties, production in large-scale, and viability during process and storage), functional (tolerance to gastric acid and bile, adhesion to the mucosal surface), and physiological (cholesterol metabolism, lactose metabolism, and properties of antimutagenic and anticarcinogenic) (Tripathi and Giri, 2014). The genera of *Bifidobacterium* and *Lactobacillus* strains are the probiotic microorganisms most extensively studied and widely used, due to the beneficial effects attributed (Bedani et al., 2014).

The viability of probiotic microorganisms in different food products is influenced by many factors during production, processing, and storage. According to Tripathi and Giri (2014), the factors include microbiological parameters (probiotics strains and inoculation proportion), food ingredients (pH, acidity, water activity, presence of salt, sugar, and chemicals), fermentation conditions (dissolved oxygen and incubation temperature), and process applied (scale of production, heat treatment, product cooling rate, packaging materials, and storage condition).

Probiotic Strains Used in Cereal Foods

Applications of probiotic cultures in nondairy products represent a great challenge. The survival and stability during production and storage of cereal probiotic foods have been shown to depend not only on the food matrix, water activity, and pH of the final products, but also on the strain of probiotic bacteria (Aspri et al., 2020). Also, the dairy products containing probiotics are usually stored at refrigerated temperatures, guaranteeing the shelf life of the products. Cereals contain an incredible number of vitamins, minerals, proteins, carbohydrates, fiber, and important oligosaccharides (Vinderola et al., 2017). One important advantage of using cereal matrices to incorporate probiotics is the high presence of dietary fiber and the presence of nondigestible carbohydrates, which can be used as prebiotics and can stimulate the growth of probiotic bacteria in the human colon (Kumar et al., 2015). Previous studies have reported the use of different probiotic strains for incorporation in cereal bars. Currently, the used commercial probiotic products include *Lactobacillus* spp., *Bifidobacterium*, and even a few nonlactic acid bacteria; thus the list of common probiotic microorganisms applied in foods are reported in Table 1.

Lactobacillus spp.

Lactobacillus spp. is a gram-positive bacterium, is anaerobic facultative, presents a bacilli shape and has been listed as a lactic acid bacterium (LAB). It has been widely recognized to be beneficial to human health due to its immunostimulatory activity and has been largely introduced in the functional food market, being reported as safe for consumption (Henriques, 2011).

On the other hand, it has been well documented that the genetic, biochemical, and physiological state of bacterial cells changes throughout the different growth phases; thus different *Lactobacillus* strains lead to different metabolic pathways (Bhushan et al., 2020). The most common strain applied to the functional foods has been the *L. acidophilus*, being widely reported throughout the bibliography.

One of the important aspects of probiotic strains functionality and *Lactobacillus* strains have been associated with the capability to adhere to enterocytes, blocking the potential adhesion sites for pathogens, promoting an immunomodulatory capacity, and producing antimicrobial substances. For example, the strain *L. casei* 01 has exhibited antiproliferative activity on human colon cancer cells HT-29 and reduced the cytotoxicity of 4-NQO (4-nitroquinoline 1-oxide) against human intestinal epithelial cells (Liu et al., 2011).

Bifidobacterium spp.

Bifidobacteria is an anaerobic bacterium, gram-positive, bacillus shaped, and nonmotile, which acts as a beneficial microorganism on human health, a probiotic (Turroni et al., 2019). This genus is one of the most

TABLE 1
Descriptive List of Probiotic Microorganisms Usually Applied in Foods.

***Lactobacillus* spp.**	***Bifidobacterium* spp.**	**Other Microorganisms**
L. acidophilus	*B. bifidum*	*Enterococcus faecium*
L. brevis	*B. breve*	*Enterococcus faecalis*
L. casei	*B. infantis*	*Escherichia coli* Nissle
L. rhamnosus	*B. longum*	*Streptococcus salivarius* subsp. *Thermophilus*
L. reuteri	*B. adolescentis*	*Streptococcus salivarius*
L. delbrueckii	*B. animalis*[a]	*Saccharomyces cerevisiae* var. *boulardii*
L. bulgaricus		*Bacillus coagulans*
L. plantarum		*Bacillus clausii*
L. fermentum		
L. johnsoni		
L. salivarius		
L. paracasei		
L. gaseri		

[a] *Bifidobacterium animalis* are often known commercially for *Bifidobacterium lactis*.
Adapted from Santosa, S., Farnworth, E., Jones, P.J.H., 2006. Probiotics and their potential health claims. Nutr. Rev. 64(6) 265–274. https://doi.org/10.1301/nr.2006.jun.265-274.

abundant bacteria that compose the human gut microbiota, living mainly in the colon, promoting beneficial effects on the hosts. The genus *Bifidobacterium* belongs to the *Actinobacteria phylum* (Goodfellow et al., 2012).

The genus *Bifidobacterium*, even if traditionally listed among lactic acid bacteria (LAB), is only poorly phylogenetically related to genuine LAB. This bacterium rapidly colonizes the gut of infants within the first weeks following the birth through specific mother-milk-derived oligosaccharides, commonly referred to as human milk oligosaccharides (HMOs), and its concentration and type changes throughout the growth and might vary depending on specific situations. However, the level of bifidobacterial abundance in the human gut drops with aging (Mueller et al., 2006; Eckburg et al., 2005). The *Bifidobacterium* species use a different metabolic pathway for the degradation of hexoses—fructose-6-phosphate phosphoketolase (Felis and Dellaglio, 2007). Research on this genus has focused on the capability of fermenting oligosaccharides. Between the strains of *Bifidobacterium* spp., the *B. lactis* Bb-12 strain is one of the most investigated and applied probiotics. This strain has been used in fermented products (dairy and nondairy) (Ouwehand et al., 2004).

Yeasts and Bacillus sp.

The *Saccharomyces boulardii* and probiotic *Bacillus* would be easier to add and would be more stable during storage, but both cultures need to be refrigerated to limit their growth during storage (Champagne et al., 2011). Between the different probiotics, *S. cerevisiae* var. *boulardii* presents an advantage in terms of resistance and growth rate (Singu et al., 2020). The major difference between yeast and bacteria with the probiotic application is the resistance to pH variation; while the yeast grows well at pH 7–8, optimal growth is observed between pH 4.5–6.5. Another difference that can be considered is their presence in humans; yeast's is less than 1% and bacteria is 99% (Czerucka et al., 2007). Furthermore, *S. boulardii* possesses many properties that make it a potential probiotic agent, such as the ability to survive in the transit throughout the gastrointestinal tract. In an optimum temperature (37 °C), it can inhibit the growth of different microbial pathogens and is even considered safe for infants, thus the coating using *S. boulardii* presents low cost (Czerucka et al., 2007; Singu et al., 2020).

Bacillus coagulans is a lactic-acid-producing bacterium that shows characteristics of both *Lactobacillus* and *Bacillus* genera. Additionally, it showed characteristics such as probiotic properties, high-temperature resistance, and easy storage (Sui et al., 2020). There is also evidence that probiotics prevent the diarrhea that may occur due to antibiotic use. Thus, the activity of the *S. boulardii* against *Clostridium difficile* infections is recognized, and it might be a probiotic candidate (Champagne et al., 2011).

In Table 2 is described the probiotic microorganisms applied to cereal food matrices reported throughout the years. The *Lactobacillus* spp. strains are the strains most used to turn cereal products into functional products. The cereal products vary in different types of

TABLE 2
Recent Studies Related to the Use of Cereal Matrices for Carrying Probiotic Bacteria.

Food Matrix	Probiotic Microorganism	Food Product	Reference
Cereal	*Saccharomyces boulardii*	Cornflakes	Singu et al. (2020)
Cereal	*B. animalis* subsp. *lactis* Bb-12	Cereal bar	Pereira et al. (2019)
Cereal	*Saccharomyces cerevisiae*	Cereal bar	dos Baliza et al. (2018)
Cereal	*L. plantarum*	3D printing of cereal-based food	Zhang et al. (2018)
Cereal	*L. acidophilus* *B. animalis* subsp. *lactis*	Savory cereal Bars	Bampi et al. (2016)
Cereal	*Saccharomyces boulardii* *L. acidophilus*	Cereal bar	Bastos et al. (2014)
Cereal	*L. acidophilus* La-5	Cereal bar	Rodrigues (2013)
Chocolate	*L. acidophilus* *B. lactis*	Milk and dark chocolate	Lalicic-Petronijevic et al. (2015)
Cereal and chocolate	*L. rhamnosus*	Cereal bar and dark chocolate	Raymond and Champagne (2015)
Soybean	*L. acidophilus*	Soy bar	Chen and Mustapha (2012)
Cereal	*L. rhamnosus*	Chocolate-coated breakfast cereals	Saarela et al. (2006)
Cereal	*B. lactis* Bb-12	Oat-based cereal bar	Ouwehand et al. (2004)

cereals used and applied, and in some cases, chocolate can be added to increase the flavor of the bar, as well as increase the economic value. On the other hand, other strains of probiotics might be found in the cereal products, such as *Bifidobacterium* spp. strains, followed by *Saccharomyces* spp. strains.

Commercial Probiotic Cereal Products

Nevertheless, the application of the scientific data is still controversial, since there is no clear-cut criterion about the amount of probiotic strain to be administered and which anticancer mechanisms are most relevant. Furthermore, some studies point toward a probiotic stimulation of the immune system, the growth suppression of other bacteria that convert procarcinogen metabolites, or the direct action against carcinogens present in the gut (de Moreno de LeBlanc et al., 2007).

Probiotic products have a very strict regulation, with only a few "permitted health claims" available for labeling in some countries. In Europe, the EFSA (European Food Safety Authority) demands a strain-specific approach on efficacy, which is still out of range for most manufacturers, since there are no all-inclusive biological markers, and extensive clinical trials are very expensive; so, currently no health claim is permitted related to probiotic bacteria.

In the early stages of food development, it is critical to consider consumer acceptance, from sensory evaluation to added value recognition. Intriguingly, the food industry still selects probiotic species and strains much based on their technological properties rather than their potential health benefits. The main reason seems to be the reduction of development costs and time of a new product, giving priority to safety concerns rather than specific physiological mechanisms, which require greater investments to evaluate. Nowadays in the market, several products containing probiotics can be found and are descript in Table 3.

Probiotic Cereal Products and Main Characteristics

Several studies have evaluated the cereal as carriers of probiotic cultures in beverages, bread, biscuits, cereal bars, breakfast cereals, and others (Lamsal and Faubion, 2009; Martins et al., 2013). Since probiotics are living bacteria or yeasts, something could be difficult to maintain a viable cell number throughout the storage and consumption. The cereals grains, due to their bioactive compounds list, might naturally be a good carrier for probiotics, since cereals contain native prebiotic substances that protect probiotic microorganisms from

TABLE 3
Cereal-Based Probiotic Food Available in the Market.

Food Matrix	Probiotic Microorganism	Product	Ingredients
Breakfast probiotic bars	*Bacillus coagulans* GBI-30 6086 probiotic cultures		Peanut butter dark chocolate bars: oats, cane sugar, brown rice flour, canola oil, peanut butter (peanuts, sea salt), semisweet chocolate (sugar, unsweetened chocolate, cocoa butter, soy lecithin, vanilla extract), peanuts, oat flour, gum acacia, raisin paste, water, peanut oil, peanut flour, millet, buckwheat, amaranth, honey, quinoa, sea salt, natural flavor, *Bacillus coagulans* GBI-30 6086 probiotic cultures, Vitamin E (tocopherols to maintain freshness).
Cereal bar (granola probiotic wellness bar and chocolate probiotic wellness bar)	*L. acidophilus* *L. casei* *B. lactis*		Dark chocolate coating (chocolate, sugar, cocoa powder, inulin, calcium carbonate, cocoa butter, soy lecithin (an emulsifier), vanilla), Probiotic cultures (*L. acidophilus*, *L. casei*, *B. lactis*).
Nutrition bars	*Bacillus coagulans* GBI-30 6086		Peanut butter: organic peanuts, organic peanut butter (peanuts), organic date paste, organic cocoa butter, organic sorghum flake, organic sunflower seeds, organic sesame seeds, organic cocoa beans, organic flax seeds, organic peanut oil, organic pumpkin seeds, natural flavors, organic coconut oil, organic chia seeds, sea salt, *Bacillus coagulans* GBI-30 6086.
Bar	*Bacillus coagulans* GBI-30 6086		Gluten free oats, brown rice syrup, brown rice crisps, natural cashew butter, vegetable glycerin, coconut chips, chicory root fiber, sea salt, natural flavor, *Bacillus coagulans* GBI-30 6086 (probiotic).
Cereal bar	*L. acidophilus* *L. plantarum* *Streptococcus thermophilus* *B. lactis*	 	Oat flakes 26%, yogurt coating 20% (sugar, fully hydrogenated palm fat, yogurt powder 4%, lactose, emulsifier: soya lecithin, acid: citric acid, flavoring), glucose syrup, fructose syrup, extruded cereals 12% (wheat flour, maize flour, rice flour, sugar, salt), dried plums 13% (contain antioxidant: sulfur dioxide), cornflakes, rapeseed oil, cranberry component 3% (cranberries 65%, sugar, elderberry fruity concentrate, sunflower oil, acid: citric acid), sugar, maltodextrin, inulin 1.5% (fiber), barley malt extract, palm fat, humectant: sorbitol, acidity regulator: citric acid, salt, emulsifier: soya lecithin, flavoring, dairy cultures 0.065% (*L. acidophilus*, *L. plantarum*, *B. lactis*, *Streptococcus thermophilus*).

Continued

TABLE 3
Cereal-Based Probiotic Food Available in the Market—cont'd

Food Matrix	Probiotic Microorganism	Product	Ingredients
Bar	*B. lactis* Bb-12		Goodbelly honey almond butter probiotic bars: almonds, almond butter (almonds), organic honey, buckwheat grits, maple sugar, organic coconut oil, puffed quinoa, buckwheat flour, pumpkin seeds, palm oil, whole grain quinoa, salt, cultures (*B. lactis* Bb-12), baking soda, vanilla extract.
Bar	*Bacillus coagulans*		Coconut, prebiotic dietary tapioca fiber, erythritol, tapioca syrup, unsweetened chocolate, chicory root fiber, cocoa butter, chia seed, extra virgin coconut oil, sea salt, vanilla extract, purified Stevia extract, probiotics (*Bacillus coagulans*), rosemary extract.
Cereal	*B. lactis Bb*-12		Whole grain oats, whole grain corn, sugar, peanut butter (peanuts, salt), dextrose, corn starch, whole grain rice, corn bran, whole grain sorghum, whole grain millet, brown sugar syrup, contains 2% or less of: canola oil, salt, tripotassium phosphate, cultures (*B. lactis*), color (caramel color and annatto extract). Vitamin E (mixed tocopherols) added to preserve freshness.
Cereal	*L. casei*		Wheat bran, rice, degerminated yellow corn meal, evaporated cane juice crystals, kashi seven whole grains and sesame flour (whole: oats, hard red winter wheat [Graham flour], rye, triticale, long grain brown rice, barley, buckwheat, sesame seeds), oat fiber, vanilla coating (evaporated cane juice crystals, fractionated palm kernel oil, nonfat dry milk, nonfat yogurt [cultured nonfat milk, yogurt is heat-treated after culturing], soy lecithin, vanilla extract), crisp rice (rice flour, salt), evaporated cane juice syrup, corn bran, chicory root fiber, expeller pressed canola oil, calcium carbonate, natural flavor, soy protein concentrate, salt, broccoli extract, ginger, annatto (color), *L. casei* probiotic culture, Vitamin D.

Muesli	*Bacillus coagulans*		Rolled oats, barley flakes, rolled wheat, triticale flakes, golden syrup, dark compound probiotic chocolate chips [sugar, vegetable fat (certified sustainable palm), cocoa powder, probiotic culture (1.5%) (organic agave inulin, *Bacillus coagulans* GBI-30 6086), emulsifiers (soy lecithin, 476, 492), natural flavors], almonds (4%), pear manuka honey pieces (4%) {fruit [pear puree (33%), concentrated apple puree (31%)], invert sugar, sugar, glycerol, manuka honey (2.5%), wheat fiber, pectin, natural flavor, citric acid}, honey, almond paste, coconut (coconut, preservative 220, 223), sunflower oil, linseed, pepitas, sunflower kernels, cinnamon, natural flavor, antioxidant (307b).
Granola	*Bacillus coagulans*		Gluten-free oats, coconut sugar, raw virgin coconut oil, fair trade dark chocolate chunks (cane sugar, chocolate liquor, cocoa butter, vanilla extract), sunflower seeds, puffed amaranth, cacao powder, quinoa flakes, cinnamon, sea salt, chia seeds, probiotic cultures (inulin, palm oil, *Bacillus coagulans* GBI-30 6086).

the adverse conditions of the gastrointestinal tract digestion.

Nevertheless, there is another alternative to guarantee the maintenance of the probiotics in the cereal bars. Pereira et al. (2019) have described the immobilization of probiotics and prebiotics in edible films that can coat the cereal bar. Such solution appears to be a promising way to protect the probiotics, mainly from the processing and storage of the cereal bars, as well as to protect other sensitive probiotics that cannot be incorporated by direct inoculation. On the other hand, the immobilization of probiotic strains in the film might be also useful to protect the microorganisms from the action of the gastrointestinal tract, since it is crucial to take the probiotic strains until the human gut, where they may exert the beneficial effect (Pereira et al., 2018).

CEREAL BARS' MAIN INGREDIENTS

Granola

Granola has emerged as an important breakfast cereal mixture product. It is considered a ready-to-eat (RTE) cereal, produced from the mixture of grain components with other ingredients such as chopped nuts and fruit pieces through extensive processing, i.e., granulation. The principal grains used in the manufacture of granola include corn, rice, wheat, oats, and barley, which are overall considered healthy ingredients (Pathare et al., 2012). In this respect, ready-to-eat cereals are an especially promising alternative to dairy products for the incorporation of probiotics. They are a well-known fortified product which can be given to people with lactose intolerance.

The expansion of the market range for these kinds of granulated products can be associated with life-style trends, such as fitness and well-being, and with their many already known health benefits, such as the reduction of satiety and improvement of bowel movement. Several multinational companies are investing in intensive research for new convenient healthy products, and much of the attention is now being drawn to cereal bars fortified with nutritional ingredients. Their portable and modern daily-routine concepts may carry out a larger perception of high added value (Annunziata and Vecchio, 2011).

Chocolate

Chocolate is one of the most well-known ingredients, and it is extensively consumed, not only because of its flavor and taste but also because it promotes good digestibility and accelerates metabolisms. Thus, the term functional chocolate has been brought to the table as an important ingredient for balanced and nutritional cereal bars. Chocolate is rich in polyphenols, flavonoids, minerals, proteins, and carbohydrates, and numerous studies have suggested that chocolate might be a good substrate for probiotic bacteria, as a lipid fraction on the cereal foods, since cocoa butter protects probiotics during storage and in the upper gastrointestinal tract passage (Kemsawasd et al., 2016; Aspri et al., 2020). Therefore, as can be seen in Table 2, several cereal bars are also combined with chocolate and probiotics, which will improve the quality and viability of the functional bars.

Cereal Grains

Oat

Oat (*Avena sativa*) is a type of cereal grain grown from its seeds, usually cultivated in temperate regions. In 2018, its production reached around 23 million tons and is slowly growing since 2010 (FAOSTAT, 2019). Oats have numerous uses in foods but can be crushed into oatmeal, a very well-known breakfast, or can be fine ground into powder. On the other hand, oats are an important ingredient of muesli and granola.

Oat grain contains a mixture of insoluble and soluble (beta-glucan) fibers and provides the best of both worlds. While soluble oat dietary fiber is an important source to lower cholesterol levels, the insoluble dietary fiber or oat fiber has been described as a possible promoter of healthier gut microbiota (Jefferson and Adolphus, 2019).

Abbeele et al. (2018) have reported the effect of oats intake in human health, and results have shown that different oat ingredients promoted a stimulation of health-related microbial metabolites in the gut microbiome, which suggests that these ingredients can be used as prebiotic substrates upon digestion. In the same study, authors associated the results with the amount of soluble dietary fiber in the oats content (Abbeele et al., 2018). Oats own bioactive compounds, such as lipids and beta-glucan, which are associated to improve the diet quality by reducing cholesterol, regulate satiety, and protect against carcinogenesis of the colon. Korczak et al. (2020) have described that the best amount of oats intake in the day should be 40–100 g per day, which provide between 2.5 and 2.9 g of beta-glucan per day, which has shown to decrease fecal pH and alter beneficially the fecal bacteria.

Wheat

Wheat is a grass that is widely cultivated since its seed, an important source of carbohydrates, is consumed by humans,. Wheat is eaten as whole grain, as a source of multiple nutrients, vitamins, and soluble and insoluble dietary fiber. In 2018, 734 million tons of wheat was produced in all world, being the highest production in Asia (FAOSTAT, 2019).

Wheat is used widely to produce flours and it is a base of a huge number of occidental meals, and attached to huge consumption by humans caused some problems associated with refined flours, which could negatively affect the positive effect of this seed.

Costabile et al. (2008) have described that the whole grain wheat in cereals has a prebiotic effect upon the human gut when compared with the refined wheat, and results have shown different impacts in the human gut. The consumption of whole grain acted as a prebiotic agent on the growing promotion of bifidobacteria and lactobacilli, comparing with the starting levels; also in the same study, the antioxidant content was evaluated, and the presence of specific polyphenols promoted an antioxidant effect on the human gut, promoting the beneficial environment.

Maize

The maize production has achieved 1147 million tons all over the world in 2018 as per records. The biggest producer is the American continent, representing 52.5% of the total production (FAOSTAT, 2019). The maize is a cereal grain domesticated about 10,000 years ago, but a very small amount of the total maize production is consumed directly by humans, which is mainly applied to other ends. The consumption of 100 g of maize per day provided a good source of the daily intake of the required vitamins. Furthermore, a maize-based soluble fiber is a versatile source of dietary fiber with potential selective fermentation properties as determined in vitro and in vivo. Costabile et al. (2016) have shown the prebiotic effect of dietary fiber coming from maize grains since the stimulation of bifidobacteria in the overall gut microbiota was demonstrated following the consumption of 6 g fiber/delivered/day. Konings et al. (2014) described the positive effect of ingredients coming from maize, such as starch, reporting beneficial impact regarding the metabolic pathways from the human gastrointestinal tract.

Rice

Besides oat and maize cereals, rice grain is also one of the most applied into cereal-based meals and is well recognized due to its richness of fiber, minerals, vitamins, and antioxidants. Rice is a seed originally from *Oryza glaberrima* or *Oryza sativa*. Rice is an agricultural commodity, being one of the most harvested. In 2017, 770 million tons of rice were produced worldwide, the Asian continent being the biggest producer, with 90.7% of the total production (FAOSTAT, 2019). Of course, the nutritional value of rice is recognized, and the nutrient analysis will vary between the different varieties of rice and the conditions of growing (FAOSTAT, 2019).

Rice has been widely applied in the production of cereal bars as an alternative source of value. On the other hand, it has been applied to give an extra crunch and softness to the taste of the cereal bars. Pham et al. (2017) have studied the isolated impact of rice in the human gut microbiota, and conclusions taken have shown that the short-chain fatty acids produced by white rice have been similar to the ones produced by the control (fructooligosaccharides). White rice promoted a different fermentation, showing a different modulation effect upon butyrate-producing bacteria.

PREBIOTIC

The oldest concept of prebiotic was introduced by Gibson and Roberfroid (1995), where it states that a prebiotic is a nondigestible food ingredient that promotes benefits in human health with a positive modulation of the human gut. Later, in 2016, ISAPP (International Scientific Association for Probiotics and Prebiotics) recommended that a prebiotic was "a substrate that is selectively utilized by host microorganisms conferring a health benefit." Prebiotics are nutrients that, while not easily digested by humans, beneficially modulate the gut microbiota for the host. They differ from most dietary fibers that stimulate the growth of a wide variety of intestinal microorganisms and not necessarily its beneficial constituents (Gibson et al., 2017; Kerry et al., 2018). Thus, the prebiotic molecule can be an inorganic or organic substance stimulating specific microorganisms in the microbiota (Tomasik and Tomasik, 2020).

There are many types of prebiotics, and the majority of them are part of the carbohydrate group. Prebiotics can be obtained naturally from sources like vegetables, fruits, and grains consumed in our daily life; these are generally nondigestible soluble fibers such as inulin, oligofructose, fructooligosaccharides or galactooligosaccharides (GOS), and pectin derivatives (Kerry et al., 2018). According to the new definition of prebiotics, polyphenols, fatty acids, and some peptides can be included in the prebiotic group since they can be metabolized by bacteria into active ingredients (Tomasik and Tomasik, 2020). In Table 4 is described the category of prebiotics already applied in foods and nutraceuticals.

TABLE 4
Category of the Prebiotics With Demonstrated Effect Upon Probiotics.

Category of Prebiotics	Types of Prebiotics
Fructans	Inulin and fructooligosaccharide
Galactooligosaccharides	Galactooligosaccharides with excess galactose at C_3, C_4, or C_6 galactooligosaccharides manufactured from lactose through enzymatic transglycosylation
Starch- and glucose-derived oligosaccharides	Starch and polydextrose
Other oligosaccharides	Oligosaccharides originated from polysaccharide and pectic oligosaccharide (POS)
Noncarbohydrate oligosaccharides	Compounds that are not classified as carbohydrates but are recommended to be classified as prebiotics: cocoa-derived flavanols

Prebiotic Cereal Bars

Characteristics of prebiotic cereal products

Prebiotics are a subcategory of functional food ingredients that can be added to many foods including products from cereals, such as cereal bars. Also, in the literature, there are reports of the use of prebiotics in other products based on cereal grains, such as cookies and bread. In Table 5 are described the recent studies where prebiotic compounds have been studied in matrices containing cereal.

Addition of Natural Prebiotic Into Cereal Products

The addition of natural prebiotic into cereal bars has been a rising trend. In Table 6 are described the products in the market, applying natural ingredients with prebiotic potential. More than 20 natural ingredients have been described, but in this chapter are described the most usually applied, not only by the prebiotic promotion but also due to another characteristic's enhancement. The chosen ingredients can vary between roots, leaves, flowers, vegetables, and fruits.

Chicory root and dandelion greens

Chicory is a grown plant native from Europe, commonly known as the plant of the blue daisy and its roots have been described as a coffee substitute. The roots of chicory are baked, roasted, ground, and used as an additive. In the 70s, it was found that chicory roots contain high levels of dietary fiber especially inulin, a well-recognized prebiotic. In average, the roots might contain around 20% of inulin. Thus, chicory root extract has been produced as a dietary supplement or food additive, containing by dry weight, 98% of inulin and 2% of other compounds. On the other hand, dandelion is a big genus of flower plants and has been used by humans for food preparations and treatment with herbal extracts mixed with medicinal purposes. The entire dandelion plant is edible, including leaves, stems, flowers, and roots. Dandelion greens can be used in

TABLE 5
Recent Studies Related to the Use of Cereal Matrices for Carrying Prebiotic.

Food Product	Prebiotic	Reference
Cereal bar	Inulin	Pereira et al. (2019)
Cereal bar	Inulin, oligofructose, and gum acacia	Dutcosky et al. (2006)
Cereal bar	Inulin and gum acacia	Srebernich et al. (2016)
Gluten-free bread	Fructooligosaccharides and inulin	Morais et al. (2014)
Gluten-free bread	Inulin	Morais et al. (2014)
Cookies	Burdock root flour	Moro et al. (2018)
Gluten-free chocolate cookies	Inulin and oligofructose	da Silva and Conti-Silva (2018)

TABLE 6
Examples of Cereal-Based Prebiotic Food Available on the Market.

Food Matrix	Prebiotic Ingredient	Product	Ingredients
Baked breakfast bars	Chicory root fiber		Mixed berry filling (pear juice concentrate, tapioca syrup, cane sugar, blueberry puree concentrate, apple powder, corn starch, vegetable glycerin, blackberry puree concentrate, blueberry juice concentrate, strawberry puree concentrate, natural flavors), Kashi seven whole grain flour (oats, hard red wheat, brown rice, rye, triticale, barley, buckwheat), whole wheat flour, whole grain oats, invert cane syrup, expeller pressed canola oil, honey, chicory root fiber, vegetable glycerin, tapioca syrup, leavening (sodium acid pyrophosphate, baking soda), soy lecithin, xanthan gum, natural flavors.
Cereal	Inulin		Brown rice flour, whole grain corn meal, cane sugar, corn meal, yellow corn flour, inulin, honey, quinoa, flax seeds, flavor, buckwheat flour, sea salt, amaranth, molasses, tocopherols (vitamin E).
Cereal bar	Oligofructose Syrup		Cereals (40%) (whole oats, rice, whole wheat), milk chocolate (22%) (sugar, cocoa butter, skimmed milk powder, cocoa mass, full cream milk powder, milk fat, emulsifier: soya lecithin), glucose syrup, raisins (11%), oligofructose syrup, grape extract, roasted chopped nuts (3%) (hazelnuts, almonds), humectant: glycerol, sunflower oil, sugar, flavoring, emulsifier: soya lecithin, salt.
Granola bar	Inulin		Granola (whole grain oats, organic cane sugar, canola oil, mixed tocopherols), tapioca syrup, rice crisp (whole grain brown rice flour), organic palm oil, organic semisweet chocolate chips (organic cane sugar, organic chocolate, organic cocoa butter, organic vanilla extract), organic tapioca syrup solids, organic inulin, organic invert cane syrup, rice fiber, organic cocoa, sea salt, sunflower lecithin, natural flavor.

salads as a great source of fiber, like chicory root, and contain a great amount of inulin, ca. 4g of fiber per 100g of serving (Clare et al., 2009).

Jerusalem artichoke

The Jerusalem artichoke (*Helianthus tuberosus* L.) is also called earth apple, and this plant is a sunflower species. It has been used as a feedstock for ethanol production and in the food industry due to its interesting nutritional and functional properties, being used as a supplement for sugar as high-fructose syrup. A fresh Jerusalem artichoke tuber contains 80% water, 15% of sugar, and around 2% of protein. Tubers contain a great amount of inulin instead of starch and sucrose found in other tubers, being also a great source of soluble and insoluble dietary fiber (Brkljača et al., 2014).

Bananas and apples

The fruit has been recognized as an excellent source of prebiotic molecules, bananas being the most popular, since they are rich in vitamins, minerals, and fiber. Unlike the aforementioned examples, bananas contain small amounts of inulin, but when unripe, bananas have a high content of resistant starch that has been recognized with prebiotic effects. On the other hand, apples are an incredible source of pectin, counting to at least 50% of the total fiber content (Licht et al., 2010). Pectin is also recognized with prebiotic benefits, coupled to this the fact it contains a high amount of polyphenols, that combined with pectin improves the digestive health of the human gut.

Bananas and apples are the most-applied fruits to cereal bars, which improve the overall health benefit of the cereal bar since they will have a synergistic effect on the human gut microbiota.

Cocoa beans

Cocoa beans also called cacao is a dried and fully fermented seed, being the basis for the production of chocolate. Cocoa contains various metabolites, such as flavonols (including epicatechin) and procyanidins (Davison and Howe, 2015). The highest levels of cocoa flavanols are found in raw cocoa, which is lost during chocolate production. Several in vitro and in vivo studies underlying cocoa polyphenol interactions with the gut microbiota have been conducted. Flavanols positively influence the growth rate of *Bifidobacterium* and *Lactobacillus* but reduce the number of *Clostridium histolyticum*, as demonstrated by Tzounis et al. (2008), when they incubated fecal bacteria with flavanol. They also noted that flavanol does not have any considerable effect on the growth of *Clostridium coccoides* and *Eubacterium rectale*. Further, over the incubation period, a decrease in flavanol forms was observed, indicating their degradation by fecal microbiota (Tzounis et al., 2008; Sorrenti et al., 2020).

CONCLUSIONS AND FUTURE PERSPECTIVE

In recent years, people's consciousness regarding health and diet has been increasing, making people consume healthier foods, looking for meals with functional properties. Thus, cereal-based products and ready-to-eats have been a suitable alternative to the quick meals and snacks that adults and children have been adopting. The search for higher quality cereal products has also increased due to health and well-being concerns, and the food industry has been developing new products to keep up with the actual trends. As can be seen throughout the chapter, several cereal bars products have been developed to fulfill the market needs of different niches. Products with probiotics or prebiotics have emerged and created an impact on the snacks market. On the other hand, the future will bring revolution and challenge for the human feeding and the cereal bars will not be different on this aspect, since by 2050, it is estimated that worldwide population will have 9.7 billion humans, and that will bring challenges and opportunities to the food industry and market. Cereals bars might be an alternative to deliver not only important drivers of health, such as probiotics and prebiotics, but also, natural bioactive molecules and natural flours coming from new sources of value.

Certainly, studies on the impact of cereal bars improved with functional ingredients on human gut microbiota will keep going as a scientific target of research, which will also bring more knowledge and application to the end products in the market.

ACKNOWLEDGMENTS

The authors would like to thank to "MULTIBIOREFINERY", supported by Programa Operacional Competitividade e Internacionalização e pelo Programa Operacional Regional de Lisboa, na sua componente FEDER, e pela Fundação para a Ciência e Tecnologia and project UID/Multi/50016/2013. And to the project: Co-CerealValue: Valorização sustentável de subprodutos do processamento e armazenamento de cereais (POCI-01-0247-FEDER-033607), supported by Fundo Europeu de Desenvolvimento Regional (FEDER), through the Programa Operacional Competitividade e Internacionalização (POCI).

REFERENCES

Abbeele, P.V., et al., 2018. Different oat ingredients stimulate specific microbial metabolites in the gut microbiome of three human individuals in vitro. ACS Omega 3 (10), 12446–12456.

Angelov, A., et al., 2006. Development of a new oat-based probiotic drink. Int. J. Food Microbiol. 112 (1), 75–80. https://doi.org/10.1016/j.ijfoodmicro.2006.05.015.

Annunziata, A., Vecchio, R., 2011. Functional foods development in the European market: a consumer perspective. J. Funct. Foods 3 (3), 223–228. https://doi.org/10.1016/j.jff.2011.03.011.

Aspri, M., Papademas, P., Tsaltas, D., 2020. Review on non-dairy probiotics and their use in non-dairy based products. Fermentation 6 (1), 30.

Bampi, G.B., et al., 2016. Spray chilling microencapsulation of *Lactobacillus acidophilus* and *Bifidobacterium animalis* subsp. lactis and its use in the preparation of savory probiotic cereal bars. Food Bioprocess Technol. 9 (8), 1422–1428. https://doi.org/10.1007/s11947-016-1724-z.

Bastos, G.A., Paulo, E.M., Chiaradia, A.C.N., 2014. Aceitabilidade de barra de cereais potencialmente probiótica. Braz. J. Food Technol., 113–120. Available from http://www.scielo.br/scielo.php?script=sci_arttext&pid=S1981-67232014000200003&lang=pt.

Bedani, R., et al., 2014. Influence of daily consumption of synbiotic soy-based product supplemented with okara soybean by-product on risk factors for cardiovascular diseases. In: Food Research International. Elsevier Ltd, https://doi.org/10.1016/j.foodres.2014.11.006.

Bhushan, B., et al., 2020. Characterization of riboflavin-producing strains of lactobacillus plantarum as potential probiotic candidate through in vitro assessment and principal component analysis. Probiotics Antimicrob. Proteins, 9–12. https://doi.org/10.1007/s12602-020-09696-x.

Brkljača, J., et al., 2014. Quantification of inulin content in selected accessions of Jerusalem artichoke (*Helianthus tuberosus* L.). Helia 37 (60), 105–112. https://doi.org/10.1515/helia-2014-0009.

Brüssow, H., 2019. Probiotics and prebiotics in clinical tests: an update [version 1; peer review: 2 approved]. F1000Res. 8, 1–9. https://doi.org/10.12688/f1000research.19043.1.

Centurión, M., Machín, L., Ares, G., 2019. Relative impact of nutritional warnings and other label features on cereal bar healthfulness evaluations. J. Nutr. Educ. Behav. 51 (7), 850–856. https://doi.org/10.1016/j.jneb.2019.01.021.

Champagne, C.P., Aachary, A.A., Prapulla, S.G., 2011. Technology and stability of probiotics in probiotic and prebiotic vegetable food. In: Probiotic and Prebiotic Foods: Technology, Stability and Benefits to Human Health, pp. 365–404.

Chen, M., Mustapha, A., 2012. Survival of freeze-dried microcapsules of α-galactosidase producing probiotics in a soy bar matrix. Food Microbiol. 30 (1), 68–73. https://doi.org/10.1016/j.fm.2011.10.017.

Clare, B., Conroy, R., Spelman, K., 2009. The diuretic effect in human subjects of an extract of *Taraxacum officinale* folium over a single day. J. Altern. Complement. Med. 15 (8), 929–934. https://doi.org/10.1089/acm.2008.0152.

Costabile, A., et al., 2016. Prebiotic potential of a maize-based soluble fibre and impact of dose on the human gut microbiota. PLoS ONE 11 (1), 1–12. https://doi.org/10.1371/journal.pone.0144457.

Costabile, A., Klinder, A., Fava, F., Napolitano, A., Fogliano, V., Leonard, C., Gibson, G.R., Tuohy, K.M., 2008. Whole-grain wheat breakfast cereal has a prebiotic effect on the human gut microbiota: a double-blind, placebo-controlled, crossover study. Br. J. Nutr. 99 (1), 110–120. https://doi.org/10.1017/S0007114507793923.

Czerucka, D., Piche, T., Rampal, P., 2007. Review article: yeast as probiotics—*Saccharomyces boulardii*. Aliment. Pharmacol. Ther. 26 (6), 767–778. https://doi.org/10.1111/j.1365-2036.2007.03442.x.

da Silva, T.F., Conti-Silva, A.C., 2018. Potentiality of gluten-free chocolate cookies with added inulin/oligofructose: chemical, physical and sensory characterization. LWT Food Sci. Technol. 90, 172–179. https://doi.org/10.1016/j.lwt.2017.12.031.

da Silva, E.P., et al., 2016. Physicochemical and sensory characteristics of snack bars added of jerivá flour (*Syagrus romanzoffiana*). Food Sci. Technol. 36 (3), 421–425. https://doi.org/10.1590/1678-457X.08115.

Davison, K., Howe, P., 2015. Potential implications of dose and diet for the effects of cocoa flavanols on cardiometabolic function. J. Agric. Food Chem. 63 (45), 9942–9947.

Day, L., 2013. Proteins from land plants—potential resources for human nutrition and food security. Trends Food Sci. Technol. 32 (1), 25–42. https://doi.org/10.1016/j.tifs.2013.05.005.

de Moreno de LeBlanc, A., Matar, C., Perdigon, G., 2007. The application of probiotics in cancer. Br. J. Nutr. 98 (Suppl 1), S105–S110.

Dey, G., 2018. Non-dairy probiotic foods: innovations and market trends. In: Innovations in Technologies for Fermented Food and Beverage Industries, pp. 159–173, https://doi.org/10.1007/978-3-319-74820-7_9.

Donkor, O.N., et al., 2007. Survival and activity of selected probiotic organisms in set-type yoghurt during cold storage. Int. Dairy J. 17, 657–665. https://doi.org/10.1016/j.idairyj.2006.08.006.

dos Baliza, D.D.M.S., Da Silva, J.F.M., Pimenta, R.S., 2018. Evaluation of the applicability of a probiotic strain of *Saccharomyces cerevisiae* in cereal bars. Braz. J. Food Technol. 21. https://doi.org/10.1590/1981-6723.14817.

D'Souza, A.L., Rajkumar, C., Cooke, J., Bulpitt, C.J., 2002. Probiotics in prevention of antibiotic associated diarrhoea: meta-analysis. BMJ 324 (7350), 1361.

Dutcosky, S.D., et al., 2006. Combined sensory optimization of a prebiotic cereal product using multicomponent mixture experiments. Food Chem. 98 (4), 630–638. https://doi.org/10.1016/j.foodchem.2005.06.029.

Eckburg, P.B., et al., 2005. Diversity of the human intestinal microbial flora. Science 308 (5728), 1635–1638.

FAOSTAT, 2019. Crops/World Regions/Production Quantity Pick Lists. Food and Agriculture Organization, Statistics Division. Retrieved 14 August 2020.

Felis, G.E., Dellaglio, F., 2007. Taxonomy of lactobacilli and bifidobacteria. Curr. Issues Intest. Microbiol. 8 (2), 44–61.

Garcia, A., et al., 2018. Physical-sensory evaluation of a cereal bar with quinoa: a preliminary study. J. Biomed. Biopharm. Res. 15 (1), 25–36. https://doi.org/10.19277/bbr.15.1.171.

Gibson, G.R., Roberfroid, M.B., 1995. Dietary modulation of the human colonic microbiota: introducing the concept of prebiotics. J. Nutr. 125 (6), 1401–1412. https://doi.org/10.1093/jn/125.6.1401.

Gibson, G.R., et al., 2017. Expert consensus document: the International Scientific Association for Probiotics and Prebiotics (ISAPP) consensus statement on the definition and scope of prebiotics. Nat. Rev. Gastroenterol. Hepatol. 14 (8), 491–502. https://doi.org/10.1038/nrgastro.2017.75.

Goodfellow, M., et al., 2012. Bergey's Manual of Systematic Bacteriology, second ed. Springer, New York, NY, USA.

Henriques, S., 2011. Incorporation of Probiotics in Cereal Bars. Universidade Católica Portuguesa.

Jefferson, A., Adolphus, K., 2019. The effects of intact cereal grain fibers, including wheat bran on the gut microbiota composition of healthy adults: a systematic review. Front. Nutr. 6 (33). https://doi.org/10.3389/fnut.2019.00033.

Kemsawasd, V., Chaikham, P., Rattanasena, P., 2016. Survival of immobilized probiotics in chocolate during storage and with an in vitro gastrointestinal model. Food Biosci. 16, 37–43.

Kerry, R.G., et al., 2018. Benefaction of probiotics for human health: a review. J. Food Drug Anal., 1–13. https://doi.org/10.1016/j.jfda.2018.01.002.

Khalesi, S., et al., 2014. Effect of probiotics on blood pressure: a systematic review and meta-analysis of randomized, controlled trials. Hypertension 64 (4), 897–903. https://doi.org/10.1161/HYPERTENSIONAHA.114.03469.

Konings, E., et al., 2014. Effect of polydextrose and soluble maize fibre on energy metabolism, metabolic profile and appetite control in overweight men and women. Br. J. Nutr. 111 (1), 111–121. https://doi.org/10.1017/S0007114513002183.

Korczak, R., Kocher, M., Swanson, K.S., 2020. Effects of oats on gastrointestinal health as assessed by in vitro, animal and human studies. Nutr. Rev. 78 (5), 343–363.

Kumar, B.V., Vijayendra, S.V.N., Reddy, O.V.S., 2015. Trends in dairy and non-dairy probiotic products—a review. J. Food Sci. Technol. 52 (10), 6112–6124. https://doi.org/10.1007/s13197-015-1795-2.

Lalicic-Petronijevic, J., et al., 2015. Viability of probiotic strains *Lactobacillus acidophilus* NCFM® and *Bifidobacterium lactis* HN019 and their impact on sensory and rheological properties of milk and dark chocolates during storage for 180 days. J. Funct. Foods 15, 541–550. https://doi.org/10.1016/j.jff.2015.03.046.

Lamsal, B.P., Faubion, J.M., 2009. The beneficial use of cereal and cereal components in probiotic foods. Food Rev. Int. 25 (2), 103–114. https://doi.org/10.1080/87559120802682573.

Licht, T.R., et al., 2010. Effects of apple and specific apple components on the cecal environment of conventional rats: role of apple pectin. BMC Microbiol. 10, 13.

Liu, C.T., et al., 2011. Antiproliferative and anticytotoxic effects of cell fractions and exopolysaccharides from Lactobacillus casei 01. Mutat. Res. Genet. Toxicol. Environ. Mutagen. 721 (2), 157–162. https://doi.org/10.1016/j.mrgentox.2011.01.005.

Markets and Markets, 2020. www.marketsandmarkets.com. (Accessed 6 July 2020).

Martinez, R.C.R., Bedani, R., Saad, S.M.I., 2015. Scientific evidence for health effects attributed to the consumption of probiotics and prebiotics: an update for current perspectives and future challenges. Br. J. Nutr. 114, 1993–2015. https://doi.org/10.1017/S0007114515003864.

Martins, E.M.F., et al., 2013. Products of vegetable origin: a new alternative for the consumption of probiotic bacteria. Food Res. Int. 51 (2), 764–770. https://doi.org/10.1016/j.foodres.2013.01.047.

McSweeney, P.L.H., Day, L., 2016. Food products and ingredients. In: Reference Module in Food Science. Elsevier, https://doi.org/10.1016/b978-0-08-100596-5.03345-x.

Moayyedi, P., Ford, A.C., Talley, N.J., Cremonini, F., Foxx-Orenstein, A.E., Brandt, L.J., Quigley, E.M.M., 2010. The efficacy of probiotics in the treatment of irritable bowel syndrome: a systematic review. Gut 59 (3), 325–332.

Mondragón-bernal, O.L., et al., 2017. Stability and functionality of synbiotic soy food during shelf-life. J. Funct. Foods 35, 134–145. https://doi.org/10.1016/j.jff.2017.05.021.

Morais, E.C., et al., 2014. Prebiotic gluten-free bread: sensory profiling and drivers of liking. LWT Food Sci. Technol. 55 (1), 248–254. https://doi.org/10.1016/j.lwt.2013.07.014.

Moro, T.M.A., et al., 2018. Use of burdock root flour as a prebiotic ingredient in cookies. LWT Food Sci. Technol. 90, 540–546. https://doi.org/10.1016/j.lwt.2017.12.059.

Mueller, S., et al., 2006. Differences in fecal microbiota in different European study populations in relation to age, gender, and country: a cross-sectional study. Appl. Environ. Microbiol. 72 (2), 1027–1033.

Nazzaro, F., et al., 2008. Synbiotic potential of carrot juice supplemented with *Lactobacillus* spp. and inulin or fructooligosaccharides. J. Sci. Food Agric. 88, 2271–2276. https://doi.org/10.1002/jsfa.3343.

Ogunremi, O.R., Banwo, K., Sanni, A.I., 2017. Starter-culture to improve the quality of cereal-based fermented foods: trends in selection and application. Curr. Opin. Food Sci. 13, 38–43. https://doi.org/10.1016/j.cofs.2017.02.003.

Ouwehand, A.C., Kurvinen, T., Rissanen, P., 2004. Use of a probiotic Bifidobacterium in a dry food matrix, an in vivo study. Int. J. Food Microbiol. 95 (1), 103–106. https://doi.org/10.1016/j.ijfoodmicro.2004.03.015.

Panghal, A., et al., 2018. Potential non-dairy probiotic products—a healthy approach. Food Biosci. 21, 80–89. https://doi.org/10.1016/j.fbio.2017.12.003.

Parvez, S., et al., 2006. Probiotics and their fermented food products are beneficial for health. J. Appl. Microbiol. 100 (6), 1171–1185. https://doi.org/10.1111/j.1365-2672.2006.02963.x.

Pathare, P.B., Baş, N., Fitzpatrick, J.J., Cronin, K., Byrne, E.P., 2012. Production of granola breakfast cereal by fluidised bed granulation. Food Bioprod. Process. 90, 549–554.

Pereira, J., et al., 2018. Impact of whey protein coating incorporated with Bifidobacterium and Lactobacillus on sliced ham properties. Meat Sci. 139, 125–133. https://doi.org/10.1016/j.meatsci.2018.01.016.

Pereira, J.O., et al., 2019. Cereal bars functionalized through: bifidobacterium animalis subsp. Lactis BB-12 and inulin incorporated in edible coatings of whey protein isolate or alginate. Food Funct. 10 (10), 6892–6902. https://doi.org/10.1039/c9fo00370c.

Pham, T., et al., 2017. In vitro fermentation patterns of rice bran components by human gut microbiota. Nutrients 9 (11), 1–15. https://doi.org/10.3390/nu9111237.

Qi, D., Nie, X.L., Zhang, J.J., 2020. The effect of probiotics supplementation on blood pressure: a systemic review and meta-analysis. Lipids Health Dis. 19 (1), 1–11. https://doi.org/10.1186/s12944-020-01259-x.

Raymond, Y., Champagne, C.P., 2015. The use of flow cytometry to accurately ascertain total and viable counts of *Lactobacillus rhamnosus* in chocolate. Food Microbiol. 46, 176–183. https://doi.org/10.1016/j.fm.2014.07.002.

Rivera-Espinoza, Y., Gallardo-Navarro, Y., 2010. Non-dairy probiotic products. Food Microbiol. 27 (1), 1–11. https://doi.org/10.1016/j.fm.2008.06.008.

Roberfroid, M.B., 1999. 'Nutritional and health benefits of inulin and oligofructose concepts in functional foods: the case of inulin and oligofructose. J. Nutr. 129, 1398–1401.

Roberfroid, M., Gibson, G.R., Hoyles, L., McCartney, A.L., Rastall, R., Rowland, I., et al., 2010. Prebiotic effects: metabolic and health benefits. Br. J. Nutr. 104 (S2), S1–S63. https://doi.org/10.1017/s0007114510003363.

Rodrigues, C.S., 2013. Desenvolvimento de barras de cereais com ingredientes prebióticos e probiótico. Universidade Estadual de Campinas. Available from: http://repositorio.unicamp.br/jspui/bitstream/REPOSIP/256000/1/Rodrigues_CinthiaSoares_M.pdf.

Saarela, M., et al., 2006. Fibres as carriers for *Lactobacillus rhamnosus* during freeze-drying and storage in apple juice and chocolate-coated breakfast cereals. Int. J. Food Microbiol. 112 (2), 171–178. https://doi.org/10.1016/j.ijfoodmicro.2006.05.019.

Salmerón, I., Thomas, K., Pandiella, S.S., 2015. Effect of potentially probiotic lactic acid bacteria on the physicochemical composition and acceptance of fermented cereal beverages. J. Funct. Foods 15, 106–115. https://doi.org/10.1016/j.jff.2015.03.012.

Salminen, S., Bouley, C., Boutron-Ruault, M.C., Cummings, J.H., Franck, A., Gibson, G.R., Isolauri, E., Moreau, M.C., Roberfroid, M., Rowland, I., 1998. Functional food science and gastrointestinal physiology and function. Br. J. Nutr. 80 (Suppl 1), S147–S171.

Santosa, S., Farnworth, E., Jones, P.J.H., 2006. Probiotics and their potential health claims. Nutr. Rev. 64 (6), 265–274. https://doi.org/10.1301/nr.2006.jun.265-274.

Singu, B.D., Bhushette, P.R., Annapure, U.S., 2020. Thermotolerant *Saccharomyces cerevisiae* var. boulardii coated cornflakes as a potential probiotic vehicle. Food Biosci. 36, 100668. https://doi.org/10.1016/j.fbio.2020.100668.

Sorrenti, V., et al., 2020. Cocoa polyphenols and gut microbiota interplay: bioavailability, prebiotic effect and impact on human health. Nutrients 12, 1908. https://doi.org/10.3390/nu12071908.

Srebernich, S.M., et al., 2016. Physico-chemical, sensory and nutritional characteristics of cereal bars with addition of acacia gum, inulin and sorbitol. Food Sci. Technol. 36 (3), 555–562. https://doi.org/10.1590/1678-457X.05416.

Sui, L., et al., 2020. In vitro assessment of probiotic and functional properties of *Bacillus coagulans* T242. Food Biosci. 36 (1), 100675. https://doi.org/10.1016/j.fbio.2020.100675.

Thushara, R.M., et al., 2016. Cardiovascular benefits of probiotics: a review of experimental and clinical studies. Food Funct. 7 (2), 632–642. https://doi.org/10.1039/c5fo01190f.

Tomasik, P., Tomasik, P., 2020. Probiotics, non-dairy prebiotics and postbiotics in nutrition. Appl. Sci. 10 (4), 1470.

Tripathi, M.K., Giri, S.K., 2014. Probiotic functional foods : survival of probiotics during processing and storage. J. Funct. Foods 9, 225–241. https://doi.org/10.1016/j.jff.2014.04.030.

Turroni, F., et al., 2019. Bifidobacterium bifidum: a key member of the early human gut microbiota. Microorganisms 7 (11), 544. https://doi.org/10.3390/microorganisms7110544.

Tzounis, X., et al., 2008. Flavanol monomer-induced changes to the human faecal microflora. Br. J. Nutr. 99, 782–792.

Vinderola, G., Burns, P., Reinheimer, J., 2017. Probiotics in non-dairy products. In: Vegetarian and Plant-Based Diets in Health and Disease Prevention. Academic Press, pp. 809–835.

Wang, L., et al., 2018. The effects of probiotics on total cholesterol. Medicine (United States) 97 (5). https://doi.org/10.1097/MD.0000000000009679.

Zendeboodi, F., et al., 2020. ScienceDirect Probiotic: conceptualization from a new approach. Curr. Opin. Food Sci. 32, 103–123. https://doi.org/10.1016/j.cofs.2020.03.009.

Zhang, Q., et al., 2018. Changes in antioxidant activity of Alcalase-hydrolyzed soybean hydrolysate under simulated gastrointestinal digestion and transepithelial transport. J. Funct. Foods 42, 298–305. https://doi.org/10.1016/j.jff.2018.01.017.

CHAPTER 12

Fruit Juice Added With Prebiotics and Probiotics

S. CHAKKARAVARTHI • S. MITHUL ARAVIND
Department of Basic and Applied Science, National Institute of Food Technology, Entrepreneurship and Management, Sonipat, Haryana, India

INTRODUCTION

Fruits and vegetables are the most important natural food sources with various nutrients present. Fruits generally contain a large amount of sugar, dietary fiber, and many other bioactive compounds. According to various research findings, fruit and vegetable sources can be categorized as functional foods (Patel, 2017; Chakkaravarthi et al., 2005; Singla and Chakkaravarthi, 2017). Due to increased health consciousness, consumers give high preference to fruit-based products. Although fruits are processed into various forms, one of the most accepted products is fruit juice. Also, fruit and vegetable juices serve effectively as a media for probiotic growth (Mattila-Sandholm et al., 2002). The consumption of functional fruit juices modulates the action of complex gut microbiota system. The gut floral population, popularly termed probiotics, includes various microbial species, especially bacteria, that have the large intestinal tract as their primary habitat. The majority of the gut population includes species of *Bifidobacterium, Lactobacillus, Bacteroides, Ruminococcus,* and *Faecalibacterium*. These microbes utilize fruit juice ingredients such as organic acids and dietary fibers, which promote the viability of probiotics (Horackova et al., 2018).

Recently, Zendeboodi et al. (2020) redefined probiotics as viable or inviable (vegetative or spore; intact or ruptured) microbial cells that significantly confirm the host health benefits. According to the International Scientific Association for Probiotics and Prebiotics (ISAPP), prebiotics are defined as a substrate that is selectively used by the host microorganisms, conferring a health benefit. This definition was modified to consider the noncarbohydrates other than food sources that do not function solely in the gastrointestinal tract (Gibson et al., 2017). The most prominent active prebiotic compounds in functional foods include dietary fibers such as oligosaccharides (fructans and glucans, etc.) and polyphenols such as catechin, epigallocatechin, quercetin, resveratrol, etc. (Arena et al., 2014; Gómez et al., 2014; Javier et al., 2018; Mithul Aravind et al., 2021). However, prebiotics and probiotics are interrelated, which almost maintains the health of an individual. The resulting change in the consumed host can either be positive or negative. This statement is explained by the gut microbial action on functional foods. Functional foods such as pomegranate extract contain a fermentable compositional matrix such as punicalagin and ellagic acid that is responsible for stimulating the growth of gut microbiota; these are either beneficial (several species of lactobacilli and bifidobacteria) or pathogenic (*E. coli* O157:H7) (Bialonska et al., 2009). All these findings support the consumption of natural fruit juices. However, in the fast forwarded world, the consumption of fresh fruit juice is quite limited. Addressing this along with the active health consciousness of consumers, food/beverage technologists are producing more healthy probiotic and synbiotic fruit juices. Supporting these findings and observations, this chapter mainly focuses on the addition, activity, and effect of prebiotics and probiotics added to fruit juice.

NATURAL FRUIT JUICES AND THEIR PROBIOTIC EFFICIENCY

Fresh fruit juices are prepared by squeezing and/or crushing the fruits. The resulting fruit juice has liquid and some fleshy fruit parts such as pomace or the fruit sac. Natural fruit juice from fresh fruits is enriched with an abundance of nutrients. These include vitamins, minerals, sugars, dietary fibers such as oligosaccharides, polysaccharides, amino acids, and bioactive compounds such as lycopene, lutein, stilbenes, phenolic acids, aromatic, volatile compounds, and other phytochemicals. Thus, they are healthy when consumed. Though the fruit matrix has various nutrients,

Probiotics and Prebiotics in Foods. https://doi.org/10.1016/B978-0-12-819662-5.00013-6

they are not completely extracted in the juice. Also, the nutrient pool from the juice may not be completely absorbed instantly due to the presence of fibers and certain secondary metabolites. The latter compounds are metabolized and absorbed in the gut with the help of gut microbes. However, there are some compounds such as antinutrients (tannins) and certain pathogenic microbes present naturally in the fruit matrix that interrupt and restrict the beneficial effects of natural and fresh fruit juices. According to the US Food and Drug Administration (FDA), the microbial limit of fruit juices lies around 25 CFU/mL. In natural fruit juice, the probiotic bacterial population includes *Lactobacillus* and acetic acid bacteria along with pathogens such as *E. coli*, the *Clostridium* species, and *Salmonella*. The presence of these pathogenic bacteria species degrades the quality of juice by spoilage, causing disease to the consumers (Ahmed et al., 2019). Thus, the probiotic efficiency of natural fruit and vegetable juices is less significant and effective when compared with probiotic-fortified fruit and vegetable juices.

EFFECT OF FRUIT JUICES ON GUT MICROBIOTA

When food is ingested, most of the nutrients get absorbed and metabolized in the digestive system before reaching the gut. Certain nutritive compounds cannot be digested and are ultimately utilized by the complex gut microbiota. The undigested fruit nutrients reach the gut, then interact and modulate the gut microbiota. Fruits act as an important resource for many nutrients and bioactive compounds. When fresh fruit juices are compared with processed fruit juices, the availability of prebiotics and probiotics is limited. This can be explained by stating that, the quantity of pulp in fresh juice is much high when compared with processed juices. For example, an in vivo study was done using male Wistar rats that were fed with nonprocessed pure apple juices (with pulp). Fecal sample analysis was done using the plate count method (specific agar was used). The results showed that apple juice enhanced the growth of *Lactobacillus* and *Bifidobacterium* spp. in rat feces (Sembries et al., 2006). A similar outcome was achieved with orange juice, which has several nutrients and other bioactive compounds such as ascorbic acid, carotenoids, and flavonoids that influence the gut microbial population.

Duque et al. (2016) reported that orange juice influences the gut microbial population. They proved that orange juice possesses selective prebiotic effects in the host. In this study, the experiment compared the viability of the gut microbiota of fresh and pasteurized orange juice using the Simulator of the Human Intestinal Microbial Ecosystem (SHIME) model. These results showed that fresh orange juice increased *Lactobacillus* spp., *Enterococcus* spp., *Bifidobacterium* spp., and *Clostridium* spp. and reduced the species of *Enterobacteria* in the gut. However, pasteurized orange juice increased the counts of *Lactobacillus* spp. and reduced *Enterobacteria*. Both fresh and pasteurized orange juice increased the concentration of short chain fatty acids such as butyric, acetic, and propionic acid because the production of ammonium due to gut bacterial fermentation was significantly reduced. Thus, it was suggested that the increase in beneficial gut microbiota was due to the available bioactive compounds such as polyphenols that interact and control the gut bacterial growth (Pereira-Caro et al., 2015).

PREPARATION OF PROCESSED PROBIOTIC FRUIT JUICES

In order to avoid any spoilage or deterioration, juice needs to be processed. The existing pathogens and their spores are destroyed through different processing techniques, thereby extending the shelf life of natural juices with minimal or no nutrient loss. The other aspect of processing is to increase the health benefits by fortification. Fermented and fortified apple juice (pH 3.4 to 3.6) with protocatechuic acid and ethanol extract from roselle calyx inhibited the growth of pathogenic microbes such as *S. aureus*, *L. monocytogenes*, *B. cereus*, and *E. coli* O157:H7 during 3 days of storage at 25 °C (Chao and Yin, 2009).

The current trend in fruit juice processing focuses more on the fortification of health-supporting microbes that are native to the gut, that is, probiotic fortification. One of the most important and significant ways to boost probiotic-fortified fruit juices is to incorporate them with prebiotics. The most popular probiotics added to fruit juices include species of *Lactobacillus* such as *L. acidophilus*, *L. casei*, *L. plantarum*, *L. paracasei*, *L. reuteri*, *L. fermentum*, etc.; species of *Bifidobacterium* such as *B. bifidum*, *B. lactum*, *B. infantis*, *B. lactis*; *Weissella* spp.; *Pediococcus* spp.; *and Saccharomyces cerevisiae* (Patel, 2017; Tripathi and Giri, 2014). Among these bacteria, *L. casei* is one of the most important probiotics because these bacteria were found in vitro to increase the length of intestinal villi cells and restrict the growth of *Trichinella spiralis* and *Plasmodium chabaudi* in mice (Bautista-Garfias et al., 1999; Martínez-Gómez et al., 2006). The preparation of processed probiotic fruit juice is shown in Fig. 1.

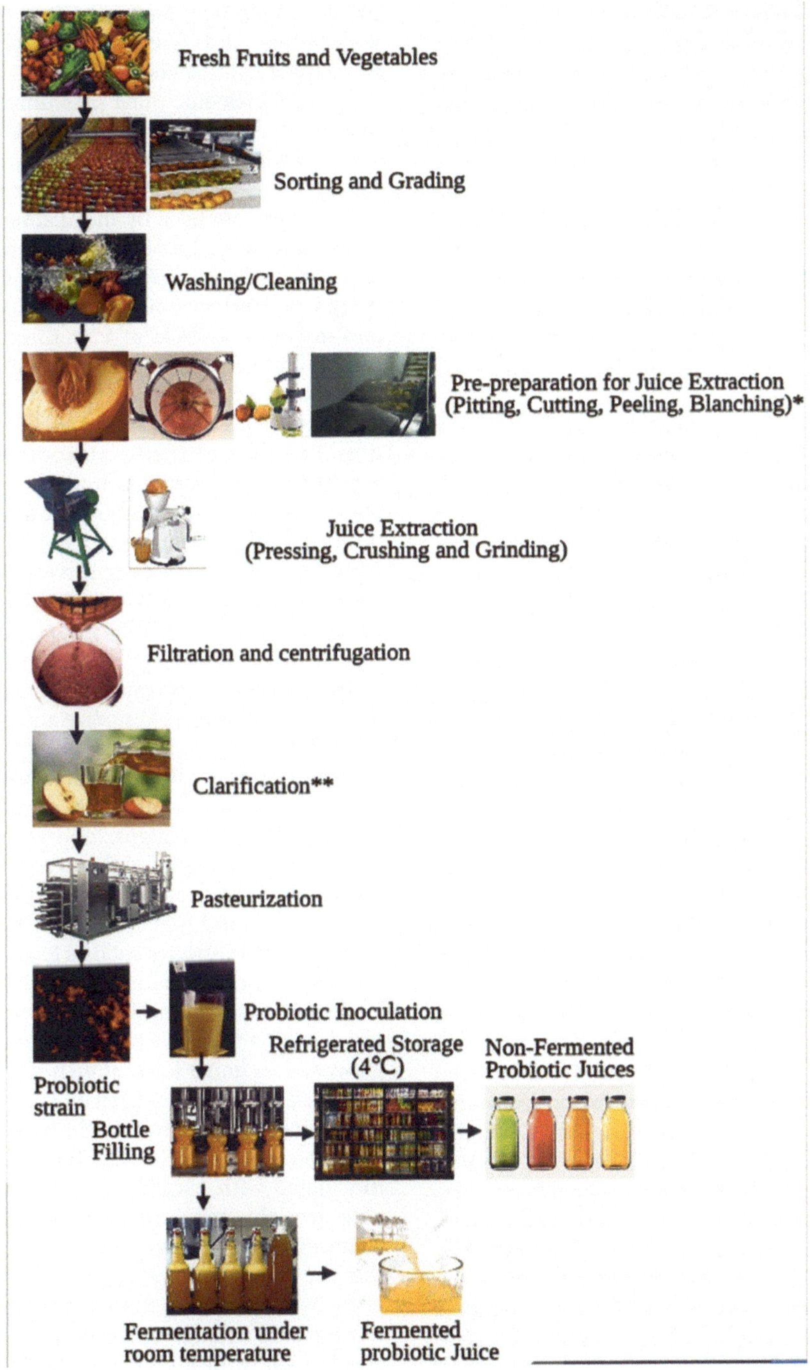

FIG. 1 Preparation of processed probiotic fruits and vegetables juice. * Generally, hard fruits and vegetables such as pineapples, apples, and carrots were cut into pieces for maximum juice extraction. Pitting is done for fruits such as mangos and peaches to remove the seed. Almost all fruit juices were prepared after peeling. Blanching is done for many vegetables to inactivate enzymes that cause spoilage. Vegetables such as cabbage are blanched to destroy some antibacterial compounds. ** Fruits such as a cashew apple was clarified by making use of gelatin to precipitate tannin.

ADDITION OF PROBIOTICS IN FRUIT AND VEGETABLE JUICES

Globally, several dairy-based commercial probiotic beverages are available. It is also important to produce probiotic fortified fruit juice, as it is an alternative efficient carrier in comparison to dairy-fortified probiotic beverages. This is because unlike dairy probiotic beverages, fruit probiotic juices don't have starter cultures such as *Streptococcus thermophilus* to compete with added probiotics for nutrients (Ranadheera et al., 2014). Moreover, the probiotic efficiency of fortified/added probiotic strains was found to be more viable and active than the native strains in the fruits and vegetables (Garcia et al., 2018).

Nagpal et al. (2012) explained the fortification and fermentation of fruit juices with *L. plantarum* and *L. acidophilus*. These strains, when inoculated in fruit juices, have the tendency to survive and utilize the nutrients in the fruit juices. Yoon et al. (2004) tested tomato juice for probiotic addition with *L. casei* A4, *L. delbrueckii* D7, *L. plantarum* C3, and *L. acidophilus* LA 39. The results showed that *L. casei* and *L. delbrueckii* utilized the tomato juice for their growth. After 72 h of fermentation, the pH decreased from 4.1 to 3.5. *L. plantarum* majorly contributed to sugar consumption and pH drop. After 4 weeks of storage at 4 °C, *L. acidophilus* and *L. delbrueckii* were not affected while *L. casei* and *L. plantarum* decreased slightly. However, when skim milk was added to the tomato juice, *L. acidophilus* grew well in a shorter period of time. This may be due to the addition of skim milk with tomato juice (Babu et al., 1992).

The optimization of cantaloupe juice fermentation by probiotic *L. casei* NRRL B-442 was done by Fonteles et al. (2012). After about 8 h of fermentation, the highest viability (8.98 log CFU/mL) was observed at pH 6.1. The viable cell count was decreased after storage of 42 days. Luckow and Delahunty (2004) found that there were more differences between the conventional and probiotic blackcurrant juice varieties. Generally, conventional blackcurrant juices possess a natural characteristic color, flavor, and other optical properties. But the probiotic juice possesses a perfumery and dairy odor as well as a sour and savory flavor due to the formation of metabolites such as lactic acid as well as decreased pH and increased turbidity. These changes were formed by the inoculation of probiotics such as *L. plantarum 299v*. The overall preference seems to be almost the same. They also found that age and gender showed a significant difference in the sensory evaluation of the probiotic blackcurrant juice. Females older than 40 highly preferred the probiotic blackcurrant juice. This could be because of their lower sensitivity as well as the fact that older consumers were highly exposed to a wide variety of different aromas and flavors compared to younger consumers.

Pomegranate juice after pasteurization and 10% inoculation of *L. plantarum* and *L. bulgaricus* was found to survive well with 10^9 CFU/mL, as it is enriched with nutrients (Thakur and Joshi, 2017). Mousavi et al. (2011) researched on pomegranate juice by inoculating four strains of lactic acid bacteria—*Lactobacillus plantarum, L. delbrueckii, L. paracasei,* and *L. acidophilus*—to find the viability and stability, when they were allowed to ferment for 72 h at 30 °C and stored for 4 weeks at 4 °C. Citric acid, the major organic compound in the pomegranate, acted as a good carbon source for the fermenting bacteria. *L. plantarum* and *L. delbrueckii* possessed higher viability during the storage time. Also, these two bacteria accounted for the major citric acid and sugar consumption and consecutively increased the pH in the stored pomegranate juices. It was likely recommended by researchers that pomegranate juice was highly suitable for probiotic survival, even during storage. However, *L. paracasei* was the most affected species due to a change in pH and its viability was highly decreased; *L. acidophilus* accounted for very little sugar consumption. Thus, *L. paracasei* and *L. acidophilus* exhibited very poor stability and viability during cold storage and are generally not recommended for high-acid fruit juices.

The probiotic efficiency of three different juices—apple, orange, and grape—was compared in two different experimental designs, one with supplementation of probiotics directly with juices while the other had fermentation of the fruit juice with that probiotic species. It was found that apple juice fermented with various lactobacilli species such as *L. acidophilus* strains L10 and 1674, *L. casei* strains L26 and 667, *L. paracasei* 672, *L. plantarum* 299v, and *L. rhamnosus* LGG exhibited more counts than in supplemented juice. *L. casei* L26 grew well in grape juice at 48 h of fermentation. Also, strains of *L. rhamnosus* such as LGG and 607 grew well and reached a maximum limit in all juice varieties when fermented for 48 h. The fermented apple juice inoculated with *Lactobacilli* was found to be more viable than the pomegranate juice fermented by *L. planatarum, L. acidophilus,* and *L. paracasei*. It was found that the viability of the lactobacilli strains was more in fermented fruit juices than in supplemented ones, even during cold storage (Espirito-Santo et al., 2015). In another study, probiotic apple juice was also prepared by fermenting it with *L. plantarum* and *S. thermophilus* for 72 h; the effects were studied for 4 weeks at 4 °C. Although apple is the best-suited medium with abundant nutrients for microbial growth, *L. plantarum* survived better than *S. thermophilus*. A sensory evaluation conducted by the authors showed that the fermentation was not influenced much in their characteristics (Thakur and Joshi, 2017). The tendency of probiotic strains such as *L. rhamnosus* 6224 and *Enterococcus faecium* 21,605 was to adsorb toxins such as patulin in apple juice without influencing the functional quality of the juice. The peptidoglycan and polysaccharides of the bacterial cell wall composition were the major reason for this activity (Hatab et al., 2012).

Pakbin et al. (2014) produced probiotic peach juice by inoculating three lactobacilli strains: *L. casei*, *L. delbrueckii*, and *L. plantarum*. This was stored at 4°C for 4 weeks after 72h of fermentation. It was found that *L. delbrueckii*, which was optimum to produce probiotic peach juice, utilized the maximum amount of sugar in the peach juice. The consumed sugar was fermented to lactic acid, thereby the pH was reduced. The viability of *L. casei* was below the deduction limit as it could not survive at low pH. However, *L. delbrueckii* and *L. plantarum* grew well from the storage time of the third week. The initial microbial population after fermentation influenced the viability of probiotics after storage. This was clearly understood because the *L. plantarum* count was very low compared to *L. delbrueckii* after fermentation. Thus, they couldn't survive after 4 weeks of storage at 4°C.

Pediococcus pentosaceus LBF2 was inoculated in soursop juice and stored for 3 weeks at 4°C. The pH and TSS of the probiotic-fortified juice decreased throughout storage. Maximum viability was observed as 1.85×10^7 CFU/mL (Akpeji et al., 2017). Prickly pear juice was allowed to ferment with *L. fermentum*-ATCC 9338. The acidity increased throughout storage. Tartaric acid showed a major increase to 1.7g/100mL from 0.5g/100mL with a significant decrease in ascorbic acid and sugar from 3.95 to 1.7g/100mL. The prebiotic rich juices found to support strongly for the growth of probiotics (Panda et al., 2017).

Yoon et al. (2006) formulated probiotic cabbage juice. They inoculated *L. casei*, *L. plantarum*, and *L. delbrueckii* in cabbage juice. They found that all three species grew well in the juice after 48h. The principal probiotics responsible for an increase in acidity were found to be *L. plantarum* and *L. delbrueckii*. It was observed that *L. casei* cannot survive the low pH and high acidity, even under low-temperature storage.

Rafiq et al. (2016) developed probiotic carrot juice by using *L. acidophilus*, *L. plantarum*, *L. casei*, and *B. bifidum* subsp. *longum*. They found that the optimum pH and temperature for the production of probiotic carrot juice was 6°C and 30°C, respectively. There was a greater increase in protein content and a greater decrease in ash, fat, and carbohydrate content in probiotic carrot juice over fresh carrot juice. However, the overall energy content was higher in fresh carrot juice over probiotic juice. It was also found that the health benefits of probiotic carrot juice were more than fresh juice. When a sensory test was conducted, the overall acceptance was higher for fresh juice than probiotic juice. This might be due to the acidity and aroma developed during fermentation.

Yang et al. (2020) researched the probiotic efficiency of vegetable juice using a mixture of 10% carrot juice, 60% purple cabbage juice, and 20% tomato juice incubated with *L. plantarum*. The authors blanched the cabbage to eliminate the antibacterial compounds. They also found that the fermentation time required for the juice mixture was less than 48h. Also, the maximum count reached 9.13 ± 0.19 log CFU/mL after 16h of fermentation and the volatile compounds increased to 55.

The production of fermented probiotic strain (*L. casei* ATCC 393)-inoculated potato juice with different potato varieties (Haryoung, Hongyoung, Jayoung, and Superior) was studied by Kim et al. (2012). The fermented juices were stored for 4 weeks at 4°C. Due to the increased production of lactic acid during fermentation, the pH reduced to 4.63 in the Haryoung variety juice. The strain was more viable in the Haryoung variety, even after treatment with gastric juice and bile salt; the strain did not survive well in the Hongyoung variety. Thus, the authors concluded that the Haryoung variety was good for probiotic drink production.

Under extremely low temperature and pH, the probiotic bacteria were highly sensitive for growth in any media. It is mandatory to choose the specific probiotic strains before inoculation, as cold storage resists many of the viable cells to be dormant (Heenan et al., 2002).

ADDITION OF PREBIOTICS AND PROBIOTICS (SYNBIOTIC) IN FRUIT AND VEGETABLE JUICES

The other method for the optimal survival of incorporated probiotics in fruit and vegetable juices was making them adhere to prebiotic fibers. The fruit juices fortified with probiotics and prebiotics as well as the viability of probiotics at different storage durations are summarized in Table 1. Reports from Saarela et al. (2006) found that oat flour containing 20% β-glucan was able to protect probiotics such as *L. rhamnosus* when inoculated and stored in apple juice.

Nguyen et al. (2019) explored the fermentation of pineapples with probiotic strains of lactobacilli (*L. acidophilus* and *L. plantarum*) and *Bifidobacterium* (*B. lactis*) with and without the addition of the prebiotic raftiline, a fructo-oligosaccharide. Although pineapple juice serves as an excellent growth medium for both

TABLE 1
Probiotic and Prebiotic Fortified Fruit, Vegetable Juices and Viability of Probiotics at Different Storage Durations.

Sl. No	Name of the Fruit Juice	Fortified Prebiotics	Name of the Probiotic Inoculated	Maximum Viability of Cells	Storage Duration	Reference
1.	Apple juice	–	*L. acidophilus*L10 *L. casei* L26 *L. rhamnosus* LGG *L. paracasei* L33 *L. plantarum* 299v *L. acidophilus* (CIRMBIA 1674) *L. casei* (CIRMBIA 667) *L. paracasei* (CIRMBIA 672) *L. plantarum* (CIRMBIA 466) *L. rhamnosus* (CIRMBIA 607)	2.01 ± 0.05 2.20 ± 0.01 0.64 ± 0.01 2.94 ± 0.10 1.66 ± 0.09 2.36 ± 0.02 2.77 ± 0.01 3.76 ± 0.03 3.55 ± 0.05 4.01 ± 0.06 log CFU/mL	28 days at 5°C	Espirito-Santo et al. (2015)
2.	Apple juice	–	*L. plantarum* (MTCC 1407) *S. thermophilus* (NCDC 074)	$0.2 \pm 0.2 \times 10^9$ CFU/mL	4 weeks at 4	Thakur and Joshi (2017)
3.	Cabbage juice	–	*L. casei* A4 *L. delbrueckii* D7 *L. plantarum* C3	$1.1 \pm 5.3 \times 10^6$ $4.5 \pm 3.32 \times 10^5$ $4.1 \pm 0.16 \times 10^7$ CFU/mL	1 week 4 weeks 4 weeks at 4°C	Yoon et al. (2006)
4.	Cantaloupe juice	–	*L. casei* B-442	7.5 log CFU/mL	42 days at 4°C	Fonteles et al. (2012)
5.	Cashew apple juice	Sucrose	*L. casei* NRRL B-442	8.66 ± 0.02 log CFU/mL	42 days at 4°C	Horackova et al. (2018); Pereira et al. (2013)
6.	Grape juice	–	*L. acidophilus* L10 *L. casei* L26 *L. rhamnosus* LGG *L. paracasei* L33 *L. plantarum* 299v *L. acidophilus* (CIRMBIA 1674), *L. casei* (CIRMBIA 667), *L. paracasei* (CIRMBIA 672), *L. plantarum* (CIRMBIA 466) *L. rhamnosus* (CIRMBIA 607)	8.0 – 9.8 log CFU/mL	48 hours at 5°C	Espirito-Santo et al. (2015)
7.	Grape juice	–	*L. plantarum*, *L. acidophilus*	6.8×10^8 8.4×10^8 CFU/mL	72 hours at 37°C	Nagpal et al. (2012)
8.	Grape juice	Short Chain Inulin Fibre.	*L. rhamnosus* GR-1	$1.73 \pm 1.72 \times 10^7$ CFU/mL	72 hours at 4°C	White and Hekmat (2018)
9.	Guava juice	4% Honey	*S. boulardii* *S. cerevisiae* – 1 *S. cerevisiae* – 2 Yeast isolate from guava	5.5×10^7 5.5×10^7 5.20×10^7 5.20×10^7 CFU/mL	45 days at 4°C	Saba et al. (2018)

TABLE 1
Probiotic and Prebiotic Fortified Fruit Juices and Viability of Probiotics at Different Storage Durations—cont'd

Sl. No	Name of the Fruit Juice	Fortified Prebiotics	Name of the Probiotic Inoculated	Maximum Viability of Cells	Storage Duration	Reference
10.	Orange juice	–	*L. acidophilus* L10 *L. casei* L26 *L. rhamnosus* LGG *L. paracasei* L33 *L. plantarum* 299v *L. acidophilus* (CIRMBIA 1674) *L. casei* (CIRMBIA 667) *L. paracasei* (CIRMBIA 672) *L. plantarum* (CIRMBIA 466) *L. rhamnosus* (CIRMBIA 607)	7.9 – 8.4 log CFU/mL	48 hours at 5°C	Espirito-Santo et al. (2015)
11.	Orange juice	–	*L. plantarum* *L. acidophilus*	6.9×10^8 8.1×10^8 CFU/mL	72 hours at 37°C	Nagpal et al. (2012)
12.	Orange juice	Long Chain Inulin Fiber	*L. rhamnosus* GR-1	$5.91 \pm 4.69 \times 10^7$ CFU/mL	72 hours at 4°C	White and Hekmat (2018)
13.	Peach juice	–	*L. plantarum* DSMZ 20179 *L. delbrueckii* DSMZ 15996 *L. casei* DSMZ 20011	7.2×10^5 1.7×10^7 log CFU/mL	28 days at 4°C	Pakbin et al. (2014)
14.	Pine apple juice	Cells encapsulated with milk protein matrix	*B. animalis* subsp. *lactis* BB-12	7 log CFU/mL (for free cells) 8 log CFU/mL (for encapsulated cells)	28 days and 8°C	Horackova et al. (2018); Pereira et al. (2013)
15.	Pomegranate juice	–	*L. acidophilus* DSMZ 20079 *L. plantarum* DSMZ 20174 *L. delbrueckii* DSMZ 20006 *L. paracasei* DSMZ 15996	$2.8 \times 10^5 \pm 0.05 \times 10^5$ $2.8 \times 10^5 \pm 0.05 \times 10^5$ $1.5 \times 10^5 \pm 0.26 \times 10^5$ $1.3 \times 10^5 \pm 0.1 \times 10^5$ CFU/mL	1 week 2 weeks 2 weeks 1 week at 4°C	Mousavi et al. (2011)
16.	Potato juice "superior" variety	–	*L. casei* ATCC 393	$2.5 \pm 0.3 \times 10^7$ CFU/mL	4 weeks at 4°C	Kim et al. (2012)
17.	Potato juice "haryoung" variety			$1.4 \pm 0.4 \times 10^9$ CFU/mL		
18.	Potato juice "hongyoung" variety			$9.8 \pm 1.6 \times 10^5$ CFU/mL		
19.	Potato juice "jayoung" variety			$6.0 \pm 0.0 \times 10^7$ CFU/mL		

Continued

TABLE 1
Probiotic and Prebiotic Fortified Fruit Juices and Viability of Probiotics at Different Storage Durations—cont'd

Sl. No	Name of the Fruit Juice	Fortified Prebiotics	Name of the Probiotic Inoculated	Maximum Viability of Cells	Storage Duration	Reference
20.	Strawberry-apple juice	Cells encapsulated with milk protein matrix	*B. animalis* subsp. *lactis* BB-12	5.2 log CFU/mL (for free cells) 6.1 log CFU/mL (for encapsulated cells)	28 days at 8°C	Horackova et al. (2018); Pereira et al. (2013)
21,	Sweet apple cider	Long Chain Inulin Fiber	*L. rhamnosus* GR-1	$7.58 \pm 5.12 \times 10^7$ CFU/mL	72 hours at 4°C	White and Hekmat (2018)
22.	Tomato juice	–	*L. plantarum* *L. acidophilus*	7.7×10^8 9.0×10^8 CFU/mL	72 hours at 37°C	Nagpal et al. (2012)
23.	Tomato juice	–	*L. acidophilus* *L. casei* *L. plantarum* *L. delbrueckii*	$1.4 \pm 0.4 \times 10^9$ $1.7 \pm 0.5 \times 10^8$ $3.4 \pm 1.0 \times 10^6$ $8.1 \pm 3.5 \times 10^8$ CFU/mL	4 weeks at 4°C	Yoon et al. (2004)

bacterial strains, the growth of the *L. casei* NRRL B442 strain was abundant. It was observed that when prebiotics were added in pineapple juice, the production of lactic acid was greater by *B. lactis*. The probiotic cell viability was high in fermented pineapple juice when compared with fresh juice. This was due to a higher buffering capacity of pineapple juice, which made it remain intact in the first month of storage. It was found that the addition of prebiotics did not influence the existence of probiotics in the first month. The observed results showed that the viability of *L. plantarum* was far better than *L. acidophilus*, which survived for more than 80 days. The presence of light, oxygen, and probiotic strains reduced the total phenolic content. Similarly, sonicated pineapple juice obtained the viability of *L. casei* B-442 within a short time of 12h and sustained its survival for 21 days (Costa et al., 2013).

White and Hekmat (2018) developed novel fruit juices from apple cider, grape, and orange juice by inoculating *L. rhamnosus* GR-1, which was then fortified with long and short chain inulin fibers. The average viability was found to be higher in short chain inulin fiber-fortified apple cider juice than in short chain inulin fiber-fortified grape juice. At the beginning of fermentation, the viability was high in long chain inulin fiber-fortified apple cider juice while at 24h of fermentation, the count was more in long chain inulin fiber-fortified orange juice. However, at the end of 72h of fermentation, the count was high in long chain inulin fiber-fortified apple cider juice ($7.58 \pm 5.12 \times 10^7$ CFU/mL). Poor outcomes were observed in long chain inulin fiber-fortified grape juice ($0.50 \pm 0.33 \times 10^7$ CFU/mL). Under refrigerated storage conditions, the viability count was found to be more in long chain inulin fiber-fortified orange juice when stored over 30 days. The overall acceptance was given to short chain inulin fiber-fortified orange juice than the control apple cider juice, which was due to the presence of residue in the apple cider juice. But where flavor was concerned, orange was not preferred as it possessed an increased acidic flavor, long chain fortified apple cider juice was preferred. Where texture was concerned, long chain fortified apple cider juice earned a higher preference. Thus, the addition of prebiotics fairly enhanced probiotic survival. Also, probiotic survival was dependent on the strain and type of fruit juice.

Saba et al. (2018) enriched guava juice with 4% honey (as a prebiotic) and inoculated it with four different yeast strains (probiotic): *Saccharomyces boulardii*, *Saccharomyces cerevisiae* 1, *S. cerevisiae* 2, and yeast isolate from guava. After 7 days of fermentation, the pH dropped from 3.5 to 2.92 (the least in guava juice fortified with yeast isolates from guava). The juice fermented with added honey showed more sugar utilization. The yeast count was high in honey-fortified guava juice compared with the plain fermented guava juice. The count reached the maximum until 30 days of fermentation. Consumer acceptance was high in guava juice added with honey, especially inoculated with the yeast strain *S. boulardii*.

Fruit juice enriched with prebiotics and probiotics was prepared by Fujita et al. (2017) for tube-fed patients to increase the micronutrient concentration in their body and to serve other health benefits. Extracts from whole, unripe *Carica papaya* Linn. juice were allowed to ferment aerobically with *Enterococcus faecalis* and *Aspergillus oryzae* for 1 year and 6 months, respectively. The polymerized polyphenols were fermented to low molecular weight phenolic acids, which acted as prebiotic sources. This juice greatly decreased the populations of Firmicutes, *Eggerthella lenta*, and *Clostridium scindens* in the fecal samples of the tube-fed patients when supplemented with a dose of 9 g/day. The prebiotic source from the prepared extract had no significant effect on bifidobacteria but decreased the concentration of *Bacteroidetes*, *C. perfringens*, and *C. difficile*.

SURVIVAL OF PROBIOTICS IN FRUIT JUICES AND THEIR QUALITY ATTRIBUTES

Major issues in the addition of probiotics to fruit juices are the off-flavor and poor surviving nature. The added probiotics struggle to survive in fruit and vegetable juices as the pH is low, presence of oxygen and other native compounds in it. Below are some techniques to enhance the survival of probiotics and increase consumer acceptance.

Addition of Sucrose

Pereira et al. (2013) researched the viability of the probiotic strain *L. casei* NRRL B-442 on fermented cashew apple juice. They found that the viability was more in sucrose incurred juice throughout the storage period of 42 days. The count of *L. casei* remained above the mean limit of 8 log CFU/mL. An increase in lactic acid in the cashew apple juice content proved that the added probiotics can produce acids as a result of metabolism, even under low temperatures (Pereira et al., 2013). Similar outcomes were obtained in orange, grapefruit, black currant, and pineapple juices fermented with *B. longum* NCIMB 8809 after 42 days of storage with added sucrose. It was clearly observed that an increase in the the viability of the probiotic *L. casei* in cashew apple juice was found tied to a proportionate decrease in sugar content throughout storage.

As far as appearance was concerned, fermented cashew juice was turbid and slightly yellow. Fermented cashew apple juice samples possess a stable color in nature. They found that the antioxidant activity and ascorbic acid content of fermented cashew apple juice were higher than nonfermented juice. Ascorbic acid in cashew apple juice was decreased from its initial concentration during storage. The reduction of ascorbic acid was more in fermented juice than unfermented juice. During the storage of probiotic cashew apple juice, the polyphenol oxidase (PPO) and peroxidase (POD) activities were decreased, which directly reduced the browning reaction and nutritional breakdown. Also, the stored product gains greater preference where the overall preference was concerned in sensory analysis (Pereira et al., 2011).

Inoculation of Spores

Over the decades, the use of spores mainly of the Bacillus species, especially *B. coagulans*, in food products was proven safe and healthy for the host. An example is the use of *Bacillus* sp. spores in date paste (Marcial-Coba et al., 2019). Spores of *B. coagulans* MTCC 5856 in strawberry preserves were also studied (Majeed et al., 2016). Spores of *B. coagulans* produced xylanase, which was mostly used to process fruit juices, like in the process of clarification (Konuray and Erginkaya, 2018). With individual inoculation of different Bacillus strains (*B. coagulans* MTCC 5856; *B. coagulans* GBI-306086, and *B. subtilis* PXN 21) 7–8 log CFU/g in orange juice and simulated under gastrointestinal tract (GIT) conditions, it was observed that *B. coagulans* spores were the most resistant with a final count of ~ 7–8 log CFU/mL. Therefore, it was concluded that spore-forming probiotic strains can be an alternative to normal probiotic strains, as they were more resistant against juice processing and GIT conditions (Soares et al., 2015). Although these findings were taken, research with spores in the field of fruit and vegetable juices was limited.

Microencapsulation of Probiotics

The probiotics that efficiently survive in fruit juices were enhanced by the microencapsulation of the desired probiotic strains (Ding and Shah, 2008; Nualkaekul et al., 2012). Simple microbial viability experiments on the microencapsulation of *L. rhamnosus* and *L. acidophilus* in orange juice showed that there was an increase in the viability of these probiotic strains and a decreased acidity of orange juice (Sohail et al., 2012). The encapsulation of *L. casei* 01 with alginate and Thai herbal cashew flower extract, which were inoculated in mulberry and melon juices, increased their viability under refrigerated conditions (Chaikham, 2015).

The immobilization of probiotics in fruit juice using a calcium alginate complex enhanced their survival over normal free cells added in fruit juices. This technique was adopted in banana puree (Tsen et al., 2004) and tomato juice preparation (King et al., 2007). When both mobilized and immobilized probiotics were concerned, the latter were found to be more viable than the former. This can be explained by the effect of oxygen, which is high due to its exposure in mobilized probiotics rather than immobilized. It is difficult to maintain the viability of probiotic strains in fruit and vegetable juices during cold storage; however, this can be overcome by probiotic encapsulation. The overall acceptance of these juices by customers was high, as they were independent of probiotic incorporation. It was suggested that at poor conditions such as low pH and high acidity, the survival of probiotics in such fruit juices was possible by the microencapsulation of probiotic bacteria, which was comparatively more beneficial than raw fruit juices (Wang et al., 2012).

A study was done by Horackova et al. (2018) on pineapple and strawberry-apple juice by making them with probiotics. Commercial strains of *B. animalis* subsp. *lactis* BB-12 were inoculated after encapsulation. The oil-in-water emulsion-based encapsulation was done with a reconstituted skim milk protein matrix in which rennet and $CaCl_2$ were added; the emulsification was done with canola oil. This was then centrifuged and added in the juice. They found that pineapple juice showed better results than strawberry-apple juice, that is, the stability was found to be better in pineapple juice than in strawberry- apple juice. Strawberry-apple juice was found to be a very poor source for probiotics. Also, the viability was greater in pineapple juice than in strawberry-apple juice. The stability was greater in encapsulated probiotics than in free cells. After 14 days, it was found that at $22 \pm 1\,°C$, the viability of such existing cells decreased in strawberry-apple juice.

Flavor Masking

The addition of certain suitable flavors and aromatic ingredients hides the presence of probiotics. This can be explained by, if such volatile ingredients are added to the probiotic-incorporated juice, they mask the fermented aroma, which is due to the action of microbiota in the juice. This ultimately increases consumer preference (Rakin et al., 2007; Seelam et al., 2014). Luckow et al. (2006) observed that the addition of tropical juice concentrates of pineapple, mango, and passion fruit in pasteurized orange juice samples mask the off-flavor produced by probiotic microorganisms such as *Lactobacillus paracasei* ssp. *paracasei* NFBC 43338. The practical application was found to be very effective. The willingness of consumers was much higher for masked samples than the control and unmasked samples. Also, the crew found that the customer preference of probiotic juice with added tropical juice flavors in pasteurized orange juice was greater when exposed for 7 days. Proper exposure and adequate information about prepared juice samples also increased the acceptance of consumers.

FACTORS AFFECTING PROBIOTIC SURVIVAL IN FRUIT JUICES

Many juices have been used as carriers for the preparation of probiotic fruit juices. Probiotics are incorporated in fruit juices, and these may undergo biochemical reactions resulting in the development of compounds that may be desirable and undesirable. Thus, it may affect the acceptance of such probiotic-incorporated fruit juices by customers (de Souza Neves Ellendersen et al., 2012). There are a few factors that limit the survival of probiotics in fruit juices. It is necessary to consider all the internal and external factors as they directly influence the growth of probiotics in fruit juices. The major concern is the acidity or pH of the juice in which the probiotics are inoculated. Further, the presence of oxygen, other nutrients, inhibitor compounds, temperature, and the specific chemical composition of juice varieties may influence the growth. Hence, the capability of probiotic bacterial strains to survive in such fruits and vegetable juices differs according to these factors.

pH

The natural and acidic conditions of fruit juice limit the growth of probiotic bacteria. Thus, pH is more important (Yáñez et al., 2008). It was noted that, if fruit juice has extremely high pH, acidophilic bacteria cannot grow. If the temperature is too hot (during pasteurization), the bacteria die off before consumption. The optimal pH range for the survival of lactobacilli in juice lies between 3.7 and 4.3. However, bifidobacteria species cannot survive such low pH. They can be active between pH 4.8 and 6.8. The survival of probiotic lactobacilli in acidic environments was high because the proton translocating ATPase was found to play a major role in controlling the pH homeostasis between the intracellular and extracellular environments. This enzyme action generated the energy for proton extrusion from the cytoplasm of the probiotic cell (Corcoran et al., 2005). Thus, strains of lactobacilli and bifidobacteria can survive longer in orange juice and pineapple juice than in cranberry juice with the increasing order of pH 3.5, 4.5, and 5.5, respectively (Sheehan et al., 2007; Tripathi and Giri, 2014). Ranadheera et al. (2014) explained that lactic acid bacteria in high acidic fruit juices increased the resistance of bacteria to subsequent stressful conditions.

Presence of Oxygen

Another major concern is the availability of oxygen in the package during storage. When a beverage with added probiotics was stored in the presence of oxygen, the probiotic cells received oxidative damage due to the formation of reactive oxygen species (ROS), which kills probiotic organisms (Nag and Das, 2013). Due to this factor, the viability of probiotic dosage in fruit juices can be extremely affected by storage. The count of probiotic bacterial species was almost extinct when pomegranate and cranberry juice were stored in the presence of oxygen for 1 week and 4 weeks, respectively (Nualkaekul and Charalampopoulos, 2011). This can be overcome by filling with CO_2 or with the addition of antioxidants such as epigallocatechin gallate (EGCG), which is a known prebiotic (Corbo et al., 2014; Gaudreau et al., 2013).

Presence of Other Compounds

Apart from pH and oxygen, the presence of other compounds may also affect the survival of probiotics in fruit juices. A comparison was done among the effects of pH, citric acid, and ascorbic acid for the growth of probiotics in fruit juices. The outcome of the experiment showed that pH and citric acid were found to be important factors influencing the survival of probiotics. Ascorbic acid had a negligible or no role to play in the survival of probiotics in fruit juices. Nualkaekul and Charalampopoulos (2011) reported that organic acids act as antimicrobials and had a positive effect on probiotic survival in fruit juices. In orange, pineapple, and blackcurrant juices, the surviving capacity of probiotic *Lactobacillus plantarum* NCIMB 8826 was found to be excellent. Among these three, orange juices had higher cell survival capacity. That might be through the maintained optimum pH of 3.8 with a combination of 11 g/L citric and malic acid and 2.8 g/L of total dietary fiber. Although the other two juices could maintain an optimum pH of 3.8, their total dietary fibers were found to 1.6 g/L and 0.2 g/L in pineapple and blackcurrant juice, respectively. It was clear from this study that the surviving capacity depends on pH provided organic acids and dietary fiber. It was also suggested that dietary fiber protected probiotics during processing and storage, as fiber supports the immobilization of probiotics in a liquid juice matrix.

The survival of *L. plantarum* in lemon and grapefruit juices was supportive, although the pHs of these two juices were observed to be 2.5 and 3.2, respectively. The reason for this supportive survival in these juices might be due to the presence of dietary fiber of 1.9 and 2.8 g/L in lemon and grape juices, respectively. Apart from this, it was also suggested that the presence of protein (6 g/L) in grape juice contributes to the survival of *L. plantarum*. In the case of cranberry juice, the concentration of organic acid is low (approximately 6 g/L of malic plus citric acid), due to which a low pH (~2.5) was observed. Moreover, high levels of phenolic acids such as coumaric and benzoic acid (1.5 g/L) in this fresh juice have been suggested to reduce the survival of probiotic lactobacilli. The viable probiotic cell concentration in pomegranate juice decreased sharply and reached zero in 4 weeks of storage. This was due to the low pH of about 3.25 and the high concentration of polyphenols (4.3 g/L) such as tannins (Nualkaekul and Charalampopoulos, 2011).

CONCLUDING REMARKS

Fresh and natural fruit juices are nutritious, but some antinutrients (tannins) and certain pathogenic microbes restrict the beneficial effects. However, the presence of prebiotic oligosaccharides in these fresh fruit juices enables probiotic efficiency. The fermentation of probiotic bacteria in fresh fruit juices produces various organic acids that lead to a reduction in pH and affect the viability of other probiotic strains. Recent studies have overcome this issue by adopting various methods such as the encapsulation of probiotics to enhance their viability in fruit juices. Meanwhile, many explorations on prebiotics suggested that the consumption of prebiotic fiber-based foods is better than probiotic- supplemented foods. Nevertheless, the continuous consumption of these kinds of probiotic-enabled fruit juices definitely maintains the gut microbiota and host health. The probiotic research on fruit juices will be explored in all different edible fruit and vegetable juices with different colon microbes. However, the selection of juice-specific probiotics would be challenging as it depends on several factors such as the availability of juice prebiotics, optimization of the fermentation condition, acceptability of the probiotic juice, etc. The outcome of such studies will help to provide society with healthy probiotic juices any time throughout the year.

REFERENCES

Ahmed, T., Das, K., Uddin, M., 2019. The microbiological quality of commercial fruit juices-current perspectives. Bangladesh J. Microbiol. 35 (2), 128–133.

Akpeji, S., Adebayo-Tayo, B., Sanusi, J., et al., 2017. Production and properties of Probioticic soursop juice using Pediococcus pentosaceus LBF2 as starter. Int. J. Biochem. Res. Rev. 17 (2), 1–10.

Arena, P.M., Caggianiello, G., Fiocco, D., et al., 2014. Barley β-glucans-containing food enhances probiotic performances of beneficial bacteria. Int. J. Mol. Sci. 15 (2), 3025–3039.

Babu, V., Mital, B.K., Garg, S.K., 1992. Effect of tomato juice addition on the growth and activity of Lactobacillus acidophilus. Int. J. Food Microbiol. 17 (1), 67–70.

Bautista-Garfias, C.R., Ixta, O., Orduña, M., et al., 1999. Enhancement of resistance in mice treated with Lactobacillus casei: effect on Trichinella spiralis infection. Vet. Parasitol. 80 (3), 251–260.

Bialonska, D., Kasimsetty, S.G., Khan, S.I., et al., 2009. Urolithins, intestinal microbial metabolites of pomegranate ellagitannins, exhibit potent antioxidant activity in a cell-based assay. J. Agric. Food Chem. 57 (21), 10181–10186.

Chaikham, P., 2015. Stability of probiotics encapsulated with Thai herbal extracts in fruit juices and yoghurt during refrigerated storage. Food Biosci. 12, 61–66.

Chakkaravarthi, S., Philomina, D., Bharathi, P., 2005. Fruit flavonoids and cardiovascular health. Indian J. Pharm. Sci. 67 (4), 399–403.

Chao, C.Y., Yin, M.C., 2009. Antibacterial effects of roselle calyx extracts and protocatechuic acid in ground beef and apple juice. Foodborne Pathog. Dis. 6 (2), 201–206.

Corbo, M.R., Bevilacqua, A., Petruzzi, L., et al., 2014. Functional beverages: the emerging side of functional foods. Compr. Rev. Food Sci. Food Saf. 13 (6), 1192–1206.

Corcoran, B.M., Stanton, C., Fitzgerald, G.F., et al., 2005. Survival of probiotic lactobacilli in acidic environments is enhanced in the presence of metabolizable sugars. Appl. Environ. Microbiol. 71 (6), 3060–3067.

Costa, M.G.M., Fonteles, T.V., de Jesus, A.L.T., et al., 2013. Sonicated pineapple juice as substrate for L. casei cultivation for probiotic beverage development: process optimisation and product stability. Food Chem. 139 (1), 261–266.

de Souza Neves Ellendersen, L., Granato, D., Bigetti Guergoletto, K., et al., 2012. Development and sensory profile of a probiotic beverage from apple fermented with Lactobacillus casei. Eng. Life Sci. 12 (4), 475–485.

Ding, W.K., Shah, N., 2008. Survival of free and microencapsulated probiotic bacteria in orange and apple juices. Int. Food Res. J. 15 (2), 219–232.

Duque, A.L.R.F., Monteiro, M., Adorno, M.A.T., et al., 2016. An exploratory study on the influence of orange juice on gut microbiota using a dynamic colonic model. Food Res. Int. 84, 160–169.

Espirito-Santo, A.P., Carlin, F., Renard, C.M.G.C., 2015. Apple, grape or orange juice: which one offers the best substrate for lactobacilli growth?—a screening study on bacteria viability, superoxide dismutase activity, folates production and hedonic characteristics. Food Res. Int. 78, 352–360.

Fonteles, T.V., Costa, M.G.M., de Jesus, A.L.T., et al., 2012. Optimization of the fermentation of cantaloupe juice by Lactobacillus casei NRRL B-442. Food Bioprocess Technol. 5 (7), 2819–2826.

Fujita, Y., Tsuno, H., Nakayama, J., 2017. Fermented papaya preparation restores age-related reductions in peripheral blood mononuclear cell cytolytic activity in tube-fed patients. PLoS One 12 (1), e0169240.

Garcia, E.F., de Oliveira, A.A., Luciano, W.A., et al., 2018. The performance of five fruit-derived and freeze-dried potentially probiotic Lactobacillus strains in apple, orange, and grape juices. J. Sci. Food Agric. 98 (13), 5000–5010.

Gaudreau, H., Champagne, C.P., Remondetto, G.E., et al., 2013. Effect of catechins on the growth of oxygen-sensitive probiotic bacteria. Food Res. Int. 53 (2), 751–757.

Gibson, G.R., Hutkins, R., Sanders, M.E., et al., 2017. Expert consensus document: the International Scientific Association for Probiotics and Prebiotics (ISAPP) consensus statement on the definition and scope of prebiotics. Nat. Rev. Gastroenterol. Hepatol. 14 (8), 491–502.

Gómez, B., Gullón, B., Remoroza, C., et al., 2014. Purification, characterization, and prebiotic properties of pectic oligosaccharides from orange peel wastes. J. Agric. Food Chem. 62 (40), 9769–9782.

Hatab, S., Yue, T., Mohamad, O., 2012. Removal of patulin from apple juice using inactivated lactic acid bacteria. J. Appl. Microbiol. 112 (5), 892–899.

Heenan, C.N., Adams, M.C., Hosken, R.W., et al., 2002. Growth medium for culturing probiotic bacteria for applications in vegetarian food products. LWT Food Sci. Technol. 35 (2), 171–176.

Horackova, S., Kristýna, R., Veselá, K., et al., 2018. Fruit juices with probiotics—new type of functional foods. Czech J. Food Sci. 36 (4), 284–288.

Javier, A.L.-P., Anna, A., Eva Nordberg, K., 2018. Structural considerations on the use of endo-xylanases for the production of prebiotic Xylooligosaccharides from biomass. Curr. Protein Pept. Sci. 19 (1), 48–67.

Kim, N.J., Jang, H.L., Yoon, K.Y., 2012. Potato juice fermented with Lactobacillus casei as a probiotic functional beverage. Food Sci. Biotechnol. 21 (5), 1301–1307.

King, V.A.E., Huang, H.Y., Tsen, J.H., 2007. Fermentation of tomato juice by cell immobilized Lactobacillus acidophilus. Mid-Taiwan J. Med. 12, 1–7.

Konuray, G., Erginkaya, Z., 2018. Potential use of *Bacillus coagulans* in the food industry. Foods 7 (6), 92.

Luckow, T., Delahunty, C., 2004. Which juice is 'healthier'? A consumer study of probiotic non-dairy juice drinks. Food Qual. Prefer. 15 (7), 751–759.

Luckow, T., Sheehan, V., Fitzgerald, G., et al., 2006. Exposure, health information and flavour-masking strategies for improving the sensory quality of probiotic juice. Appetite 47 (3), 315–323.

Majeed, M., Majeed, S., Nagabhushanam, K., et al., 2016. Evaluation of the stability of *Bacillus coagulans* MTCC 5856 during processing and storage of functional foods. Int. J. Food Sci. Technol. 51 (4), 894–901.

Marcial-Coba, M.S., Pjaca, A.S., Andersen, C.J., et al., 2019. Dried date paste as carrier of the proposed probiotic *Bacillus coagulans* BC4 and viability assessment during storage and simulated gastric passage. LWT 99, 197–201.

Martínez-Gómez, F., Ixta-Rodríguez, O., Aguilar-Figueroa, B., et al., 2006. Lactobaciilus casei ssp. rhamnosus enhances non specific protection against Plasmodium chabaudi AS in mice. Salud Publica Mex. 48 (6), 498–503.

Mattila-Sandholm, T., Myllärinen, P., Crittenden, R., et al., 2002. Technological challenges for future probiotic foods. Int. Dairy J. 12 (2), 173–182.

Mithul Aravind, S., Wichienchot, S., Tsao, R., Ramakrishnan, S., Chakkaravarthi, S., 2021. Role of dietary polyphenols on gut microbiota, their metabolites and health benefits. Food Res. Int. 142, 110189.

Mousavi, Z.E., Mousavi, S.M., Razavi, S.H., et al., 2011. Fermentation of pomegranate juice by probiotic lactic acid bacteria. World J. Microbiol. Biotechnol. 27 (1), 123–128.

Nag, A., Das, S., 2013. Improving ambient temperature stability of probiotics with stress adaptation and fluidized bed drying. J. Funct. Foods 5 (1), 170–177.

Nagpal, R., Kumar, A., Kumar, M., 2012. Fortification and fermentation of fruit juices with probiotic lactobacilli. Ann. Microbiol. 62 (4), 1573–1578.

Nguyen, B.T., Bujna, E., Fekete, N., et al., 2019. Probiotic beverage from pineapple juice fermented with. Front. Nutr. 6, 54.

Nualkaekul, S., Charalampopoulos, D., 2011. Survival of Lactobacillus plantarum in model solutions and fruit juices. Int. J. Food Microbiol. 146 (2), 111–117.

Nualkaekul, S., Lenton, D., Cook, M.T., et al., 2012. Chitosan coated alginate beads for the survival of microencapsulated *Lactobacillus plantarum* in pomegranate juice. Carbohydr. Polym. 90 (3), 1281–1287.

Pakbin, B., Razavi, S., Mahmoudi, R., et al., 2014. Producing probiotic peach juice. J. Biotechnol. Health Sci. 1, 1–5.

Panda, S.K., Behera, S.K., Witness Qaku, X., et al., 2017. Quality enhancement of prickly pears (Opuntia sp.) juice through probiotic fermentation using *Lactobacillus fermentum*—ATCC 9338. LWT 75, 453–459.

Patel, A., 2017. Probiotic fruit and vegetable juices- recent advances and future perspective. Int. Food Res. J. 24 (5), 1850–1857.

Pereira, A.L.F., Maciel, T.C., Rodrigues, S., 2011. Probiotic beverage from cashew apple juice fermented with Lactobacillus casei. Food Res. Int. 44 (5), 1276–1283.

Pereira, A.L.F., Almeida, F.D.L., de Jesus, A.L.T., et al., 2013. Storage stability and acceptance of probiotic beverage from cashew apple juice. Food Bioprocess Technol. 6 (11), 3155–3165.

Pereira-Caro, G., Borges, G., Ky, I., et al., 2015. In vitro colonic catabolism of orange juice (poly)phenols. Mol. Nutr. Food Res. 59 (3), 465–475.

Rafiq, S., Sharma, V., Nazir, A., et al., 2016. Development of probiotic carrot juice. J. Nutr. Food Sci. 6 (4), 534.

Rakin, M., Vukasinovic, M., Siler-Marinkovic, S., et al., 2007. Contribution of lactic acid fermentation to improved nutritive quality vegetable juices enriched with brewer's yeast autolysate. Food Chem. 100 (2), 599–602.

Ranadheera, S., Prasanna, P.H.P., Vidanarachchi, J., 2014. Fruit juice as probiotic carriers. In: Elder, K.E. (Ed.), Fruit Juices: Types, Nutritional Composition and Health Benefits. Nova Science Publishers, Hauppauge, NY, USA, pp. 253–267.

Saarela, M., Virkajärvi, I., Nohynek, L., et al., 2006. Fibres as carriers for *Lactobacillus rhamnosus* during freeze-drying and storage in apple juice and chocolate-coated breakfast cereals. Int. J. Food Microbiol. 112 (2), 171–178.

Saba, N., Swamy, B.N., Nalini, B.S., et al., 2018. Bioenrichment of guava juice with prebiotic and probiotics. Int. J. Curr. Microbiol. App. Sci. 7 (7), 1307–1315.

Seelam, D.N.S., Katike, U., Obulam, V.S., 2014. A comparative study on probiotication of mixed watermelon and tomato juice by using probiotic strains of lactobacilli. Int. J. Curr. Microbiol. App. Sci. 3 (11), 977–984.

Sembries, S., Dongowski, G., Mehrländer, K., et al., 2006. Physiological effects of extraction juices from apple, grape, and red beet pomaces in rats. J. Agric. Food Chem. 54 (26), 10269–10280.

Sheehan, V.M., Ross, P., Fitzgerald, G.F., 2007. Assessing the acid tolerance and the technological robustness of probiotic cultures for fortification in fruit juices. Innovative Food Sci. Emerg. Technol. 8 (2), 279–284.

Singla, V., Chakkaravarthi, S., 2017. Applications of prebiotics in food industry: a review. Food Sci. Technol. Int. 23 (8), 649–667.

Soares, M.B., Martinez, R.C.R., Da Silva, B.S., et al., 2015. Survival of probiotic lactic acid bacteria and bacillus spores in orange juice and creamy curd cheese through simulated gastrointestinal tract conditions. In: The Latin American Symposium on Food Science, Brazil.

Sohail, A., Turner, M.S., Prabawati, E.K., et al., 2012. Evaluation of *Lactobacillus rhamnosus* GG and Lactobacillus acidophilus NCFM encapsulated using a novel impinging aerosol method in fruit food products. Int. J. Food Microbiol. 157 (2), 162–166.

Thakur, A., Joshi, V.K., 2017. Preparation of probiotic apple juice by lactic acid fermentation. Int. J. Food Fermen. Technol. 7 (1), 67–85.

Tripathi, M.K., Giri, S.K., 2014. Probiotic functional foods: survival of probiotics during processing and storage. J. Funct. Foods 9, 225–241.

Tsen, J.-H., Lin, Y.-P., An-Erl King, V., 2004. Fermentation of banana media by using κ-carrageenan immobilized Lactobacillus acidophilus. Int. J. Food Microbiol. 91 (2), 215–220.

Wang, J.-C., Zhang, W.-Y., Zhong, Z., et al., 2012. Transcriptome analysis of probiotic Lactobacillus casei Zhang during fermentation in soymilk. J. Ind. Microbiol. Biotechnol. 39 (1), 191–206.

White, J., Hekmat, S., 2018. Development of probiotic fruit juices using *Lactobacillus rhamnosus* GR-1 fortified with short chain and long chain inulin fiber. Fermentation 4 (2).

Yáñez, R., Marques, S., Gírio, F.M., et al., 2008. The effect of acid stress on lactate production and growth kinetics in *Lactobacillus rhamnosus* cultures. Process Biochem. 43 (4), 356–361.

Yang, F., Wang, Y.-P., Zhao, H., 2020. Quality enhancement of fermented vegetable juice by probiotic through fermented yam juice using *Saccharomyces cerevisiae*. Food Sci. Technol. 40 (1), 26–35.

Yoon, K.Y., Woodams, E.E., Hang, Y.D., 2004. Probiotication of tomato juice by lactic acid bacteria. J. Microbiol. 42 (4), 315–318.

Yoon, K.Y., Woodams, E.E., Hang, Y.D., 2006. Production of probiotic cabbage juice by lactic acid bacteria. Bioresour. Technol. 97 (12), 1427–1430.

Zendeboodi, F., Khorshidian, N., Mortazavian, A.M., et al., 2020. Probiotic: conceptualization from a new approach. Curr. Opin. Food Sci. 32, 103–123.

CHAPTER 13

Probiotics and Prebiotics in Animal Feed

MOHAMED ZOMMITI[a,b] • MOUNIR FERCHICHI[a]
[a]Unit of Functional Proteomics and Nutraceutical Potential of Biodiversity in Tunisia, Higher Institute of Applied Biological Sciences of Tunis, University of Tunis El Manar, Tunis, Tunisia, [b]Laboratory of Microbiology, Signals and Microenvironment (LMSM), University of Rouen, Evreux, France

INTRODUCTION

In the livestock sector, it has been shown that the animal intestinal microbiota represents a key organ, playing a critical role in the harvesting, storage, and expenditure of energy obtained from the diet reflecting a direct influence on the health and weight of the animal (Krajmalnik-Brown et al., 2012). The link between nutrition and the microbiota is well established. Diet is considered one of the main drivers in shaping the gut microbiota over a lifetime. Thus, a healthy gut is key to a healthy animal. An infected gut (i.e., by coccidiosis or by necrotic enteritis) is not a healthy gut and is not efficient in digesting and transporting nutrients (Choct, 2009). Thus, a homeostatic balanced and diverse microbial composition is salient for optimal digestion and nutrient uptake. The key tool for good gut health is to provide the best feed possible that meets the nutritional needs of the age, category, and stage of production. An outstanding report has been announced by FAO (2013), in which this organization stated the possibility of probiotic application for animal nutrition as gut ecosystem enhancers. Likewise, two pilot investigations conducted by Seal et al. (2018) and Yirga (2015) reported and argued on the use of probiotics, particularly LAB-probiotics, in boosting the growth and reproduction performances, promoting the survival rate and health status of different pools of animals in livestock husbandry. Among the main patented probiotic microorganisms known, species of lactic acid bacteria (LAB) such as *Lactobacillus* (*plantarum, paracasei, acidophilus, casei, rhamnosus, crispatus, gasseri, reuteri, bulgaricus*) are mostly used, as well as *Bifidobacterium* (*longum, catenulatum, breve, animalis, bifidum*) and *Saccharomyces boulardii*. The application of microorganisms like *Enterococcus faecium* and *Bacillus* (*coagulans, subtilis, laterosporus*) are also described (Dixit et al., 2016). Probiotic strains belonging to the genera *Lactobacillus* and *Bifidobacterium* are more common among commercial strains (de Simone, 2019).

Much is known about the benefaction of probiotic microorganisms and/or the prebiotic feed ingredients on the livestock field. Nevertheless, consistent data regarding the effects promoted by these beneficial microbes and substrates on the parameters associated with performance parameters, breeding, livestock production, and welfare still need clarification and accuracy. Thus, the aim of this chapter is to review the salient effects of probiotics, potential probiotic microbes, and prebiotics on nonhuman fields for the purpose to assess the impact of the supplementation with probiotics and/or prebiotics on various factors and parameters related to the animal field. A glance at the market of probiotics and prebiotics for animal feeding has also been discussed additionally to the main selective criteria, and requirements for probiotic strains and the feed ingredients associated. To wrap up, a recap and sum-up of the concluding remarks and future outlook have been stated to shed light on the more efficient and salient pre- and probiotics in terms of their application on livestock as a vital nonhuman sector. When summing up the advantages of pro-and prebiotic use, one should emphasize on their role in the protection of animals against pathogens, enhancement of immunological response, reduced need for antibiotic-based growth stimulants, and high safety of those formulas.

Historical Notes About Probiotics

According to Hill et al. (2014), the International Scientific Association for Probiotics and Prebiotics (ISAPP) suggested a consensus statement on the proper use of the term probiotic: "live microorganisms that, when administered in adequate amounts, confer a health benefit on the host." Before explaining the mechanisms and benefactions provided by these beneficial microbes, it is crucial to specify why, if a microorganism is to be considered a probiotic, it should meet a critical package of requirements: (i) not pathogenic; (ii) high adhesive potential to epithelial cells; (iii) can

Probiotics and Prebiotics in Foods. https://doi.org/10.1016/B978-0-12-819662-5.00011-2

colonize and reproduce itself in the host; (iv) able to survive passage through the harsh gastrointestinal tract (GIT) conditions; (v) high resistance to gastric acidity and bile content; (vi) produces metabolites to inhibit or kill pathogenic bacteria; and (vii) has undergone trials in vitro and in vivo that demonstrate its benefits. Finally, a probiotic or a potential probiotic microbe should remain viable under processing, production, and storage conditions (Kabir, 2009). According to Syngai et al. (2016), administration of probiotics should provide the following beneficial effects: (i) stimulation of the development of beneficial microbiota; (ii) reduction and prevention of colonization by enteric pathogens; (iii) aid in the maturation of the intestinal epithelium; (iv) stimulation of epithelial health; (v) improvement of digestive functionality; and (vi) immunomodulation of the immune system and immunological activity. Probiotic microbes can affect the immune system both directly and indirectly. Direct influence is exerted by numerous species of *Lactobacillus* that increase cytokine and antibody levels (Haghighi et al., 2006; Brisbin et al., 2011). In livestock, probiotics have shown tremendous effects in terms of improvement of body weight, growth performance, and a wide range of various parameters. The growing concern about the spread of microorganisms resistant to antibiotics in humans has determined the ban on the use of antibiotics as growth promoters in livestock (Schwarz et al., 2001) leaving the way open in front of promising natural tools also known as probiotics (Fig. 1).

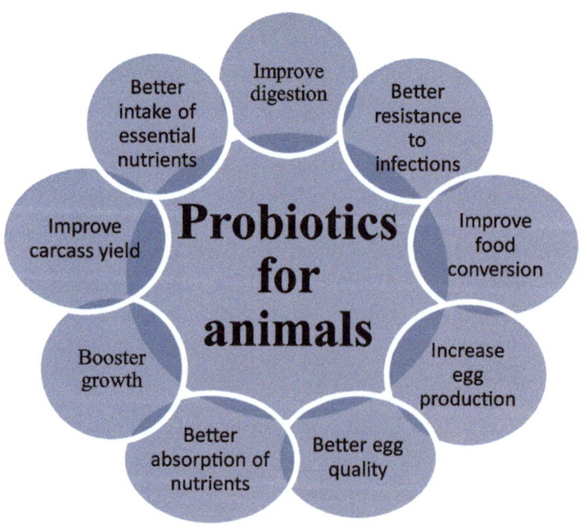

FIG. 1 The main effects of probiotics on animal nutrition, health, and breeding (Bouchard et al., 2015; Zommiti et al., 2019).

Table 1 recaps the list of the microbial strains, related to LAB and non-LAB probiotics, that are potentially usable in animal husbandry because of their multifaceted functions. Various investigations and studies have demonstrated the claimed beneficial effects of probiotics on animals including improving the growth performance of chickens (Lei et al., 2015), ameliorating growth performance rates in ruminants (Rigobelo and Ávila, 2012), boosting growth parameters in pigs (Bajagai et al., 2016), and increasing disease resistance, survival, and final weight (Jha et al., 2015). The following paragraphs describe the main beneficial effects of probiotics on farm animals and their limitations.

Poultry

According to Farrell (2013), poultry husbandry is one of the most important contributors to the food provision industry in many different countries around the globe. Scientific literature has supplied a nonstop investigation about probiotics and their application in poultry production in terms of enhancement of the broiler chicken growth rates and performance parameters (Afsharmanesh and Sadaghi, 2014; Mookiah et al., 2014; Zhang and Kim, 2014; Lei et al., 2015). Additionally, various studies have demonstrated the astonishing potential of probiotics in inhibiting pathogens, thus, fighting and/or preventing enteric diseases such as campylobacteriosis (Stern et al., 2006), salmonellosis (Biloni et al., 2013), coccidiosis (Dalloul et al., 2003), and necrotic enteritis (Jayaraman et al., 2013).

Probiotics for poultry may contain one or more strains of microorganisms and may be given either alone or in combination with other additives in feed or water. Among these beneficial microbes, a variety of bacterial species (*Bacillus*, *Bifidobacterium*, *Enterococcus*, *Lactobacillus*, *Lactococcus*, and *Streptococcus* spp.) and yeast species (*Saccharomyces* spp.) have been tested and used as probiotics in the field of poultry production. It is pertinent to record that the efficiency of probiotic microorganisms firmly depends on various factors, encompassing the origin of the microorganisms, the composition of the mixed rations, and the time of their administration. It seems that the effectiveness of probiotic cultures is greater when they contain a larger number of genera (Chambers and Gong, 2011). Likewise, the source strictly affects effectiveness, as strains that come directly from chicken intestines are more efficient than those from other sources. Moreover, the probiotic composition may be beneficial for one breed of chicken but not for others. The application of probiotic microbes in poultry had many features and touched many sectors. For instance, in 2015, Cox and Dalloul

TABLE 1
Probiotic Microbes Used in Animal Husbandry as Gut Ecosystem Enhancers, Antibiotic Replacers, and Feed Additives (Kabir, 2009; Seo et al., 2010; Hai, 2015; Banerjee and Ray, 2017; Carnevali et al., 2017; Liao and Nyachoti, 2017).

Animal and Its Products	BACTERIA				
	LAB	Non-LAB	Yeast	Fungi	Microalgae
Poultry and eggs	*Lactobacillus* *Enterococcus* *Pediococcus* *Streptococcus* *Weissella*	*Bacillus* *Bifidobacterium*	*Candida* *Kluyveromyces* *Saccharomyces*	*Aspergillus*	–
Pig	*Lactobacillus* *Pediococcus* *Enterococcus* *Weissella*	*Bacillus* *Bifidobacterium* *Clostridium*	*Kluyveromyces* *Saccharomyces*	–	–
Ruminant	*Lactobacillus* *Enterococcus*	*Bacillus* *Bifidobacterium* *Megasphaera* *Prevotella* *Propionibacterium*	*Saccharomyces* *Kluyveromyces* *Trichosporon*	*Aspergillus*	–
Aquaculture	*Lactobacillus* *Lactococcus* *Leuconostoc* *Enterococcus* *Pediococcus* *Carnobacterium* *Weissella*	*Bacillus* *Clostridium* *Alteromonas* *Enterobacter* *Pseudomonas* *Streptomyces* *Brevibacterium* *Eubacterium* *Microbacterium* *Micrococcus* *Roseobacter* *Staphylococcus* *Streptomyces*	*Saccharomyces* *Debaryomyces*	*Aspergillus*	*Tetraselmis* *Phaeodactylum*

performed a pilot study in which they investigated the in ovo spraying of the probiotic on chicks or embryonated eggs as a novel strategy. However, the outcomes from probiotic use are not consistent.

In terms of egg production and quality, probiotics can affect the production, feed use efficiency, and quality of eggs in laying hens, but these effects have been very inconsistent. Table 2 lists the examples of results of investigations on the effects of probiotics and probiotic microorganisms in animal nutrition.

Swine industry

One astonishing criterion of antibiotic use in the porcine production field is their imminent potency to promote animal growth potential, reflecting their potential as "AGPs" or antibiotic growth promoters (Dibner and

TABLE 2
Probiotic Effects on the Performance of Poultry and Egg Production (Zommiti et al., 2019).

Probiotic Microorganisms	Effects	References
Poultry		
L. acidophilus, *B. subtilis*, *S. cerevisiae; A. oryzae*	Remarkable improvement in growth performances Severe decrease in feed conversion ratio parameter	Shim et al. (2012)
E. faecium	Increase in final body weight; the histology of the intestinal mucosa is positively affected.	Abdel-Rahman et al. (2013)
L. fermentum, *S. cerevisiae*	Increase in final body weight and improvement in feed intake No significant effects on feed conversion ratios	Bai et al. (2013)
E. faecium	Increase in the villus height and the villus crypt ratio in the intestinal mucosa Improvement in growth rate performances; reduction in *E. coli* K88 counts	Cao et al. (2013)
L. acidophilus, *B. subtilis* DSM 17299, *C. butyricum*	Increase in final body weight with no effects of this probiotic complex on the feed conversion ratio and feed intake parameters	Zhang and Kim (2014)
L. acidophilus, *L. casei*, *B. bifidum*, *E. faecium*	Enhancing growth rate performances and food intake parameters; significant decrease in FCR criterion	Landy and Kavyani (2013)
L. salivarius 15 K	High antimicrobial activity against *Klebsiella* and *E. coli*	Bujnakova et al. (2014)
Probiotic complex of 11 *Lactobacillus* strains	Improvement of the growth rate parameter; remarkable decline in FCR	Mookiah et al. (2014)
L. paracasei J.R + *L. rhamnosus* 15b + *L. lactis* Y + *L. lactis* FOa	High level of in vivo reduction in *Campylobacter jejuni* (5 log reduction) in broiler chicken	Cean et al. (2015)
L. plantarum P-8	Induction of the beneficial immune response of broiler chickens	Wang et al. (2015)
Wickerhamomyces anomalus LV-6	Improvement of broiler production	García-Hernández et al. (2012)
B. subtilis	Significant increase in final body weight of chicken in grower-finisher phase and/or in all growing phases	Abdel-Rahman et al. (2013)
B. subtilis	An important increase in final body weight of chicken in all growing phases with improvement in feed intake criterion; increase in the absorption of nutrients	Afsharmanesh and Sadaghi (2014)
B. subtilis B2A	Reduction in intestinal *Salmonella* burden (linear and quadratic, $P < .05$)	Park and Kim (2014)
B. amyloliquefaciens	Boost in growth rate performances with the feed intake parameter; A decrease in feed conversion ratio	Ahmed et al. (2014)
B. amyloliquefaciens	Increase in final body weight parameters in grower-finishing phase; significant decline in FCR in all growing phases	Jahromi et al. (2014)
Streptomyces sp. JD9 (KF878075)	Enhancing the broiler production	Latha et al. (2015)
Bacillus spp. *L. salivarius* subsp. *salivarius*, *L. salivarius* subsp. *salicinius*	Significant in vivo reduction in the pathogen *Campylobacter jejuni* counts (1 to 2 log) in comparison to controls	Arsi et al. (2015)

TABLE 2
Probiotic Effects on the Performance of Poultry and Egg Production—cont'd

Probiotic Microorganisms	Effects	References
B. subtilis	Significant reduction in the *Salmonella gallinarum* level in chicken	Upadhaya et al. (2016)
Egg		
L. acidophilus	Significant increase in egg production and the FCR; the quality of eggs is ameliorated with a decrease in yolk cholesterol	Fajardo et al. (2012)
L. acidophilus, *L. casei; B. bifidum*, *A. oryzae*, *S. faecium*, *Torulopsis* spp.	High improvement in the egg production parameters with an increase in the eggshell thickness and a significant drop in yolk cholesterol	Hung et al. (2012)
L. acidophilus D2/CSL	Boosts egg production when supplemented to laying hens' diets with significant improvement in albumin viscosity parameters	Gallazzi et al. (2009)
P. acidilactici	Significant effects on the quality of eggs, boosting the weight and reducing the yolk cholesterol but no effects on the egg production	Mikulski et al. (2012)
B. licheniformis *B. subtilis*	Improvement in the egg production parameter and reduction in the yolk cholesterol; a significant decrease in the FCR criterion	Kurtoglu et al. (2004)
R. capsulatus	No effects on egg production and egg quality parameters except a significant reduction in cholesterol and triglycerides concentration in serum (15% and 11%) as well as in egg-yolks (13% and 16%). Improvement in yolk color with 0.04%	Salma et al. (2007)

Richards, 2005; Guerra and Castro, 2009). In the swine industry, the use of antibiotics is the most studied of all livestock species. This is due to the fact that the subtherapeutic use of these agents can tremendously improve pig growth rate, boost production and reproduction performances, and remarkably reduce morbidity and mortality of piglets (Cromwell, 2002; Thacker, 2013).

The knowledge that the use of probiotics may replace antibiotic therapies with less aggressive methods has resulted in a new vision in the pharmaceutical industry. Although banned in some areas, encompassing the European Union, subtherapeutic use of antibiotics in feed to prevent diarrhea and improve performance is still common in the swine industry. Hence, the substitution of AGPs with probiotic microbes to address the issue of antibiotic resistance is critical in pig production. For monogastric animals, this switch has been more extensively studied in poultry than in pigs. As with other livestock, it is not an easy task to make generalizations because of the drastic variation in the microorganisms used, origins, doses, duration of treatment, and husbandry practices (Kenny et al., 2011).

To wrap up, the use of different strains and doses of potential probiotics and beneficial microbes and the differences in husbandry practices (nutrition, housing, etc.), age of pigs, and feed type may explain contrasting results with the same probiotic microorganisms. There is increasing research on the use of probiotics to intervene in decreasing pathogen load and in ameliorating gastrointestinal disease symptoms in pigs. Besides the in vitro test to identify the best potential probiotics, several studies are conducted in vivo utilizing different probiotic microorganisms. For a summary of probiotic applications in swine, refer to Table 3.

Cattle

According to Oltjen and Beckett (1996), ruminants have the potential to convert agricultural by-products and fibrous materials, which are not edible for humans, into edible high-quality food for humans. This category of animals is able to produce massive meat and milk reflecting a salient role in the agricultural output systems of every country. The continuously growing human population around the world makes scientific

TABLE 3
Effects of Probiotics on the Growth Performance of Swine (Zommiti et al., 2019).

Probiotic Microorganism	Effects	References
Lactobacillus casei	Reduces the number of the first causal agent of severe diarrhea in gnotobiotic piglets, *E. coli*, colonizing their jejunal mucosa by 1 logarithm (4.95 log $10/cm^2$) in comparison with the control group (5.96 log $10/cm^2$) *L. casei* administered in combination with maltodextrin decreased the number of *E. coli* colonizing the jejunum of conventional piglets by more than 2.5 log (4.75 log $10/cm^2$) compared to the control (7.42 log $10/cm^2$)	Bomba et al. (1999)
L. reuteri BSA131	Improvement in average daily gain, body weight, and feed conversion ratio; significant reduction of the number of fecal coliforms	Chang et al. (2000)
Enterococcus faecium EF1	Induces strong cytokines and chemokines production (antiinflammatory factors) in the small intestine	Huang et al. (2012)
L. plantarum ZJ316	Improves ADG and FCR parameters; reduces the frequency of diarrhea in growing-finishing pigs and improves meat organoleptic characteristics and quality	Suo et al. (2012)
L. fermentum I5007	Increases average daily gain in neonatal piglets; Significant improvement in intestinal immunity; Decreases *Clostridium* spp. and *Escherichia* spp. levels	Liu et al. (2014)
L. johnsonii XS4	The supplementation of this probiotic in diets during the end of pregnancy and lactation has positive effects including an increase in the litter weight at birth, the number of piglets at weaning, and the weaning litter weight	Wang et al. (2014)
L. acidophilus	Significant increase in the average daily gain in weaned piglets	Lv et al. (2015)
L. plantarum ATCC 4336 *L. fermentum* DSM 20016 *E. faecium* ATCC 19434	Improvement of the ADG in weaned piglets	Bajagai et al. (2016)
Bifidobacterium longum AH1206	No significant impacts of *Bifidobacterium longum* AH1206 intake on ADG and FCR parameters in neonatal piglets	Herfel et al. (2013); Bajagai et al. (2016)
E. faecium NCIMB 11181	Significant decrease in fecal shedding of *E. coli* counts	Pajarillo et al. (2015)
E. faecium	Inhibitory activity against *Clostridium perfringens*	Hanczakowska et al. (2016)
Bacillus subtilis *C. butyricum*	An important increase in the ADG criterion in growing-finishing pigs; No effect on feed conversion ratios	Meng et al. (2010)
Bacillus subtilis KN-42	Reduces the number of *E. coli* bacteria and improves ADG and FCR	Hu et al. (2014)
Bacillus toyonensis	The addition of this Bacillus has significantly improved the growth rate of ADG and food intake	Kantas et al. (2015)
B. licheniformis *B. subtilis*	A remarkable increase in ADG criterion but no effects of the Bacillus added, even a decrease in the feed conversion ratio criterion in growing-finishing pigs	Alexopoulos et al. (2004); Bajagai et al. (2016)
B. subtilis MA139	No significant effects on ADG with a decrease in FCR parameters	Bajagai et al. (2016)
S. cerevisiae subsp. *boulardii* CNCM I-1079	Severe decrease in feed conversion ratio parameter in weaned piglets	Le Bon et al. (2010)

ADG, average daily gain; *FCR*, feed conversion ratio.

investigators and researchers pursue a nonstop investigation to improve the yield of animal production to supply enough food sources for human beings. As a result, the accomplishment of this aim needs urgently the modulation of the GIT of animals, in particular ruminants, representing one of the most important and reliable strategies.

Ruminants have a complex ecosystem called the rumen. This has a complex microbial ecology, where polysaccharides and proteins ingested by the host are degraded by different rumen microbial populations, resulting in the synthesis of short-chain fatty acids (SCFAs) and microbial protein, which are used by the host as energy and protein sources. There is an increasing global interest in manipulating the rumen compartment, particularly the rumen ecosystem, to increase the effectiveness of the ruminal fermentation processes to enhance animal productivity and reduce the unwanted by-products, such as methane. This has put scientists into the obligation to urgently characterize the digestive ecosystem of the rumen in terms of the microbial communities' composition and the specific functionality of each microbial strain. According to Rigobelo and Ávila (2012), probiotics, with their key impact on the composition and functionalities of the GIT microbial communities, could promote health and ameliorate growth performance parameters in ruminants. It has been reported that the use of bacterial probiotics is more efficient in the case of chickens, pigs, and young calves, whereas administration of probiotic yeast strains (*Saccharomyces cerevisiae*) and fungi (*Aspergillus oryzae*) offers better results in adult ruminants (Fuller, 1999). Yeast (*S. cerevisiae*) is a commonly used probiotic in ruminants (Chaucheyras-Durand et al., 2008), affecting mainly the microbial community dynamics in the rumen and the breakdown of nutrients. Lactic acid-producing bacteria are another important group of probiotics. The application of probiotics in ruminants represents a field with limited recent publications. Studies in this sector have been conducted in the preruminant's life and adult ruminants, taking into account both the economic parameters and the health status of the animals. Despite the number of investigations about the role of probiotics in ruminants, the findings are somehow inconsistent due to the interference of several factors including the probiotic strains, doses, in vivo experiments and their duration, the animals, and their genotypes, and the environmental surroundings (Table 4).

Horses

There are very few data and limited studies regarding the use of probiotics in horses, and these data are contentious mainly due to the small number of observations and recorded cases. Their administration is mainly related to the control of intestinal diseases. For instance, common management practices such as high-concentrate diets, low forage quality, meal feeding, and confinement housing have a direct impact on intestinal function, precisely on large intestinal fermentation. The microbiome of the equine large intestine is a complex and diverse ecosystem. The disturbance of microflora and their environment can lead to an imminent dysbiosis reflecting a high gastrointestinal disorder.

Although there are still many unanswered questions about the mode of action of successful probiotics, evidence indicates competitive inhibition and enhanced immunity. Manufacturers and industrials of prebiotics and probiotics suggest that the incorporation of these supplements in horse diets may help them in maintaining, even restoring, the whole equine intestinal microbiome. However, the unsolved question is, "Does science support the use of prebiotics and probiotics in horses?" According to recent data and limited reviews available on the subject, the answer is, unfortunately, not exact (Costa and Weese, 2012; Coverdale, 2016). Only a very limited number of studies support the use of probiotics in the equine field. The bulk of the data shows that probiotics have no reliable impact on diarrhea and other gastrointestinal diseases. Two preeminent researchers in the field of the equine intestinal microbiome, Costa and Weese, have described the evidence for probiotic efficacy in the treatment of equine gastrointestinal disease as "disappointing" (Costa and Weese, 2012).

According to Coverdale (2016), a wide range of microbial species have been tested in the equine sector as probiotics, such as *Lactobacillus* spp., *Enterococcus* spp., *Bifidobacterium* spp., and *Saccharomyces* spp. For instance, diets containing *L. acidophilus* had limited effects either on reducing the risk of acidosis associated with feeding high-starch concentrates to horses or on nutrient digestibility. Live yeasts have been demonstrated to elicit an increase in fiber digestibility in the colon and to modulate the balance of hindgut bacterial communities, with a decreased risk of lactic acidosis (Jouany et al., 2008). The use of these agents has resulted in ameliorated fiber digestibility in horses offered both high-starch and high-fiber diets. Although probiotics have shown to be a promising tool in the treatment of selected diseases in humans, the evidence that they can be used to control diseases in the equine sector is so far weak. Table 5 summarizes some of the most relevant investigations of probiotic application in the equine field.

TABLE 4
Probiotic Effects on Ruminants, Beneficial Effects, and Limitations (Zommiti et al., 2019).

Probiotic Microorganism	Observed Effect	References
L. acidophilus 15, *S. cerevisiae* NCDC49	Significant reduction in the incidence of diarrhea in calves fed on milk fermented	Agarwal et al. (2002)
Colicinogenic probiotic *E. coli* strains	Reduces carriage and shedding of *E. coli* O157:H7 in adult cattle	Tkalcic et al. (2003)
Colicin E7-producing *E. coli*	Daily addition of 10^8 of colicin E7-producing *E. coli* per gram of feed: reduce the fecal shedding of serotype O157:H7 in cattle	Schamberger et al. (2004)
L. acidophilus NP51	A high reduction in *E. coli* O157:H7 fecal shedding by 57%	Younts-Dahl et al. (2004)
Live yeasts	Supplementation of live yeasts into the grain reduced the number of days with diarrhea in young calves	Galvao et al. (2005)
Probiotic complexes of six *Lactobacillus* spp. of bovine and human origin	Significant reduction in the incidence of diarrhea, fecal coliform counts, and the overall mortality in veal calves	Timmerman et al. (2005)
E. coli strain Nissle 1917	Beneficial effects on the prophylaxis and treatment of neonatal calf diarrhea	Von Buenau et al. (2005)
Saccharomyces cerevisiae	Improvement of performance parameters; Increase in dry matter intake and milk production; Increase of average milk yield per day by 14% in Saanen dairy goats receiving *S. cerevisiae* at the rate of 4×10^9 CFU/day/animal	Jouany (2006); Stella et al. (2007)
Several bacterial and yeast strain supplementation	Amelioration of weight gain and rumen development in young calves	Adams et al. (2008)
L. acidophilus NP51	Daily administration of this probiotic strain decreases *E. coli* O157:H7 fecal shedding by 35% in beef cattle	Peterson et al. (2007)
Saccharomyces cerevisiae	Increases fiber digestibility rates	Guedes et al. (2008)
L. acidophilus NP51, *Propionibacterium freudenreichii* NP 24	Boosts some growth parameters when fed to finishing cattle; Effects on growth performances were inconsistent	Vasconcelos et al. (2008)
Saccharomyces cerevisiae	High levels of dry milk intake, increased dry matter and milk yield by about 0.44 g/kg and 1.2 g/kg body weight, respectively, rumen pH, rumen volatile fatty acids concentration, and organic matter digestibility	Desnoyers et al. (2009)
L. plantarum PCA 236	Significant reduction of *Clostridium* spp. counts in goats	Maragkoudakis et al. (2010)
L. johnsonii, *L. salivarius*, *L. murinus*, *L. mucosae*, *L. amylovorus*, *L. mucosae*	High in vitro microbial antagonism	Maldonado et al. (2012)
L. reuteri DDL 19, *L. alimentarius* DDL 48, *E. faecium* DDE 39, *B. bifidum* DDBA	Increased milk production in goats and polyunsaturated fatty acids; Improvement in the average body weight by 9%; Significant antimutagenic activity	Apás et al. (2014)
Propionibacterium freudenreichii *L. acidophilus* NP51	Lowers the fecal shedding of *E. coli* O157:H7 levels in beef cattle	Wisener et al. (2015)
B. amyloliquefaciens	Fed to dairy calves at the rate of 3.16 × 108 CFU per kg dietary dry matter (DM); an improvement in the growth rate by 39% (551 vs 767 g/day), Increased feed use efficiency by 14% (2.5 vs 2.9 kg milk + starter DM/kg weight gain)	Le et al. (2017)

TABLE 5
Probiotic Use in Equine Field: Success and Limitations of Clinical Trials (Zommiti et al., 2019).

Probiotic Microorganism	Effect	References
A mixture of six *lactobacilli* and two *Bifidobacterium* strains	No significant differences between two probiotic and two placebo groups in the study of the prevalence of *Salmonella* shedding, length of antimicrobial therapy, diarrhea, and stay in hospital in colic surgery horse patients	Parraga et al. (1997)
L. lactis; *Streptococcus faecium*; *S. cerevisiae*	No significant effects of probiotic application compared to the placebo group on the occurrence of *Salmonella* shedding and the intensity of diarrhea in horses hospitalized due to colic	Kim et al. (2001)
L. casei; *L. acidophilus*; *E. faecium*	High reduction in the frequency of *Salmonella* shedding by 65% in the probiotic-treated group of horses with gastrointestinal disease	Ward et al. (2004)
L. salivarius; L. reuteri; L. crispatus; L. johnsonii	Significant decrease of diarrhea incidence; body weight gain in probiotic-treated foals	Yuyama (2004)
S. boulardii	Significant reduction in the severity and duration of diarrhea in a hospital setting in probiotic-treated horses of clinical signs in horses with enterocolitis	Desrochers et al. (2005)
L. pentosus WE7	A negative effect of probiotic supplementation is reflected in a high increase in the occurrence of diarrhea and an urgent need for veterinary intervention in the probiotic group	Weese and Rousseau (2005)
S. cerevisiae; L. acidophilus; E. faecium	The study on the effect of probiotic/psyllium on fecal sand clearance in healthy horses showed a remarkable increase in sand output after 4 weeks of treatment compared to the baseline	Landes et al. (2008)
S. boulardii	No beneficial effects of the probiotic yeast on the duration of diarrhea, heart and respiratory rates, amelioration of appetite, and other parameters compared to the placebo group	Boyle et al. (2013)

Companion Animals

For the canine sector, *Lactobacillus* spp. and *Enterococcus* spp. were the most studied microbes as probiotics for dogs, and *Bifidobacterium* spp. was the most investigated microbes in the feline field.

SELECTION CRITERIA AND REQUIREMENTS FOR PROBIOTIC STRAINS

Safety, functionality, and technological utility represent the crucial criteria for the selection of probiotic microbes. This drastic selection was set up according to the World Health Organization (WHO), Food and Agriculture Organization (FAO), and EFSA (the European Food Safety Authority). Hill and coworkers (2014) revealed that probiotic potential is directly connected to a particular species, not to the genus or species of a microorganism. Regarding the safety of a bacterial strain, its origin, its antibiotic resistance profile, and the total absence of association with pathogenic cultures correspond to the main basis of the safety profile. As for the functional aspects, viability represents a prerequisite for probiotic functionality; it enhances several mechanisms including adherence to epithelial cells, reduction of mucosal gut permeability, and immunomodulatory effects, and this constitutes an industrial challenge (Galdeano and Perdigón, 2004; Kosin and Rakshit, 2006).

Lee (2009) showed that probiotic strains should answer to the requirements, in direct link with the technology of their manufacture, reflecting their ability to survive and maintain their biotechnological properties throughout the storage and distribution processes. However, the value of these parameters is still under debate as there are matters of significance, in vitro and in vivo discrepancies, and a lack of standardization to operate the procedures to be considered. As there is an obvious absence of common standards to all probiotic applications, the most relevant approach to establish the properties of a microbial strain is to target populations and target specific physiologic functions via specific studies (Saarela et al., 2000). Or, according to Ouwehand et al. (1999), it is not necessary for a potential probiotic microbial strain to fulfill all such drastic selective criteria.

Probiotics in Animal Feed: A Commercial Glance

Farm animals are exposed to the environment-related stress encompassing diet, rearing methods, weaning, and other factors. Numerous factors may cause a critical dysbiosis in the intestinal ecosystem reflecting a high risk for pathogenic infections. Regardless of the species, animal health represents a salient issue in the production chain. Probiotics and their uses in animal nutrition and feeding are directly related to their verified efficacy in maintaining the equilibrium of the intestinal microbiota, also known as intestinal homeostasis. The intake of probiotic strains may have an important impact on absorption and utilization of the feed, on metabolism, and several parameters such as daily and total body weight gain of a wide range of farm animals including poultry (chicken and turkeys), cattle (sheep and goats), pigs, and horses (Table 6).

Ready-made probiotic preparations for the animal world usually contain one, two, or a combination of bacterial strains and yeasts (Rachwał, 2003). Such formulas and their effectiveness are directly associated

TABLE 6
Examples of Probiotic Formulas Used in the Nutrition of Livestock.

Commercial Name of the Formula (Manufacturer)	Microorganisms in the Preparation
Poultry	
Avian PAC (Soluble Loveland Industries)	*Streptococcus faecium*, *Lactobacillus acidophilus*
Biogen D (Bio-Gen)	*Bifidobacterium bifidum*, *Lactobacillus acidophilus*, *Pediococcus faecium*
B.I.O.Sol (Biochem)	*Enterococcus faecium*
Cerbiogalli	*Lactobacillus*: *acidophilus*, *casei*, *plantarum*
Ecobiol (Norel animal nutrition)	*Bacillus amyloliquefaciens*
Enviva Pro (Danisco Animal Nutrition)	*Bacillus subtilis*
Farmafore soluble (Farm'apro)	*Lactobacillus*: *rhamnosus*, *farciminis*
FloraMax-B11 (Pacific Vet Group)	*Lactobacillus salivarius*, *Pediococcus parvulus*
GalliPro (Evonik Industries)	*Bacillus subtilis*
Galvit Probiotyk (Galvit)	*Enterococcus faecium*
Lavipan (JHJ)	*Lactobacillus*: *plantarum*, *casei*; *Lactococcus lactis*, *Carnobacterium divergens*, *Saccharomyces cerevisiae*
Propoul (International Company s.r.o.)	*Lactobacillus fermentum*
Pigs	
Anta Pro EF (Dr. Eckel)	*Enterococcus faecium*
Biogen N (Bio-Gen)	*Bifidobacterium bifidum*, *Lactobacillus acidophilus*, *Pediococcus faecium*
Biogen T (Bio-Gen)	*Bifidobacterium bifidum*, *Lactobacillus acidophilus*, *Enterococcus faecium*
BioPlusYC (Evonik Industries)	*Bacillus licheniformis*, *Bacillus subtilis*
Bro-biofair (Vitality Co)	*Saccharomyces cerevisiae*
Cerbiopor	*Lactobacillus*: *acidophilus*, *brevis*, *casei*, *fermentum*, *lactis*, *plantarum*; *Bacillus*: *subtilis*, *megaterium*, *pumilus*; *Enterococcus faecium*, *Cellulomonas* sp., *Saccharomyces cerevisiae*

TABLE 6
Examples of Probiotic Formulas Used in the Nutrition of Livestock—cont'd

Commercial Name of the Formula (Manufacturer)	Microorganisms in the Preparation
MicroSource S (Agtech Products Inc.)	*Bacillus*: *subtilis*, *licheniformis*
Toyocerin (Rubinum S.A.)	*Bacillus toyonensis*
EnvivaMPI (DANISCO Animal Nutrition)	*Lactobacillus*: *farciminis*, *rhamnosus*
LSP 122 (Alpharma)	*Bacillus licheniformis*
Cattle (calves)	
Probiosacc C-I	*Saccharomyces cerevisiae*
Yea Sacc (Alltech)	*Lactobacillus rhamnosus*, *Enterococcus faecium*
Poultry and pigs	
Acid-Pak-4-Way (Alltech)	*Lactobacillus acidophilus*, *Enterococcus faecium*
Calsporin (ORFFA)	*Bacillus subtilis*
Lavipan (JHJ)	*Lactobacillus*: *plantarum*, *casei*; *Lactococcus lactis*, *Carnobacterium divergens*, *Saccharomyces cerevisiae*
Microguard (PeterLab Holdings)	*Bacillus*: *licheniformis*, *megaterium*, *mesentricus*, *polymyxa*, *subtilis*; *Saccharomyces boulardii*; *Bifidobacterium bifidum*; *Lactobacillus*: *acidophilus*, *bulgaricus*, *plantarum*; *Streptococcus faecium*
Probion (Woogene B&G Co. Ltd.)	*Bacillus subtilis*, *Clostridium butyricum*, *Lactobacillus acidophilus*
Probios (Chr Hansen)	*Lactobacillus*: *acidophilus*, *casei*, *plantarum*, *lactis*; *Enterococcus faecium*; *Bacillus subtilis*
Poultry and calves	
Probiomix	*Bifidobacterium bifidum*; *Lactobacillus amylovorus Enterococcus faecium*
UltraCruz (Santa Cruz Animal Health)	*Enterococcus faecium*, *Lactobacillus*: *acidophilus*, *casei*, *plantarum*
Calves and pigs	
Cernivet LBC (Cerbios)	*Enterococcus faecium*
Provita LE (Schaumann)	*Lactobacillus rhamnosus*, *Enterococcus faecium*
Poultry, pigs, and calves	
Bio Plus2B (Chr. Hansen)	*Bacillus subtilis*, *Bacillus licheniformis*
Cylactin (DSM)	*Enterococcus faecium*
Doctor Em (Biotron)	*Lactobacillus*: *paracasei*, *plantarum*; *Lactococcus lactis*, *Saccharomyces cerevisiae*
Lactiferm	*Enterococcus faecium*
Oralin (Chevita GmbH)	*Enterococcus faecium*
Poultry, pigs, calves, and horses	
Pro-Biotyk em15 (ProBiotics)	*Bacillus subtilis*, *Bifidobacterium*: *animalis*, *bifidum*, *longum*, *Lactobacillus*: *acidophilus*, *casei*, *delbrueckii* subsp. *bulgaricus*, *fermentum*, *plantarum*; *Lactococcus lactis* subsp. lactis; *Saccharomyces cerevisiae*; *Streptococcus thermophilus*
Super-CyC (Choong Ang Biotech Co. Ltd.)	*Bacillus subtilis*, *Saccharomyces cerevisiae*

(Continued)

TABLE 6
Examples of Probiotic Formulas Used in the Nutrition of Livestock—cont'd

Commercial Name of the Formula (Manufacturer)	Microorganisms in the Preparation
Poultry, pigs, sheep, and cattle	
Protexin (Protexin Probiotics International Ltd.)	*Lactobacillus*: *acidophilus*, *delbrueckii* subsp. *bulgaricus*, *plantarum*, *rhamnosus*; *Bifidobacterium bifidum*; *Streptococcus salivarius* subsp. *thermophilus*; *Enterococcus faecium*; *Aspergillus oryzae*; *Candida pintolepesii*
Poultry, pigs, beef, dairy, horses, and deer	
PrimaLac (Star Labs, Inc.)	*Bifidobacterium*: *bifidum*, *thermophilus*; *Enterococcus faecium*; *Lactobacillus*: *acidophilus*, *casei*

Reproduced with permission from Markowiak, P., Śliżewska, K., 2018. The role of probiotics, prebiotics and synbiotics in animal nutrition. Gut Pathog. 10, 21 (with slight modifications).

with two major key factors: the accurate selection of the strain(s) and the unitary dose containing an appropriate count of viable cells. To preserve the properties of probiotic preparations and formulas, these beneficial microbes have to be stored and used according to the instructions of the manufacturers. Due to the presence of viable microorganisms in the commercial preparations, the latter is highly vulnerable to critical conditions such as temperature and light. It is pertinent to know that no other substances should be added while probiotics are administered, and the water used for dilution should contain no chlorine or other disinfectants (Szeleszczuk, 2005). Formulas containing many ingredients, in terms of the highest number of microbial species, are the most effective. Commercial probiotic formulas available in the market are listed in Table 7.

LABELING OF PROBIOTICS USED IN ANIMAL FEED

The packaging of commercial probiotic products should be labeled with clear and accurate information in terms of content, positive effects of the products, date of expiry, dose rates, contraindications (if any), and other information. However, commercial probiotics are often inadequately or incorrectly labeled. In 2003, a pilot study carried out by Weese in which he suggested that an ideal probiotic label "should state the organisms that are present to the strain level, correctly spell and identify the contents, state the number of live organisms, and guarantee that the stated number would be present at the time of expiry." In the same line, Weese (2003) emphasized on the dose rate that should be used for different categories of animals; crucial information is often neglected on the labels.

TABLE 7
Examples of Prebiotic Formulas Available in the Market and Intended for Livestock.

Commercial Name of the Preparation (Producer)	Prebiotic Ingredients	Animal Receptor
Bacto CS1000	Polysaccharides, oligosaccharides	Poultry
BionatStart	MOS, β-glucans	Calves
DOLSORB DN (Dolfos)	MOS, β-glucans	Poultry
MetSac MOS (VITTRA)	MOS, β-glucans	Calves, pigs, and poultry
Mycocyd forte (Herbiline)	β-glucans	Poultry
Mycostop (Extra-vit)	MOS, β-glucans	Poultry, pigs
PROFEED (Beghin Meiji)	scFOS	Calves, horses, pigs, and poultry

FOS, fructo-oligosaccharides; *MOS*, malto-oligosaccharides; *scFOS*, short-chain fructo-oligosaccharides.
Data from Markowiak, P., Śliżewska, K., 2018. The role of probiotics, prebiotics and synbiotics in animal nutrition. Gut Pathog. 10, 21.

In fact, a limited number of investigations have examined the quality and authenticity of probiotic labeling. In 2011, Weese and Martin showed that the labeling of commercial probiotics was so tenuous. In this field, the most common errors in the labeling were failing to mention specific names of the microorganisms in the product, failing to give the number of viable microorganisms in the product, giving contradictory information, not mentioning the expiry date, and misspelling the microbial name (Weese, 2003; Weese and Martin, 2011). Weese (2003) demonstrated in his same investigation that the majority of pharmacies, manufacturers, health food stores, grocery stores, companion pet stores, and veterinary clinics use vaguely descriptive terms on labels of commercial probiotics such as "dried Lactobacillus," "Lactobacillus cultures," "probiotic cultures," "fermentation products," and so on, instead of specific names of the microorganisms in the product.

A significant number of producers misspelled the name of microbes, including using obsolete names and even listing the names of microbes that did not exist (Weese, 2003). This latter reported that very few of the products had labels with the name of microorganisms to the strain level. Likewise, not all products had enough data about the number of viable bacteria, and even if the information was present, it was not clear whether the specified quantity was at the time of manufacture or at the time of expiry (Weese, 2003). Only about 8% of the studied products had both a satisfactory label and the number of viable bacteria as claimed on the label (Weese and Martin, 2011).

The most critical probiotic labeling errors occurred from wrong information, such as labeling the product as yeast instead of *Lactobacillus* sp., or claiming to have bacteria present that were not detected in the product, or claiming to have more bacteria than were actually present in the products (Weese, 2002; Lata et al., 2006). In the same context, the addition of microorganisms with no proven probiotic effects and the inclusion of potentially pathogenic microorganisms in commercial products were other solemn issues revealed (Weese, 2002).

"KILLED" PROBIOTICS—BETWEEN MYTH AND REALITY

Despite the safety of probiotic microorganisms, there is always skepticism on whether the administration of live organisms has zero risks, particularly in certain populations (von Wright, 2005; Boyle et al., 2006). Therefore, it is of considerable interest to determine whether the health benefits of probiotics can be attained without the risks associated with the administration of a live microorganism (Adams, 2010). An excellent review has been written by Kataria et al. (2009) in which it has been reported that heat-killed, ultraviolet-inactivated *L. rhamnosus* GG, better known as LG, and even components of these agents may be just as effective and considerably safer for the host. These constituents also seem to induce milder immunological reactions to the host. In the world of animals, Zhang et al. (2005) demonstrated that the administration of *S. cerevisiae* cell components to broilers significantly enhanced growth performance parameters, meat tenderness, and ileal villus development. Nonetheless, there is a substantial argument on whether dead microbes or components of their carcasses "may be a safer alternative."

PREBIOTICS: FROM CONCEPT TO APPLICATIONS

According to the ISAPP, a prebiotic is currently defined as a "substrate that is selectively used by host microorganisms conferring a health benefit" (Gibson et al., 2017). Prebiotic components can also be added to different food formulations to develop various products with functional claims that would attract consumers keen seeking healthy foods and health (Hutkins et al., 2016). Their presence, these nondigestible carbohydrates, in the GIT may induce the development and/or metabolic activation of beneficial microbes residing in the intestinal microflora through the selectivity of the substrate (Martinez et al., 2015).

The most identified prebiotics are carbohydrates and oligosaccharides with various molecular structures normally occurring in the human and animal diet. Dietary carbohydrates, such as fibers, are candidate prebiotics, but the most promising are nondigestible oligosaccharides (NDOs). Actually, the main well-known and the most used prebiotics are nondigestible carbohydrates such as galacto-oligosaccharides (GOS) (Fan et al., 2019), fructo-oligosaccharides (FOS), inulin (Xavier dos Santos et al., 2019; Xavier-Santos et al., 2019), and lactulose (Zeng et al., 2019). Numerous other nondigestible carbohydrates have been assessed for their prebiotic potential, encompassing xylo-oligosaccharides (XOS) (Madhukumar and Muralikrishna, 2012), soybean oligosaccharides (Ma et al., 2017), isomalto-oligosaccharides (IMO) (Wu et al., 2017), xylo-polysaccharide (XPS) (Ho et al., 2018), polydextrose (Costa et al., 2019), arabinoxylan (Chen et al., 2019), and β-glucans (Velikonja et al., 2019). However, the majority of the data available in the scientific literature on prebiotics and their effects are directly related to FOS and inulin (Martinez et al., 2015). These nonabsorbable feed ingredients promote the growth and metabolic activity of the presumably beneficial gut bacteria, particularly species of *Bifidobacterium* and *Lactobacillus*. According to Sartor (2004), prebiotic components may confer numerous health benefits via the production of SCFAs, encompassing butyrate and lactate, reflecting a significant cytokine production within the intestinal mucosa.

Currently, in the animal world, a wide list of the nondigestible oligosaccharides is added to animal feed crosswise various animal species. Among them, we can find FOS), mannose oligosaccharides (MOS), lactulose and GOS, chito-oligosaccharides (COS), arabinoxylan oligosaccharides (AXOS), XOS, trans-galacto-oligosaccharide (TOS), glucan, inulin, levan, polydextrose, peptidoglycan, chitin, yeast cell wall, inactivated yeast bacteria, galacto-mannan, galacto-mannan-oligosaccharides (GMOs), galacto-glucomannan-oligosaccharide-arabinoxylan complex (GGMO-AX), acidic oligosaccharides (AOS), arabino-galactan, phosphorylated mannans (MAN), arabino-xylan, and mannobiose (Anadón et al., 2016). These feed ingredients have a direct impact on the intestinal ecosystem, particularly via improving and boosting lactic acid fermentation.

Diverse investigations have determined the major physiological effects promoted by prebiotic supplementation, particularly nondigestible oligosaccharides, in terms of their chemical structure, the nature of the glycosidic bonds, degree of polymerization, fermentability, level of solubility, and viscosity (Rastall, 2010; Rastall and Gibson, 2015; Singh et al., 2017; Chen and Karboune, 2019). In the same line, the level of efficiency of a prebiotic component is directly associated with the composition of the resulting products from its metabolization by the colon bacteria (Chen and Karboune, 2019). The latter proved that the design of nondigestible carbohydrates as prebiotics demands a wide knowledge about its distinct metabolic pathways especially among probiotic bacteria reflecting a tight connection between the type of prebiotic biomolecule and the extent of its therapeutic efficacy (Chen and Karboune, 2019).

In humans, the addition of prebiotics and prebiotic feed ingredients to the diet has brought positive aspects to the gut microbial homeostasis. However, the use of prebiotics in animal nutrition and production, as a promising alternative to antimicrobial growth promoters, has given conflicting consequences, whereas their use in the modulation of the gut microbial equilibrium is worthwhile. They contribute to the establishment of a "healthier" microflora where bifidobacteria and/or lactobacilli become predominant and exert possible health-promoting effects at the expense of more harmful species.

The "probiotic approach" adds one or two species to a wide spectrum of various species in the gut flora; nevertheless, the "prebiotic approach" aims at fertilization of the intestinal ecosystem. Two pilot studies conducted by Patterson and Burkholder (2003) and Gaggìa et al. (2010) demonstrated that techno-functional imitations of the naturally occurring prebiotics GOS and FOS, enhance intestinal growth of bifidobacteria as a marker of probiotics in a dose-dependent manner.

PREBIOTICS: WHAT ARE THEIR BENEFIT(S) FOR ANIMALS?

Prebiotics have been widely investigated and used for companion animals and animal husbandry, encompassing livestock, poultry, and aquaculture. According to Stevens and Hume (1998) and Gibson et al. (2017), the intrinsic differences among animal species in terms of the living environment, anatomy and physiology, nutritional composition, and reliance on the gut microbiota for energy must be taken into consideration when assessing the impact of prebiotics on animal health.

The majority of prebiotic components seem to stimulate lactic acid and bifidogenic bacteria. In animal husbandry, these nondigestible ingredients may confer various functions including their attachment to pathogens, improvement of osmosis in the lumen of the intestine, serving as substrates for fermentation, a significant modulation of the immune system reflecting a high response of macrophages, and a notable production of SCFAs (Patel and Goyal, 2012).

Poultry

Poultry, which is used primarily for the production of meat or eggs, include land fowl species (i.e., chickens, turkeys, and quail) and waterfowl species (i.e., ducks and geese), which respond to prebiotics although most have a fairly short midgut and hindgut that includes a short, straight colon, and twin ceca. Although the concept of functional foods has been introduced decades ago, scientific data for the use of prebiotic ingredients in animal nutrition exists from the late 1990s essentially for poultry and pigs (Hajati and Rezaei, 2010). The majority of investigations conducted in prebiotic components has been performed in the field of poultry, as this is the most studied monogastric animal. Most of the experiments conducted with these feed components have focused on improving the microbial health, performance parameters, and decreasing carcass contamination of young meat birds.

Prebiotics were found to increase the stool volume of chicken via regulating intestinal microbial flora through selective stimulation of beneficial bacterial populations such as bifidobacteria and inhibiting pathogenic bacteria, including *Salmonella* and Clostridia (Sims et al., 2004; Cao et al., 2005; Totton et al., 2012; Park et al., 2013). According to Park et al. (2013), the most common prebiotics used in monogastric animals is inulin, FOS, mannan-oligosaccharides (MOS), and GOS, with GOS being the least investigated in the poultry industry compared to FOS. In 2013, two pilot investigations performed by Allaart et al. and Samanta et al. reported the critical situation in terms of drawing final conclusions on the prebiotic effects in animals from published studies, due to the wide assortment of these investigations concerning subjects, age, diet, weight, performance parameters, substances tested, dose, duration of the trials, and outcomes. Such discrepancies in terms of different reported results have been mostly recorded after the use of MOS to reduce the intestinal numbers of *Clostridium perfringens* in poultry and after the use of inulin as a prebiotic ingredient to boost the growth performance parameters of layers and broilers, reflecting that the effects are both dose- and diet-dependent (Yusrizal and Chen, 2003a,b; Biggs et al., 2007; Ortiz et al., 2009; Allaart et al., 2013).

Prebiotic research on poultry has been performed since 1990 and, as a result, a large database of research is accessible in this area. For instance, prebiotics in broiler diets have been revealed to be so important to increase the beneficial bacterial population count, in particular, the lactobacilli count in the GIT (Xu et al., 2003; Yusrizal and Chen, 2003b; Baurhoo et al., 2007). Some authors recorded a significant reduction in the counts of Salmonella and Coliforms (Spring et al., 2000), whereas others observed a decrease in *Escherichia coli* (Xu et al., 2003; Zduńczyk et al., 2005; Baurhoo et al., 2007). Numerous other pathogenic bacteria such as bacilli, staphylococci, streptococci, and even yeast, have also been announced to decrease with prebiotic supplementation (Samarasinghe et al., 2003; Cao et al., 2005).

In terms of assessment of performance parameters in broilers, the supplementation of prebiotic dietary ingredients has shown an obvious improvement in body weight (Yusrizal and Chen, 2003a; Sims et al., 2004; Zduńczyk et al., 2005). In the same line, body weight gain, feed conversion, and carcass weight were significantly boosted (Samarasinghe et al., 2003; Xu et al., 2003; Yusrizal and Chen, 2003a; Sims et al., 2004; Józefiak et al., 2008; Yang et al., 2008). In 2003, two separate studies carried out by Samarasinghe et al. and Yusrizal and Chen showed that MOS and inulin supplementation increased carcass weight and abdominal fat weight. According to other investigations, contradictory results have been revealed showing a reduction in feed intake and feed:gain ratios (F:G) with the supplementation of fructans and MOS (Samarasinghe et al., 2003; Xu et al., 2003; Yusrizal and Chen, 2003a; Baurhoo et al., 2007).

In the pathway of fermentation profiles, the prebiotic supplementation of poultry diets seemed to play a key role. For instance, increased butyrate levels have been recorded (Zduńczyk et al., 2004; Yang et al., 2008). According to Cao and collaborators (2005), nutrition supplemented with FOS reduced cecal indole and phenol concentrations. A significant correlation has been reported between the improvement of the total SCFA and lactic acid concentrations, and the intestinal pH decreased when prebiotics were supplemented (Yang et al., 2008; Zduńczyk et al., 2005). As regards the intestinal morphology, and increased intestinal villus height was reported when prebiotics were included in the broiler nutrition (Baurhoo et al., 2007). Other changes in intestinal characteristics have also been observed, encompassing increased gut length (Yusrizal and Chen, 2003a). In a pilot investigation conducted by Świątkiewicz and Arczewska-Włosek (2011), the study on the effect of prebiotics on mineral utilization and bone quality in broilers showed negative results. Table 8 summarizes the main effects of prebiotics on poultry production.

To wrap up, the application of prebiotics in animal feed is a relatively recent effort. Although the outcomes are promising, numerous critical concerns must be resolved, such as the establishment of the efficacy of prebiotics in the routine diet of livestock. Advanced approaches such as next-generation sequencing could be very useful to corroborate any potential prebiotic impact on animal microbiota whereas, at the same time, future research on prebiotic feed ingredients in livestock should focus on the immunological aspects, changes at the gut epithelium, and livestock product quality (Samanta et al., 2013).

Pigs

In swine production, different types of chemically defined or undefined dietary compounds are added to the diet of pigs to test their influence on gastrointestinal microbiota and/or on health status improvement during the challenge with pathogens. Prebiotics are known to influence the GIT microflora in pigs. For instance, in a trial conducted by Smiricky-Tjardes et al. (2003), they demonstrated that the supplementation

TABLE 8
Effects of Prebiotics on Growth, Health, and Performance Parameters in Poultry.

Subjects	Prebiotic	Main Results	References
Absorption and utilization of feed, diarrhea, body weight gain			
Broiler chickens	FOS	Significant positive effects on the daily growth of studied animals; improvement in the growth of *Bifidobacterium* and *Lactobacillus* bacteria; high inhibition of *Escherichia coli* in experimental animals gastrointestinal tract	Xu et al. (2003)
Broiler chickens	Fructans from chicory	Boosting body weight gain, feed turnover, and reduction in serum cholesterol	Yusrizal and Chen (2003a)
Turkeys	MOS	Improvement in the growth of experimental animals	Sims et al. (2004)
Chickens	MOS	Improved daily weight gain, feed intake, and feed conversion ratios	Flemming et al. (2004)
Broiler chickens	Bio-MOS	Enhancement of growth performance parameters of birds	Hooge (2004)
Turkeys	MOS, inulin	No increase in feed consumption or enhancement in the bodyweight of experimental animals. A higher SCFA concentration was found in animals fed with prebiotics, compared to the control	Stanczuk et al. (2005)
Turkeys	FOS	No effect on the growth and productivity of experimental animals. Reduction in the intestinal pH was noted in the case of FOS administration at a concentration of 2%	Juśkiewicz et al. (2006)
Chickens	Inulin, oligofructose, MOS, short-chain oligosaccharide, TOS	No significant body weight gain. An excessively high prebiotic dose may lead to a negative impact on the gastrointestinal system reflecting an obvious delay in the process of growth in animals	Baurhoo et al. (2007)
Chickens	FOS	Improved broiler's weight gain about 5%–8% and a feed conversion ratio of about 2%–6%	Yang et al. (2009)

TABLE 8
Effects of Prebiotics on Growth, Health, and Performance Parameters in Poultry—cont'd

Subjects	Prebiotic	Main Results	References
Intestinal ecosystem imbalance, pathogenic infections			
Chickens	Fructans from artichoke	Reduced *Clostridium perfringens* count and bacterial endotoxin levels	Kleessen et al. (2003)
Broiler chickens	Fructans from chicory	An improvement in the *Lactobacillus* genus bacteria count and reduction in the count of potential pathogens, such as *Salmonella* and *Campylobacter*	Yusrizal and Chen (2003a)
Broiler chickens infected with *Salmonella typhimurium*	IMO	A significant reduction in the *Salmonella typhimurium* count. Digestion and effectiveness of the administered nutrition were not significantly different from the control group. An important loss of weight in the case of animals fed with 1% IMO compared to the control group. The supplementation with IMO caused an increase in the *Bifidobacterium* count in the GI system of experimental animals	Thitaram et al. (2005)
Improved quality of meat, milk, eggs			
Eggs from Ross 308 broiler	DiNovo (DN; laminarin and fucoidan), Bi2tos (BI; nondigestive TOS)	Summing up, the administration of prebiotics in ovo resulted in an improvement of many parameters in terms of increase of BW, FCR, carcass weight, the weight of the myocardium, and weight of the breast significant for the commercial production of poultry	Maiorano et al. (2017)

BW, body weight; *FCR*, feed conversion ratio; *FOS*, fructo-oligosaccharides; *GOS*, galacto-oligosaccharides; *IMO*, isomalto-oligosaccharides; *MOS*, manno-oligosaccharides; *TOS*, transgalacto-oligosaccharides.

of TOS at 35 g/kg in a diet for growing pigs led to a significant augmentation in fecal bifidobacteria and lactobacilli without any improvement of growth performance parameters. However, MOS isolated from the *S. cerevisiae* cell wall showed a beneficial effect on the intestinal microbiota, as well as on animal growth (Shim et al., 2005). It has been suggested that they suppress the growth of several sturdy pathogens such as *Clostridium botulinum*, and *Clostridium sporogenes*, *E. coli*, and *Salmonella typhimurium* and conversely, stimulate the growth of *B. longum*, *L. casei*, *L. acidophilus*, and *Lactobacillus delbrueckii*. The use of a novel blend of GOS by Tzortzis et al. (2005) at 40 g/kg feed in a three-step system of continuous culture resulted in an important increase in *bifidobacteria* and acetate concentration, with a simultaneous reduction in intestinal pH, compared with the control group and a control diet supplemented with inulin. Additionally, the investigation of this oligosaccharide mixture strongly inhibited the adhesion of enterotoxigenic *E. coli* (ETEC) and *Salmonella enterica* serotype *typhimurium* to HT29 cells in in vitro studies (Tzortzis et al., 2005). Prebiotics such as oligosaccharides of fructose, mannose, and chitin protect piglets against high environmental stressors (i.e., transport, intensive breeding system, and antibiotics) and sturdy pathogen loads, encompassing fecal *E. coli* shedding and reducing infection-associated responses to *S. enterica* serovar *typhimurium* infection or porcine reproductive and respiratory syndrome virus (Liu et al., 2008; Che et al., 2011).

In 2003, Mikkelsen and collaborators experimented with the incorporation of several oligosaccharides into swine diets at levels oscillating from 5 to 40 g/kg diet. They found contradictory results in terms of no significant effects regarding the beneficial modulation of microbial communities determined in various intestinal segments and feces of swine. Similar results have been reported with Mountzouris et al. (2006), who illustrated that the dietary treatment with FOS or TOS showed no influence on the beneficial bacterial population but led to an obvious promotion of saccharolytic activities in the porcine colon based on the value of the total volatile fatty acids (VFA), acetate level and glycolytic activities. In a separate trial performed by Modesto et al. in 2009, it has been reported that GOS from milk whey and sugar beet fructo-oligosaccharides (sbFOS) added to the diet of weaned pigs in different amounts had no effect on the hindgut microbiota, except for sbFOS at a concentration of 40 g/kg, which tended to increase the endogenous bifidobacteria; whereas, the growth performance was not influenced.

Prebiotics, similar to probiotics, give various and contradictory results in swine production. They sometimes modulate the microbiota toward beneficial bacteria, such as bifidobacteria and lactobacilli, enhancing the intestinal defense systems in terms of immunomodulation, pathogen displacement, and bacteriocin production. These nondigestible feed ingredients rarely affect growth performance positively.

Ruminants

According to Quigley et al. (1997), calves are born in a preruminant state and function as nonruminants until the rumen and other compartments of the stomach fully develop. During the first few weeks of life, or longer, in the case of veal calves maintained on low-roughage diets (low in fibers and fibrous ingredients), prebiotics can be used to enhance growth, improve feed conversion ratio (FCR), reduce the frequency and severity of scours in terms of diarrhea and/or reduce the occurrence of respiratory diseases (Quigley et al., 1997; Ghosh and Mehla, 2012; Roodposhti and Dabiri, 2012).

Two popular oligosaccharides used in domestic livestock are MOS and FOS (Verdonk et al., 2005; Hill et al., 2008; Morrison et al., 2010). The use of prebiotics in ruminants, particularly in cattle, has been limited by the ability of this category of animals to degrade most prebiotics. Nevertheless, the enhancement of rumen-protective technologies may allow these feed substances to be used in the feedlot and dairy cattle (Callaway et al., 2008), reflecting the presence of various classes of nondigestible oligosaccharides in the plant cell wall in nature including the plants normally used for livestock nutrition. Conversely, the main prebiotic feed ingredients used in preweaned calves are cello-oligosaccharides, galactosyl lactose, manno-oligosaccharides, and yeast cell wall extracts.

The outcomes of Heinrichs and coworkers in 2003 showed that the addition of MOS to the diet of Holstein calves ameliorated fecal scores just as for antibiotic treatment when compared to the control milk replacer, whereas any improvement of body weight has not been noticed. According to de Vaux et al. (2002), the supplementation of sorbitol, L-arabinose, trehalose, and rhamnose to the cattle rumen medium displaced *E. coli* O157:H7 within 72 h. The overall studies on the impact of forage and concentrate diets on fecal shedding and colonization of the gut by *E. coli* O157:H17 are still unclear and little data are available as shown by Lema et al. (2002). However, the manipulation of the fiber content could bring a new outlook in terms of maintaining the animals on a concentrate diet without sacrificing cattle weight gain. In the same line, the use of prebiotics and prebiotic feed ingredients can reduce the adhesive potential of certain pathogenic species to the intestinal epithelium and can either prevent this dysbiosis in the intestinal tract and stop the incidence of diarrhea cases (Callaway et al., 2003), or reduce its prevalence in calves (Morrison et al., 2010). In 2008, Hill and collaborators suggested that supplementation of prebiotics have a prophylactic potential to diarrhea when they are first administered to a healthy animal, and specifically, they stated that there are no studies showing successful treatment of diarrhea using prebiotics.

The potential role of prebiotics in improving the health and performance of calves is of increasing interest because of the public concern about the use of antimicrobials in cattle production (Terré et al., 2007). It is important to know that the use of prebiotics has been augmented as a substitute therapeutical approach that prevents the use of antibiotics and, thus, reduces the emergence and spread of antibiotic-resistant bacteria and residual antibiotics in dairy foods, meat, and milk (Hill et al., 2008). Inversely, an accord has not been reached as to whether prebiotics may be effective in improving animals' growth performance or reducing the prevalence of gastrointestinal diseases in calves.

Horses

Horses are large nonruminant herbivores that rely principally on microbial fermentation for energy, with more than half of their maintenance energy requirement coming from microbial fermentation occurring in their enlarged caecum and colon (Stevens and Hume, 1998). As their typical diet is high in roughage, fibrous materials, and feedstuffs that are consumed all through the day, prebiotic interventions might improve the effectiveness of fermentation (Morgan et al., 2007; Respondek et al., 2011). It is pertinent to know that the use of prebiotics has stimulatory potential on LAB encompassing *Lactobacillus*, *Bifidobacterium*, and *Enterococcus*. Many of these have been used successfully as efficient equine probiotics. Prebiotic feed ingredients may play a crucial role as a substrate for LAB, in terms of improving the microflora of the large intestine. For instance, the supplementation of *S. cerevisiae* fermentation products in conjunction with low-quality forage has led to a significant improvement in terms of apparent digestibility of dry matter, crude protein, and neutral detergent fiber, indicating that prebiotic supplementation is more effective when high-starch, high-fiber, or low-quality forage diets are fed.

Starch intake after supplementing the diet of the horse with short-chain fructo-oligosaccharides (scFOS) (Respondek et al., 2011) or MOS have beneficial effects in preventing digestive disorders associated with both the prebiotics. Furthermore, the use of scFOS in horses has shown a significant reduction in disruptions in colonic microbial communities after an abrupt change in the diet and altered fecal VFA concentrations toward propionate and butyrate (Coverdale, 2016). The lack of investigations in terms of equine digestibility using prebiotics as digestive aids had led to inconsistency in results. Similar to the studies on the use of probiotics, the results are variable.

COMPANION ANIMALS: CATS AND DOGS

Prebiotics may be considered as functional food ingredients. They are attracting considerable interest from pet owners and pet food manufacturers. Dogs and cats evolved as Carnivora eating diets high in protein and fat but low in fibrous ingredients. According to Stevens and Hume (1998), these nonruminant animals with short, simple GIT have little capacity to ferment nondigestible components, which occurs mainly in the colon. However, numerous health benefits have been achieved with prebiotic administration in terms of reduction in infections, enhancement of insulin sensitivity, and better fecal consistency (Respondek et al., 2008; Verbrugghe et al., 2009). Several studies have investigated the impact of prebiotics as dietary supplements on the microflora in healthy cats and dogs. Two pilot studies performed by Sparkes and coworkers (1998a,b) have shown that the supplementation of FOS at 0.75% dry matter produced drastic qualitative and quantitative variations in the fecal flora of healthy cats. Compared with the samples from cats fed a basal diet, increased numbers of lactobacilli and Bacteroides species, and decreased numbers of *E. coli* were observed and associated with the FOS diet. However, bacteriological examination of the duodenal juice in these same cats showed significant variation in the composition of the duodenal flora, across sampling periods, which was not affected by the same prebiotic supplementation. In a separate experiment, Kanakupt and his team (2011) demonstrated that healthy cats fed a diet containing short-chain GOS and FOS had greater fecal *Bifidobacterium* species populations and butyrate concentrations versus the control. There is a single report assessing the effects of GOS on the fecal microflora in healthy cats and cats with inflammatory bowel disease (IBD) (Biagi et al., 2013). Using a randomized, double-blinded, crossover feeding trial, oligonucleotide probes targeting specific bacterial populations showed no significant differences in the fecal microbiota of IBD cats and healthy cats fed on the same diet. Overall, interanimal variation was moderately high whereas a trend of increased *Bifidobacterium* species levels was observed with the GOS supplementation. In cases of trials on dogs, Willard et al. (2000) revealed that healthy Beagle dogs fed on a 1% FOS diet over a 3-month experiment showed inconsistent fecal excretion of both *Lactobacillus* and *Bifidobacterium* species.

PREBIOTICS IN AQUACULTURE

All fish have a quite simple and short GIT. The short length and simple structure in terms of the absence of special adaptations and segments of the fish gut, results in the rapid transit of digested material, limiting the time available for microbial or prebiotic activity.

The prebiotic feed ingredients play a key role in increasing the beneficial gut resident commensal bacterial populations, resulting in the improvement of the host health. These indigestible fibers are found to stimulate the growth of intestinal bacterial species in aquatic species. The most efficient doses of prebiotic components in aquatic host species are oscillating between 1% and 3% (weight/volume) of the diet (Li et al., 2007; Hoseinifar et al., 2013, 2014). The benefaction of prebiotics results from the by-products generated from their fermentation by gut commensal bacteria. FOS, MOS, inulin, and β-glucans are called immuno-saccharides. In farmed aquatic fish, a wide range of beneficial effects has been observed encompassing a direct boost of the innate immune responses with phagocytic activation, neutrophil activation, activation of the alternative complement system, and an increase in lysozyme activity (Table 9). According to Kyu Song et al. (2014), immuno-saccharides directly activate the innate immune system via interacting with pattern recognition receptors (PRR) expressed on innate immune cells. They can also associate with microbe-associated molecular patterns (MAMPs) to activate innate immune cells. Appropriate immune responses are crucial for combating pathogens and weight gain. Various investigations have indicated that immuno-saccharides are beneficial to both fish and shellfish. In 2009, Xu et al. reported that the prebiotic MOS boosts growth and enhances digestive enzymes such as protease and amylase.

Prebiotic Selection Criteria

The investigation of the prebiotic potential of a component is based on numerous criteria encompassing its origin, source, chemical composition, purity, and structure. Prebiotics have to cover safety guidelines required by all the countries worldwide, including the possession of the generally recognized as safe (GRAS) status, adequate dose, absence of any kind of contaminants and/or impurities, and nonalteration of intestinal microflora reflecting the absence of negative effects on the host and side effect assessment. In 2007, FAO experts showed that the use of the term prebiotic is sternly related to the microbiota modulation in a specific site reflecting its beneficial health effects on the host.

The concept of prebiotics has been developed over time. For a dietary substrate to be classed as an effective component, it is proposed that a candidate prebiotic must fulfill numerous crucial criteria, specifically the five basic criteria, which have to be proven via in vitro and in vivo investigations (Wang, 2009) (Fig. 2). For instance, (i) the substrate must not be hydrolyzed or absorbed in the upper sections of the alimentary tract (the stomach or small intestine); (ii) prebiotics must be selective for beneficial commensal bacteria in the large intestine such as the bifidobacteria; (iii) the fermentation of the substrate should introduce beneficial luminal (systemic) effects within the host (variation in metabolic processes and enhancement of the immunological system); (iv) the prebiotic ingredient may lead to selective stimulation of growth of the probiotic bacteria, and (v) the importance of the prebiotic ingredient in food/feed processing conditions in terms of its techno-functional features, stability, and availability for bacterial metabolism in the intestine. The effects of dietary fibers on upper and lower GIT are summarized in Table 10.

TABLE 9
Prebiotics as Immunostimulants in Aquaculture (Anadón et al., 2016).

Oligosaccharides	Polysaccharides
Arabinoxylan-oligosaccharide (AXOS)	Inulin
Fructo-oligosaccharides (FOS)	β-Glucan
Galacto-oligosaccharide (GOS)	Chitin/chitosan
Mannan-oligosaccharide (MOS)	
Mannan-oligosaccharide (MOS) + β-glucan	

CONCLUSIVE REMARKS AND FUTURE OUTLOOK

Numerous studies in farm animals and healthy or stressed animals have demonstrated the effects of these feed additives, probiotics, and/or potential probiotics in poultry, pigs, and ruminants, via enhancing the number of beneficial bacteria and reducing the potential pathogen load. A promising path of probiotics and their beneficial effects on animal performance and welfare has been reported. Besides, the blocking limitation for the widespread and reliable use of probiotics is the uncertainty in the reproducibility of their impact, with a wide diversity of probiotic species and strains, livestock species, and agricultural practices highlighting the complexity of the interactions in animal production systems. The study of the probiotics' mechanisms of action is a complicated network where, from human studies, it is possible to increase knowledge and extrapolate on the animal field and vice versa.

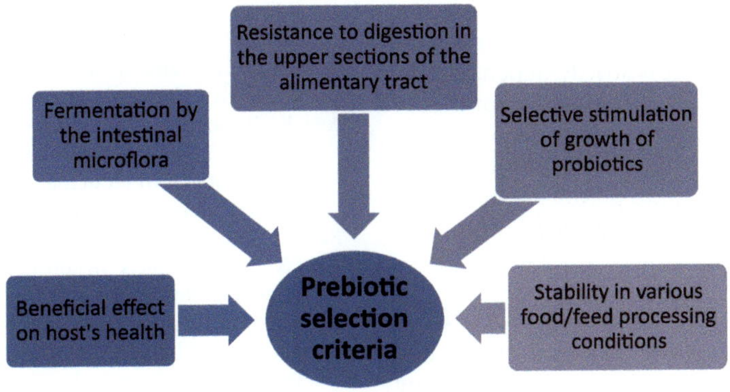

FIG. 2 Requirements for potential prebiotics (Wang, 2009).

TABLE 10
Intestinal Functions Assigned to Prebiotics and Prebiotic Ingredients.

Dietary Feed Ingredients and Gastrointestinal Functions	
Outcomes on upper GIT	Resistance to digestion
	Delay of gastric emptying
	Increase of the oro-cecal transit time
	Reduction in glucose absorption and low glycemic index
	Hyperplasia of the small intestinal epithelium
	Stimulation of secretion of intestinal hormonal peptides
Outcomes on lower GIT	Acting as food for colonic microflora and as substrates for colonic fermentation
	Production of fermentation end products (mainly SCFAs)
	Stimulation of saccharolytic fermentation
	Acidification of the colonic content
	Hyperplasia of the colonic epithelium
	Stimulation of secretion of colonic hormonal peptides
	Bulking effects on stool production
	Regularization of stool production (frequency and consistency)
	Acceleration of the ceco-anal transit

From Gaggìa, F., Mattarelli, P., Biavati, B., 2010. Probiotics and prebiotics in animal feeding for safe food production. Int. J. Food Microbiol. 141, S15–S28 (slightly modified).

Zero risk does not exist. This leads to continuing uncertainty about the efficacy and safety of pre- and probiotics in accordance with the little evidence that these beneficial microbes and the agents associated are absolutely safe. In a nutshell, not all problems can be addressed via the administration of probiotics. New horizons and alternatives should be explored.

All these mentioned various effects reflect an impression of inconsistency in terms of results; thus, some investigators remain skeptical about their reported beneficial outcomes. However, it is pertinent to record that inconsistencies in the results represent a salient feature of every new development. Our belief is that probiotics and/or prebiotics are not a panacea and should be used after critical thinking. They can be introduced in cases similar to those described as successful in the scientific literature. If novel applications are sought, a large number of target animals should be used, and the product should be drastically tested. Failure does not necessarily mean that probiotics and prebiotics or their association do not work. Their effect is strictly dependent on several parameters and factors. A trial-and-error theory should be practiced at least in the first years of investigating their effects. Their potential to substitute for antibiotics in simple illnesses and to produce "ecological" meat products may also guide practitioners' decisions.

COMPLIANCE WITH ETHICAL STANDARDS

Conflict of Interest Statement: The authors declare that the research was conducted in the absence of any commercial or financial relationships that could be construed as a potential conflict of interest.

REFERENCES

Abdel-Rahman, H., Shawky, S., Ouda, H., Nafeaa, A., Orabi, S., 2013. Effect of two probiotics and bioflavonoids supplementation to the broilers diet and drinking water on the growth performance and hepatic antioxidant parameters. Glob. Vet. 10 (6), 734–741.

Adams, C.A., 2010. The probiotic paradox: live and dead cells are biological response modifiers. Nutr. Res. Rev. 23, 37–46.

Adams, M.C., Luo, J., Rayward, D., King, S., Gibson, R., Moghaddam, G.H., 2008. Selection of a novel direct-fed microbial to enhance weight gain in intensively reared calves. Anim. Feed Sci. Technol. 145, 41–52.

Afsharmanesh, M., Sadaghi, B., 2014. Effects of dietary alternatives (probiotic, green tea powder and Kombucha tea) as antimicrobial growth promoters on growth, ileal nutrient digestibility, blood parameters, and immune response of broiler chickens. Comp. Clin. Pathol. 23 (3), 717–724.

Agarwal, N., Kamra, D.N., Chaudhary, L.C., Agarwal, L., Sahoo, A., Pathak, N.N., 2002. Microbial status and rumen enzyme profile of crossbred calves on different microbial feed additives. Lett. Appl. Microbiol. 34, 329–336.

Ahmed, S.T., Islam, M.M., Mun, H.S., Sim, H.J., Kim, Y.J., Yang, C.J., 2014. Effects of Bacillus amyloliquefaciens as a probiotic strain on growth performance, cecal microflora, and fecal noxious gas emissions of broiler chickens. Poult. Sci. 93 (8), 1963–1971.

Alexopoulos, C., Georgoulakis, I.E., Tzivara, A., Kritas, S.K., Siochu, A., Kyriakis, S.C., 2004. Field evaluation of the efficacy of a probiotic containing Bacillus licheniformis and *Bacillus subtilis* spores, on the health status and performance of sows and their litters. J. Anim. Physiol. Anim. Nutr. 88, 381–392.

Allaart, J.G., van Asten, A.J.A.M., Gröne, A., 2013. Predisposing factors and prevention of *Clostridium perfringens*-associated enteritis. Comp. Immunol. Microbiol. Infect. Dis. 36 (5), 449–464.

Anadón, A., Martínez-Larrañaga, M.R., Aresi, M.M.A., 2016. Prebiotics: safety and toxicity considerations. In: Gupta, R.C. (Ed.), Nutraceuticals: Efficacy, Safety and Toxicity. Academic Press, Amsterdam, pp. 757–775 (Chapter 54).

Apás, A.L., González, S.N., Arena, M.E., 2014. Potential of goat probiotic to bind mutagens. Anaerobe 28, 8–12.

Arsi, K., Donoghue, A.M., Woo-Ming, A., Blore, P.J., Donoghue, D.J., 2015. The efficacy of selected probiotic and prebiotic combinations in reducing Campylobacter colonization in broiler chickens. J. Appl. Poultry Res. 24, 327–334.

Bai, S.P., Wu, A.M., Ding, X., Lei, Y., Bai, J., Zhang, K.Y., Chio, J.S., 2013. Effects of probiotic-supplemented diets on growth performance and intestinal immune characteristics of broiler chickens. Poult. Sci. 92 (3), 663–670.

Bajagai, Y.S., Klieve, A.V., Dart, P.J., Bryden, W.L., 2016. Probiotics in animal nutrition, production, impact and regulation. In: Makkar, H.P.S. (Ed.), FAO Animal Production and Health Paper No. 179. Food and Agriculture Organization of the United Nation, Rome.

Banerjee, G., Ray, A.K., 2017. The advancement of probiotics research and its application in fish farming industries. Res. Vet. Sci. 115, 66–77.

Baurhoo, B., Letellier, A., Zhao, X., et al., 2007. Cecal populations of lactobacilli and bifidobacteria and *Escherichia coli* populations after in vivo *Escherichia coli* challenge in birds fed diets with purified lignin or mannan-oligosaccharides. Poult. Sci. 86 (12), 2509–2516.

Biagi, G., Cipollini, I., Bonaldo, A., et al., 2013. Effect of feeding a selected combination of galacto-oligosaccharides and a strain of *Bifidobacterium pseudocatenulatum* on the intestinal microbiota of cats. Am. J. Vet. Res. 74, 90–95.

Biggs, P., Parsons, C.M., Fahey, G.C., 2007. The effects of several oligosaccharides on growth performance, nutrient digestibilities, and cecal microbial populations in young chicks. Poult. Sci. 86 (11), 2327–2336.

Biloni, A., Quintana, C., Menconi, A., Kallapura, G., Latorre, J., Pixley, C., Layton, S., Dalmagro, M., Hernandez-Velasco, X., Wolfenden, A., 2013. Evaluation of effects of Early Bird associated with FloraMax-B11 on *Salmonella enteritidis*, intestinal morphology, and performance of broiler chickens. Poult. Sci. 92 (9), 2337–2346.

Bomba, A., Nemcova, R., Gancarcikova, S., Herich, R., Kastel, R., 1999. Potentiation of the effectiveness of *Lactobacillus casei* in the prevention of *E. coli* induced diarrhea in conventional and gnotobiotic pigs. Adv. Exp. Med. Biol. 473, 185–190.

Bouchard, D.S., Seridan, B., Saraoui, T., Rault, L., Germon, P., GonzalezMoreno, C., Nader-Macías, F.M., Baud, D., François, P., Chuat, V., Chain, F., Langella, P., Nicoli, J., Le Loir, Y., Even, S., 2015. Lactic acid bacteria isolated from bovine mammary microbiota: potential allies against bovine mastitis. PLoS One 10 (12), e0144831.

Boyle, R.J., Robins-Browne, R.M., Tang, M.L.K., 2006. Probiotic use in clinical practice: what are the risks? Am. J. Clin. Nutr. 83, 1256–1264.

Boyle, A.G., Magdesian, K.G., Durando, M.M., et al., 2013. *Saccharomyces boulardii* viability and efficacy in horses with antimicrobial-induced diarrhea. Vet. Rec. 172, 128.

Brisbin, J.T., Gong, J., Orouji, S., et al., 2011. Oral treatment of chickens with lactobacilli influences elicitation of immune responses. Clin. Vaccine Immunol. 18 (9), 1447–1455.

Bujnakova, D., Strakova, E., Kmet, V., 2014. *In vitro* evaluation of the safety and probiotic properties of lactobacilli isolated from chicken and calves. Anaerobe 29, 118–127.

Callaway, T.R., Anderson, R.C., Edrington, T.S., et al., 2003. Preslaughter intervention strategies to reduce food-borne pathogens in food animals. J. Anim. Sci. 81 (Suppl. 2), E17–E23.

Callaway, T.R., Edrington, T.S., Brabban, A.D., Anderson, R.C., Rossman, M.L., Engler, M.J., Carr, M.A., Genovese, K.J., Keen, J.E., Looper, M.L., Kutter, E.M., Nisbet, D.J., 2008. Bacteriophage isolated from feedlot cattle can reduce *Escherichia coli* O157:H7 populations in ruminant gastrointestinal tracts. Foodborne Pathog. Dis. 5 (2), 183–191.

Cao, B.H., Karasawa, Y., Guo, Y.M., 2005. Effects of green tea polyphenols and fructooligosaccharides in semi-purified diets in broilers' performance and caecal microflora and their metabolites. Asian Australas. J. Anim. Sci. 18, 85–89.

Cao, G.T., Zeng, X.F., Chen, A.G., Zhou, L., Zhang, L., Xiao, Y.P., Yang, C.M., 2013. Effects of a probiotic, *Enterococcus faecium*, on growth performance, intestinal morphology, immune response, and cecal microflora in broiler chickens challenged with *Escherichia coli* K88. Poult. Sci. 92, 2949–2955.

Carnevali, O., Maradonna, F., Gioacchini, G., 2017. Integrated control of fish metabolism, wellbeing and reproduction: the role of probiotic. Aquaculture 472, 144–155.

Cean, A., Stef, L., Simiz, E., Julean, C., Dumitrescu, G., Vasile, A., Pet, E., Drinceanu, D., Corcionivoschi, N., 2015. Effect of human isolated probiotic bacteria on preventing *Campylobacter jejuni* colonization of poultry. Foodborne Pathog. Dis. 12, 122–130.

Chambers, J.R., Gong, J., 2011. The intestinal microbiota and its modulation for Salmonella control in chickens. Food Res. Int. 44 (10), 3149–3159.

Chang, Y.H., Kinm, J.K., Kim, H.J., Kim, W.Y., Kim, Y.B., Park, Y.H., 2000. Probiotic effects of *Lactobacillus reuteri* BSA-131 on piglets. San'oeb Misaengmul Haghoeji 28, 8–13.

Chaucheyras-Durand, F., Walker, N.D., Bach, A., 2008. Effects of active dry yeasts on the rumen microbial ecosystem: past, present and future. Anim. Feed Sci. Technol. 145, 5–26.

Che, T.M., Johnson, R.W., Kelley, K.W., et al., 2011. Mannan oligosaccharide improves immune responses and growth efficiency of nursery pigs experimentally infected with porcine reproductive and respiratory syndrome virus. J. Anim. Sci. 89, 2592–2602.

Chen, L., Karboune, S., 2019. Prebiotics in food and health: properties, functionalities, production, and overcoming limitations with second-generation levan-type fructo-oligosaccharides. In: Varelis, P. (Ed.), Encyclopedia of Food Chemistry. vol. 3. Elsevier Inc, pp. 271–279.

Choct, M., 2009. Managing gut health through nutrition. Br. Poultry Sci. 50, 9–15.

Costa, M.C., Weese, J.S., 2012. The equine intestinal microbiome. Anim. Health Res. Rev. 13 (1), 121–128.

Costa, G.M., Paula, M.M., Barão, C.E., Klososki, S.J., Bonafe, E.G., Visentainer, J.V., Pimentel, T.C., 2019. Yoghurt added with *Lactobacillus casei* and sweetened with natural sweeteners and/or prebiotics: implications on quality parameters and pro-biotic survival. Int. Dairy J. 97, 139–148.

Coverdale, J.A., 2016. Can the microbiome of the horse be altered to improve digestion? J. Anim. Sci. 94 (6), 2275–2281.

Cox, C.M., Dalloul, R.A., 2015. Immunomodulatory role of probiotics in poultry and potential in ovo application. Benefic. Microbes 6 (1), 45–52.

Cromwell, G.L., 2002. Why and how antibiotics are used in swine production. Anim. Biotechnol. 13, 7–27.

Dalloul, R., Lillehoj, H., Shellem, T., Doerr, J., 2003. Enhanced mucosal immunity against *Eimeria acervulina* in broilers fed a Lactobacillus-based probiotic. Poult. Sci. 82 (1), 62–66.

de Simone, C., 2019. The unregulated probiotic market. Clin. Gastroenterol. Hepatol. 17, 809–817.

de Vaux, A., Morrison, M., Hutkins, R.W., 2002. Displacement of *Escherichia Coli* O157:H7 from rumen medium containing prebiotic sugars. Appl. Environ. Microbiol. 68 (2), 519–524.

Desnoyers, M., Giger-reverdin, S., Bertin, G., Duvaux-Ponter, C., Sauvant, D., 2009. Meta analysis of the influence of *Saccharomyces cerevisiae* supplementation on ruminal parameters and milk production of ruminants. J. Dairy Sci. 92, 1620–1632.

Desrochers, A.M., Dolente, B.A., Roy, M.F., Boston, R., Carlisle, S., 2005. Efficacy of *Saccharomyces boulardii* for treatment of horses with acute enterocolitis. J. Am. Vet. Med. Assoc. 227, 954–959.

Dibner, J.J., Richards, J.D., 2005. Antibiotic growth promoters in agriculture: history and mode of action. Poult. Sci. 84, 634–643.

Dixit, Y., Wagle, A., Vakil, B., 2016. Patents in the field of probiotics, prebiotics, synbiotics: a review. J. Food Microbiol. Saf. Hyg. 1 (2), 1–13.

Fajardo, P., Pastrana, L., Mendez, J., Rodriguez, I., Fucinos, C., Guerra, N.P., 2012. Effects of feeding of two potentially probiotic preparations from lactic acid bacteria on the performance and faecal microflora of broiler chickens. Sci. World J. 2012. https://doi.org/10.1100/2012/562635, 562635.

Fan, Y., Ji-Dong, W., Yu-Fan, L., Ya-Lun, S., Qia, W., Rui-Ling, Z., 2019. Galacto-oligosaccharides modulate gut microbiota dysbiosis and intestinal permeability in rats with alcohol withdrawal syndrome. J. Funct. Foods 60, 103423.

FAO, 2007. Technical Meeting on Prebiotics: Food Quality and Standards Service (AGNS). Food and Agriculture Organization of the United Nations (FAO) 15-16.09.2007. FAO Technical Meeting Report,.

FAO, 2013. Poultry Development Review. FAO, Rome.

Farrell, D., 2013. The Role of Poultry in Human Nutrition. http://wwwfaoorg/docrep/013/al709e/al709e00pdf. (Accessed 10 April 2015).

Flemming, J.S., Freitas, J.R.S., Fontoura, P., Montanhini, N.R., Arruda, J.S., 2004. Use of mannanoligosaccharides in broiler feeding. Braz. J. Poult. Sci. 6 (3), 159–161.

Fuller, R., 1999. Probiotics for farm animals. In: Tannock, G.W. (Ed.), Probiotics: A Critical Review. Horizon Scientific Press, Wymondham, UK, pp. 15–22.

Gaggìa, F., Mattarelli, P., Biavati, B., 2010. Probiotics and prebiotics in animal feeding for safe food production. Int. J. Food Microbiol. 141, S15–S28.

Galdeano, C.M., Perdigón, G., 2004. Role of viability of probiotic strains in their persistence in the gut and in mucosal immune stimulation. J. Appl. Microbiol. 97 (4), 673–681.

Gallazzi, D., Giardini, A., Mangiagalli, M.G., Marelli, S., Ferrazzi, V., Orsi, C., Cavalchini, L.G., 2009. Effects of *Lactobacillus acidophilus* D2/CSL on laying hen performance. Ital. J. Anim. Sci. 7 (1), 27–38.

Galvao, K.N., Santos, J.E., Coscioni, A., Villasenor, M., Sischo, W.M., Berge, A.C., 2005. Effect of feeding live yeast products to calves with failure of passive transfer on performance and patterns of antibiotic resistance in fecal *Escherichia coli*. Reprod. Nutr. Dev. 45, 427–440.

García-Hernández, Y., Rodríguez, Z., Brandão, L.R., Rosa, C.A., Nicoli, J.R., Elías Iglesias, A., Peréz-Sanchez, T., Salabarría, R.B., Halaihel, N., 2012. Identification and in vitro screening of avian yeasts for use as probiotic. Res. Vet. Sci. 93, 798–802.

Ghosh, S., Mehla, R.K., 2012. Influence of dietary supplementation of prebiotics (mannanoligosaccharide) on the performance of crossbred calves. Tropl. Anim. Health Prod. 44, 617–622.

Gibson, G.R., Hutkins, R., Sanders, M.E., Prescott, S.L., Reimer, R.A., Salminen, S.J., et al., 2017. Expert consensus document: the International Scientific Association for Probiotics and Prebiotics (ISAPP) consensus statement on the definition and scope of prebiotics. Nat. Rev. Gastroenterol. Hepatol. 14, 491–502.

Guedes, C.M., Gonçalves, D., Rodrigues, M.A.M., Dias-Da-Silva, A., 2008. Effects of a *Saccharomyces cerevisiae* yeast on ruminal fermentation and fibre degradation of maize silages in cows. Anim. Feed Sci. Technol. 145, 27–40.

Guerra, N.P., Castro, L.P., 2009. Probiotics: Production, Evaluation and Uses in Animal Feed. Research Signpost, Trivandrum.

Haghighi, H.R., Gong, J., Gyles, C.L., et al., 2006. Probiotics stimulate production of natural antibodies in chickens. Clin. Vaccine Immunol. 13 (9), 975–980.

Hai, N.V., 2015. The use of probiotics in aquaculture. J. Appl. Microbiol. 119, 917–935.

Hajati, H., Rezaei, M., 2010. The application of prebiotics in poultry production. Int. J. Poult. Sci. 9 (3), 298–304.

Hanczakowska, E., Świątkiewicz, M., Natonek-Wiśniewska, M., Okoń, K., 2016. Medium chain fatty acids (MCFA) and/or probiotic *Enterococcus faecium* as a feed supplement for piglets. Livest. Sci. 192, 1–7.

Heinrichs, A.J., Jones, C.M., Heinrichs, B.S., 2003. Effects of mannan oligosaccharide or antibiotics in neonatal diets on health and growth of dairy calves. J. Dairy Sci. 86 (12), 4064–4069.

Herfel, T.M., Jacobi, S.K., Lin, X., Jouni, Z.E., Chichlowski, M., Stahl, C.H., Odle, J., 2013. Dietary supplementation of *Bifidobacterium longum* strain AH1206 increases its cecal abundance and elevates intestinal interleukin-10 expression in the neonatal piglet. Food Chem. Toxicol. 60, 116–122.

Hill, T.M., Bateman, H.G.I.I., Aldrich, J.M., et al., 2008. Oligosaccharides for dairy calves. Prof. Anim. Sci. 24, 460–464.

Hill, C., Guarner, F., Reid, G., Gibson, G.R., Merenstein, D.J., Pot, B., et al., 2014. The International Scientific Association for Probiotics and Prebiotics consensus statement on the scope and appropriate use of the term probiotic. Nat. Rev. Gastroenterol. Hepatol. 11, 506–514.

Ho, A.L., Kosik, O., Lovegrove, A., Charalampopoulos, D., Rastall, R.A., 2018. *In vitro* fermentability of xylo-oligosaccharide and xylo-polysaccharide fractions with different molecular weights by human faecal bacteria. Carbohydr. Polym. 179, 50–58.

Hooge, D., 2004. Meta-analysis of broiler chicken pen trials evaluating dietary mannan oligosaccharide. Int. J. Poult. Sci. 3, 163–174.

Hoseinifar, S.H., Khalili, M., Rostami, H.K., et al., 2013. Dietary galactooligosaccharide affects intestinal microbiota, stress resistance, and performance of Caspian roach (*Rutilus rutilus*) fry. Fish Shellfish Immunol. 35, 1416–1420.

Hoseinifar, S.H., Soleimani, N., Ringø, E., 2014. Effects of dietary fructo-oligosaccharide supplementation on the growth performance, haemato-immunological parameters, gut microbiota and stress resistance of common carp (*Cyprinus carpio*) fry. Br. J. Nutr. 112, 1296–1302.

Hu, Y., Dun, Y., Li, S., Zhao, S., Peng, N., Liang, Y., 2014. Effects of *Bacillus subtilis* KN-42 on growth performance, diarrhea and faecal bacterial flora of weaned piglets. Asian Australas. J. Anim. Sci. 27, 1131–1140.

Huang, Y., Li, Y.L., Cui, Z.W., Yu, D.Y., Rajput, I.R., Hu, C.H., Li, W.F., 2012. Effect of orally administered *Enterococcus faecium* EF1 on intestinal cytokines and chemokines production of suckling piglets. Pak. Vet. J. 32, 81–84.

Hung, A.T., Lin, S.Y., Yang, T.Y., Chou, C.K., Liu, H.C., Lu, J.J., Wang, B., Chen, S.Y., Lien, T.F., 2012. Effects of *Bacillus coagulans* ATCC 7050 on growth performance, intestinal morphology, and microflora composition in broiler chickens. Anim. Prod Sci. 52 (9), 874–879.

Hutkins, R.W., Krumbeck, J.A., Bindels, L.B., Cani, P.D., Fahey Jr., G., Goh, J.Y., Sanders, M.E., 2016. Prebiotics: why definitions matter. Curr. Opin. Biotechnol. 37, 1–7.

Jahromi, M.F., Liang, J.B., Ebrahimi, R., Shokryazdan, P., 2014. Inventors; Universiti Putra Malaysia, assignee: composition for reducing metal toxicity and a method thereof. Malaysia Patent PI2014702845.

Jayaraman, S., Thangavel, G., Kurian, H., Mani, R., Mukkalil, R., Chirakkal, H., 2013. *Bacillus subtilis* PB6 improves intestinal health of broiler chickens challenged with *Clostridium perfringens*-induced necrotic enteritis. Poult. Sci. 92 (2), 370–374.

Jha, D.K., Bhujel, R.C., Anal, A.K., 2015. Dietary supplementation of probiotics improves survival and growth of rohu (*Labeo rohita* Ham.) hatchlings and fry in outdoor tanks. Aquaculture 435, 475–479.

Jouany, J.P., 2006. Optimizing rumen functions in the close-up transition period and early lactation to drive dry matter intake and energy balance in cows. Anim. Reprod. Sci. 96, 250–264.

Jouany, J.P., Gobert, J., Medina, B., Bertin, G., Julliand, V., 2008. Effect of live yeast culture supplementation on apparent digestibility and rate of passage in horses fed a high-fiber or high-starch diet. J. Anim. Sci. 86, 339–347.

Józefiak, D., Kaczmarek, S., Rutkowski, A., 2008. A note on the effects of selected prebiotics on the performance and ileal microbiota of broiler chickens. J. Anim. Feed Sci. 17, 392–397.

Juśkiewicz, J., Jankowski, J., Zduńczyk, Z., et al., 2006. Performance and gastrointestinal tract metabolism of turkeys fed diets with different contents of fructo-oligosaccharides. Poult. Sci. 85, 886–891.

Kabir, S.M.L., 2009. The role of probiotics in the poultry industry. Int. J. Mol. Sci. 10 (8), 3531–3546.

Kanakupt, K., Vester Boler, B.M., Dunsford, B.R., et al., 2011. Effects of shortchain fructooligosaccharides and galactooligosaccharides, individually and in combination, on nutrient digestibility, fecal fermentative metabolite concentrations, and large bowel microbial ecology of healthy adult cats. J. Anim. Sci. 89, 1376–1384.

Kantas, D., Papatsiros, V.G., Tassis, P.D., Giavasis, I., Bouki, P., Tzika, E.D., 2015. A feed additive containing *Bacillus toyonensis* (Toyocerin (R)) protects against enteric pathogens in postweaning piglets. J. Appl. Microbiol. 118 (3), 727–738.

Kataria, J., Li, N., Wynn, J.L., et al., 2009. Probiotic microbes: do they need to be alive to be beneficial? Nutr. Rev. 67 (9), 546–550.

Kenny, M., Smidt, H., Mengheri, E., Miller, B., 2011. Probiotics—do they have a role in the pig industry? Animal 5 (3), 462–470.

Kim, L.M., Morley, P.S., Traub-Dargatz, J.L., Salman, M., Gentry Weeks, C., 2001. Factors associated with Salmonella shedding among equine colic patients at a veterinary teaching hospital. J. Am. Vet. Med. Assoc. 218, 740–748.

Kleessen, B., Elsayed, N.A.A.E., Loehren, U., Schoedl, W., Krueger, M., 2003. Jerusalem artichokes stimulate growth of boiler chickens and protect them against endotoxins and potential cecal pathogens. J. Food Prot. 11, 2171–2175.

Kosin, B., Rakshit, S.K., 2006. Criteria for production of probiotics. Food Technol. Biotechnol. 44 (3), 371–379.

Krajmalnik-Brown, R., Ilhan, Z.E., Kang, D.W., DiBaise, J.K., 2012. Effects of gut microbes on nutrient absorption and energy regulation. Nutr. Clin. Pract. 27, 201–214.

Kurtoglu, V., Kurtoglu, F., Seker, E., Coskun, B., Balevi, T., Polat, E., 2004. Effect of probiotic supplementation on laying hen diets on yield performance and serum and egg yolk cholesterol. Food Addit. Contam. 21 (9), 817–823.

Kyu Song, S., Ram Beck, B., Kim, D., et al., 2014. Prebiotics as immunostimulants in aquaculture: a review. Fish Shellfish Immunol. 40, 40–48.

Landes, A.D., Hassel, D.M., Funk, J.D., Hill, A., 2008. Fecal sand clearance is enhanced with a product combining probiotics, prebiotics, and psyllium in clinically normal horses. J. Equine Vet. Sci. 28, 79–84.

Landy, N., Kavyani, A., 2013. Effects of using a multi-strain probiotic on performance, immune responses and caecal microflora composition in broiler chickens reared under cyclic heat stress condition. Iran. J. Appl. Anim. Sci. 3 (4), 703–708.

Lata, J., Juránková, J., Doubek, J., Příbramská, V., Frič, P., Dítě, P., Kolář, M., Scheer, P., Kosakova, D., 2006. Labelling and content evaluation of commercial veterinary probiotics. Acta Vet. Brno 75 (1), 139–144.

Latha, S., Vinothini, G., John Dickson Calvin, D., Dhanasekaran, D., 2015. *In vitro* probiotic profile based selection of indigenous actinobacterial probiont Streptomycessp. JD9 for enhanced broiler production. J. Biosci. Bioeng. 121, 124–131.

Le Bon, M., Davies, H.E., Glynn, C., Thompson, C., Madden, M., Wiseman, J., et al., 2010. Influence of probiotics on gut health in the weaned pig. Livest. Sci. 133, 179–181.

Le, O., Dart, P., Harper, K., Zhang, D., Schofield, B., Callaghan, M., Lisle, A., Klieve, A., McNeill, D., 2017. Effect of probiotic *Bacillus amyloliquefaciens* strain H57 on productivity and the incidence of diarrhoea in dairy calves. Anim. Prod. Sci. 57 (5), 912–919.

Lee, Y.K., 2009. Selection and maintenance of probiotic microorganisms. In: Lee, Y.K., Salminen, S. (Eds.), Handbook of Probiotics and Prebiotics. Wiley-VCH, Weinheim, pp. 177–187.

Lei, X., Piao, X., Ru, Y., Zhang, H., Péron, A., Zhang, H., 2015. Effect of *Bacillus amyloliquefaciens*-based direct-fed microbial on performance, nutrient utilization, intestinal morphology and cecal microflora in broiler chickens. Asian Australas. J. Anim. Sci. 28 (2), 239–246.

Lema, M., Williams, L., Walker, L., Rao, D.R., 2002. Effect of dietary fiber on *E. coli* O157: H7 shedding in lambs. Small Rumin. Res. 43, 249–255.

Li, P., Burr, G.S., Gatlin, D.M., et al., 2007. Dietary supplementation of short chain fructooligosaccharides influences gastrointestinal microbiota composition and immunity characteristics of Pacific white shrimp. *Litopenaeus vannamei*, cultured in a recirculating system. J. Nutr. 137, 2763–2768.

Liao, S.F., Nyachoti, C.M., 2017. Using probiotics to improve swine gut health and nutrient utilization. Anim. Nutr. 3, 331–343.

Liu, P., Piao, X.S., Kim, S.W., et al., 2008. Effects of chito-oligosaccharide supplementation on the growth performance, nutrient digestibility, intestinal morphology, and fecal shedding of *Escherichia coli* and *Lactobacillus* in weaning pigs. J. Anim. Sci. 86, 2609–2618.

Liu, H., Zhang, J., Zhang, S., Yang, F., Thacker, P.A., Zhang, G., Qiao, S., Ma, X., 2014. Oral administration of *Lactobacillus fermentum* I5007 favors intestinal development and alters the intestinal microbiota in formula-fed piglets. J. Agric. Food Chem. 62, 860–866.

Lv, C.H., Wang, T., Regmi, N., Chen, X., Huang, K., Liao, S.F., 2015. Effects of dietary supplementation of selenium-enriched probiotics on production performance and intestinal microbiota of weanling piglets raised under high ambient temperature. J. Anim. Physiol. Anim. Nutr. 99, 1161–1171.

Ma, Y., Wu, X., Giovanni, V., Meng, X., 2017. Effects of soybean oligosaccharides on intestinal microbial communities and immune modulation in mice. Saudi J. Biol. Sci. 24, 114–121.

Madhukumar, M.S., Muralikrishna, G., 2012. Fermentation of xylo-oligosaccharides obtained from wheat bran and Bengal gram husk by lactic acid bacteria and bifidobacteria. J. Food Sci. Technol. 49 (6), 745–752.

Maiorano, G., Stadnicka, K., Tavaniello, S., et al., 2017. *In ovo* validation model to assess the efficacy of commercial prebiotics on broiler performance and oxidative stability of meat. Poult. Sci. 96 (2), 511–518.

Maldonado, N.C., de Ruiz, C.S., Otero, M.C., Sesma, F., Nader-Macías, M.E., 2012. Lactic acid bacteria isolated from young calves—characterization and potential as probiotics. Res. Vet. Sci. 92, 342–349.

Maragkoudakis, P.A., Mountzouris, K.C., Rosu, C., Zoumpopoulou, G., Papadimitriou, K., Dalaka, E., Hadjipetrou, A., Theofanous, G., et al., 2010. Feed supplementation of *Lactobacillus plantarum* PCA 236 modulates gut microbiota and milk fatty acid composition in dairy goats—a preliminary study. Int. J. Food Microbiol. 141, 109–116.

Martinez, R.C.R., Bedani, R., Saad, S.M.I., 2015. Scientific evidence for health effects attributed to the consumption of probiotics and prebiotics: an update for current perspectives and future challenges. Br. J. Nutr. 114, 1993–2015.

Meng, Q.W., Yan, L., Ao, X., Zhou, T.X., Wang, J.P., Lee, J.H., et al., 2010. Influence of probiotics in different energy and nutrient density diets on growth performance, nutrient digestibility, meat quality, and blood characteristics in growing-finishing pigs. J. Anim. Sci. 88, 3320–3326.

Mikkelsen, L.L., Jakobsen, M., Jensen, B.B., 2003. Effects of dietary oligosaccharides on microbial diversity and fructo-oligosaccharide degrading bacteria in faeces of piglets postweaning. Anim. Feed Sci. Technol. 109, 133–150.

Mikulski, D., Jankowski, J., Naczmanski, J., Mikulska, M., Demey, V., 2012. Effects of dietary probiotic (*Pediococcus acidilactici*) supplementation on performance, nutrient digestibility, egg traits, egg yolk cholesterol, and fatty acid profile in laying hens. Poult. Sci. 91 (10), 2691–2700.

Modesto, M., D'Aimmo, M.R., Stefanini, I., 2009. A novel strategy to select Bifidobacterium strains and prebiotics as natural growth promoters in newly weaned pigs. Livest. Sci. 122, 248–258.

Mookiah, S., Sieo, C.C., Ramasamy, K., Abdullah, N., Ho, Y.W., 2014. Effects of dietary prebiotics, probiotic and synbiotics on performance, caecal bacterial populations and caecal fermentation concentrations of broiler chickens. J. Sci. Food Agric. 94 (2), 341348.

Morgan, L.M., Coverdale, J.A., Froetschel, M.A., et al., 2007. Effect of yeast culture supplementation on digestibility of varying forage quality in mature horses. J. Equine Vet. Sci. 27, 260–265.

Morrison, S.J., Dawson, S., Carson, A.F., 2010. The effects of mannan oligosaccharide and *Streptococcus faecium* addition to milk replacer on calf health and performance. Livest. Sci. 131, 292–296.

Mountzouris, K., Xypoleas, I., Kouseris, I., Kostas, F., 2006. Nutrient digestibility, faecal physicochemical characteristics and bacterial glycolytic activity of growing pigs fed a diet supplemented with oligofructose or trans-galactooligosaccharides. Livest. Sci. 105 (1), 168–175.

Oltjen, J.W., Beckett, J.L., 1996. Role of ruminant livestock in sustainable agricultural systems. J. Anim. Sci. 74 (6), 1406–1409.

Ortiz, L.T., Rodriguez, M.L., Alzueta, C., Rebole, A., Trevino, J., 2009. Effect of inulin on growth performance, intestinal tract sizes, mineral retention and tibial bone mineralisation in broiler chickens. Br. Poultry Sci. 50 (3), 325–332.

Ouwehand, A.C., Kirjavainen, P.V., Shortt, C., Salminen, S., 1999. Probiotics: mechanisms and established effects. Int. Dairy J. 9, 43–52.

Pajarillo, E.A.B., Chae, J.P., Balolong, M.P., Kim, H.B., Park, C.S., Kang, D.K., 2015. Effects of probiotic *Enterococcus faecium* NCIMB 11181 administration on swine fecal microbiota diversity and composition using barcoded pyrosequencing. Anim. Feed Sci. Technol. 201, 80–88.

Park, J.H., Kim, I.H., 2014. Supplemental effect of probiotic *Bacillus subtilis* B2A on productivity, organ weight, intestinal Salmonella microflora, and breast meat quality of growing broiler chicks. Poult. Sci. 93, 2054–2059.

Park, S.H., Hanning, I., Perrota, A., Bench, B.J., Alm, E., Ricke, S.C., 2013. Modifying the gastrointestinal ecology in alternatively raised poultry and the potential for molecular and metabolomic assessment. Poult. Sci. 92 (2), 546–561.

Parraga, M.E., Spier, S.J., Thurmond, M., Hirsh, D., 1997. A clinical trial of probiotic administration for prevention of Salmonella shedding in the postoperative period in horses with colic. J. Vet. Intern. Med. 11, 36–41.

Patel, S., Goyal, A., 2012. The current trends and future perspectives of prebiotics research: a review. 3 Biotech 2 (1), 1–15.

Patterson, J.A., Burkholder, K.M., 2003. Application of prebiotics and probiotics in poultry production. Poult. Sci. 82, 627–631.

Peterson, R.E., Klopfenstein, T.J., Erickson, G.E., Folmer, J., Hinkley, S., Moxley, R.A., Smith, D.R., 2007. Effect of *Lactobacillus acidophilus* strain NP51 on *Escherichia coli* O157:H7 fecal shedding and finishing performance in beef feedlot cattle. J. Food Prot. 70 (2), 287–291.

Quigley, J.D., Drewry, J.J., Murray, L.M., et al., 1997. Body weight gain, feed efficiency, and fecal scores of dairy calves in response to galactosyllactose or antibiotics in milk replacers. J. Dairy Sci. 80, 1751–1754.

Rachwał, A., 2003. Naturalne promotory wzrostu. Hod. Drobiu 8, 31–32.

Rastall, R.A., 2010. Functional oligosaccharides: application and manufacture. Annu. Rev. Food Sci. Technol. 1, 305–339.

Rastall, R.A., Gibson, G.R., 2015. Recent developments in prebiotics to selectively impact beneficial microbes and promote intestinal health. Curr. Opin. Biotechnol. 32, 42–46.

Respondek, F., Swanson, K.S., Belsito, K.R., et al., 2008. Short-chain fructo-oligosaccharides influence insulin sensitivity and gene expression of fat tissue in obese dogs. J. Nutr. 138, 1712–1718.

Respondek, F., Myers, K., Smith, T.L., et al., 2011. Dietary supplementation with short-chain fructo-oligosaccharides improves insulin sensitivity in obese horses. J. Anim. Sci. 89, 77–83.

Rigobelo, E.C., Ávila, F.A., 2012. Protective effect of probiotics strains in ruminants. In: Rigobelo, E.C. (Ed.), Probiotic in Animals. In Tech, Croatia, pp. 455–466.

Roodposhti, P.M., Dabiri, N., 2012. Effects of probiotic and prebiotic on average daily gain, fecal shedding of *Escherichia coli*, and immune system status in newborn female calves. Asian Australas. J. Anim. Sci. 25, 1255–1261.

Saarela, M., Mogensen, G., Fondén, R., Mättö, J., Mattila-Sandholm, T., 2000. Probiotic bacteria: safety, functional and technological properties. J. Biotechnol. 84 (3), 197–215.

Salma, U., Miah, A., Tareq, K., Maki, T., Tsujii, H., 2007. Effect of dietary *Rhodobacter capsulatus* on egg-yolk cholesterol and laying hen performance. Poult. Sci. 86 (4), 714–719.

Samanta, A.K., Jayapal, N., Senani, S., Kolte, A.P., Sridhar, M., 2013. Prebiotic inulin: useful dietary adjuncts to manipulate the livestock gut microflora. Braz. J. Microbiol. 44 (1), 1–14.

Samarasinghe, K., Wenk, C., Silva, K.F.S.T., Gunasekera, J.M.D.M., 2003. Turmeric (*Curcuma longa*) root powder and mannanoligosaccharides as alternatives to antibiotics in broiler chicken diets. Asian Australas. J. Anim. Sci. (10), 495–1500.

Sartor, R.B., 2004. Therapeutic manipulation of the enteric microflora in inflammatory bowel diseases: antibiotics, probiotics, and prebiotics. Gastroenterology 126, 1620–1633.

Schamberger, G.P., Ronald, L., Phillips, R.L., Jennifer, L., Jacobs, J.L., Diez-Gonzalez, F., 2004. Reduction of *Escherichia coli* O157:H7 populations in cattle by addition of colicin E7-producing *E. coli* to feed. Appl. Environ. Microbiol. 70, 6053–6060.

Schwarz, S., Kehrenberg, C., Walsh, T.R., 2001. Use of antimicrobial agents in veterinary medicine and food animal production. Int. J. Antimicrob. Agents 17, 431–437.

Seal, B.S., Drider, D., Oakley, B.B., Brüssow, H., Bikard, D., Rich, J.O., et al., 2018. Microbial-derived products as potential new antimicrobials. Vet. Res. 49, 66.

Seo, J.K., Kim, S.W., Kim, M.H., Upadhaya, S.D., Kam, D.K., Ha, J.K., 2010. Direct-fed microbials for ruminant animals. Asian Australas. J. Anim. Sci. 23, 1657–1667.

Shim, S.B., Verstegen, W.A., Kim, I.H., et al., 2005. Effects of feeding antibiotic-free creep feed supplemented with oligofructose, probiotics or synbiotics to suckling piglets increases the preweaning weight gain and composition of intestinal microbiota. Arch. Anim. Nutr. 59, 419–427.

Shim, Y., Ingale, S., Kim, J., Kim, K., Seo, D., Lee, S., Chae, B., Kwon, I., 2012. A multi-microbe probiotic formulation processed at low and high drying temperatures: effects on growth performance, nutrient retention and caecal microbiology of broilers. Br. Poult. Sci. 53 (4), 482–490.

Sims, M.D., Dawson, K.A., Newman, K.E., Spring, P., Hooge, D.M., 2004. Effects of dietary mannan oligosaccharide, bacitracin methylene disalicylate, or both on the live performance and intestinal microbiology of turkeys. Poult. Sci. 83, 1148–1154.

Singh, S.P., Jadaun, J.S., Narnoliya, L.K., Pandey, A., 2017. Prebiotic oligosaccharides: special focus on fructo-oligosaccharides, its biosynthesis and bioactivity. Appl. Biochem. Biotechnol. 183, 613–635.

Smiricky-Tjardes, M.R., Grieshop, C.M., Flickinger, E.A., et al., 2003. Dietary galactooligosaccharides affect ileal and total tract nutrient digestibility, ileal and fecal bacterial concentrations, and ileal fermentative characteristics of growing pigs. J. Anim. Sci. 81, 2535–2545.

Sparkes, A.H., Papasouliotis, K., Sunvold, G., et al., 1998a. Bacterial flora in the duodenum of healthy cats, and effect of dietary supplementation with fructo-oligosaccharides. Am. J. Vet. Res. 59, 431–435.

Sparkes, A.H., Papasouliotis, K., Sunvold, G., et al., 1998b. Effect of dietary supplementation with fructo-oligosaccharides on fecal flora of healthy cats. Am. J. Vet. Res. 59, 436–440.

Spring, P., Wenk, C., Dawson, K.A., Newman, K.E., 2000. The effects of dietary mannanoligosaccharides on cecal parameters and the concentrations of enteric bacteria in the ceca of Salmonella challenged broiler chicks. Poult. Sci. 79 (2), 205–211.

Stanczuk, J., Zdunczyk, Z., Juskiewicz, J., et al., 2005. Indices of response of young turkeys to diets containing mannan-oligosaccharide or inulin. Vet. Zootec. 31, 98–101.

Stella, A.V., Paratte, R., Valnegri, L., Cigalino, G., Soncini, G., Chevaux, E., Dell'Orto, V., avoini, G., 2007. Effect of administration of live *Saccharomyces cerevisiae* on milk production, milk composition, blood metabolites, and faecal flora in early lactating dairy goats. Small Rumin. Res. 67, 7–13.

Stern, N.J., Svetoch, E.A., Eruslanov, B.V., Perelygin, V.V., Mitsevich, E.V., Mitsevich, I.P., Pokhilenko, V.D., Levchuk, V.P., Svetoch, O.E., Seal, B.S., 2006. Isolation of a *Lactobacillus salivarius* strain and purification of its bacteriocin, which is inhibitory to *Campylobacter jejuni* in the chicken gastrointestinal system. Antimicrob. Agents Chemother. 50, 3111–3116.

Stevens, C.E., Hume, I.D., 1998. Contributions of microbes in vertebrate gastrointestinal tract to production and conservation of nutrients. Physiol. Rev. 78, 393–427.

Suo, C., Yin, Y., Wang, X., Lou, X., Song, D., Wang, X., Gu, Q., 2012. Effects of *Lactobacillus plantarum* ZJ316 on pig growth and pork quality. BMC Vet. Res. 8, 89.

Świątkiewicz, S., Arczewska-Włosek, A., 2011. Effect of inulin and oligofructose on performance and bone characteristics of broiler chickens fed on diets with different concentrations of calcium and phosphorus. Br. Poultry Sci. 52, 483–491.

Syngai, G.G., Gopi, R., Bharali, R., et al., 2016. Probiotics: the versatile functional food ingredients. J. Food Sci. Technol. 53, 921–933.

Szeleszczuk, P., 2005. Weterynaryjne aspekty stosowania żywych kultur mikroorganizmów w praktyce drobiarskiej. Cz I Prakt. klin. 11 (99), 56–58.

Terré, M., Calvo, M.A., Adelantado, C., et al., 2007. Effects of mannan oligosaccharides on performance and microorganism fecal counts of calves following an enhanced-growth feeding program. Anim. Feed Sci. Technol. 137, 115–125.

Thacker, P.A., 2013. Alternatives to antibiotics as growth promoters for use in swine production: a review. J. Anim. Sci. Biotechnol. 4, 35.

Thitaram, S.N., Chung, C.H., Day, D.F., Hinton, A., Bailey, J.S., Siragusa, G.R., 2005. Isomaltooligosaccharide increases cecal Bifidobacterium population in young broiler chickens. Poult. Sci. 84, 998–1003.

Timmerman, H.M., Mulder, L., Everts, H., van Espen, D.C., van der Wal, E., Klaassen, G., et al., 2005. Health and growth of veal calves fed milk replacers with or without probiotics. J. Dairy Sci. 88, 2154–2165.

Tkalcic, S., Zhao, T., Harmon, B.G., Doyle, M.P., Brown, A., Zhao, P., 2003. Fecal shedding of entero-hemorrhagic *Escherichia coli* in weaned calves following treatment with probiotic *Escherichia coli*. J. Food Prot. 66, 184–1189.

Totton, S.C., Farrar, A.M., Wilkins, W., Bucher, O., Waddell, L.A., Wilhelm, B.J., McEwen, S.A., Rajic, A., 2012. The effectiveness of selected feed and water additives for reducing *Salmonella* spp. of public health importance in broiler chickens: a systematic review, meta-analysis, and meta-regression approach. Prev. Vet. Med. 106 (3–4), 197–213.

Tzortzis, G., Goulas, A.K., Gee, J.M., et al., 2005. A novel galactooligosaccharide mixture increases the bifdobacterial population numbers in a continuous *in vitro* fermentation system and in the proximal colonic contents of pigs *in vivo*. J. Nutr. 135 (7), 1726–1731.

Upadhaya, S.D., Hossiendoust, A., Kim, I.H., 2016. Probiotics in Salmonella-challenged Hy-Line brown layers. Poult. Sci. 95, 1894–1897.

Vasconcelos, J.T., Elam, N.A., Brashears, M.M., Galyean, M.L., 2008. Effects of increasing dose of live cultures of *Lactobacillus acidophilus* (strain NP 51) combined with a single dose of *Propionibacterium freudenreichii* (strain NP 24) on performance and carcass characteristics of finishing beef steers. J. Anim. Sci. 86, 756–762.

Velikonja, A., Lipoglavsek, L., Zorec, M., Orel, R., Avgustin, G., 2019. Alterations in gut microbiota composition and metabolic parameters after dietary intervention with barley beta glucans in patients with high risk for metabolic syndrome development. Anaerobe 55, 67–77.

Verbrugghe, A., Hesta, M., Gommeren, K., et al., 2009. Oligofructose and inulin modulate glucose and amino acid metabolism through propionate production in normal-weight and obese cats. J. Nutr. 102, 694–702.

Verdonk, J.M.A.J., Shim, S.B., van Leeuwen, P., et al., 2005. Application of inulin-type fructans in animal feed and pet food. Br. J. Nutr. 93 (Suppl. 1), S125–S138.

Von Buenau, R., Jaekel, L., Schubotz, E., Schwarz, S., Stroff, T., Krueger, M., 2005. *Escherichia coli* strain Nissle 1917: significant reduction of neonatal calf diarrhea. J. Dairy Sci. 88, 317–323.

von Wright, A., 2005. Regulating the safety of probiotics—the European approach. Curr. Pharm. Des. 11, 17–23.

Wang, Y., 2009. Prebiotics: present and future in food science and technology. Food Res. Int. 42, 8–12.

Wang, J., Ji, G.F., Hou, C.L., Wang, S.X., Zhang, D.Y., Liu, H., Shan, D.C., Wang, Y.M., 2014. Effects of *Lactobacillus johnsonii* XS4 supplementation on reproductive performance, gut environment, and blood biochemical and immunological index in lactating sows. Livest. Sci. 164, 96–101.

Wang, L., Liu, C., Chen, M., Ya, T., Huang, W., Gao, P., Zhang, H., 2015. A novel *Lactobacillus plantarum* strain P-8 activates beneficial immune response of broiler chickens. Int. Immunopharmacol. 29, 901–907.

Ward, M.P., Alinovi, C.A., Couetil, L.L., et al., 2004. A randomized clinical trial using probiotics to prevent Salmonella fecal shedding in hospitalized horses. J. Equine Vet. Sci. 24, 242–247.

Weese, J.S., 2002. Microbiologic evaluation of commercial probiotics. J. Am. Vet. Med. Assoc. 220 (6), 794–797.

Weese, J.S., 2003. Evaluation of deficiencies in labeling of commercial probiotics. Can. Vet. J. 44 (12), 982–983.

Weese, J.S., Martin, H., 2011. Assessment of commercial probiotic bacterial contents and label accuracy. Can. Vet. J. 52 (1), 43–46.

Weese, J.S., Rousseau, J., 2005. Evaluation of *Lactobacillus pentosus* WE7 for prevention of diarrhea in neonatal foals. J. Am. Vet. Med. Assoc. 226, 2031–2034.

Willard, M.D., Simpson, R.B., Cohen, N.D., et al., 2000. Effects of dietary fructooligosaccharide on selected bacterial populations in feces of dogs. Am. J. Vet. Res. 61, 820–825.

Wisener, L.V., Sargeant, J.M., O'Connor, A.M., Faires, M.C., GlassKaastra, S.K., 2015. The use of direct-fed microbials to reduce shedding of *Escherichia coli* O157 in beef cattle: a systematic review and meta-analysis. Zoonoses Public Health 62, 75–89.

Wu, Y., Pan, L., Shang, Q.H., Ma, X.K., Long, S.F., Xu, Y.T., Piao, X.S., 2017. Effects of isomalto-oligosaccharides as potential prebiotics on performance, immune function and gut microbiota in weaned pigs. Anim. Feed Sci. Technol. 230, 126–135.

Xavier dos Santos, D., Casazza, A.A., Aliakbarian, B., Bedani, R., Saad, S.M.I., Perego, P., 2019. Improved probiotic survival to *in vitro* gastrointestinal stress in a mousse containing Lactobacillus acidophilus La-5 microencapsulated with inulin by spray drying. LWT Food Sci. Technol. 99, 404–410.

Xavier-Santos, D., Bedani, R., Perego, P., Converti, A., Saad, S.M.I., 2019. *L. acidophilus* La-5, fructooligosaccharides and inulin may improve sensory acceptance and texture profile of a synbiotic diet mousse. LWT Food Sci. Technol. 105, 329–335.

Xu, Z.R., Hu, C.H., Xia, M.S., et al., 2003. Effects of dietary fructooligosaccharide on digestive enzyme activities, intestinal microflora and morphology of male broilers. Poult. Sci. 82 (6), 1030–1036.

Xu, Q., Chao, Y., Wan, Q., 2009. Health benefit application of functional oligosaccharides. Carbohydr. Polym. 77, 435–441.

Yang, Y., Iji, P.A., Kocher, A., Thomson, E., Mikkelsen, L.L., Choct, M., 2008. Effects of mannanoligosaccharide in broiler chicken diets on growth performance, energy, energy utilization, nutrient digestibility and intestinal microflora. Br. Poultry Sci. 49, 186–194.

Yang, Y., Iji, P.A., Choct, M., 2009. Dietary modulation of gut microflora in broiler chickens: a review of the role of six kinds of alternatives to in-feed antibiotics. Worlds Poult. Sci. J. 65, 97–114.

Yirga, H., 2015. The use of probiotics in animal nutrition. J. Probiotics Health 3, 132.

Younts-Dahl, S.M., Galyean, M.L., Lonergan, G.H., Elam, N.A., Brashears, M.M., 2004. Dietary supplementation with Lactobacillus–Propionibacterium-based direct-fed with microbials and prevalence of *Escherichia coli* O157 in beef feedlot cattle and on hides at harvest. J. Food Prot. 67, 889–893.

Yusrizal, Chen, T.C., 2003a. Effect of adding chicory fructans in feed on broiler growth performance, serum cholesterol and intestinal length. Int. J. Poult. Sci. 2 (3), 214–219.

Yusrizal, Chen, T.C., 2003b. Effect of adding chicory fructans in feed on fecal and intestinal microflora and excreta volatile ammonia. Int. J. Poult. Sci. 2, 188–194.

Yuyama, T., 2004. Evaluation of a host-specific Lactobacillus probiotic in neonatal foals. J. Appl. Res. Vet. Med. 2, 26–32 (Yusa S, ed.).

Zduńczyk, Z., Juśkiewicz, J., Jankowski, J., Koncicki, A., 2004. Performance and caecal adaptation of turkeys to diets without or with antibiotic and with different levels of mannan-oligosaccharide. Arch. Anim. Nutr. 58, 367–378.

Zduńczyk, Z., Jankowski, J., Juśkiewicz, J., 2005. Performance and intestinal parameters of turkeys fed diet with inulin and oligofructose. J. Anim. Feed Sci. 14, 511S–516S.

Zeng, J., Song, M., Jia, T., Gao, H., Zhang, R., Jiang, J., 2019. Immunomodulatory influences of sialylated lactuloses in mice. Biochem. Biophys. Res. Commun. 514, 351–357.

Zhang, Z., Kim, I., 2014. Effects of multistrain probiotics on growth performance, apparent ileal nutrient digestibility, blood characteristics, cecal microbial shedding, and excreta odor contents in broilers. Poult. Sci. 93 (2), 364–370.

Zhang, A.W., Lee, B.D., Lee, S.K., et al., 2005. Effects of yeast (*Saccharomyces cerevisiae*) cell components on growth performance, meat quality, and ileal mucosa development of broiler chicks. Poult. Sci. 84 (7), 1015–1021.

Zommiti, M., Chikindas, M.L., Ferchichi, M., 2019. Probiotics—live biotherapeutics: a story of success, limitations, and future prospects—not only for humans. Probiotics Antimicrob. Proteins. https://doi.org/10.1007/s12602-019-09570-5 (Online ahead of print).

CHAPTER 14

Probiotics and Prebiotics in Fish Aquaculture

ELIANA DE FÁTIMA MARQUES DE MESQUITA[a] • FLÁVIA ALINE ANDRADE CALIXTO[a,b,c] • RODRIGO TAKATA[c,d]
[a]Departamento de Tecnologia de Alimentos, Universidade Federal Fluminense/UFF, Rio de Janeiro, Brazil, [b]Mestrado Profissional em Ciência e Tecnologia de Alimentos (PCTA)/IFRJ, Rio de Janeiro, Brazil, [c]Fundação Instituto de Pesca do Estado do Rio de Janeiro/Fiperj, Rio de Janeiro, Brazil, [d]Programa de Pós-graduação em Aquicultura e Recursos Aquáticos Tropicais/UFRA, Pará, Brazil

INTRODUCTION

The fish production from aquaculture has increased its representativeness in world market, mainly for human consumption. Brazil has favorable environmental and geographical conditions for aquaculture development. An increase in the aquatic organism production from different systems used in the country has been observed, in recent years. This growth occurred mainly due to research advances and adaptations of technological packages for the growth of species in production systems and public policies implemented for the field. Therefore, food and nutrition have been widely explored for intensifying production, and for this reason, aquaculture activity has been a reality as a source of income for rural producers. Meat production has been intensified by significant advances in the studies of genetics, nutrition, ambience, and reproduction. In pigs, specifically, from the aspects of nutrition, antimicrobial growth promoters (antibiotics and chemotherapeutics) were and are fundamental, bringing great benefits in terms of performance and feed efficiency when used in subtherapeutic doses as ingredients in their diets. Currently, world meat producers are working to drastically reduce the use of antibiotics, following the guidelines of the World Health Organization (WHO): "We recommend a general reduction in the use of all important classes of antimicrobial drugs in food-producing animals." Brazilian Decree No. 6.296/07 and Brazilian Normative Instruction No. 13/04 regularize the manufacture, fractionation, import, export, commercialization, and use of additives for animal feed. Therefore, the improvement of antibiotic substitutes is important, and functional foods and/or food additives stand out among these. In 2002, the WHO and FAO considered probiotics as "live microorganisms when administered in adequate amounts confer a health benefit on the host." Since most of the chapters in this book refer to food for humans, it is important to explain the concept of additive used throughout this chapter. In animal feed, according to IN 13/04, additive is a "substance, microorganism or formulated product, intentionally added to products, which is not used as an ingredient, whether or not nutritional value and which improves the characteristics of products for animal feed or animal products, improve the performance of healthy animals or meet the nutritional needs." However, as food for human, food additive is, according to Ordinance SVS/MS 540/1997, any and all ingredients intentionally added to foods without the purpose of nourishing, with the aim of modifying physical, chemical, biological, or sensory characteristics, during the manufacture, processing, preparation, treatment, packaging, conditioning, storage, transport, or handling of a food.

Fish gut microbiota is formed by a complex ecosystem of microorganisms that interact with each other; its relationships and predispositions with the predominant bacterial group are associated, among other factors, with host feeding and health. Despite knowledge advances on the subject, there are still gaps that need to be filled in order for a direct applicability in aquaculture, more specifically in aquatic organisms feeding and nutrition. Thus, the ingredients that act stimulating the beneficial bacteria for the proper functioning of the gastrointestinal tract are viable and feasible alternatives for the good performance of fish production. These ingredients can be a chemical, a stressor, or any other substance that acts by increasing the nonspecific or specific defense mechanisms of organisms.

In general, it is considered that useful microorganisms are those that act in aiding digestion and absorption of nutrients, produce vitamins that can be used by the host animal, and can attenuate the proliferation of pathogens by competitive exclusion. Harmful microorganisms can cause inflammation in the intestinal

Probiotics and Prebiotics in Foods. https://doi.org/10.1016/B978-0-12-819662-5.00010-0

mucosa, generate toxic metabolites, and weaken the animal in face of production systems challenges, providing diseases appearance and can lead to the individual's death. Under normal and health conditions, these populations are in balance; however, under stress conditions (diet and climate change, high stocking density, low aeration, or any other unfavorable situation), useful populations decrease and harmful populations proliferate, which is negatively reflected on animal health and performance.

In aquaculture, there is also an intensification of production systems as a way to optimize the resources for the area, and this has made aquatic animals increasingly dependent on the formulated diet for the specific dietary requirements. Studies in the area of nutrition were and are increasingly important for the nutritional improvement of the species under production and food management adopted. Currently, the nutrition of aquatic organisms aims that the formulated diet provides all the nutrients required for their growth and, at the same time, provides an improvement in their health. In this manner, the functional ingredients, i.e., probiotics, prebiotics, and synbiotics ingredients (prebiotic and probiotic junction), are used to improve the immune systems and, as a consequence, promote better performance of animals.

PROBIOTICS

Probiotics are food ingredients or supplements consisting of living microorganisms that can benefit the animal by balancing the intestinal microbiota. These microorganisms should exert beneficial effects on the host animal, increasing its performance or its resistance to diseases. They act as viable cells, capable of surviving and metabolizing themselves in the intestinal environment, resisting the stomach low pH and organic acids, being stable and able to remain viable for long periods under field storage conditions, and finally should not be pathogens or toxics. The definition of the term "probiotic" has been changed and broadened over the years. Although, the definition developed by WHO and FAO is accepted by the majority of researchers, but during recent years, new definitions are added to the probiotic terminology such as "paraprobiotics" (dead/inactivated cells of probiotics) and "postbiotics" (healthful metabolites of probiotics). In such way, the probiotic can be defined as viable or inviable microbial cell (vegetative or spore; intact or ruptured) that is potentially healthful to the host (Zendeboodi et al., 2020). One of the most important properties of probiotics is adherence and proliferation in specific locations to maximize their potential by host species. To achieve the maximum benefit, probiotics must reach specific locations where their action is required. Probiotics can act in several ways, including production of inhibitory compounds, competition for adhesion sites, modulation of host immune responses, competition for chemicals or available energy, competition for nutrients, improvement of water quality of production systems, in which case probiotics can be linked to the diet to enter the system indirectly through feces or added directly into the aquatic environment, interference in *quorum* detection (cell-to-cell communication between bacteria), antibacterial, antiviral, and antifungal activities.

Among the most common mechanisms of action of probiotics were the competitive exclusion of pathogenic bacteria, the improvement of the nutrition of the host animal, the enzymatic contribution to nutrient digestion, and the stimulation of the host immune response. For fishes, some positive actions have already been described, such as improving performance, carcass composition, utilization of nutrients and energy from formulated diets; and the action of digestive enzymes, health status, disease resistance, antioxidant activity, intestinal morphology, intestinal microbial composition, reduction of stress, and malformations.

The effect of probiotics depends on some factors, such as the form of supplementation, vector administration, dosage level, and duration of application. More commonly, in aquaculture-related studies, live crops are sprayed on or added to mixtures in dietary formulations. However, other compounds derived from these living cultures have already been tested, such as freeze-dried, encapsulated cells, cellular components, cell-free supernatants, and spores. In the midst of positive results or not, these are the main lines used to convey probiotics in fish feeding.

A probiotic can be derived from intestinal bacteria, isolated from the host itself, known as autochthonous. The idea is that this type of bacteria is already well adapted to the intestinal environment of the fish, and therefore, its effect can be further enhanced. Therefore, these bacteria should not be pathogenic and, at the same time, need to be apt to resist the action of bile salts, extreme pH values, and digestive tract proteases, besides ensuring intestinal colonization and greater safety as a dietary supplement. Some antimicrobial substances produced by bacilli isolated from the fish intestine have been described as biological control agents.

Lactic acid bacteria are among the main groups of microorganisms used as probiotics, as they have important characteristics, such as rapid growth and colonization in hosts, production of antimicrobial compounds (such as lactic acid, hydrogen peroxide, organic acids, and bactericides), and stimulate nonspecific immune response. These bacteria can be isolated from the intestinal environment, as already described for "acará-bandeira" *Pterophyllum scalare*, Nile tilapia *Oreochromis niloticus*, and "pacu" *Piaractus mesopotamicus*. Some research lines argue that bacteria isolated from the host itself have the advantages of not being pathogenic, being able to resist the action of bile salts, extreme pH values, and digestive tract proteases, in addition to ensuring intestinal colonization and greater safety as a food supplement. However, the other groups of bacteria have positive results and should also be considered effective in promoting fish health and growth.

The promotion of fish health using probiotics can lead to an improvement in their growth and in this case, considered as growth promoters. Some of these examples can be seen in Table 1. Studies have shown that the application of probiotics can improve feed conversion, growth rates, weight gain, and/or dietary use by fish,

TABLE 1
Some relevant results with the use of probiotics in the feeding of fish species of interest for Brazilian aquaculture.

Species (Developmental phase)	Parameters	Probiotics	Results	Reference
Pacu *Piaractus mesopotamicus* (juvenile)	Isolation and selection of potentially beneficial autochthonous bacteria	A number of 522 isolates from different anatomical components (30 lactic acid bacteria and 8 *Bacillus* strains). Only 8 were selected, since they have probiotic characteristics	Selected bacteria: *Enterococcus faecium (KJ740155), Pediococcus acidilactici (KJ740154), Enterococcus faecium (KJ740156, KJ740152, KJ740153), Bacillus subtilis (KJ754388, KJ740157, KJ740158)*	Guidoli et al. (2015)
Pacu *Piaractus mesopotamicus* (juvenile)	Challenge to infection by *Aeromonas hydrophila*	2, 4, 8, and 16 g/kg of *Bacillus cereus* and *Bacillus subtilis* 1:1, 10^8 CFU/g	Recommended: 8 g/kg of *Bacillus cereus* and *Bacillus subtilis* 1:1, 10^8 CFU/g, with higher survival rates and immune responses	Farias et al. (2016)
Pacu *Piaractus mesopotamicus* (larval)	Performance and development of the digestive system	Autochthonous strains of *Bacillus subtilis* (A252, A253, and A254)	The gastrointestinal development was favored (appearance of gastric glands), however, with no effect on performance	Guidoli et al. (2016)
Pacu *Piaractus mesopotamicus* (larval)	Performance and survival	8 autochthonous bacteria strains (6×10^4 and 6×10^7, depending on strains)	Improved performance	Guidoli et al. (2018)
Tambaqui *Colossoma macropomum* (juvenile)	Performance, body composition, hematological parameters, and challenge to infection by *Streptococcus agalactiae*	*Bacillus subtilis* (10^9 CFU/g) and *Saccharomyces cerevisiae* (10^9 CFU/g)	No difference in performance, body composition, and mortality in challenge to infection. Improvement in hematological parameters (red and white series) with *Saccharomyces cerevisiae*	Paixão et al. (2017)

Continued

TABLE 1
Some relevant results with the use of probiotics in the feeding of fish species of interest for Brazilian aquaculture—cont'd

Species (Developmental phase)	Parameters	Probiotics	Results	Reference
Nile tilapia *Oreochromis niloticus* (juvenile)	Performance, body composition, use of nutrients, and energy	Commercial probiotic Biogen (0; 0.5; 1.5; 2.0; and 2.5% of diet)	Improved performance and use of nutrients and energy from diets. Lipids and body energy decreased with 0.5% of probiotic	El-Haroun et al. (2006)
Nile tilapia *Oreochromis niloticus* (juvenile)	Hematological parameters	Autochthonous bacteria (*Lactobacillus plantarum* and *Lactobacillus brevis*, 1×10^8 CFU/g)	*Lactobacillus plantarum* improved the immune system	Jatobá et al. (2008)
Nile tilapia *Oreochromis niloticus* (juvenile)	Acute inflammatory response induced by carrageenan injection	*Lactobacillus plantarum*	The probiotic did not influence the total number of erythrocytes, lymphocytes, basophils, and hematocrit. The number of neutrophils was higher in fish supplemented with probiotics and injected with carrageenan. The glucose in fish fed with probiotic and noninjected fish was higher than in those injected with saline solution	Dotta et al. (2011)
Nile tilapia *Oreochromis niloticus* (juvenile)	Performance and immune response	*Bacillus amyloliquefaciens* and *Lactobacillus* sp.	Improved the evaluated parameters	Ridha and Azad (2012)
Nile tilapia *Oreochromis niloticus* (juvenile)	Performance, digestibility, and resistance to infection by *Aeromonas hydrophila*	*Lactobacillus plantarum* (22,708 Cpqba DRM) and *Saccharomyces cerevisiae* (1×10^8 CFU/mL, with 100 mL/kg of diet)	Improved performance, protein and energy retention rates, and resistance to infection by *Aeromonas hydrophila*. Supplementation with *Saccharomyces cerevisiae* improved digestibility of protein, energy, and dry matter	Cornélio et al. (2013)
Nile tilapia *Oreochromis niloticus* (juvenile)	Performance, hematology, and vaccination against *Streptococcus agalactiae*	Cell wall of *Saccharomyces cerevisiae* (0 and 0.3% of diet)	Improved performance. The interaction between diet and vaccination resulted in higher rates for hematocrit, hemoglobin, and leukocytes	Salvador et al. (2013)

TABLE 1
Some relevant results with the use of probiotics in the feeding of fish species of interest for Brazilian aquaculture—cont'd

Species (Developmental phase)	Parameters	Probiotics	Results	Reference
Tilapia hybrid *Oreochromis niloticus* and *Oreochromis aureus* (juvenile)	Performance, adhesive intestinal microbiota, immune response and resistance to infection by *Aeromonas hydrophila*	2 strains of *Lactobacillus*: highly adhesive *Lactobacillus brevis* JCM 1170 (HALB) and with low stickiness *Lactobacillus acidophilus* JCM 1132 (LALB)	No effects on performance. The strain HALB 10^9 cel/g of diet provided protection from infection by *Aeromonas hydrophila*	Liu et al. (2013)
Nile tilapia *Oreochromis niloticus* (adults)	Challenge to infection by *Streptococcus agalactiae*	*Saccharomyces cerevisiae* (Sc, 0.3% of diet), *chromium carbquelate* (18 mg of Cr/kg of diet) or in association (Sc + Cr)	The probiotic did not change the variables analyzed in the face of the challenge. Chromium carbquelate was more efficient in an acute response to infection	Castro et al. (2014)
Nile tilapia *Oreochromis niloticus* (juvenile and adult)	Performance, immunity, and resistance to disease	Autochthonous bacteria *Bacillus subtilis* NCBI 1701438; commercial probiotic *Lactobacillus acidophilus*; commercial additive containing *Lactobacillus sporogenes*, *Lactobacillus acidophilus*, *Bacillus subtilis*, *Bacillus licheniformis*, and *Saccharomyces cerevisiae*	The autochthonous probiotic and the mixture of bacteria improved performance, immunity, and resistance to diseases. The development phase (fish size) showed a different response to probiotics	Ridha and Azad (2015)
Nile tilapia *Oreochromis* sp. (juvenile)	Viability of the probiotic in the diet after drying	*Lactobacillus plantarum* Lab9 (CPQBA 144-09 DRM 03)	The viability of bacteria in the diet was less than 10^6 CFU/g after 43 days of drying	Londoño et al. (2015)
Nile tilapia *Oreochromis niloticus* (juvenile)	Performance, morphology and intestinal microbiota	Diets containing enzymes: phytase, protease, and xylanase; probiotics: *Bacillus subtilis*, *Bacillus licheniformis*, and *Bacillus pumilus* and the combination of enzymes and probiotics	Enzymes and probiotics in diet were able to improve performance and intestinal morphology features, with no effect on the composition of the intestinal microbiota	Adeoye et al. (2016)
Nile tilapia *Oreochromis niloticus* (juvenile)	Performance and lead intoxication	*Lactobacillus plantarum* CCFM8661 (10^8 CFU/g of diet)	Improved performance and prevented the fish death by lead intoxication	Zhai et al. (2017)

Continued

TABLE 1
Some relevant results with the use of probiotics in the feeding of fish species of interest for Brazilian aquaculture—cont'd

Species (Developmental phase)	Parameters	Probiotics	Results	Reference
Nile tilapia *Oreochromis niloticus* (juvenile)	Performance, enzymatic and antioxidant activities, immunity, and hormone modulation activities	*Paenibacillus polymyxa* HGA4C (1×10^6 and 1×10^8 CFU/g)	Improved performance, enzyme, and antioxidant activities, immunity, and modulated the expression of GHR-I, GHR-II, IGF-I, and IGF-II hormones	Midhun et al. (2019)
Nile tilapia *Oreochromis niloticus* (juvenile)	Performance, immunity, and disease resistance	Host-associated probiotics (*Lactobacillus plantarum* N11, 10^8 CFU/g of diet and *Bacillus velezensis* H3.1, 10^7 CFU/g of diet). Used separately and combined	The separate or combined use of probiotics was beneficial for the parameters evaluated	Doan et al. (2018)
Nile tilapia *Oreochromis niloticus* (juvenile)	Oxidative state, heat shock protein, and immune system-related gene expression in hypoxia challenge	*Aspergillus oryzae* (ASP, 1×10^6 and 1×10^8/g of diet)	The addition of ASP to the diet improved the fish immune system-related against hypoxia stress	Dawood et al. (2018)
Rainbow trout *Oncorhynchus mykiss* (juvenile)	Control lactococcosis and streptococcosis (*Lactococcus garvieae* and *Streptococcus iniae*)	*Aeromonas sobria* (10^7 cel/g of diet)	The probiotic was beneficial for controlling lactococcosis and streptococcosis	Brunt and Austin (2005)
Rainbow trout *Oncorhynchus mykiss* (juvenile)	Challenge to infection by *Yersinia ruckeri*	Subcellular components of *Aeromonas sobria* GC2 and *Bacillus subtilis* JB-1	Infection control	Abbass et al. (2010)
Rainbow trout *Oncorhynchus mykiss* (juvenile)	Challenge to infection by *Vibrio anguillarum*	Allochthonous bacteria Kocuria SM1 (10^8 cel/g of diet)	The probiotic showed beneficial effects on defense mechanisms	Sharifuzzaman and Austin (2010)
Rainbow trout *Oncorhynchus mykiss* (juvenile)	Performance, well-being, meat quality, and intestinal microbiota	Multistrains of probiotic bacteria *Bacillus* sp., *Pediococcus* sp., *Enterococcus* sp., *Lactobacillus* sp. (0.2% of diet, 2×10^9 CFU/kg of diet)	No difference in performance. It improved the health of the animals, without interfering in the quality of the final product	Ozório et al. (2015)
Rainbow trout *Oncorhynchus mykiss* and brown trout *Salmo trutta* (juvenile)	Performance, immune response, and intestine morphology	*Bacillus subtilis* and *Bacillus cereus toyoi* [0; 0.03 (6×10^3 CFU/g of diet); and 0.06% (1.5×10^6 CFU/g of diet)]	No effect on performance, body composition, or immune response (plasma lysozyme, alternative complement, and peroxidase activities). Increase in inflammatory cells in groups fed with probiotics	Ramos et al. (2016)

TABLE 1
Some relevant results with the use of probiotics in the feeding of fish species of interest for Brazilian aquaculture—cont'd

Species (Developmental phase)	Parameters	Probiotics	Results	Reference
Rainbow trout *Oncorhynchus mykiss* (juvenile)	Adsorption capacity of aflatoxin B1 from yeasts isolated from the intestine	Strains of bacteria isolated from the intestine *Debaryomyces hansenii* and *Kazaschtania exigua*	Improved fish health, inhibiting pathogens and the adsorption of aflatoxin B1	Martínez et al. (2018)
Jundiá *Rhamdia quelen* (larval)	Performance	*Bacillus cereus* var. *toyoi* (~ 5×10^8 CFU/mL) and *Saccharomyces boulardii* (~ 2×10^9 CFU/mL)	The probiotics had an inhibitory effect on *Vibrio carchariae*. No effect on performance	Souza et al. (2012)
Pirarucu *Arapaima gigas* (juvenile)	Performance and fish health	*Enterococcus faecium* (1×10^6 and 1×10^8 CFU/g)	The concentration of 1×10^8 CFU/g of diet modified the intestinal microbiota, improved performance and hematology, and reduced the quantity of parasites	Sousa et al. (2019)
Matrinxã *Brycon amazonicus* (adults)	Reproduction, hematological and immunological parameters, and concentrations of lipids in muscle and liver tissue	*Bacillus subtilis* (strain C-3102-Calsporin, 10 g/kg of diet)	Females fed probiotics had an increase in the number of oocytes and higher rates of fertilization and hatching of larvae. Improved immune system and decreased lipids in muscle and liver	Dias et al. (2010)
Piavuçu *Leporinus macrocephalus* (juvenile)	Performance and hematological response to stress	*Saccharomyces cerevisiae* (9×10^5 CFU/g), garlic (10 g/Kg), and cinnamon (10 g/Kg)	The probiotic did not improve performance or reduce stress	Lima et al. (2015)
Lambari *Astyanax bimaculatus* (juvenile)	Performance and immune system	Allochthonous bacteria (*Lactobacillus* spp. strains CPQBA 1168-15 DRM-01, $\geq 1.0\times10^7$ CFU/g)	Improved performance and immune system	Moraes et al. (2018)
Sea bass *Centropomus undecimalis* (juvenile)	Performance, body composition, immune response, and hematological parameters, in alternating and continuous feeding regimes	*Bacillus subtilis* (C-3102) at concentrations of 5×10^9 CFU/Kg of diet	Addition of the probiotic did not promote growth. However, it provided an immunostimulating action when supplied alternately	Noffs et al. (2015)

Continued

TABLE 1
Some relevant results with the use of probiotics in the feeding of fish species of interest for Brazilian aquaculture—cont'd

Species (Developmental phase)	Parameters	Probiotics	Results	Reference
Cardinal tetra, neon, *Paracheirodon axelrodi* (juvenile)	Transport	Efinol (10 mg/L)	Reduction of mortality and maintenance of water quality (reducing metabolic waste)	Gomes et al. (2009)
Acara-bandeira *Pterophyllum scalare* (juvenile)	Performance and resistance to pathogens	Allochthonous bacteria *Enterococcus faecium* (2×10^8 and 2×10^6 CFU mL/g)	Improved performance and in vitro tests showed good resistance to the pathogens evaluated	Dias et al. (2019)

CFU, colony-forming unit.

and we can mention, for example, *Oncorhynchus mykiss* rainbow trout, *Oreochromis niloticus* Nile tilapia, *Brycon amazonicus* "matrinxã," *Piaractus mesopotamicus* "pacu," *Colossoma macropomum* "tambaqui," *Arapaima gigas* "pirarucu," *Astyanax bimaculatus* "lambari," *Centropomus undecimalis* sea bass, *Paracheirodon axelrodi* "cardinal tetra neon," and *Pterophyllum scalare* "acará bandeira."

Another positive point regarding the use of probiotics in fish feed is the possibility of improving the reproductive performance of animals, as observed by adults of "matrinxã." The authors observed that the use of the probiotic *Bacillus subtilis* (strain C-3102-Calsporin) provided an increase in the number of oocytes, and in the rates of fertilization and larvae hatching. These results probably occurred due to the better availability of nutrients for females, which led to an increase in oocytes and probably aiding in vitellogenesis. Specifically for larvae, the endogenous microbial population depends on both the egg and the environmental microbial composition.

Altricial larvae, at the beginning of their exogenous feeding, have an immature digestive tract, still developing, which hinders digestion and assimilation of nutrients. The lack of acid digestion and an active gallbladder makes colonization of probiotic beneficial bacteria facilitated. However, the inoculation route may be somewhat unstable, due to the need to work with live food, dry formulated diet, or for aquaculture water. The success of the techniques employed depends on several factors, from the species to the stage of larval development.

PREBIOTICS

As stated in Chapter 1 of this book, prebiotic is "a substrate that is used selectively by host microorganisms that confer a health benefit." In IN 13/04, which regulates additives for products intended for animal feed, it is stated that prebiotics are ingredients that are not digested by the host's digestive enzymes, but that are fermented by the microbiota of the animals' digestive tract, contributing to their balance.

These compounds have been used as an alternative to growth promoters in order to maintain the beneficial balance of the intestinal microbiota, especially in young animals or in imminent stress conditions. These carbohydrates can be classified according to the molecular size or the degree of polymerization of the carbon chain (number of monosaccharide units), such as monosaccharides, polysaccharides, or oligosaccharides.

A prebiotic has a characteristic to be resistant to digestion by acids, salts, and enzymes produced by the animal organism and is potentially fermentable (cellulose, hemicelluloses, resistant starch oligosaccharides, phenolic compounds, among others). However, not all of them act as stimulators in the development of beneficial microorganisms in the digestive tract. That is, the fact that they are indigestible, but fermentable, does not mean that they act as prebiotics. Nondigestible oligosaccharides have been preferably used as prebiotics due to their greater fermentative selectivity.

The action of prebiotics is modulating the profile of the native intestinal microbiota present in the host, favoring the action of beneficial bacteria; competing for the same glycoconjugates on the surface of epithelial cells; preferring the nonspecific barrier function; and improving the production of mucus, short-chain fatty acids, and cytokines. A working line directs to the hypothesis that one of the beneficial actions of prebiotics is specific intestinal translocation of pathogenic bacteria, causing no adhesion to epithelial cells and acting directly in the activation of the acquired immune response. The direct result of prebiotics is the development of beneficial bacteria populations to the host, by the improvement in luminal conditions, in the morphology of the digestive tract, in the organism defense system, and, as a consequence, in the improvement of fish performance. The potential action of prebiotics in diets is linked to their specificity in fermentation, stimulating the growth and stability of microbial populations producing organic acids, especially lactic and acetic acids. These compounds have the functions of reducing the pH of the intestinal lumen and inhibiting the proliferation of harmful agents such as *Escherichia coli*, *Clostridium* sp., and *Salmonella*, which are sensitive to acids. These actions occur through antibacterial substances and enzymes produced by this microbiota favored by the adoption of prebiotics.

Pattern recognition receptors, such as β-glucan receptors and dectin-1 receptors expressed in macrophages, have direct action on the immunomodulatory activities of prebiotics. This type of interaction activates high-level reactions that stimulate the immune system. Saccharides can also interact with pattern knowledge receptors to act associated with microbial agents such as teichoic acid, peptidoglycan, glycosylated protein, or the bacteria capsular polysaccharide, leading to an immune response.

The most common prebiotics, used in fish and shrimp, includes inulin, fructooligosaccharides (FOS), and mannanoligosaccharides (MOS). Other well-known prebiotics are galactooligosaccharides (GOS), xylooligosaccharides (XOS), arabinoxylan-oligosaccharides (AXOS), isomaltooligosaccharides (IMO), and GroBiotic (commercial mixture containing partially self-lysed yeast, dairy derivatives, and dry fermentation products).

The most studied prebiotics in aquaculture are MOS, FOS, inulin, and β-glucans. MOS are oligosaccharides which have complex mannoproteins derived from the cell walls of the yeast *Saccharomyces cerevisiae*. The cell wall of these yeasts is composed of mannoproteins and chitin. FOS are nondigestible oligosaccharides composed of fructose and glucose units. Inulin is a type of fructose sugar fiber, classified as fructooligosaccharides, composed of β-D-fructofuranoside. The β-glucans are polysaccharides having glucose in their molecular structure, with β-glycosidic linkages. In nature, this prebiotic component can be found in the cell wall of many plants, such as wheat, rye, barley, and oats, within yeast for bread and beer production (genus *Saccharomyces*). Another source of β-glucans is algae, such as *Laminaria* sp., which in particular is very interesting, because it is part of a natural source in which not only prebiotics are found as additives capable of improving the health and performance of animals, but also a number of other products that make macroalgae an interesting source to be explored in aquaculture (Table 2).

Algae are a natural alternative of prebiotic source, and can be considered in the formulation of diets for organic aquaculture. So, several studies have been conducted to evaluate its potential for fish in dietary formulations. Algae generally contain 37.3% of fiber, 84.8% of which are considered soluble, and the insoluble is essentially composed of cellulose. Ergosan, a dietary supplement consisted of dried seaweed and plant extracts of *Laminaria digitata* and *Ascophyllum nodosum*, was tested in tilapia showing no gain in feed conversion. Algae such as *Ascophyllum nodosum* and *Gracilaria edulis* were tested in tilapia feeding but without significant results. The same was observed in a study testing species of algae, *Fucus vesiculosus*, and *Ulva rigida*, in zebrafish; and the use of spirulina microalgae (*Spirulina maxima*) tested as prebiotic in tambaqui feeding. Despite efforts in recent years to find and elect algae as a potential prebiotic, the results indicate that much research is still needed.

SYNBIOTICS

The joint action of probiotics and prebiotics is known as "synbiotic." This interaction in vivo can be favored by an adaptation of the probiotic to the prebiotic substrate prior to consumption, resulting in competitive advantages. The combined application is based on the principle that the prebiont provides the probiont competitive advantage (energy source for fermentation) over endogenous populations, improving survival and implantation of the live microbial food supplement in the host gastrointestinal tract, selectively stimulating growth, and/or activating the metabolism of one or a limited number of health-promoting bacteria, thereby improving fish welfare.

The synbiotic effect can be directed to the target regions of the gastrointestinal tract, such as the small and large gut. Many beneficial effects such as competition among pathogens for nutrients, or for adhesion sites and modulation of the immune system, can be expected from probiotics. The beneficial effects of synbiotics may also be related to the production of acetic, lactic, propionic, and butyric acid, which are typical final products of prebiotic fermentation in the gut by probiotic bacteria. However, since all the desired characteristics in

TABLE 2
Some relevant results with the use of prebiotics in the feeding of fishes species of interest for Brazilian aquaculture.

Species (Developmental phase)	Parameters	Prebiotics	Results	Reference
Pacu *Piaractus mesopotamicus* (juvenile)	Performance and digestibility	Autolysed yeast *Saccharomyces cerevisiae* in substitution to fish meal (0—control, 30%, 35%, 50%, 70%, and 100%)	Replacing 50% of fishmeal with yeast improved feed efficiency, performance and reduced nitrogen loss	(Ozório et al., 2010)
Pacu *Piaractus mesopotamicus* (juvenile)	Hematological parameters	*Mananoligosaccharide* (MOS, 0%; 0.2%; 0.4%; 0.6%; 0.8%; 1.0%; 1.5%; and 2.0% of diet)	The MOS levels did not minimize the effects of stress on hematological and biochemical parameters	Sado et al. (2014)
Pacu *Piaractus mesopotamicus* (juvenile)	Performance and intestine morphology	*Mananoligosaccharide* (MOS, 0%; 0.2%; 0.4%; 0.6%; 0.8%; 1.0%; 1.5%; and 2.0% of diet)	The MOS levels did not provide differences in the performance and morphometry of the intestinal villi	Sado et al. (2014)
Pacu *Piaractus mesopotamicus* (juvenile)	Stress challenge: hypoxia, and infection by *Aeromonas hydrophila* (1×10^6 CFU)	*β-Glucans (BG1) and products based on β-glucans: 1.3* and *1.6 β-glucans (BG2). Concentrations:* 0.1% and 0.5% of diet	BG1 at 0.5% for 10 days and 0.1% for 15 days improved the leukocyte respiratory activity. Both BG1 and BG2 produced similar responses compared with the conventional BG product	Sabioni et al. (2019)
Tilapia hybrid (*Oreochromis niloticus* × *Oreochromis aureus*) (juvenile)	Performance, body composition, and histology of the intestine and liver	*Mananoligosaccharide* (MOS: 0; 1.5; 3; 4.5 g/Kg of diet)	No effect on performance and body rates. The increased concentration of MOS in the diet provided an increase in dry matter and body protein. The average villus length was greater in fish fed 1.5 g of MOS/Kg of diet	Genc et al. (2007)
Nile tilapia *Oreochromis niloticus* (juvenile)	Performance and hematological parameters	*Mananoligosaccharide* (MOS—ActiveMOS and Biorigin—0%; 0.2%; 0.4%; 0.6%; 0.8%; and 1.0%)	No change in hematological parameters and a decrease in dietary intake	Sado et al. (2008)
Tilapia hybrid (*Oreochromis niloticus* × *Oreochromis aureus*) (juvenile)	Intestinal bacteria communities	Autolysate of yeasts and short-chain fructooligosaccharides	There was a change in the intestinal bacterial community, with different patterns for each prebiotic	Zhou et al., (2009)

Nile tilapia *Oreochromis niloticus* (juvenile)	Performance, hematological parameters, immunity, and resistance	Inulin (5 g/Kg of diet) and vitamin C (500 mg/kg of diet)	Vitamin C was more efficient than inulin for the evaluated parameters	Ibrahem et al. (2010)
Nile tilapia *Oreochromis niloticus* (larval)	Performance and intestine morphology	*Mananoligosaccharide* (MOS, 0%; 0.15%; 0.30%; 0.45%; 0.60%; and 0.75%)	The MOS at 0.34% in the diet improved feed conversion and promoted an increase in intestine length, villus height, and intestinal villus density	Schwarz et al. (2011)
Nile tilapia *Oreochromis niloticus* (juvenile)	Performance, blood parameters, and resistance to infection by *Aeromonas hydrophila*	Prebiotic Power top (based on *mananoligosaccharide*—MOS and *β-glucans*). *Concentrations:* 0%; 0.05%; 0.1%; 0.15%; and 0.2%)	The prebiotic altered blood parameters and was efficient in promoting resistance to infection by *Aeromonas hydrophila*. The concentration of 0.15% Power top/kg improved performance	Ahmad et al. (2015)
Nile tilapia *Oreochromis niloticus* (juvenile)	Performance, survival, intestinal microbiota, and extracellular production of superoxide anion	GroBiotic-A (1% and 2%)	The concentration of 2% improved the evaluated parameters	Peredo et al. (2015)
Nile tilapia *Oreochromis niloticus* (juvenile)	Performance, immune system, and gut morphology	*Mananoligosaccharide* (MOS, 0%; 0.2%; 0.4%; and 0.6%)	Performance and immune system have not changed. The MOS concentration of 0.4% provided an increase in the height of intestinal folds and 0.4% and 0.6% of MOS led to an increase in the thickness of the intestinal muscle layer	Sado et al. (2015)
Nile tilapia *Oreochromis niloticus* (juvenile)	Performance, resistance to *Yersinia ruckeri* infection, and intestinal morphology	*β-Glucans* (0; 1.5 and 3.0 g/kg of diet) and *mananoligosaccharide* (MOS, 180 g/kg of diet)	The prebiotics improved performance, survival, and the immune system. There were an increase in protein and lipids in fish body, and also in the height of the villi, the number of goblet cells, and the number of intraepithelial lymphocytes	Selim and Reda (2015)

Continued

TABLE 2
Some relevant results with the use of prebiotics in the feeding of fishes species of interest for Brazilian aquaculture—cont'd

Species (Developmental phase)	Parameters	Prebiotics	Results	Reference
Nile tilapia *Oreochromis niloticus* (juvenile)	Performance and parameters of blood, immunological and histological	*Inulin*: 0; 2.5 and 5 g/kg of diet. *Jerusalem artichoke tuber* (rich in inulin and fructooligosaccharides): 5 and 10 g/kg of diet	Inulin 5 g/kg and the *Jerusalem artichoke tuber* at 10 g/kg were effective in promoting performance and health	Tiengtam et al. (2015)
Nile tilapia *Oreochromis niloticus* (juvenile)	Performance, survival, use of nutrient, and body composition	Prebiotic Fermacto (0, 1, 2, 3, and 4 g/kg of diet)	The concentration of 3 g of Fermacto/kg of diet showed the best cost/benefit for the evaluated parameters	Hassan (2016)
Nile tilapia *Oreochromis niloticus* (juvenile)	Performance, immunity, and resistance to *Pseudomonas flourescens*	*Prebiotic* β-polo (based on *β-glucans*, 0; 0.5; 1; 1.5 mL/kg of diet)	The prebiotic improved performance, immunity, and resistance to infection, especially at a concentration of 1.5 mL/kg	Tawab et al. (2016)
Nile tilapia *Oreochromis niloticus* (juvenile)	Performance and resistance to infection by *Streptococcus agalactiae*	*β-Glucans* (*Biorigin:* 0.1 g of BG01/kg of diet and 0.1 of BG02/kg of diet)	Both sources of β-glucans were effective in promoting resistance. The BG02 improved performance	Pilarski et al. (2017)
Nile tilapia *Oreochromis niloticus* (juvenile)	Hematological parameters, serum lysozyme, and intestinal microbiology	*Mananoligosaccharide* (MOS: 1, 8, and 15 g/Kg of diet)	MOS modulate the intestinal microbiota, increasing the number of beneficial bacteria and stimulating immunity, giving rise to white blood cells and lysozymes	Levy-Pereira et al. (2018)
Rainbow trout *Oncorhynchus mykiss* (juvenile)	Performance and blood parameters	*Prebiotic* Profeed (*fructooligosaccharides*): 1 g/kg of diet	The prebiotic improved performance and provided positive results for blood parameters	Řehulka et al. (2011)
Rainbow trout *Oncorhynchus mykiss* (juvenile)	Performance	*Concentrated autolysate of yeast cell wall Saccharomyces cerevisiae:* 0; 0.5; 1; 1.5; and 2 g/kg of diet	The prebiotic at a concentration of 0.5 g/kg improved performance	Ghorbani et al. (2012)
Rainbow trout *Oncorhynchus mykiss* (juvenile)	Performance and survival	*Prebiotic Saccharomyces cerevisiae* (A-max, United States): 0.2%. Acidifying Acesol Plus, Italy): 0.15%	The prebiotic alone and together with the acidifier showed positive effects on the evaluated parameters	Tabrizi et al. (2012)

Rainbow trout *Oncorhynchus mykiss* (juvenile)	Performance, survival, blood parameters, and challenge to alkalinity	*Prebiotic* A-Max (mix of *mananoligosaccharide, fructooligosaccharides, and β-glucans*): 0; 0.5; 1; and 1.5 g/kg of diet	The inclusion of 1.5 g of the prebiotic A-Max/kg improved performance and survival to the challenge of alkalinity. No change in blood parameters	Akrami et al. (2015)
Rainbow trout *Oncorhynchus mykiss* (juvenile)	Performance, intestinal microbiota, body composition, and blood parameters	*Mananoligosaccharide* (MOS, 0, 1.0, 2.5, and 4.0 g/kg of diet)	The inclusion of 1.0 g of MOS/kg of diet improved performance, body composition, intestinal microbiota profile, and some blood parameters	Denji et al. (2015)
Rainbow trout *Oncorhynchus mykiss* (juvenile)	Performance, immunity and survival	*Mananoligosaccharide* (MOS, 0.2%)	Improved performance and reduced mortality	Mínguez et al. (2016)
Rainbow trout *Oncorhynchus mykiss* (juvenile)	Performance, survival, immunity, enzyme activity, and resistance to *Yersinia ruckeri*	*Mananoligosaccharide* (MOS, 0.2%) and poly-β-hydroxybutyrate (2%)	No change in performance. The enzymatic activity was not altered (except for pepsin and amylase with 2% poly-β-hydroxybutyrate). Increased resistance to infection	Najdegerami et al. (2016)
Rainbow trout *Oncorhynchus mykiss* (juvenile)	Immunological indicators, enzyme responses, stress tolerance, and resistance infection by *Aeromonas hydrophila*	Immunogen (2 g/kg of diet)	The prebiotic changed the hematological and immunological parameters when the fish was exposed to *Aeromonas hydrophila*. Increased stress tolerance. Histopathological examinations showed a potential protective effect on liver tissue, which resulted in a lower level of serum enzyme in fish fed with immunogen after challenge with *Aeromonas hydrophila*	Yarahmadi et al. (2016)
Rainbow trout *Oncorhynchus mykiss* (juvenile)	Performance, survival, microbiota profile, and amino acid metabolism	*Mananoligosaccharide* (MOS, 0; 0.4; 1 g/kg of diet) and GroBiotic A (1%)	Change in microbiota profile associated with the metabolism of nonessential amino acids	Betiku et al. (2017)

CFU, colony-forming unit.

the selection of probiotic strains are hardly achieved, prebiotics can be used to maximize the effects expected by probiotics. The use of prebiotics benefits the growth and fixation of the desirable beneficial microbiota in the intestinal tract of aquatic animals, and the joint application is an example of synbiotic product.

The concept of synbiotic became public for aquaculture with the publication of an article in which the authors combined the effect of *Enterococcus faecalis*, mananoligosaccharides, and polyhydroxybutyrate acid, with positive responses to the performance and immune response of rainbow trout. After this research, a series of studies have already been conducted with the objective of achieving the synergistic effect between probiotics and prebiotics, and standardizing their use in formulated diets. Achieving this synbiotic relationship in a host is still a challenge that deserves more research efforts for this technology to be used in formulated diets. The literature shows us that there are already positive results, mainly with the modulation of the intestinal microbiota, improvement of health and immune response and resistance to diseases. As a consequence of an improvement in the organic systems of fish, a greater development of these animals appears. In a study with the golden fish *Cyprinus carpio koi*, the combined use of chitosan oligosaccharides and *Bacillus coagulans* led to an increase in weight and specific growth rate, and also promoting immunity and disease resistance (Table 3). Therefore, investigations on probiotics with plant-derived products, such as algae, deserve greater attention, mainly due to their prebiotic potential already reported in the literature. The inclusion of these plant-derived products may be an alternative to optimize their use in formulated diets; however, there is a very fine line between the beneficial potential of these plant food components and the antinutritional effects presented by them. Thus, studies to adapt the levels of inclusion to potentiate the beneficial effects of these products together with probiotics, directing to a better modulation of the intestinal microbiota, should be carried out to promote the health and performance of fish.

FINAL NOTES AND PERSPECTIVES

Sustainability is a current theme for all productions, and aquaculture does not escape the rule. Production, increasingly close to a balance with the environment and social development, brings new perspectives for aquaculture, mainly due to the challenges that production systems bring every day. Despite this tendency to balance production, fishes are increasingly exposed to biotic and abiotic variations, which leads animals to stress, prejudicing their development and may lead to death.

Nutrition, more specifically food additives, i.e., probiotics, prebiotics, and synbiotics, brought a new perspective to improve fish health facing aquaculture challenges, strengthening the animal's immune system, and providing greater diseases control. Thus, treatment is avoided and prevention is implemented. The performance of these additives is well described in Tables 1–3, where are shown the actions of the main bacteria, autochthonous or not, used as a source of probiotics. In relation to prebiotics, the most studied and verified are inulin, MOS, and FOS. However, other nondigestible saccharides deserve special attention and more specific studies to determine their beneficial potential to the host, especially those from natural products, such as seaweed.

The high point of these food additives is synbiotics, and their potential emerged in the 2000s, due to the first studies. The selection of probiotic strains with prebiotics that show an interaction, more specifically a synbiotic synergisms, can gives a direction to the beneficial actions these additives. In fact, better than using bacterial strains in the diets or in the systems production water so that they can reach the gastrointestinal tract is the capacity to their colonization using efficient substrates and substances capable of improving the physiology of the digestive tract.

Considering the importance and effective results of symbiotic action for probiotics and prebiotics in aquaculture, greater efforts should be directed for using these additives for breeders and larvae of Neotropical fish species, mainly because they know the importance of nutrition in these development stages and their influence on the quality of eggs and larvae. Furthermore, larva of altricial species has incipient digestive tract at the beginning of their exogenous feeding. The colonization of beneficial bacteria during organic structures development of the digestive tract could help in nutrient absorption, accelerating the physiological aspects related to digestion. This fact can help to introduce early formulated diet in larvae feeding, one of the major goals of larviculture. Then these food additives should be used increasingly frequent to improve the fish health during production and, as consequence, to ensure the quality of the fish product.

ACKNOWLEDGMENTS

The authors would like to thank the researcher and colleague of the Federal Institute of Rio de Janeiro (IFRJ) Adriano Gomes da Cruz, for the guidelines and suggestions in the preparation of this work.

TABLE 3
Some relevant results with the use of synbiotics in the feeding of fishes species of interest for Brazilian aquaculture.

Species (Developmental phase)	Parameters	Synbiotic	Results	Reference
Tambaqui *Colossoma macropomum* (juvenile)	Performance, food utilization, survival, and economic analysis	*Bacillus subtilis* and *Mananoligosaccharide* (MOS)	The evaluated parameters improved with the synbiotic supplementation	Azevedo et al. (2016)
Pacu *Piaractus mesopotamicus* (juvenile)	Performance, immune response, and morphological changes in the intestine	β-Glucans (240 mg/kg), *Bacillus subtilis* (106 CFU/kg), simbiótico (240 mg of β-glucans/kg and 106 CFU/kg of *Bacillus subtilis*)	No evidence of the individual or symbiotic action of pre- and probiotics on immune response or performance. Small changes in intestinal morphology were observed	Cerozi et al. (2017)
Nile tilapia *Oreochromis niloticus* (juvenile)	Performance and challenge to infection by *Aeromonas hydrophila*	*Bacillus subtilis* and *Bacillus amyloliquefaciens* subsp. *plantarum* (AP193) (0.028% of diet, 4.2×10^7 CFU/g) and Previda (0.5%)	The synbiotic combination did not improve fish performance. The use of probiotics alone or in combination with prebiotics has been shown to be effective in reducing mortality from bacterial infection	Addo et al. (2017)
Rainbow trout *Oncorhynchus mykiss* (juvenile)	Performance and immune response	*Enterococcus faecalis* (1%), *Mananoligosaccharide* (MOS, 0.4%), and polyhydroxybutyrate acid (1%)	Improved performance and immune response	Rodriguez-Estrada et al. (2009)
Rainbow trout *Oncorhynchus mykiss* (juvenile)	Performance, body composition, and enzyme activity	GroBiotic-A (1% and 2%). Aqualase (0.1% and 0.2%). Evaluation separately	The performance was better with the prebiotic at a concentration of 2% of the diet. Prebiotic and probiotic improved enzyme activity	Azari et al. (2011)
Rainbow trout *Oncorhynchus mykiss* (juvenile)	Antioxidant defense and resistance to infection by *Streptococcus iniae*	*Galactooligosaccharides* (GOS, 1%), *Pediococcus acidilactici* (7.57 log CFU/g of diet), *and the* synbiotic *(GOS and Pediococcus acidilactici)*	The synbiotic had a positive effect on antioxidant capacity and resistance to infection by *Streptococcus iniae*	Hoseinifar et al. (2017)

CFU, colony-forming unit.

REFERENCES

Abbass, A., Sharifuzzaman, S.M., Austin, B., 2010. Cellular components of probiotics control *Yersinia ruckeri* infection in rainbow trout, *Oncorhynchus mykiss* (Walbaum). J. Fish Dis. 33, 31–37. https://doi.org/10.1111/j.1365-2761.2009.01086.x.

Addo, S., Carrias, A.A., Williams, M.A., Liles, M.R., Terhune, J.S., Davis, D.A., 2017. Effects of *Bacillus subtilis* strains and the prebiotic Previda® on growth, immune parameters and susceptibility to *Aeromonas hydrophila* infection in Nile tilapia, *Oreochromis niloticus*. Aquac. Res. 48 (9), 4798–4810. https://doi.org/10.1111/are.13300.

Adeoye, A.A., Yomla, R.Y., Jaramillo-Torres, A., Rodiles, A., Merrifield, D.L., Davies, S.J., 2016. Combined effects of exogenous enzymes and probiotic on Nile tilapia (*Oreochromis niloticus*) growth, intestinal morphology and microbiome. Aquaculture 463, 61–70.

Ahmad, M.H., El-Mousallamy, A., Zein, N., El-Naby, A.S.A., Mohamed, S.Z., 2015. Evaluation of prebiotic as natural additives on growth performance and blood biochemistry for Nile tilapia (*Oreochromis niloticus*). Middle East. J. Appl. Sci. 5 (2), 526–533.

Akrami, R., Mansour, M.R., Chitsaza, H., Alamdar, O.G., Denji, K.A., 2015. Prebiotic (A-Max) and growth of juvenile rainbow trout (*Oncorhynchus mykiss*). J. Appl. Aquac. 27, 61–71. https://doi.org/10.1111/jwas.12356.

Azari, A.H., Rezaei, M.H., Baei, M.S., Najafpour, S., Roohi, A., Darvishi, M., 2011. The effects of commercial probiotic and prebiotic usage on growth performance, body composition and digestive enzyme activities in juvenile rainbow trout (*Oncorhynchus mykiss*). World Appl. Sci. J. (Special Issue of Food and Environment) 14, 26–35. ISSN 1818-4952.

Azevedo, R.V., Filho, J.C.F., Pereira, S.L., Cardoso, L.D., Vidal Júnior, M.V., Andrade, D.R., 2016. Suplementação com prebiótico, probiótico e simbiótico para juvenis de tambaqui a duas densidades de estocagem. Pesq. Agrop. Brasileira 51 (1), 9–16. Brasília. jan https://doi.org/10.1590/S0100-204X2016000100002.

Betiku, O.C., Yeoman, C.J., Gaylord, T.G., Duff, G.C., Hamerly, T., Bothner, B., Block, S.S., Sealey, M., 2017. Differences in amino acid catabolism by gut microbes with/without prebiotics inclusion in GDDY-based diet affect feed utilization in rainbow trout. Aquaculture 490, 108–119. https://doi.org/10.1016/j.aquaculture.2017.09.006.

Brunt, J., Austin, B., 2005. Use of a probiotic to control lactococcosis and streptococcosis in rainbow trout, *Oncorhynchus mykiss* (Walbaum). J. Fish Dis. 28, 693–701.

Castro, M.P., Claudiano, G.S., Petrillo, T.R., Shimada, M.T., Belo, M.A.A., Marzocchi-Machado, C.M., Moraes, J.R.E., Manrique, G.W., Moraes, F.R., 2014. Acute aerocystitis in Nile tilapia bred in net cages and supplemented with chromium carbochelate and *Saccharomyces cerevisiae*. Fish Shellfish Immunol. 36, 284–290. https://doi.org/10.1016/j.fsi.2013.11.012.

Cerozi, B.S., Zanon, R.B., Silva, T.S.C., Cyrino, J.E.P., 2017. Combination of dietary β-glucan and *Bacillus subtilis* as a potential synbiotic for pacu *Piaractus mesopotamicus* (Holmberg, 1887). J. Appl. Ichthyol. 33 (3), 563–565. https://doi.org/10.1111/jai.13340.

Cornélio, F.H.G., Cargnin-Ferreira, E., Borba, M.G., Mouriño, J.L.P., Fernandes, V.A.G., Fracalossi, D.M., 2013. Crescimento, digestibilidade e resistência à infecção por patógeno em tilápia-do-nilo alimentada com probióticos. Pesq. Agrop. Brasileira 48 (8), 863–870. Brasília. ago https://doi.org/10.1590/S0100-204X2013000800008.

Dawood, M.A.O., Koshio, S., Esteban, M.A., 2018. Beneficial roles of feed additives as immunostimulants in aquaculture: a review. Rev. Aquac. 10 (4), 950–974. https://doi.org/10.1111/raq.12209.

Dias, D.C., Leonardo, A.F.G., Tachibana, L., Correa, C.F., Bordon, I.C.A.C., Romagosa, E., Ranzani-Paiva, M.J.T., 2010. Effect of incorporating probiotics into the diet of matrinxã (*Brycon amazonicus*) breeders. J. Appl. Ichthyol. 28, 40–45. https://doi.org/10.1111/j.1439-0426.2011.01892.x.

Denji, K.A., Mansour, M.R., Akrami, R., Ghobadi, S., Jafarpour, S.A., Mirbeygi, S.K., 2015. Effect of dietary prebiotic mannan oligosaccharide (MOS) on growth performance, intestinal microflora, body composition, haematological and blood serum biochemical parameters of rainbow trout (*Oncorhynchus mykiss*) juveniles. J. Fish. Aquat. Sci. 104, 255–265. https://doi.org/10.3923/jfas.2015.255.265.

Dias, J.A.R., Abe, H.A., Sousa, N.C., Silva, R.D.F., Cordeiro, C.A.M., Gomes, G.F.E., Ready, J.S., Mouriño, J.L.P., Martins, M.L., Carneiro, P.C.F., Maria, A.N., Fujimoto, R.Y., 2019. *Enterococcus faecium* as potential probiotic for ornamental neotropical cichlid fish, *Pterophyllum scalare* (Schultze, 1823). Aquac. Int. https://doi.org/10.1007/s10499-019-00339-9.

Doan, H.V., Hoseinifar, S.H., Khanongnuch, C., Kanpiengjai, A., Unban, K., Kim, V.V., Srichaiyo, S., 2018. Host-associated probiotics boosted mucosal and serum immunity, disease resistance and growth performance of Nile tilapia (*Oreochromis niloticus*). Aquaculture. https://doi.org/10.1016/j.aquaculture.2018.03.019.

Dotta, G., Mouriño, J.L.P., Jatobá, A., Morán, R.E.B., Pilati, C., Martins, M.L., 2011. Acute inflammatory response in Nile tilapia fed probiotic *Lactobacillus plantarum* in the diet. Acta Sci. Biol. Sci. 33 (3), 239–246. https://doi.org/10.4025/actascibiolsci.v33i3.8011.

El-Haroun, E.R., Goda, A.M.A-S., Chowdhury, M.A.K., 2006. Effect of dietary probiotic Biogen® supplementation as a growth promoter on growth performance and feed utilization of Nile tilapia *Oreochromis niloticus* (L.). Aquac. Res. 37 (14), 1473–1480. https://doi.org/10.1111/j.1365-2109.2006.01584.x.

Farias, T.H.V., Levy-Pereira, N., Alves, L.O., Dias, D.C., Tachibana, L., Pilarski, F., Belo, M.O.A., Ranzani-Paiva, M.J.T., 2016. Probiotic feeding improves the immunity of pacus, *Piaractus mesopotamicus*, during *Aeromonas hydrophila* infection. Anim. Feed Sci. Technol. 211, 137–144. https://doi.org/10.1016/j.anifeedsci.2015.11.004.

Genc, M.A., Yilmaz, E., Genc, E., Aktas, M., 2007. Effects of dietary mannan oligosaccharides (MOS) on growth, body composition, and intestine and liver histology of the hybrid Tilapia (*Oreochromis niloticus* x *O. aureus*). Isr. J. Aquacult. Bamidgeh 59 (1), 10–16.

Ghorbani, A., Salamatdoustnobar, R., Maghami, S.S.G., Motallebi, V., 2012. The effect of different levels of prebiotic on the length of fingerling rainbow trout. Afr. J. Biotechnol. 11 (36), 8928–8931. https://doi.org/10.5897/AJB11.1905.

Gomes, L.C., Brinn, R.P., Marcon, J.L., Dantas, L.A., Brandão, F.A., Abreu, J.S., Lemos, P.E.M., McComb, D.M., Baldisserotto, B., 2009. Benefits of using the probiotic EfinolsL during transportation of cardinal tetra, *Paracheirodon axelrodi* (Schultz), in the Amazon. Aquac. Res. 40, 157–165. https://doi.org/10.1111/j.1365-2109.2008.02077.x.

Guidoli, M.G., Santinón, J.J., Pasteris, S.E., Sánchez, S., Nader-Macías, M.E.F., 2015. Isolation and selection of potentially beneficial autochthonous bacteria for *Piaractus mesopotamicus* aquaculture activities. J. Bioprocess. Biotech. 5, 9. https://doi.org/10.4172/2155-9821.1000254.

Guidoli, M.G., Hernández, D.R., Boehringer, S.I., Sánchez, S., Nader-Macías, M.E.F., 2016. Administration of three autochthonous *Bacillus subtilis* strains induce early appearance of gastric glands and vestiges of pylorus in *Piaractus mesopotamicus* larvae. J. Bioprocess. Biotech. 6, 3. https://doi.org/10.4172/2155-9821.1000271.

Guidoli, M.G., Mendoza, J.A., Falcón, S.L., Boehringer, S.I., Sánchez, S., Macías, M.E.F.N., 2018. Autochthonous probiotic mixture improves biometrical parameters of larvae of *Piaractus mesopotamicus* (Caracidae, Characiforme, Teleostei). Ciênc. Rural 48, 7. Santa Maria. e20170764 https://doi.org/10.1590/0103-8478cr20170764.

Hassan, A.S., 2016. Evaluation of prebiotic as a feed supplement for Nile tilapia (*Oreochromis niloticus*) fry. Abbassa Int. J. Aqua. 9 (1), 1–18.

Hoseinifar, S.H., Sun, Y.-Z., Zhou, Z., 2017. Prebiotics and synbiotics. In: Austin, B., Newaj-Fyzul, A. (Eds.), Diagnosis and Control of Diseases of Fish and Shellfish. Published by JohnWiley & Sons Ltd, pp. 185–187. 320 p. (Chapter 7).

Ibrahem, M.D., Mohamed, F., Mesalhy, S., El-Aty, A.M.A., 2010. Effect of dietary supplementation of inulin and vitamin C on the growth, hematology, innate immunity, and resistance of Nile tilapia (*Oreochromis niloticus*). Fish Shellfish Immunol. 29 (2010), 241e246. https://doi.org/10.1016/j.fsi.2010.03.004.

Jatobá, A., Vieira, F.N., Neto, C.B., Silva, B.C., Mouriño, J.L.P., Jerônimo, G.T., Dotta, G., Martins, M.L., 2008. Utilização de bactérias ácido-lácticas isoladas do trato intestinal de tilapia-do-nilo como probiótico. Pesq. Agrop. Brasileira 43 (9), 1201–1207. Brasília.

Levy-Pereira, N., Yasui, G.S., Cardozo, M.G., Neto, J.D., Farias, T.H.V., Sakabe, R., Pádua, S.B., Pilarski, F., 2018. Immunostimulation and increase of intestinal lactic acid bacteria with dietary mannan-oligosaccharide in Nile tilapia juveniles. Rev. Bras. Zootec. 47, e20170006. https://doi.org/10.1590/rbz4720170006.

Lima, K.S., Filipe, S.C., Francisco, M.O.J., William Cristiane, T.T., Ricardo Henrique, B.S., Izabela, G.P.C.S., Luís Gustavo, T.B., 2015. Performance and hematological variables of piavuçu whose diets were supplemented with phytobiotic and probiotic additives. Semina Cienc. Agrar. 36 (4), 2881–2891. jul.-ago.

Liu, W., Ren, P., He, S., Xu, L., Yang, Y., Gu, Z., Zhou, Z., 2013. Comparison of adhesive gut bacteria composition, immunity, and disease resistance in juvenile hybrid tilapia fed two different *Lactobacillus* strains. Fish Shellfish Immunol. 35, 54. e 62 https://doi.org/10.1016/j.fsi.2013.04.010.

Londoño, L., Ramírez, T., Bolívar, G.A., 2015. Effect of drying on the viability of the probiotic bacterium *Lactobacillus plantarum* Lab9 (CPQBA 144-09 DRM 03) impregnated in the feed for tilapia (*Oreochromis* sp.). Aquac. Res., 1–9. https://doi.org/10.1111/are.12911.

Martínez, M.P., Pereyra, M.L.G., Juri, M.G.F., Poloni, V., Cavaglieri, L., 2018. Probiotic characteristics and aflatoxin B1 binding ability of *Debaryomyces hansenii* and *Kazaschtania exigua* from rainbow trout environment. Aquac. Res., 1–10. https://doi.org/10.1111/are.13614.

Midhun, S.J., Neethu, S., Arun, D., Vysakh, A., Divya, L., Radhakrishnan, E.K., Jyothis, M., 2019. Dietary supplementation of *Bacillus licheniformis* HGA8B improves growth parameters, enzymatic profile and gene expression of *Oreochromis niloticus*. Aquaculture 505, 289–296. https://doi.org/10.1016/j.aquaculture.2019.02.064.

Mínguez, C., Webster, P.G., Villa, M., 2016. Effect of a prebiotic supplementation of mannan oligosaccharide on growth traits and mortality of rainbow trout (*Oncorhynchus mykiss*). AACL Bioflux 9 (6), 1260–1264. http://www.bioflux.com.ro/aacl.

Moraes, A.V., Pereira, M.O., Moraes, K.N., Rodrigues-Soares, J.P., Jesus, G.F.A., Jatoba, A., 2018. Autochthonous probiotic as growth promoter and immunomodulator for *Astyanax bimaculatus* cultured in water recirculation system. Aquac. Res. 49 (8). https://doi.org/10.1111/are.13743.

Najdegerami, E.H., Tokmachi, A., Bakhshi, F., 2016. Evaluating the effects of dietary prebiotic mixture of mannan oligosaccharide and poly-β-hydroxybutyrate on the growth performance, immunity, and survival of rainbow trout, *Oncorhynchus mykiss* (Walbaum 1792), fingerlings. J. World Aquac. Soc. 48 (3), 415–425. https://doi.org/10.1111/jwas.12356.

Noffs, A.P., Tachibana, L., Santos, A.A., Ranzani-Paiva, M.J.T., 2015. Common snook fed in alternate and continuous regimens with diet supplemented with *Bacillus subtilis* probiotic. Pesq. Agrop. Brasileira 50 (4), 267–272. https://doi.org/10.1590/S0100-204X2015000400001. Brasilia.

Ozório, R.O.A., Kopecka-Pilarczyk, J., Peixoto, M.J., Lochmann, R., Santos, R.J., Santos, G., Weber, B., Calheiros, J., Ferraz-Arruda, F., Jossé, P.V.-P., Gonçalves, F.M., 2015. Dietary probiotic supplementation in juvenile rainbow trout (*Oncorhynchus mykiss*) reared under cage culture production: effects on growth, fish welfare, flesh quality and intestinal microbiota. Aquac. Res. 46 (4), 1–16. https://doi.org/10.1111/are.12724.

Ozório, R.O.A., Turini, B.G.S., Môro, G.V., Oliveira, L.S.T., Portz, L., Cyrino, J.E.P., 2010. Growth, nitrogen gain, and indispensable amino acid retention of pacu (*Piaractus mesopotamicus*, Holmberg 1887) fed different brewers yeast (*Saccharomyces cerevisie*) levels. Aquac. Nutr. 16, 276–283. https://doi.org/10.1111/j.1365-2095.2009.00662.x.

Paixão, A.E.M., Santos, J.C., Pinto, M.S., Pereira, D.S.P., Ramos, C.E.C.O., Cerqueira, R.B., Navarro, R.D., Silva, R.F., 2017. Effect of commercial probiotics (*Bacillus subtilis* and *Saccharomyces cerevisiae*) on growth performance, body composition, hematology parameters, and disease resistance against *Streptococcus agalactiae* in tambaqui (*Colossoma macropomum*). Aquac. Int. 25, 2035–2045. https://doi.org/10.1007/s10499-017-0173-7.

Peredo, A.M., Buentello, A., Gatlin, III., Hume, M.E., 2015. Evaluation of a dairy-yeast prebiotic in the diet of juvenile Nile tilapia, *Oreochromis niloticus*. J. World Aquac. Soc. 46, 92–101. https://doi.org/10.1111/jwas.12170.

Pilarski, F., Oliveira, C.A.F., Souza, F.P.B.D., Zanuzzo, F.S., 2017. Different b-glucans improve the growth performance and bacterial resistance in Nile tilapia. Fish Shellfish Immunol. 70, 25–29. https://doi.org/10.1016/j.fsi.2017.06.059.

Ramos, M.A., Gonçalves, J.F.M., Costas, B., Batista, S., Lochmann, R., Pires, M.A., Rema, P., Ozório, R.O.A., 2016. Commercial Bacillus probiotic supplementation of rainbow trout (*Oncorhynchys mykiss*) and brown trout (*Salmo trutta*): growth, immune responses and intestinal morphology. Aquac. Res., 1–12. https://doi.org/10.1111/are.13090.

Řehulka, J., Minařik, B., Cink, D., Žalak, J., 2011. Prebiotic effect of fructo-oligosaccharides on growth and physiological state of rainbow trout, *Oncorhynchus mykiss* (Walbaum). Acta Univ. Agric. Silvic. Mendel. Brun. 27 (5), 227–236.

Ridha, M.T., Azad, I.S., 2012. Preliminary evaluation of growth performance and immune response of Nile tilapia *Oreochromis niloticus* supplemented with two putative probiotic bacteria. Aquac. Res. 43, 843–852. https://doi.org/10.1111/j.1365-2109.2011.02899.x.

Ridha, M.T., Azad, I.S., 2015. Effect of autochthonous and commercial probiotic bacteria on growth, persistence, immunity and disease resistance in juvenile and adult Nile tilapia *Oreochromis niloticus*. Aquac. Res., 1–11. https://doi.org/10.1111/are.12726.

Rodriguez-Estrada, U., Satoh, S., Haga, Y., Fushimi, H., Sweetman, J., 2009. Effects of single and combined supplementation of *Enterococcus faecalis*, mannan oligosaccharide and polyhydroxybutyrate acid on growth performance and immune response of rainbow trout *Oncorhynchus mykiss*. Suisan zoshoku (Aquac. Sci.) 57, 609–617.

Sabioni, R.E., Zanuzzo, F.S., Cyrino, J.E.P., 2019. Immunomodulation of juvenile Pacu, *Piaractus mesopotamicus*, by different β(1-3)(1-6)-D glucan products. Braz. Arch. Biol. Technol. 62, e19170811. https://doi.org/10.1590/1678-4324-2019170811.

Sado, R.Y., Bicudo, A.J.A., Cyrino, J.E.P., 2014. Hematology of juvenile pacu, *Piaractus mesopotamicus* (Holmberg, 1887) fed graded levels of mannan oligosaccharides (MOS). Lat. Am. J. Aquat. Res. 42 (1), 30–39. https://doi.org/103856/vol42-issue1-fulltext-3.

Sado, R.J., Bicudo, A.J.D.A., Cyrno, J.E.P., 2008. Feeding dietary mannan oligosaccharide to juvenile Nile Tilapia (*Oreochromis niloticus*) has no effect on hematological parameters and showed decreased feed consumption. J. World Aquac. Soc. 39, 821–826. https://doi.org/10.1111/j.1749-7345.2008.00219.x.

Sado, R.Y., Raulino-Domanski, F., Freitas, P.F., Baioco-Sales, F., 2015. Growth, immune status and intestinal morphology of Nile tilapia fed dietary prebiotics (mannan oligosaccharides-MOS). Lat. Am. J. Aquat. Res. 43 (5), 944–952. https://doi.org/10.3856/vol43-issue5-fulltext-14.

Salvador, R., Claudiano, G.S., Loureiro, B.A., Marcusso, P.F., Eto, S.F., Pilarski, F., Toazza, C.S., Moraes, J.R.E., Moraes, F.R., 2013. Desempenho e hematologia de tilapias-do-nilo alimentadas com *Saccharomyces cerevisiae* e vacinadas contra *Streptococcus agalactiae*. Pesq. Agrop. Brasileira 48 (8), 892–898. Brasília. ago. 2013 https://doi.org/10.1590/S0100-204X2013000800012.

Schwarz, K.K., Furuya, W.M., Natali, M.R.M., Gaudezi, M.C., Lima, P.A.G., 2011. Mannanoligosaccharides in diets for tilapia larvae. Rev. Bras. Zootec. 40 (12), 2634–2640. https://doi.org/10.1590/S1516-35982011001200003.

Selim, K.M., Reda, R.M., 2015. Beta-glucans and mannan oligosaccharides enhance growth and immunity in Nile tilapia. Beta-glucans and mannan oligosaccharides enhance growth and immunity in Nile tilapia. N. Am. J. Aquac. 77 (1), 22–30. https://doi.org/10.1080/15222055.2014.951812.

Sharifuzzaman, S.M., Austin, B., 2010. Development of protection in rainbow trout (*Oncorhynchus mykiss* Walbaum) to *Vibrio anguillarum* following use of the probiotic Kocuria SM1. Fish Shellfish Immunol. 29, 212–216. https://doi.org/10.1016/j.fsi.2010.03.008.

Sousa, N.C., Couto, M.V.S., Abe, H.A., Paixão, P.E.G., Cordeiro, C.A.M., Lopes, E.M., Ready, J.S., Jesus, G.F.A., Martins, M.L., Mouriño, J.L.P., Carneiro, P.C.F., Maria, A.N., Fujimoto, R.Y., 2019. Effects of an *Enterococcus faecium*-based probiotic on growth performance and health of Pirarucu, *Arapaima gigas*. Aquac. Res. 00, 1–9. https://doi.org/10.1111/are.14332.

Souza, D.M., Martins, G.B., Piedras, S.R.N., Pouey, J.L.O.F., Robaldo, R.B., Leite, F.P.L.L., 2012. Probiotic actions of *Bacillus cereus* var. *toyoi* and *Saccharomyces boulardii* in silver catfish (*Rhamdia quelen*) larvae culture. R. Bras. Zootec. 41 (3), 815–819. https://doi.org/10.1590/S1516-35982012000300048.

Tabrizi, J.M., Barzeghar, A., Farzampour, S., Mirzaii, H., Safarmashaei, S., 2012. Study of the effect of prebiotic (*Saccharomyces cerevisiae*) and acidifier on growth parameters in grower's rainbow trout (*Oncorhynchus mykiss*). Ann. Biol. Res. 3 (5), 2053–2057. http://scholarsresearchlibrary.com/archive.html.

Tawab, A.A.A.E., Ammar, A.M., Saleh, O.A.E-R., El-Hofy, F.I.A-A., Tolba, H.A., 2016. Effect of prebiotic on the immune status of *Oreochromis niloticus*. BVMJ 30 (1), 20–28. http://www.bvmj.bu.edu.eg.

Tiengtam, N., Khempaka, S., Paengkoum, P., Boonanuntanasarn, S., 2015. Effects of inulin and Jerusalem artichoke (*Helianthustuberosus*) as prebiotic ingredients in the diet of juvenile Niletilapia (*Oreochromis niloticus*). Anim. Feed Sci. Technol. 207 (2015), 120–129. https://doi.org/10.1016/j.anifeedsci.2015.05.008.

Yarahmadi, P., Farsani, H.G., Khazaei, A., Khodadadi, H., Rashidiyan, G., Jalali, M.A., 2016. Protective effects of the prebiotic on the immunological indicators ofrainbow trout (*Oncorhynchus mykiss*) infected with *Aeromonas hydrophila*. Fish Shellfish Immunol. 54, 589–597. https://doi.org/10.1016/j.fsi.2016.05.010.

Zendeboodi, F., Khorshidian, N., Mortazavian, A.M., Cruz, A.G., 2020. Probiotic: conceptualization from a new approach. Curr. Opin. Food Sci. 32, 103–123. https://doi.org/10.1016/j.cofs.2020.03.009.

Zhai, Q., Wang, H., Tian, F., Zhao, J., Zhang, H., Chen, W., 2017. Dietary *Lactobacillus plantarum* supplementation decreases tissue lead accumulation and alleviates lead toxicity in Nile tilapia (*Oreochromis niloticus*). Aquac. Res. 48, 5094–5103. https://doi.org/10.1111/are.13326.

Zhou, Z-G., He, S., Liu, Y., Shi, P., Huang, G., Yao, B., 2009. The effects of dietary yeast culture or short-chain fructo-oligosaccharides on the intestinal autochthonous bacterial communities in juvenile hybrid tilapia, *Oreochromis niloticus* × *Oreochromis aureus*. J. World Aquac. Soc. 40 (4), 450–459. https://doi.org/10.1111/j.1749-7345.2009.00272.x.

CHAPTER 15

Electric Technologies Applied to Probiotic and Prebiotic Food

ARTURO B. SORO[a] • ELENA S. INGUGLIA[a] • RAMON S. ROCHA[b] • ERICK A. ESMERINO[b,c] • ADRIANO G. CRUZ[b] • BRIJESH K. TIWARI[a]
[a]Department of Food Chemistry and Technology, Teagasc Food Research Centre, Dublin, Ireland, [b]Department of Food, Federal Institute of Science and Technology of Rio de Janeiro (IFRJ), Rio de Janeiro, Brazil, [c]Faculty of Veterinary Medicine, Federal Fluminense University (UFF), Niterói, Brazil

PROBIOTIC

Probiotics are live microorganisms that, when administered in suitable amounts, confer a health benefit on the host. A minimum consumption of 10^6–10^7 CFU/g or mL of the product or 10^9 in a portion is commonly suggested. However, these values may be influenced by the expected health effect, the food matrix, and the probiotic culture used (species and strain) (Hill et al., 2014).

Some criteria are used to select a potentially probiotic culture, such as safety (preferentially of a human source, isolated from the gastrointestinal tract of healthy people or animals, nonpathogenic, nontoxic, nonallergenic, nonmutagenic, or noncarcinogenic), functional (resistance to acid and bile salts, ability to adhere to intestinal epithelial cells and/or colonize the lumen of the gastrointestinal tract), and technological (genetically stable strains, viability during product processing and storage, adequate sensory properties, phage resistance, and large-scale production) (Pimentel et al., 2017).

Probiotic cultures are isolated from many sources, such as the human gastrointestinal tract, human cavities (vaginal or oral), human milk colostrum, human milk, milk and dairy products, meat, fruits and plants (Rolim et al., 2020). The associated health effect of the probiotic microorganism is more important than its origin. However, it is believed that probiotic strains have a greater functionality in environments similar to those that they have been isolated from; therefore, probiotics from human origin may sometimes have a higher functionality (Pimentel et al., 2017).

Considering the safety aspects, probiotic strains should not have a history of pathogenicity or association with diseases such as gastrointestinal disorders, negatively affecting the immune system, producing toxins, unmaking the colonocyte function, or having the ability to transfer antibiotic-resistant genes to the gut's resident microbiota (Pimentel et al., 2017). Some strains having a long period of human exposure and few reports of health problems have been classified as generally recognized as safe (GRAS). In Europe, many species that have the qualified presumption of safety (QPS) status have been evaluated for their identity, possible pathogenesis, and use (Pradhan et al., 2020).

Considering the functional aspects, the ability of the probiotic cultures to survive the acid environment of the stomach and the presence of bile salts is one of the most important criteria. Furthermore, they should be able to adhere, colonize, or multiply in the intestine, as these features facilitate the competition with pathogenic bacteria and allow their permanence for a longer period in the gut (Ashaolu, 2020). Considering the technological aspects, the probiotic strain should withstand industrial production on a large scale, such as freeze-drying and spray dryer Furthermore, it should be incorporated in food products without high decreases in viability and without compromises on the sensory aspects (Rolim et al., 2020).

The most predominant genera used as probiotic cultures are *Lactobacillus* and *Bifidobacterium*. The most used species are *Lactobacillus acidophilus, Lactobacillus gasseri, Lactobacillus johnsonii, Lactobacillus rhamnosus, Lactobacillus casei, Lactobacillus plantarum, Bifidobacterium longum, Bifidobacterium bifidum, Bifidobacterium breve, Bifidobacterium infantis, Saccharomyces cerevisiae,* and *S. boulardii* (Pradhan et al., 2020). Other species such as *Escherichia coli* Nissle, *Clostridium butyricum, Bacillus coagulans, Bacillus subtilis, Propionibacterium freudenreichii,* and *Pseudomonas fluorescens* are less frequently used. Furthermore, commensal human species are recognized as "probiotics of the future," such as *Akkermansia muciniphila, Faecalibacterium prausnitzii, Roseburia* spp., and *Eubacterium hallii* (Rolim et al., 2020). There is

Probiotics and Prebiotics in Foods. https://doi.org/10.1016/B978-0-12-819662-5.00006-9

great potential in using yeasts as probiotics, as they can grow at low pH values, water activity, and temperature, and high salt concentrations (Pimentel et al., 2017).

Many health effects are associated with the consumption of probiotic cultures, such as reduction of the risk of colon cancer and symptoms of lactose intolerance, improvements of the immune system, increases in the nutritional value of the products, decreases in the cholesterol levels, reduction of the risk of gut infections, and decreases in antibiotic-associated diarrhea (Zendeboodi et al., 2020). The mechanisms of action can be related to the alteration of the composition of the gut microbiota by competition between probiotics and the other microorganisms, production of components (short-chain fatty acids, vitamins, amino acids, and antimicrobial substances), changes in the signals of TLR4/TLR2/MyD88/NFκB and NLRP3 inflammasome, and decreases in the gut permeability (Halloran and Underwood, 2019). Furthermore, they can improve the epithelial barrier and increase the adhesion to the mucosa of the intestine (Silva et al., 2020). Each probiotic strain is associated with specific functional properties and health benefits; therefore, the health effects cannot be generalized (Hill et al., 2014).

Probiotic cultures are generally included in dairy products, such as yogurts, cheeses, fermented milk, and fermented dairy beverages. This is related to the fact that dairy products have been already recognized as healthy products, that they help in the digestion process, and are important for the bones. In addition, consumers know that microorganisms are commonly used in the preparation of dairy products (Pimentel et al., 2018).

PREBIOTICS

Prebiotics are substrates that are selectively utilized by host microorganisms, conferring a benefit to health (Gibson et al., 2017). A food ingredient to be considered prebiotic must meet some criteria: It should be safe, not be hydrolyzed or absorbed in the upper part of the gastrointestinal tract, be selectively fermented by one or a limited number of potentially beneficial bacteria in the colon, change the composition of the microbiota to a healthier composition, and, preferably, induce beneficial effects on the health of the host. Furthermore, it must be stable to the food processes, such as heating, drying, and storage; should not require refrigeration; and be easily and efficiently incorporated into processed foods (Pimentel et al., 2017).

Prebiotic component consumption are associated with several health effects such as stimulation of the immune system, decreases in the blood lipids, improvements in the insulin resistance and brain energy, function and cognition, increases in the bioavailability of minerals, and inhibition of pathogens (Gibson et al., 2017). The effect of the prebiotic component is associated with its type, dosage, and the food matrix used (Turkmen et al., 2019). A suitable dosage is associated with the observation of the prebiotic effect without adverse effects, such as nonselective utilization or gas formation, and varies according to the microbial ecosystem and the metabolic effects (Gibson et al., 2017).

Oligosaccharides, such as fructans (oligofructose, fructooligosaccharides (FOS), and inulin), galactans (galactooligosaccharides (GOS)), mannanooligosaccharides (MOS), and xylooligossacharides (XOS)) are the most recognized prebiotic components. However, polyphenols, polyunsaturated fatty acids, conjugated linoleic acid (CLA), and human milk oligosaccharides may be considered prebiotic components (Gibson et al., 2017).

Prebiotic components can be obtained from many sources, such as extraction from plants (inulin is mainly obtained from chicory roots), enzymatic hydrolysis (oligofructose is obtained from hydrolysis of inulin), synthesis (FOS is obtained from transglycosylation of sucrose and GOS from lactose), or microbial (curdlan) (Peredo-Lovillo et al., 2020).

NOVEL ELECTRIC TECHNOLOGIES

In the past decades, the new productivity, quality, and safety standards requested by the food industry, along with the increasing consumer demands for more sustainable and environmental friendly methods for food production, have led to an increased interest in the use of novel technologies for the food sector. Food operators have been focusing on the production of safer food products with the intention to maintain or even improve quality attributes, resulting in the development of novel thermal and nonthermal technologies. Such novel processing methods have had promising results at a pilot-plant scale; however, complications in the scaling-up process and the variability in consumer's acceptance of some technological application have delayed their regulatory approval for application within the food chain. Therefore, concerted efforts of food producers and researchers should aim at exploring the development and potential implementation of novel technologies on an industrial level and also improve their perception of benefits to consumers (Jermann et al., 2015; Priyadarshini et al., 2019).

Nowadays, novel technologies can be divided into thermal and nonthermal: (i) thermal technologies are based on their capacity of mild-heat generation such as ohmic, microwave, and radio frequency; (ii) other nonheat-based technologies like cold plasma, pulsed electric fields, ultrasound, ultraviolet light (continuous and pulsed), and high hydrostatic pressure, among others, use lower temperature conditions and shorter processing times. The role of these techniques has shown diverse applications within the manufacturing, pharmaceutical, medical, chemical, nutritional, and food industries (Barba et al., 2018). Currently, the antimicrobial effect has been extensively studied for both novel thermal and nonthermal in several foodborne pathogens (e.g., *Salmonella*, *E. coli*, *Campylobacter*, and *Listeria monocytogenes*) and spoilage bacteria (e.g., *Pseudomonas*, *Lactobacillus*, and *Brochothrix thermosphacta*) in different food matrices (e.g., milk, dairy products, fruit juices, meat, meat products, fish, and seafood) along with their impact on food quality attributes (Morales-de la Peña et al., 2019). Additionally, a combination of these technologies with old processes (e.g., pasteurization, freezing, refrigeration, and drying) and new approaches (e.g., natural antimicrobials and modified atmosphere packaging) has demonstrated the enhancement in the effectivity of food production and, therefore, broadened the range of application of the emerging technologies in the food sector (Knorr, 2018).

In particular, by the turn of the 19th century, a group of techniques based on the utilization of electric current was conducted as possible new technologies for food preservation. Previously, James Clerk Maxwell, mathematician and physicist, had discovered significantly key findings on the understanding of electricity and magnetism. Since then, the application of external electric fields in food has made it possible to create electronic strategies that are now employed and assessed for industrial applications. Although these techniques consist of the generation of electric fields in food products acting as semiconductor systems, their classification (such as type of electric flow, pulsed or nonpulsed technology, heat generation, etc.) varies among the literature. Ohmic, radio frequency, microwave, and pulsed electric fields have been identified as electronic technologies and are discussed throughout this chapter (Barba et al., 2018; Misra et al., 2017; Rocha et al., 2018).

Thermal Electric Technologies

In the category of thermal electric technologies, ohmic processing (OH) involves the use of an alternate electrical current that is transformed into uniform heat when it passes through the food. The transmission of electricity is highly dependent on the conductivity of the food system and is set up through metallic electrodes that are in contact with the food product (Fig. 1).

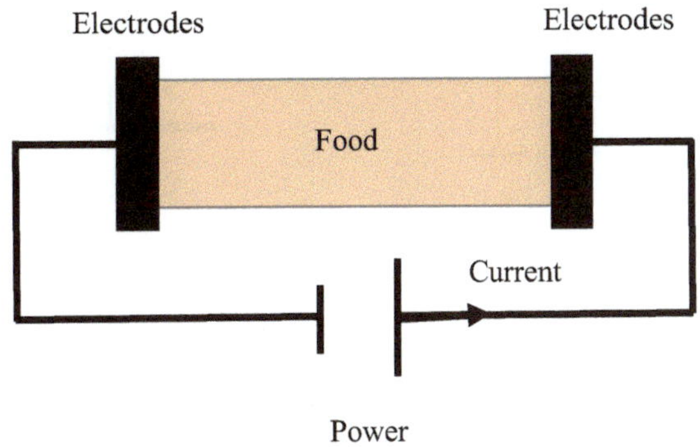

FIG. 1 Schematic working mechanism of Ohmic heating (OH).

Therefore, OH allows the generation and transference of heat directly to the food, increasing efficiency and time saving of the heating process. It is also worth noting that a phenomenon known as electroporation is present during the ohmic process but to a lesser extent. As a disadvantage, the potential erosion of the electrodes can lead to the leakage of heavy metals onto the food surface, therefore, reducing the potential perspectives of extensive ohmic application in the food industry. Nevertheless, the development of electrodes with resistant materials to erosion (like titanium) could help overcome this potential food safety concern (Jaeger et al., 2016; Knirsch et al., 2010). Overall, ohmic has been identified as a potential technology capable of replacing and/or intensifying the performance of old thermal technologies, suggesting potential application on dairy products, such as goat milk (Pereira et al., 2008), milk (Sun et al., 2008), condensed milk (Sun et al., 2011), skimmed milk and milk cream (Kim and Kang, 2015), buffalo milk (Shivmurti et al., 2014), cheeses (Rocha et al., 2020), dairy desserts (Kuriya et al., 2020), whey beverages (Ferreira et al., 2019; Coimbra et al., 2020), and infant formula (Roux et al., 2016).

Dielectric heating is based on the usage of electromagnetic waves to generate heat when high electric pulses are applied. Radio frequency and microwave are mainly grouped within the dielectric techniques and are shortly differenced (Awuah et al., 2014) (Fig. 2).

In microwave heating, domestic and industrial devices use electromagnetic waves in a frequency range from 300 MHz to 300 GHz, which are partly absorbed by the food and transformed into heat. Production

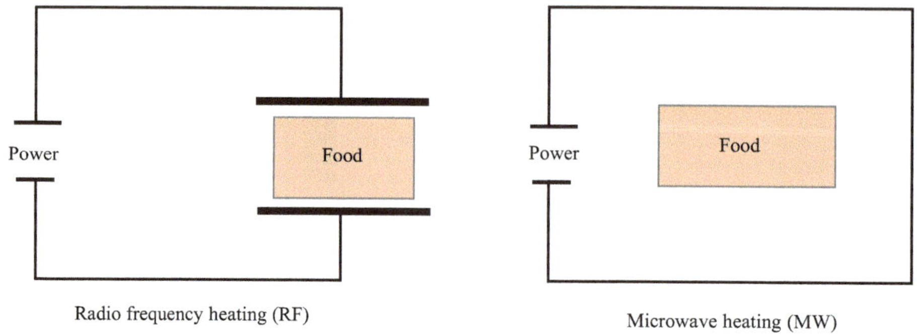

FIG. 2 Schematic working mechanism of radio frequency (RF) and microwave (MW) heating.

of internal heating has been associated with mechanisms of molecular friction (like dipolar process and ionic migration) in the presence of high electric fields. Notwithstanding the positive aspects of this technology (e.g. time-saving, handling ease, high heat production, etc.), the penetration depth, treatment parameters, and dielectric properties are contributing to a nondesirable uneven distribution of temperature within the food (Chandrasekaran et al., 2013; Regier et al., 2017). On the contrary, radio-frequency heating consists of a simpler device that produces microwaves in a frequency range of 3kHz to 3GHz. This technology has a similar mode of action and similar limitations of nonuniform heating, like the microwave. However, in the past years, the increased popularity of microwave heating has raised the interest of food researchers to develop food computational models to solve the irregular heating problem. As a matter of fact, the combination of microwave heating with other processes (e.g., compound extraction or drying) is now been explored as a result of the little effect on the quality of treated products associated with microwaves (Meda et al., 2017).

Nonthermal Electric Technology

Among the nonthermal technologies, pulsed electric fields (PEF) have been highly studied for a wide range of applications. This technique applies high electric pulses (0.1–80kV/cm) in a short treatment time (milliseconds or microseconds) to a food immersed in a liquid and placed between two electrodes (Gabrić et al., 2018). Hence, the electric current is transferred to a capacitor where it is stored and discharged in the form of electric fields through the treatment chamber. In addition, the polarity of the pulses (monopolar or bipolar) can be altered to enhance the permeabilization of cell membranes and, therefore, change the effectivity of the process (Wang et al., 2018) (Fig. 3).

The efficacy of this technology to disrupt bacterial membranes and promote mass transfer is well known; although some disadvantages of PEF such as high cost, bubble formation, electrolysis, or scaling-up problems have ruled out its implementation in few food processes, this underdevelopment technique could offer numerous possibilities as a combination and/or pretreatment within the food sector (Gomez et al., 2019).

APPLICATION OF ELECTRIC TECHNOLOGIES IN PROBIOTIC AND PREBIOTIC FOODS

The former electric-based strategies have been assessed in different applications for food purposes. Specifically, an increasing interest in fermented and probiotic food has led food researchers to investigate the impact of these technologies on probiotics and prebiotics. The well-known fermentation process consists of the transformation of complex molecules into simple compounds, carried out by molds, yeasts, and/or bacteria through enzymatic reactions. The fermentation industry employs efficiently this set of chemical reactions to develop food products with different attributes. In recent years, novel technologies have been considered by the fermentation industry to boost fermentation rates and obtain fermented products of high quality (Ojha et al., 2017). The combined effect of electroporation and mild heating in ohmic processes, for example, could increase the effectiveness of fermentation; electroporation facilitates the transport of nutrients through the cell membrane due to pore formation during the fermentation process. Additionally, ohmic technology generates and transfers heat uniformly across the medium: this mechanism helps microorganisms to reduce the lag phase and trigger an earlier start in the fermentation process when reaching the appropriate temperature in the liquid solution. In fact, the

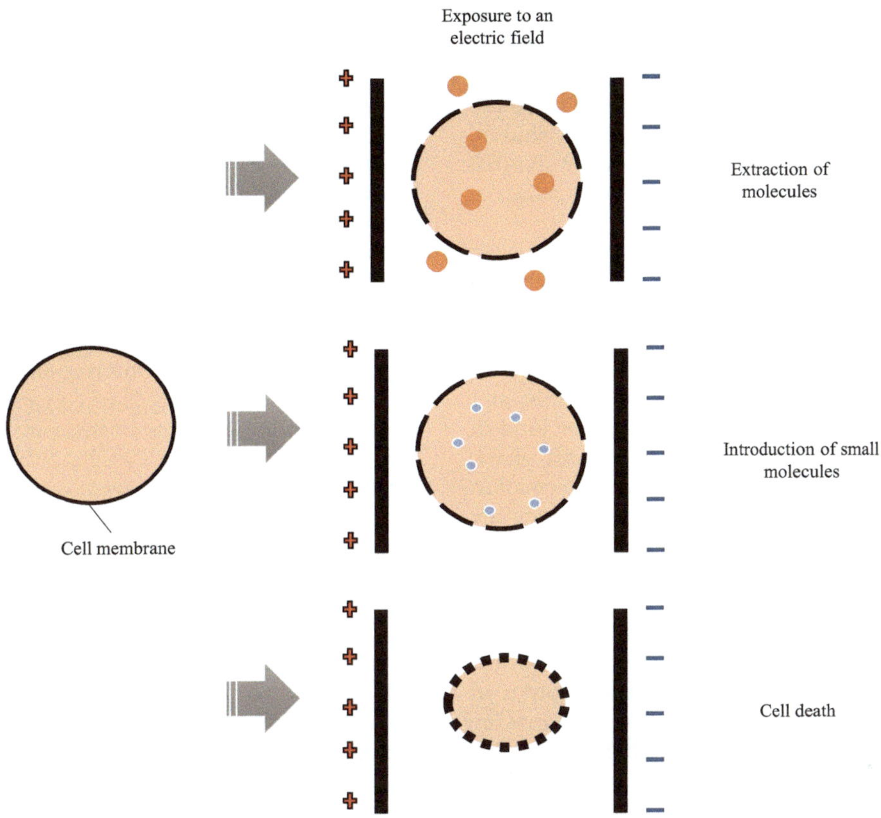

FIG. 3 Schematic working mechanism and the potential outcome of pulsed electric field (PEF) on the cell membrane.

electrical properties of ohmic heating could disrupt the permeability of cell membranes and substrates, consequently affecting the fermentative processes. However, these mechanisms are dependent on ohmic process's parameters and the conditions of the fermentation (e.g., fermentation products, type of substrate, and microorganism). Furthermore, the continuous application of electric fields may alter the productivity of the fermentation as a result of promoting the passage of inhibitory metabolites through the cell membrane at later stages of fermentation (Gavahian and Tiwari, 2020). Similarly, microwave and radio frequency promotes the random distribution of temperature within the medium, accelerating the fermentation process. Moreover, electromagnetic waves produced by these technologies contribute to the disruption of cell structures due to molecular friction mechanisms. Even though thermal electric technologies may ease fermentative processes, only a few studies have been carried out with the aim of implementation at the industrial level (Koubaa et al., 2016). For instance, Siefarth et al. (2014) compared radio frequency and convention heating after the fermentation stage in the production of yogurt as preservation technologies; the authors concluded that the radio-frequency process could be used as a postfermentation treatment because of its high heating rate and low impact on quality properties. In microwave heating, the present studies are focused on grape and wheat fermentation, and they discuss its possible use as a pretreatment. Therefore, this technique may increase the productivity of the process together with the reduction of the fermentation period (Carew et al., 2015; Melikoglu, 2012). In addition, Cho et al. (1996) and Loghavi et al. (2008, 2009) studied the benefits of ohmic heating in the lactic acid fermentation with *Lactobacillus acidophilus*. All these studies observed a reduction of the lag phase that could be associated with the transport of nutrients in the bacterial membrane through the electroporation phenomena. Other studies that investigated the administration of an ohmic treatment in the production of fermented coffee and bread have also found similar results (Gally et al., 2017;

Masure et al., 2019; Reta et al., 2017). Moreover, all these authors recommended the application of ohmic heating in the early stages of the fermentation process because of the high generation of inhibitory products. Notwithstanding the positive findings of these studies, further investigation is needed for assessing thermal electric technologies due to the lack of information in fermentative processes.

Pulsed electric fields have been identified as the most likely technique to be employed during fermentation processes. The main advantage of using PEF for fermentation is its mechanical effect on cell permeability, resulting in increased production rates. The formation of pores can be reversible or irreversible, and its severity depends on several factors like the kind of microorganism, treatment parameters, growth phase, cell density, temperature, pH, etc. The pore formation process and the effectiveness of this technology are strongly connected, and therefore, it is very important to optimize and monitor these conditions before, during, and after treatment (Galván-D'Alessandro and Carciochi, 2018). Different studies have appraised the effect of pulsed electric fields in fermentation. For example, López et al. (2008) considered to apply it in grape skin before its fermentation in the winemaking process. Results showed a significant time reduction of the maceration step and an increase of wine attributes in the final product. Mattar et al. (2015) also identified the positive effect of this technique on *Saccharomyces cerevisiae* at early stages of wine fermentation and recommended its intermittent use. In another study, PEF was applied on a starter culture (*Lactobacillus delbrueckii* subsp. *bulgaricus* and *Streptococcus thermophilus*) before the lactic fermentation of yogurt: the performance of the starter culture was improved, and the fermentation step was accelerated (Chanos et al., 2020). Continuous pulsed electric field treatment may seriously affect the viability of microorganisms because of the cell damage and temperature increase of the medium; hence, possible solutions like devices coupled with cooling systems, discontinuous treatments, or pretreatments could be taken into consideration to solve these issues. Although this may be true, more studies are required to address in detail the application of this nonthermal technology (Mota et al., 2018).

Role of Electric Technologies for Probiotics and Prebiotics

A complete understanding of the role of electromagnetic waves and electric fields in biological structures could facilitate their implementation within the food industry. In particular, there is a growing interest in the health benefits of probiotics. A key consideration in the assessment of the implementation of these novel electric technologies is the evaluation of their effects alone on microbial structures. Improvement of the viability of probiotics and production of prebiotics are two notable areas in which these technologies could be considered. In Table 1, studies focusing on the enhancement of the former characteristics are presented. For instance, Cueva and Aryana (2012) and Najim and Aryana (2013) recommend the use of pulsed electric fields to increase protease activity and bile and acid tolerances in *Lactobacillus acidophilus* and *Lactobacillus delbrueckii*. Other studies observed initial stress on the growth of the bacterial cells of *Lactobacillus delbrueckii* and *Lactobacillus plantarum* treated with this technology (Peng et al., 2020; Seratlić et al., 2013). Nevertheless, Seratlić et al. (2013) showed increased growth of *Lactobacillus plantarum* cells after the treatment and also that this new bacterial population had acquired higher resistance to pulsed electric fields. It is equally important to highlight studies that exploit the mechanism of electroporation to incorporate compounds such as magnesium, calcium, zinc, or cholesterol inside the cellular matrix of probiotics (Goral et al., 2018, 2020; Goral and Pankiewicz, 2017; Kozłowicz et al., 2019; Lye et al., 2011; Pankiewicz et al., 2020). Additionally, few studies have been carried out with pulse electric fields for enhancing the accumulation of prebiotics like trehalose in *Lactobacillus plantarum* (Vaessen et al., 2018, 2019). Lastly, microwave heating was assessed by Ambros et al. (2016) and de Paula et al. (2008) to facilitate the preservation of starter culture and probiotics through drying in *Lactobacillus paracasei* and hydrolase Levan, a product of *Zymomonas mobilis*, to generate prebiotics, respectively. Although other technologies like ohmic and radio frequency have been considered for fermentative purposes, there is little evidence to support their use in probiotic and prebiotic food. Furthermore, scarce knowledge has been provided about the impact of novel electric techniques on the structures of probiotics and prebiotics (Pereira et al., 2018). Rodrigues et al. (2020) revised the potential changes on the protein conformation of food caused by electric fields that may lead to safety and technological issues (i.e., allergenicity, loss of flavor, and cytotoxicity). These authors highlighted the lack of evidence on the development of small-scale studies at the pilot plant level that is necessary to understand and assess the effect of the electric technologies in the structure conformation of food proteins under real conditions.

Pulse electric field and ohmic have been identified by de Almada et al. (2016) as techniques for

TABLE 1
Summary of studies focused on the assessment of electric technologies for potential applications in probiotic and prebiotic food.

Technology	Application	Probiotic/prebiotic	Treatment parameters	Reference
Microwave	Drying	*Lactobacillus paracasei*	2.45 GHz and 1000 W	Ambros et al. (2016)
PEF	Improving viability	*Lactobacillus acidophilus*	25 kV/cm and 30,000 μs	Cueva and Aryana (2012)
Microwave	Improving prebiotic production	Levan produced by *Zymomonas mobilis*	650 W	de Paula et al. (2008)
PEF	Magnesium enrichment	*Lactobacillus rhamnosus* and *Lactococcus lactis*	2 kV/cm, 20 μs for 15 min	Goral et al. (2018)
PEF	Magnesium enrichment	*Lactobacillus rhamnosus*	3.5 kV/cm, 10 μs for 10 min	Goral and Pankiewicz (2017)
PEF	Calcium enrichment	*Lactobacillus rhamnosus*	3 kV/cm, 75 μs for 10 min	Goral et al. (2020)
PEF	Zinc enrichment	*Lactobacillus rhamnosus*	3 kV/cm, 20 μs for 15 min	Kozłowicz et al. (2019)
PEF	Cholesterol enrichment	*Lactobacillus acidophilus*, *Lactobacillus bulgaricus*, and *Lactobacillus casei*	7.5 kV/cm, 4 ms	Lye et al. (2011)
PEF	Improving viability	*Lactobacillus acidophilus* and *Lactobacillus delbrueckii*	3 kV/cm, 20 μs for 15 min	Najim and Aryana (2013)
PEF	Calcium enrichment	*Lactobacillus rhamnosus*	3 kV/cm, 10–150 μs for 10 min	Pankiewicz et al. (2020)
PEF	Improving viability	*Lactobacillus delbrueckii*	285 V/cm, 100 μs for 240 min	Peng et al. (2020)
PEF	Improving viability	*Lactobacillus plantarum*	31.6 kV/cm, 5 μs	Seratlić et al. (2013)
PEF	Improving prebiotic production	*Lactobacillus plantarum*/trehalose and lactose	6.3 kV/cm, 100 μs	Vaessen et al. (2019)
PEF	Improving prebiotic production	*Lactobacillus plantarum*/trehalose	7.5 kV/cm, 100 μs	Vaessen et al. (2018)

producing paraprobiotics. Some of the mechanisms (e.g., electroporation, protein coagulation, enzyme inactivation, etc.) in which these technologies are involved could be capable of promoting its production. The term paraprobiotic differs from the definition of probiotic in that it is only applicable to nonviable microbial cells with health benefits. Thus, these unviable cells are associated with the treatment of several diseases and symptoms (e.g., immunomodulation, diarrhea, respiratory diseases, etc.) (Shripada et al., 2020). However, there is currently not enough information about paraprobiotics and new technologies, and consequently, this area could be explored.

In conclusion, electric technologies could play an important role in the fermentation industry and the production of probiotic and prebiotic food. The mechanisms of action of these techniques require further investigation to take full advantage of their potential benefits and applications. In addition, microwave, ohmic, and radio-frequency heating are poorly evaluated, despite their promising results. The new field of paraprobiotics could focus its effort on assessing these electric technologies and, therefore, identifying all their advantages and disadvantages. Finally, the effect on the structures of probiotics and prebiotics may help explain in detail the mode of action of these technologies.

REFERENCES

Ambros, S., Bauer, S.A.W., Shylkina, L., Foerst, P., Kulozik, U., 2016. Microwave-vacuum drying of lactic acid bacteria: influence of process parameters on survival and acidification activity. Food Bioprocess Technol. 9 (11), 1901–1911.

Ashaolu, T.J., 2020. Immune boosting functional foods and their mechanisms: a critical evaluation of probiotics and prebiotics. Biomed. Pharmacother. 130, 110625.

Awuah, G.B., Ramaswamy, H.S., Tang, J., 2014. Radio-Frequency Heating in Food Processing: Principles and applications. CRC Press.

Barba, F.J., Ahrné, L., Xanthakis, E., Landerslev, M.G., Orlien, V., 2018. Innovative technologies for food preservation. In: Innovative Technologies for Food Preservation. Academic Press, Massachusetts, USA, pp. 25–51.

Carew, A.L., Close, D.C., Dambergs, R.G., 2015. Yeast strain affects phenolic concentration in pinot noir wines made by microwave maceration with early pressing. J. Appl. Microbiol. 118 (6), 1385–1394.

Chandrasekaran, S., Ramanathan, S., Basak, T., 2013. Microwave food processing—a review. Food Res. Int. 52 (1), 243–261.

Chanos, P., Warncke, M.C., Ehrmann, M.A., Hertel, C., 2020. Application of mild pulsed electric fields on starter culture accelerates yogurt fermentation. Eur. Food Res. Technol. 246 (3), 621–630.

Cho, H.Y., Yousef, A.E., Sastry, S.K., 1996. Growth kinetics of Lactobacillus acidophilus under ohmic heating. Biotechnol. Bioeng. 49 (3), 334–340.

Coimbra, L.O., Vidal, V.A., Silva, R., Rocha, R.S., Freitas, M.Q., Guimarães, J.T., Pollonio, M.A., 2020. Are ohmic heating-treated whey dairy beverages an innovation? Insights of the Q methodology. LWT 134, 110052.

Cueva, O., Aryana, K.J., 2012. Bile tolerance of *Lactobacillus acidophilus* LA-K as influenced by various pulsed electric field conditions. Adv. Biosci. Biotechnol. 03 (05), 620–625.

de Almada, C.N., Almada, C.N., Martinez, R.C.R., Sant'Ana, A.S., 2016. Paraprobiotics: evidences on their ability to modify biological responses, inactivation methods and perspectives on their application in foods. Trends Food Sci. Technol. 58, 96–114.

de Paula, V.C., Pinheiro, I.O., Lopes, C.E., Calazans, G.C., 2008. Microwave-assisted hydrolysis of Zymomonas mobilis levan envisaging oligofructan production. Bioresour. Technol. 99 (7), 2466–2470.

Ferreira, M.V.S., Cappato, L.P., Silva, R., Rocha, R.S., Guimarães, J.T., Balthazar, C.F., Neto, R.P., 2019. Ohmic heating for processing of whey-raspberry flavored beverage. Food Chem. 297, 125018.

Gabrić, D., Barba, F., Roohinejad, S., Gharibzahedi, S.M.T., Radojčin, M., Putnik, P., Bursać Kovačević, D., 2018. Pulsed electric fields as an alternative to thermal processing for preservation of nutritive and physicochemical properties of beverages: a review. J. Food Process Eng. 41 (1).

Gally, T., Rouaud, O., Jury, V., Havet, M., Ogé, A., Le-Bail, A., 2017. Proofing of bread dough assisted by ohmic heating. Innovative Food Sci. Emerg. Technol. 39, 55–62.

Galván-D'Alessandro, L., Carciochi, R., 2018. Fermentation assisted by pulsed electric field and ultrasound: a review. Fermentation 4 (1), 1.

Gavahian, M., Tiwari, B.K., 2020. Moderate electric fields and ohmic heating as promising fermentation tools. Innovative Food Sci. Emerg. Technol. 64, 102422.

Gibson, G.R., Hutkins, R., Sanders, M.E., Prescott, S.L., Reimer, R.A., Salminen, S.J., Verbeke, K., 2017. Expert consensus document: the International Scientific Association for Probiotics and Prebiotics (ISAPP) consensus statement on the definition and scope of prebiotics. Nat. Rev. Gastroenterol. Hepatol. 14, 491.

Gomez, B., Munekata, P.E.S., Gavahian, M., Barba, F.J., Marti-Quijal, F.J., Bolumar, T., Lorenzo, J.M., 2019. Application of pulsed electric fields in meat and fish processing industries: an overview. Food Res. Int. 123, 95–105.

Goral, M., Pankiewicz, U., 2017. Effect of pulsed electric fields (PEF) on accumulation of magnesium in lactobacillus rhamnosus B 442 cells. J. Membr. Biol. 250 (5), 565–572.

Goral, M., Kozlowicz, K., Pankiewicz, U., Goral, D., 2018. Magnesium enriched lactic acid bacteria as a carrier for probiotic ice cream production. Food Chem. 239, 1151–1159.

Goral, M., Pankiewicz, U., Sujka, M., Kowalski, R., Goral, D., Kozlowicz, K., 2020. Influence of pulsed electric field on accumulation of calcium in Lactobacillus rhamnosus B 442. J. Microbiol. Biotechnol. 30 (1), 44–53.

Halloran, K., Underwood, M.A., 2019. Probiotic mechanisms of action. Early Hum. Dev. 135, 58–65.

Hill, C., Guarner, F., Reid, G., Gibson, G.R., Merenstein, D.J., Pot, B., Calder, P.C., 2014. Expert consensus document: the International Scientific Association for Probiotics and Prebiotics consensus statement on the scope and appropriate use of the term probiotic. Nat. Rev. Gastroenterol. Hepatol. 11, 506.

Jaeger, H., Roth, A., Toepfl, S., Holzhauser, T., Engel, K.-H., Knorr, D., Steinberg, P., 2016. Opinion on the use of ohmic heating for the treatment of foods. Trends Food Sci. Technol. 55, 84–97.

Jermann, C., Koutchma, T., Margas, E., Leadley, C., Ros-Polski, V., 2015. Mapping trends in novel and emerging food processing technologies around the world. Innovative Food Sci. Emerg. Technol. 31, 14–27.

Kim, S.S., Kang, D.H., 2015. Comparative effects of ohmic and conventional heating for inactivation of *Escherichia coli* O157: H7, *Salmonella enterica* serovar Typhimurium, and *Listeria monocytogenes* in skim milk and cream. J. Food Prot. 78, 1208–1214.

Knirsch, M.C., Alves dos Santos, C., Martins de Oliveira Soares Vicente, A.A., Vessoni Penna, T.C., 2010. Ohmic heating—a review. Trends Food Sci. Technol. 21 (9), 436–441.

Knorr, D., 2018. Emerging technologies: back to the future. Trends Food Sci. Technol. 76, 119–123.

Koubaa, M., Roselló-Soto, E., Barba-Orellana, S., Barba, F.J., 2016. Novel thermal technologies and fermentation. In: Novel Food Fermentation Technologies. Springer, Cham, Switzerland, pp. 155–163.

Kozłowicz, K., Góral, M., Góral, D., Pankiewicz, U., Bronowicka-Mielniczuk, U., 2019. Effect of ice cream storage on the physicochemical properties and survival of probiotic bacteria supplemented with zinc ions. LWT 116, 1–8.

Kuriya, S.P., Silva, R., Rocha, R.S., Guimarães, J.T., Balthazar, C.F., Pires, R.P., Raices, R.S., 2020. Assessment the impact of the electric field on the quality parameters of blueberry flavored dairy desserts. Food Res. Int. 134, 109235.

Loghavi, L., Sastry, S.K., Yousef, A.E., 2008. Effect of moderate electric field frequency on growth kinetics and metabolic activity of Lactobacillus acidophilus. Biotechnol. Prog. 24 (1), 148–153.

Loghavi, L., Sastry, S.K., Yousef, A.E., 2009. Effect of moderate electric field frequency and growth stage on the cell membrane permeability of Lactobacillus acidophilus. Biotechnol. Prog. 25 (1), 85–94.

López, N., Puértolas, E., Condón, S., Álvarez, I., Raso, J., 2008. Effects of pulsed electric fields on the extraction of phenolic compounds during the fermentation of must of Tempranillo grapes. Innovative Food Sci. Emerg. Technol. 9 (4), 477–482.

Lye, H.S., Karim, A.A., Rusul, G., Liong, M.T., 2011. Electroporation enhances the ability of lactobacilli to remove cholesterol. J. Dairy Sci. 94 (10), 4820–4830.

Masure, H.G., Wouters, A.G.B., Fierens, E., Delcour, J.A., 2019. Electrical resistance oven baking as a tool to study crumb structure formation in gluten-free bread. Food Res. Int. 116, 925–931.

Mattar, J.R., Turk, M.F., Nonus, M., Lebovka, N.I., El Zakhem, H., Vorobiev, E., 2015. *S. cerevisiae* fermentation activity after moderate pulsed electric field pre-treatments. Bioelectrochemistry 103, 92–97.

Meda, V., Orsat, V., Raghavan, V., 2017. Microwave heating and the dielectric properties of foods. In: The Microwave Processing of Foods. Woodhead Publishing, Cambridge, UK, pp. 23–43.

Melikoglu, M., 2012. Solid-state fermentation of wheat pieces by Aspergillus oryzae: effects of microwave pretreatment on enzyme production in a biorefinery. Int. J. Green Energy 9 (6), 529–539.

Misra, N.N., Koubaa, M., Roohinejad, S., Juliano, P., Alpas, H., Inacio, R.S., Barba, F.J., 2017. Landmarks in the historical development of twenty first century food processing technologies. Food Res. Int. 97, 318–339.

Morales-de la Peña, M., Welti-Chanes, J., Martín-Belloso, O., 2019. Novel technologies to improve food safety and quality. Curr. Opin. Food Sci. 30, 1–7.

Mota, M.J., Lopes, R.P., Koubaa, M., Roohinejad, S., Barba, F.J., Delgadillo, I., Saraiva, J.A., 2018. Fermentation at non-conventional conditions in food- and bio-sciences by the application of advanced processing technologies. Crit. Rev. Biotechnol. 38 (1), 122–140.

Najim, N., Aryana, K.J., 2013. A mild pulsed electric field condition that improves acid tolerance, growth, and protease activity of Lactobacillus acidophilus LA-K and Lactobacillus delbrueckii subspecies bulgaricus LB-12. J. Dairy Sci. 96 (6), 3424–3434.

Ojha, K.S., Mason, T.J., O'Donnell, C.P., Kerry, J.P., Tiwari, B.K., 2017. Ultrasound technology for food fermentation applications. Ultrason. Sonochem. 34, 410–417.

Pankiewicz, U., Góral, M., Kozłowicz, K., Góral, D., 2020. Application of pulsed electric field in production of ice cream enriched with probiotic bacteria (*L. rhamnosus* B 442) containing intracellular calcium ions. J. Food Eng. 275, 1–9.

Peng, K., Koubaa, M., Bals, O., Vorobiev, E., 2020. Effect of pulsed electric fields on the growth and acidification kinetics of Lactobacillus delbrueckii Subsp. bulgaricus. Foods 9 (9), 1146.

Peredo-Lovillo, A., Romero-Luna, H.E., Jiménez-Fernández, M., 2020. Health promoting microbial metabolites produced by gut microbiota after prebiotics metabolism. Food Res. Int. 136, 109473.

Pereira, R.N., Martins, R.C., Vicente, A.A., 2008. Goat milk free fatty acid characterization during conventional and ohmic heating pasteurization. J. Dairy Sci. 91, 2925–2937.

Pereira, R.N., Teixeira, J.A., Vicente, A.A., Cappato, L.P., da Silva Ferreira, M.V., da Silva Rocha, R., da Cruz, A.G., 2018. Ohmic heating for the dairy industry: a potential technology to develop probiotic dairy foods in association with modifications of whey protein structure. Curr. Opin. Food Sci. 22, 95–101.

Pimentel, T.C., Garcia, S., Prudencio, S.H., 2017. Produtos lácteos funcionais. In: Nero, L.A., da Cruz, A.G., Bersot, L.S. (Eds.), Produção, Processamento e Fiscalização de Leite e Derivados. 1ed. vol. 1. Atheneu, pp. 205–226.

Pimentel, T.C., Marcolino, V.A., Barão, C.E., Klososki, S.J., Rosset, M., 2018. Minas frescal cheese as a probiotic carrier. In: Bioactive Molecules in Food. Springer, United Kingdom, pp. 1–32.

Pradhan, D., Mallappa, R.H., Grover, S., 2020. Comprehensive approaches for assessing the safety of probiotic bacteria. Food Control 108, 106872.

Priyadarshini, A., Rajauria, G., O'Donnell, C.P., Tiwari, B.K., 2019. Emerging food processing technologies and factors impacting their industrial adoption. Crit. Rev. Food Sci. Nutr. 59 (19), 3082–3101.

Regier, M., Knoerzer, K., Schubert, H., 2017. Introducing microwave-assisted processing of food. In: The Microwave Processing of Foods. Woodhead Publishing, Cambridge, UK, pp. 1–22.

Reta, Mursalim, Salengke, Junaedi, M., Mariati, Sopade, P., 2017. Reducing the acidity of Arabica coffee beans by ohmic fermentation technology. Food Res. 1 (5), 157–160.

Rocha, C.M.R., Genisheva, Z., Ferreira-Santos, P., Rodrigues, R., Vicente, A.A., Teixeira, J.A., Pereira, R.N., 2018. Electric field-based technologies for valorization of bioresources. Bioresour. Technol. 254, 325–339.

Rocha, R.S., Silva, R., Guimarães, J.T., Balthazar, C.F., Pimentel, T.C., Neto, R.P., Calvacanti, R.N., 2020. Possibilities for using ohmic heating in Minas Frescal cheese production. Food Res. Int. 131, 109027.

Rodrigues, R.M., Avelar, Z., Machado, L., Pereira, R.N., Vicente, A.A., 2020. Electric field effects on proteins—novel perspectives on food and potential health implications. Food Res. Int. 137, 109709.

Rolim, F.R., Neto, O.C.F., Oliveira, M.E.G., Oliveira, C.J., Queiroga, R.C., 2020. Cheeses as food matrixes for probiotics: in vitro and in vivo tests. Trends Food Sci. Technol. 100, 138–154.

Roux, S., Courel, M., Birlouez-Aragon, I., Municino, F., Massa, M., Pain, J.P., 2016. Comparative thermal impact of two UHT technologies, continuous ohmic heating and direct steam injection, on the nutritional properties of liquid infant formula. J. Food Eng. 179, 36–43.

Seratlić, S., Bugarski, B., Nedović, V., Radulović, Z., Wadsö, L., Dejmek, P., Galindo, F.G., 2013. Behavior of the surviving population of Lactobacillus plantarum 564 upon the application of pulsed electric fields. Innovative Food Sci. Emerg. Technol. 17, 93–98.

Shivmurti, S., Harshit, P., Rinkita, P., Smit, P., 2014. Comparison of chemical properties of milk when conventionally and ohmically heated. Int. Food Res. J. 21, 1425.

Shripada, R., Gayatri, A.-J., Sanjay, J., 2020. Paraprobiotics. Precision Medicine for Investigators, Practitioners and Providers. Academic Press, Massachusetts, USA, pp. 39–49.

Siefarth, C., Tran, T.B., Mittermaier, P., Pfeiffer, T., Buettner, A., 2014. Effect of radio frequency heating on yoghurt, I: technological applicability, shelf-life and sensorial quality. Foods 3 (2), 318–335.

Silva, D.R., Sardi, J.D.C.O., de Souza Pitangui, N., Roque, S.M., da Silva, A.C.B., Rosalen, P.L., 2020. Probiotics as an alternative antimicrobial therapy: current reality and future directions. J. Funct. Foods 73, 104080.

Sun, H., Kawamura, S., Himoto, J.I., Itoh, K., Wada, T., Kimura, T., 2008. Effects of ohmic heating on microbial counts and denaturation of proteins in milk. Food Sci. Technol. Res. 14, 117–123.

Sun, H., Masuda, F., Kawamura, S., Himoto, J.I., Asano, K., Kimura, T., 2011. Effect of electric current of ohmic heating on nonthermal injury to Streptococcus thermophilus in milk. J. Food Process Eng. 34, 878–892.

Turkmen, N., Akal, C., Özer, B., 2019. Probiotic dairy-based beverages: a review. J. Funct. Foods 53, 62–75.

Vaessen, E.M.J., den Besten, H.M.W., Patra, T., van Mossevelde, N.T.M., Boom, R.M., Schutyser, M.A.I., 2018. Pulsed electric field for increasing intracellular trehalose content in Lactobacillus plantarum WCFS1. Innovative Food Sci. Emerg. Technol. 47, 256–261.

Vaessen, E.M.J., den Besten, H.M.W., Esveld, E.D.C., Schutyser, M.A.I., 2019. Accumulation of intracellular trehalose and lactose in lactobacillus plantarum WCFS1 during pulsed electric field treatment and subsequent freeze and spray drying. LWT 115, 108478.

Wang, Q., Li, Y., Sun, D.W., Zhu, Z., 2018. Enhancing food processing by pulsed and high voltage electric fields: principles and applications. Crit. Rev. Food Sci. Nutr. 58 (13), 2285–2298.

Zendeboodi, F., Khorshidian, N., Mortazavian, A.M., da Cruz, A.G., 2020. Probiotic: conceptualization from a new approach. Curr. Opin. Food Sci. 32, 103–123.

CHAPTER 16

Ultrasound for Probiotic and Prebiotic Foods

HUGO SCUDINO[a] • JONAS T. GUIMARÃES[a] • DÉBORA LEMOS LINO[b] • MARIA CARMELA KASNOWSKI HOLANDA DUARTE[a] • ERICK A. ESMERINO[a] • MÔNICA Q. FREITAS[a] • ADRIANO G. CRUZ[b]
[a]Faculty of Veterinary, Fluminense Federal University (UFF), Niterói, Brazil, [b]Department of Food, Federal Institute of Science and Technology of Rio de Janeiro (IFRJ), Rio de Janeiro, Brazil

INTRODUCTION

The consumption of functional foods is increasingly common since the consumers are looking for a healthier lifestyle, which includes their diet and the health benefits they can obtain from it. Food products containing prebiotics and probiotics present many functional effects, such as the improvement of intestinal and immunological health, prevention of certain types of cancer, reduction in the risk of cardiovascular diseases, among other benefits (Guimarães et al., 2018c). As a reflex of the functional food movement, the global market of probiotics was evaluated as 62.6 billion dollars in 2014 and was estimated to reaches 96.0 billion dollars by 2020 (Espitia et al., 2016).

The probiotics are usually found in dairy products such as cheeses, dairy desserts, and ice creams, but the most popular products are fermented milk and yogurts. Currently, the nondairy probiotics have drawn attention due to the increasing interest of vegan consumers and those with dietary restrictions such as lactose intolerance, allergies to milk proteins, and even cholesterol restriction.

Besides their concern for their health, modern consumers are more conscious and want to know how food products are manufactured, because they seek processes that preserve to the maximum the natural properties of the food. In addition, the health benefits must be aligned to practicality, food safety, and sensorially pleasant products (Bleiel, 2010).

In this sense, the nonthermal emerging technologies have been studied to meet the requirements of the new trends in food production. The US is one of these technologies, mainly the high-intensity ultrasound (HIUS), which may affect food production in different ways to produce healthier and functional food.

The US can be used as a nonthermal food processing technique due to its capacity to promote physical and chemical changes in foods even at low temperatures (Guimaraes et al., 2018). This technology has been developed for decades, but in recent years, great advances have been achieved, showing it to be a process that is promising, efficient, safe, and useful. Among the several applications of the HIUS, we can highlight microbial inactivation (Piyasena et al., 2003), milk and dairy beverage homogenization (Guimaraes et al., 2018; Scudino et al., 2020), modification in lactose crystallization (Moura et al., 2016), and increase of bioactive compounds and functional activities (Lollo et al., 2015; Guimarães et al., 2019b). Currently, studies on HIUS have been carried out to develop probiotic and prebiotic products to increase their benefits and functional activities. Besides, the few studies on the subject are sufficient to show that this t is a promising technology in the field of functional food.

In this chapter, US processing will be reviewed with a focus on its application in the manufacture of prebiotic and probiotic dairy products, with special attention to the effects of acoustic cavitation to improve the properties of these products.

ULTRASOUND TECHNOLOGY

The US is defined as sound waves that exceed the range of audible waves in humans (20–20.000 Hz), which are capable of working in low temperatures. Therefore, the US may be considered a nonthermal technology and is of great interest to the processing of milk and dairy products. Although it is considered a nonthermal technology, the acoustic cavitation generated by the ultrasonic waves can heat a product, depending on the applied intensity; thus, it is considered nonthermal if

Probiotics and Prebiotics in Foods. https://doi.org/10.1016/B978-0-12-819662-5.00001-X

the temperature of the product does not exceed 55°C during the whole processing (Guimarães et al., 2018c).

Researches on the US started with Harvey and Loomis in the 1920s and then in the 1960s after the discovery that the sound waves from war submarine sonars were killing fishes. In the following years, several researches evaluating the effects of US appeared, first exploring the higher frequencies and then the higher intensities. At the same time, parallel researches were developing small devices that would allow tests in these conditions (Zisu and Chandrapala, 2015).

The applications of the US in food processing, analysis, and quality control are divided according to the intensity of the acoustic cavitation in low-intensity ultrasound (LIUS) and HIUS (Fig. 1), which may vary the frequency of sound waves and the power emitted in a specific area. The LIUS is characterized by high frequencies (> 1 MHz) and intensities lower than 1 W/cm^2, which does not cause significant changes in the material by which they propagate, being commonly used for analytical applications such as determination of the composition and structure of the foods (Mohammadi et al., 2017).

On the other hand, the HIUS is characterized by low frequencies (20–100 kHz) and intensities higher than 1 W/cm^2 capable of causing changes in the physical-chemical and microbiological properties of the foods. These effects are promising for processing, preservation, and food safety. This emerging technology has been used as an alternative to the conventional processing of foods, controlling their microstructure and the sensory characteristics of the products (Kentish and Feng, 2014). In dairy products, the capacity of the US to work with temperatures lower than that used in conventional pasteurization allows the processing of milk and its derivatives and reduces the changes caused by heat, such as protein denaturation, vitamins degradation, lactose degradation, among others (Patist and Bates, 2008).

SYSTEMS OF HIGH-INTENSITY ULTRASOUND FOR FOOD PROCESSING

The formation of US waves is possible due to devices called transducers, which produce a stable and reliable ultrasonic field, using electric energy produced by a generator. This electric energy is converted into mechanical energy in the form of ultrasonic vibrations, which are applied directly or indirectly to the food. (Arvanitoyannis et al., 2017).

The most-used systems of HIUS applied to foods are the baths and the probes (Fig. 2). The baths constitute a system of indirect application since they have the transducer attached to the base or the walls of the tank, and the ultrasonic energy is transmitted to the liquid that fills the container (usually water), and then t that ultrasonic energy is transferred to the food immersed in the liquid. Therefore, the intensity of the US inside the sample container is lower than expected; however, this method can be applied to solid foods such as vegetables, cheeses, and other ready-to-eat products (Chen et al., 2020; Azam et al., 2020). In the probe systems, the application is direct where the acoustic signal is amplified and applied directly into the food through a metallic rod, without the need for an additional liquid medium between the food and the probe. Moreover, in this method, the food must be a liquid or semisolid to produce adequate acoustic cavitation, and the US can affect the whole product. Due to the direct action, in the probe system, it is possible to apply a higher cavitation intensity (Jambrak et al., 2014; Capelo-Martínez, 2009).

During the application of the US waves, there is an energy dissipation in the form of heat. The amount of

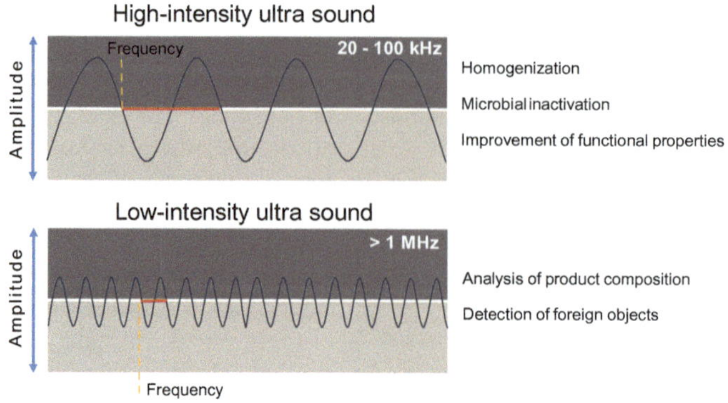

FIG. 1 Types of ultrasound waves applied to food processing and analysis.

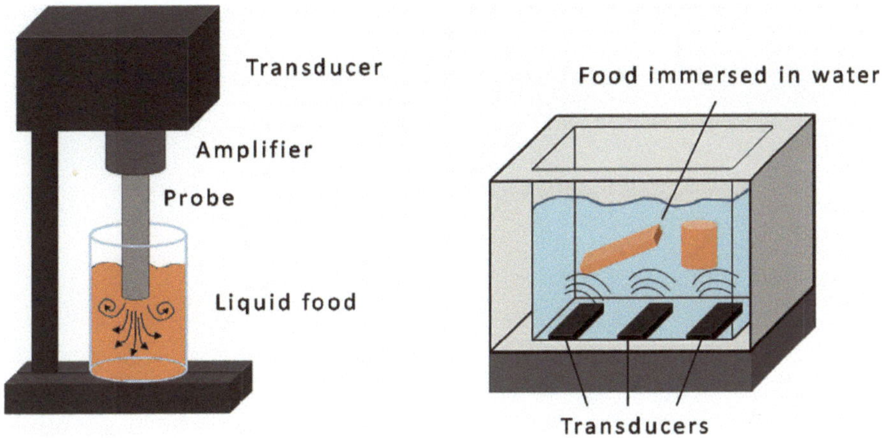

FIG. 2 Types of ultrasonic equipment used for food processing. Direct application by ultrasonic probe (left) and indirect application by ultrasonic bath (right).

dissipated energy varies in function of the wave's amplitude and frequency. The US waves find some resistance by passing through the food, which can be greater or lower depending on the material; moreover, part of the energy dissipated by the transducer is lost in the form of heat, causing an increase in the temperature of the product during the processing (Kentish and Feng, 2014).

This phenomenon is considerably important in applications of high acoustic intensity and can be determined by Eq. (1).

$$\text{Calorimetric power} = m \times C \times \frac{\Delta T}{\Delta t} \tag{1}$$

where the mass (m) of the product is multiplied by its specific heat (C) and the ratio of the temperature variation (ΔT) in a specific period of time (Δt) (Bermúdez-Aguirre, 2017).

MECHANISM OF ACTION

The main phenomenon observed in US processing and responsible for most of the physical-chemical and microbiological changes in liquid foods is acoustic cavitation. This phenomenon can be defined as the formation, growth, and collapse of microbubbles inside an aqueous solution, resulting in pressure fluctuations that occur in the applied acoustic field (Leong et al., 2011) (Fig. 3).

Acoustic cavitation can be classified in two ways: stable cavitation and unstable or transient cavitation. In the stable cavitation, usually produced in medium acoustic intensities, the bubbles inside the liquid oscillate regularly in long cycles, around its size of equilibrium, generating microagitation in the medium without collapsing. The other type is the unstable or transient cavitation that is generated in high-intensity acoustic fields where the bubbles oscillate in short cycles. During the cycle of negative pressure, the bubble expands several times its original size, and then, during the compression, the bubble collapses violently, creating microjets and shock waves (Villamiel et al., 2017).

The size of the bubbles induced by the acoustic field depends on the sound frequency. Larger bubbles (~ 100 μm) are formed in lower frequencies (20–40 kHz). Meanwhile, the number of bubbles created is influenced by the amplitude, which in turn is influenced by the provided energy; in other words, an US with higher intensity creates more bubbles (Bermúdez-Aguirre, 2017). Once formed, the bubbles tend to move in the direction of the sound waves; this movement is referred to as acoustic microstreaming and is caused by "Bjerkness" forces (Leighton et al., 1990).

Another observed effect during the acoustic cavitation is the expansion and compression of the bubbles, which happens slowly due to a process called rectified diffusion and depends on the sufficient application of a sound field and the saturation of the liquid medium with dissolved gases (Crum, 1984). When the bubbles reach a certain size they collapse, generating localized microimplosions that reach high temperatures and pressures at the locale of the event; this being the main mechanism of shear force generation in the fluid submitted to the US (Kentish and Feng, 2014). Temperatures of thousands of degree Kelvin (Suslick et al., 1999) and hundreds of pressure bars are possible in the surroundings of the bubbles (Ashokkumar and Grieser, 1999).

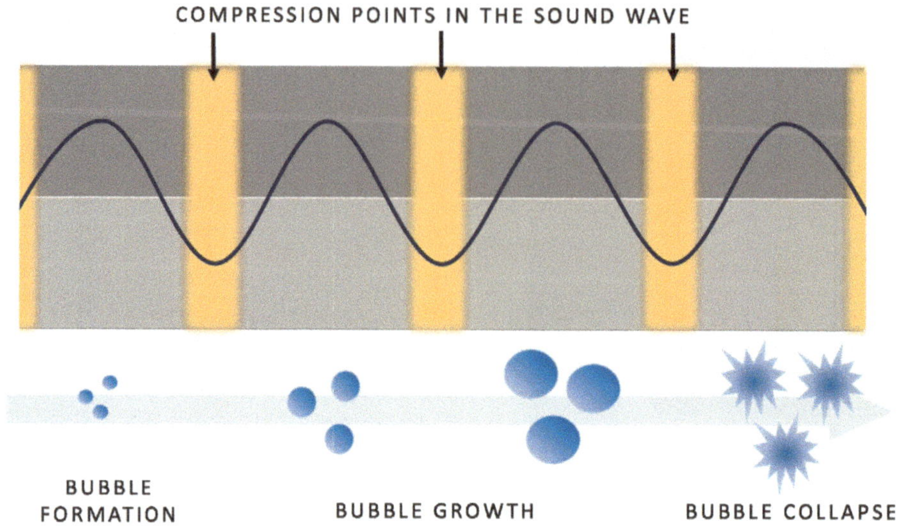

FIG. 3 Scheme of the acoustic cavitation phenomenon in liquids.

As the temperature and pressure rapidly increase inside a bubble at the end of its collapse, molecules present in the bubbles are decomposed under these extreme conditions and generate several highly reactive radicals. For example, if the medium is water, the radicals H^+ and OH^- are generated. These radicals are highly reactive and interact quickly with other radical species or chemical molecules. A common product of this reaction in water is hydrogen peroxide (Ashokkumar and Grieser, 1999; Leong et al., 2011).

EFFECT OF HIGH-INTENSITY ULTRASOUND ON MICROORGANISMS

The HIUS is widely studied in food processing aiming for microbial inactivation, with the acoustic cavitation being the main mechanism responsible for the inactivation of the microorganisms. The microimplosions generated at the collapse of the microbubbles create strong shear forces. These shock waves may be strong enough to mechanically weaken, pierce, or even rupture the cell walls, with subsequent release of intracellular material, leading it to cell death (Villamiel et al., 2017; Huang et al., 2017).

Besides the mentioned shear forces, the HIUS also produces effects related to the intracellular cavitation (Hughes and Nyborg, 1962). The effects of localized heating, production of free radicals, and microstreaming influence microbial inactivation (Bermúdez-Aguirre et al., 2011b).

The efficacy of the microbial inactivation by HIUS is influenced by the total energy received by the system; in other words, it depends on the ultrasonic wave amplitude, ultrasonic power, volume of the medium, and the time of processing. Moreover, the temperature of the treatment, the food composition, and the type of microorganism also directly affect the inactivation (Piyasena et al., 2003).

The lethal effect of the US depends on the cavitation intensity, which in turn depends on multiple physical parameters. Among the characteristics that describe the US waves, frequency and amplitude are the most important (Gao et al., 2014c). The US wave's frequency defines the size of the bubbles. Larger bubbles are formed in lower frequencies (20–40 kHz) (Bermúdez-Aguirre, 2017). Studies such as Gao et al. (2014a) compare microbial inactivation using low and high US frequencies, and the conclusion is that lower frequencies have more inactivation power. The influence of the wave's amplitude is reported to be of extreme importance for microbial inactivation. Higher inactivation rates have been observed when using higher amplitudes, which may be due to the increase in the amount of bubbles submitted to cavitation per unit of time (Chandrapala and Leong, 2015).

Another important parameter to be taken into consideration is the temperature. Studies (Salleh-Mack and Roberts, 2007; Halpin et al., 2013; Ordoñez et al., 1984; Beatty and Walsh, 2016) have compared processing by US using different temperatures and have concluded that higher temperatures increase the values of microbial inactivation in foods. In addition, thermosonication, which is the use of temperatures higher than 55 °C together with US, is very effective to inactivate microorganisms and ensure food safety (Anaya-Esparza et al., 2017).

In general, the microbial tolerance to US is increased in foods or complex mediums compared to buffered or saline solutions. This effect is probably due to the reduction in water activity and the protective effect of the food compounds. For example, Lee et al. (1989) reached inactivation of 4 log of *Salmonella* sp. in peptone water, but only 0.8 log in chocolate. In a study carried out by Bermúdez-Aguirre and Barbosa-Cánovas (2008), they concluded that the microbial resistance increases with increasing milk fat concentration.

The microorganism's characteristics also influence their resistance to the US effects. Gram-positive bacteria have a higher resistance than gram-negative ones, probably, because their cell walls are different. The cell walls of the gram-positive are thicker than those of the gram-negative, which contain a narrow layer of peptidoglycans (Ojha et al., 2017). Yeasts are more susceptible to the HIUS treatment due to their larger cells, which result in a larger surface area; similarly, the bacillus is more susceptible than the coccus. The aerobic bacteria are more resistant than the anaerobic, and the bacterial spores are extremely resistant to US processing (Bermúdez-Aguirre et al., 2011a; Gao et al. 2014b,c).

However, researchers have seen that HIUS processing can also increase the microbial count in the food. Then, US can also be used to favor the growth of beneficial microorganisms such as probiotic cultures (Huang et al., 2017). This is possible for adjusting the processing parameters. This effect will be discussed in the following sections.

ULTRASOUND TECHNOLOGY FOR FOOD PROCESSING

The HIUS has been widely studied for its applications in the dairy industry, which can be used to provide the homogenization of milk fat globules, lactose crystallization, microbial inactivation, modifications in functionality, and rheological properties (Ashokkumar et al., 2009, Chandrapala and Leong, 2015).

One of the main advantages of HIUS is to promote milk homogenization (Ashokkumar et al., 2009). In the dairy industry, homogenization is usually done with high-pressure equipment (Pereda et al., 2009). In the homogenization process by HIUS, the shear forces resulted from the strong collapse of the bubbles during cavitation, are responsible for decreasing the size of the fat globules. In this process, the fat globule membrane is ruptured, and then it is restructured spontaneously, adsorbing proteins and improving the emulsion stability. Certain dairy products such as milk, yogurt, and ice cream are homogenized to improve the product stability and avoid the creaming during the storage (Bermúdez-Aguirre et al., 2009; Villamiel and Jong, 2000; Bosiljkov et al., 2011; Soria and Villamiel, 2010).

The protein denaturation caused by US is particularly important in the development of fermented milk because it increases the water retention capacity and viscosity and decreases syneresis, which improve the texture of these products (Sfakianakis et al., 2014).

Another US effect observed in the development of dairy products is the improvement of the protein coagulation properties, especially important in the process of cheese manufacturing. Changes in the milk protein structure caused by the US may improve the coagulation time, curd firmness, and connectivity of the protein network, which can increase the cheese yield for up to 20% (Zhao et al., 2014; Bermúdez-Aguirre and Barbosa-Cánovas, 2010).

The HIUS application in dairy beverages has been studied both aiming the microbial inactivation and the kinetic stabilization of the product, by decreasing the phase separation and improving the homogenization of the ingredients (Guimaraes et al., 2018). Moreover, other positive effects may also be observed, mainly in dairy beverages with chocolate and fruit pulp additions, such as the increase in the bioactive compounds' concentration and their activities (e.g., antioxidant and antihypertensive activities) (Monteiro et al., 2018; Guimarães et al., 2019b).

In addition to its wide use in the dairy industry, the HIUS has been used in the processing of fruits and vegetables. Research has proven that, in cherry tomatoes (São José and Dantas Vanetti, 2012) and organic lettuce (Sagong et al., 2011), the US increases the efficiency of the sanitizers, due to the cavitation action that weakens the microbial cells on the product's surface making them less resistant to the sanitizer. In orange juices (Wong et al., 2012; Char et al., 2010; Ferrante et al., 2007), mangoes (Kiang et al., 2013), and apples (Gabriel, 2012), the US technology showed to be efficient for microbial inactivation, provided that it is combined with any complementary methods such as high pressure or heating. The results of microbial inactivation are excellent using combined methods since besides preserving the characteristics of the products more high temperatures are not reached.

The US is generally used in liquid foods since its effects are promoted mainly by acoustic cavitation that is more relevant in liquid media. However, some studies proved that the US could also be used for the processing of solid foods, such as meats. Siró et al. (2009)

reported that the US-assisted brine as a technique to cure meat improved the microstructure of the meat tissue, the water retention capacity, and the overall texture of the cured pork loin (*Longissimus dorsi*).

FUNCTIONAL FOODS

Prebiotics and probiotics are considered functional compounds, which can be naturally present in the food or can be isolated and used as an ingredient to develop functional foods, to promote the health and well-being of the consumers (Mohammadi et al., 2011).

The definition of prebiotic is "a substrate that is selectively utilized by host microorganisms conferring a health benefit" (Gibson et al., 2017). The mainly identified prebiotic substances are nondigestible carbohydrates, including inulin and several oligosaccharides, which are used as substrates by beneficial bacteria in the colon. The prebiotics provide many health benefits such as helping in the reduction in hypercholesterolemia, hyperglycemia, and intestinal disorders. In addition, they may reduce the production of toxins by some pathogenic bacteria, such as *Clostridium* sp. and *Escherichia coli* (Yasmin et al., 2015; Ogueke et al., 2010). Besides the health benefits, the prebiotics present technological advantages, such as the improvement of the textural and sensory properties of food products (Balthazar et al., 2017a,b; Guimarães et al., 2018b, 2020).

According to studies, it is suggested that a daily dose of 2.5–10 g is sufficient to provide some health benefits to humans, but higher consumptions may cause osmotic diarrhea (Davani-Davari et al., 2019). In some countries, there is legislation to establish the minimum quantity of prebiotics to be used in the food to exert their functional properties. These amounts would be enough to impact the colon microbiota beneficially; however, there may be unwanted effects such as flatulence and intestinal discomfort, due to the production of gases from the prebiotic fermentation in the colon. Moreover, it is recommended to consume prebiotics along with probiotic bacteria, such as *Bifidobacterium* and *Lactobacillus*, since these bacteria do not produce gases as part of their metabolism, which would prevent the occurrence of intestinal discomfort (Cruz et al., 2017).

According to Hill et al. (2014), probiotics are defined as "living microorganisms that when administered in adequate amounts confer health and well-being to the host." The reported benefits of the probiotic microorganisms are related to the improvement of the gastrointestinal transit, the ion absorption by the intestinal epithelial cells, and currently, have shown a capacity to increase the production of mucin in the epithelial cells, which act as an antimicrobial barrier that prevents the bonding of pathogenic agents. They also promote the production of immunoglobulin A (IgA) in the intestine, which bonds to the pathogenic agents and exerts an antiinflammatory effect. Moreover, they may help to prevent certain types of cancer (Bubnov et al., 2018).

The efficacy of the probiotic foods will depend on the concentration of viable and active cells at the moment of consumption. The recommended amount of probiotic microorganisms is 10^8–10^9 CFU (colony forming units) per day, which is equivalent to 100 g of a food product containing 10^6–10^7 CFU of probiotics (Settanni and Moschetti, 2010).

A microbial culture to be considered potentially probiotic must meet some criteria, such as resistance to gastric and biliary acids, adhesion to epithelial cells in the human intestine, antimicrobial activity against pathogenic bacteria, salt biliary hydrolase activity, and capacity to reduce the adhesion of pathogens to the gastrointestinal tract. The most-used genera of probiotic bacteria in dairy products are *Lactobacillus* and *Bifidobacterium* (Kolacek et al., 2017).

However, there are other ways that microorganisms can provide benefits to the host's health, which are the dead probiotics or their metabolites, even though they are not classified as probiotics. Therefore, to expand the studies about functional foods, two new terms were proposed, the paraprobiotics and the postbiotics (Champagne et al., 2018), which present some advantages over the use of probiotics in foods since they can be used in a safer way, without changing the original food characteristics (Barros et al., 2020).

Paraprobiotics are "nonviable microbial cells (intact or ruptured) or raw cellular extracts (with complex chemical composition), which when administered (orally or topically) in adequate amounts, confer a benefit to the human or animal consumer" (Taverniti and Guglielmetti, 2011). These compounds, in not using live microorganisms, present some advantages over the probiotics. Since the microbial cells are inactive, they will have lower interaction with the food compounds, exerting only minor changes in the original characteristics of the food. In addition, the food processing is simpler since there is no need to maintain the microbial viability; moreover, it can be added even before the thermal treatment. The paraprobiotics also offer higher safety to the consumer, avoiding infection and inflammatory responses. Besides that, storage and transport are simpler (Almada et al., 2016).

The paraprobiotic products may provide several health benefits to the consumers such as improvement in resistance to common infections, maintenance of a healthy state of humor (Sugahara et al., 2017), immunomodulatory properties (Murata et al., 2018), and reduction in the chronic symptoms associated with stress (Nishida et al., 2017), among others.

Different from the probiotics and paraprobiotics that are composed of microbial cells, the postbiotics, also called cell-free supernatants (CFS), are "metabolic products or by-products secreted by living microorganisms or released after microbial lysis that confer any physiological benefit to the host" (Aguilar-Toalá et al., 2018); in other words, they are bioactive soluble factors with health benefits. These by-products may exert beneficial effects similar to the probiotics but with the advantage of being safe in relation to the consumption dose. Some examples of postbiotics are short-chain fatty acids, peptides, enzymes, teichoic acids, vitamins, organic acids, among others (Barros et al., 2020).

The consumption of postbiotics is correlated with many health benefits such as the treatment of patients with the inflammatory intestinal disease (Tsilingiri et al., 2012), mitigation of the inflammatory response of the intestinal mucosa (Compare et al., 2017), reduction in the risk of mammary cancer (Tan et al., 2015), among others.

The production of paraprobiotics and postbiotics can be realized with HIUS technology since it is possible to inactivate and even to rupture the cells. A more in-depth view will be covered in the following sections.

EFFECTS OF ULTRASOUND ON PROBIOTIC AND PREBIOTIC PRODUCTS

The US technology has been standing out among the emerging technologies due to its versatility of use, and recently, the studies have advanced to other fields such as functional foods. The technology has been shown promising, with lots of evidence of its use in the improvement of the probiotic viability/survival and of the fermentative activity and higher production of fermentation by-products such as bioactive peptides.

EFFECT OF HIUS ON PROBIOTIC PRODUCTS

The effect of HIUS on probiotic products is influenced by some US parameters, such as intensity (frequency and/or power), processing time, type of probiotic culture, and the medium conditions (depends on the food characteristics) (Huang et al., 2017). The intensity and processing time are decisive and must be correctly adjusted because depending on the setting, it may stimulate the growth of microorganisms or reduce and even inactivate them (Abesinghe et al., 2019) (Fig. 4). The type of probiotic culture is also decisive for the desired effects. The time and intensity must be adjusted for the different microorganisms since different strains have different responses to the HIUS due to variations in the structure of the cell walls (thickness, composition, and size) (Ewe et al., 2012). The type of food may also interfere with the HIUS effect since some food compounds can "protect" the microorganisms from external shear forces, thereby increasing the difficulty in the action of the US (Chandrapala and Zisu, 2018).

The ultrasonic treatment when used in lower frequencies and intensities may promote microbial growth. However, the interactions between the US and the microorganisms are complex and not fully understood. The main mechanisms related to microbial growth are the disruption of cell aggregates, increased mass transfer, improvements of the media that serve as

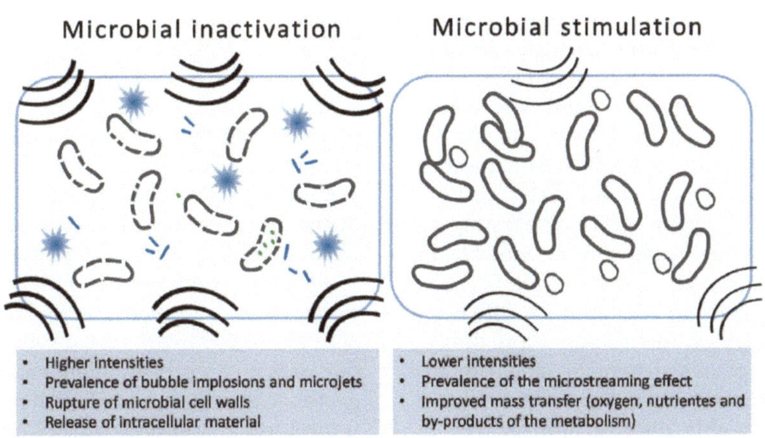

FIG. 4 Effects of the acoustic cavitation generated by different ultrasonic intensities on microorganisms.

substrates, and modulation of the physiological and genetic functions of the microorganisms (Pagnossa et al., 2020; Huang et al., 2017).

The disruption of cell aggregates increases the capacity for obtaining nutrients by the microbial cells, improves the metabolic processes, and increases the biomass (Avhad and Rathod, 2015). Another mechanism is the modulation of physiological and genetic functions. Studies (Takeuchi et al., 2008; Saura et al., 2003; Kobayashi et al., 2009) have shown that the US is able to promote microbial growth through the activation of the synthesis of DNA and proteins.

The US promotes the mass transfer between the cells and the medium, which is extremely important for microbial growth. This mechanism occurs through the microstreaming and the formation of pores in the cell membrane (Mota et al., 2018). The pressure fluctuations caused by the US create agitation in the medium. The microorganisms, when in fluid mediums, present a layer of stationary liquid that involves them and may limit the changes with the medium. The microstreaming caused by the US may remove that layer and favor the mass transfer between the microorganism and the medium (Dahroud et al., 2016; Dai et al., 2017; Tao and Sun, 2015). In addition to the microstreaming, the US may favor the mass transfer by forming reversible pores on the cell membrane. Sonoporation is the formation of cavities or transitory pores on the cell membrane due to sonication. The micropores are beneficial to the cells and allow an intense mass transfer with the medium. Moreover, there is an improvement in the permeability of the cell membranes resulting in a better transfer of substrates such as nutrients and oxygen through the membrane, besides removing the by-products of cell metabolism more efficiently (Ojha et al., 2017; Pitt and Ross, 2003). Nevertheless, if the US intensity is too high or the processing duration is long enough, there may be possible not only the formation of higher and irreversible pores but also the occurrence of cell membrane disorders and thinning, rupture of the cell wall, leakage of cell material, and damages to DNA due to the formation of free radicals (Ojha et al., 2017; Almada et al., 2016). Therefore, to reach the desired level of cell permeability and avoid cell death, the US processing parameters must be quantified and precisely controlled.

The food product wherein the probiotic is added has primordial importance in HIUS process efficiency. The impact of the HIUS on the microbiota is directly influenced by the medium's chemical and physical composition. Factors such as pH, total solid content, and the nutrients available directly influence probiotic viability (Lye et al., 2012; Dahroud et al., 2016). Environments with a lower pH (pH 4) that were sonicated had reduced probiotic viability but when the pH was more favorable (pH 7), the probiotic viability increased (Racioppo et al., 2017). The US may also change the characteristics of the medium, changing the pH, and making the environment more favorable to the growth of microorganisms (Pagnossa et al., 2020). In environments with a higher concentration of total solids, the microorganisms are more "protected" from the US action; thus, it is necessary to adjust properly the intensity and time to reach the desired effect (Gera and Doores, 2011).

One of the effects of the HIUS on the microorganisms is to stimulate the release of intracellular enzymes such as β-galactosidase and proteases (Ewe et al., 2012). In milk, the enzymes β-galactosidase hydrolyze the lactose, and the proteases hydrolyze the casein. As a result of the hydrolysis of these milk components, sugars, peptides, and amino acids are generated, which become more available for utilization by the microorganisms and then stimulate microbial growth (Lye et al., 2012).

The HIUS, when used under adequate conditions (controlling intensity and time parameters), changes the conformational structure of enzymes. It then increases the contact between the enzyme and substrate due to the cavitation and agitation in the medium (Huang et al., 2019).

Therefore, a useful application for the HIUS technology would be in the development of products with hydrolyzed lactose, highly sought after by lactose-intolerant consumers. This may be possible due to the improvement of the β-galactosidase activity, which hydrolyzes the lactose present in the product and thus making the product consumable by lactose-intolerant people (Abesinghe et al., 2019).

The most common probiotic products are fermented milk and yogurts; therefore, there is an interest in developing studies involving these products with US technology. According to some studies (Nguyen et al., 2009; Wu et al., 2000; Barukčić et al., 2015; Huang et al., 2019), it was observed that the HIUS promoted significant benefits to probiotic cultures, resulting in higher probiotic viability, fast acid development, and consequently, reduced fermentation time. One of the reasons that the US can speed up the fermentation process is that during sonication a combination of the hydrolysis of intra- and extracellular lactose may occur. The released β-galactosidase has higher hydrolysis activity than the intracellular one, which can provide higher amounts of glucose and galactose for the *Lactobacillus*; therefore, more lactic acid is produced in a shorter time (Gholamhosseinpour and Hashemi, 2019). The US also influences the production of organic acids by decreasing the proportion of acetic and propionic acids to lactic acid (Nguyen et al., 2012). Moreover, fermented milk processed by HIUS developed better sensory properties and a higher nutritional value, mainly due to the production of greater concentrations of oligosaccharides and bioactive peptides (Nguyen et al., 2009; Barukčić et al., 2015). Therefore, the processing by HIUS may result in higher probiotic viability and lower production time due to the accelerated fermentation. Products with lower lactose content have a more pleasant flavor due to the lower concentration of propionic and acetic acid and a higher concentration of oligosaccharides.

Another dairy product with a potential for the application of the HIUS is cheese. The US may promote benefits in the manufacture of these products since it can stimulate the growth of probiotic and starter cultures and the release of intracellular enzymes, thereby increasing proteolysis and lipolysis (Guimarães et al., 2018c). The increase in proteolysis can cause the development of more bioactive peptides, which promote some health benefits to the consumers. Moreover, increased proteolysis and lipolysis speed up the maturation process of cheese and may improve its flavor.

It was already discussed that adjusting the US parameters is possible to stimulate the microbial growth, but it is also possible to interrupt their multiplication, mitigating the microorganism's activity. Mitigation is a technological method used to modulate the metabolism of microorganisms. Its application is important in probiotic cultures to be added to food since their high activity can change the physicochemical and sensory properties of the food, which are often unwanted. Therefore, mitigation is usually used to avoid the acidification of the medium and to maintain probiotic viability (Racioppo et al., 2017). In a study realized by Bevilacqua et al. (2016), the HIUS was used to mitigate the probiotic microbiota, thus avoiding postacidification in a probiotic rice beverage.

In another study, also with rice beverages, Racioppo et al. (2017) concluded that the HIUS can result in the improvement of the properties associated with health benefits. After exposing *Lactobacillus reuteri* to US, they observed an increase in hydrophobicity and adhesion to Caco-2 cells. Such findings indicate that US may improve or even modulate the adhesion of the probiotic microorganism. Gholamhosseinpour and Hashemi (2019) observed that HIUS treatment could improve the antioxidant activities of dairy products because it increased the propagation capacity of the probiotic cultures in fermented milk and the release of antioxidant compounds. Lye et al. (2012) observed that the US could be used to increase the cholesterol removal capacity of the Lactobacillus (*Lactobacillus fermentum* FTDC 1311) without affecting their functionality and metabolism.

More studies are still necessary to understand the effects of HIUS on the probiotic cultures, mainly in vivo studies and clinical trials, to prove the effective improvement of probiotics on the consumer's health. Studies to the present day show that HIUS technology is promising for being used in probiotic products since it promotes higher probiotic viability and accelerates the fermentation and maturation processes. Moreover, it allows the production of dairy products with a lower concentration of lactose, a higher concentration of oligosaccharides, and a lower concentration of unpleasant organic acids (propionic and acetic acids).

EFFECTS OF HIUS ON PREBIOTIC PRODUCTS

Studies on the application of HIUS in prebiotic products are scarcer and are mainly focused on the evaluation of the physical effects of acoustic cavitation on the product's characteristics. In prebiotic products, there is always a concern with stability since the addition of nondigestible fibers can change the product's microstructure and cause instability. A possible solution to this problem could be the utilization of the US to assist the prebiotic homogenization and its incorporation into the product. In a study conducted by Guimaraes et al. (2018), they suggested that the HIUS improved the inulin crystallization forming a stronger and homogeneous gel network and thus, improving the physical stability of a whey beverage added with fruit pulp.

In the manufacturing of prebiotic products, it is important to keep the structure and concentration of the added prebiotic intact, so that it can exert its health benefits. It is important to ensure that the added prebiotic will remain intact till it reaches the intestinal tract to perform its functions. Considering the prebiotic stability, Arruda et al. (2020) studied the effect of thermal treatments (70–100°C/60s) on the inulin structure in an araticum whey beverage and observed no changes in the inulin structure, suggesting high thermal stability of this prebiotic compound. To evaluate the effect of HIUS processing on the inulin structure, Bezerra et al. (2018) studied the processing of an inulin-enriched apple juice and observed no significant effect of the US (0, 1, 4, 7, and 10kJ/g) on the inulin chemical profile. Other prebiotic compounds, such as xylooligosaccharides (XOS), were also evaluated after HIUS processing, and in general, the US treatments were not capable of changing the XOS chromatographic profiles (Silva et al., 2020). Therefore, the HIUS seems to be an effective technology for processing prebiotic products since it maintains its chemical characteristics.

In general, the HIUS has shown not to change the structure of the prebiotic compounds, but it can promote ruptures in the polymerized chains of some polysaccharides, thereby releasing compounds that are used by the probiotic cultures. Therefore, small changes may also be desirable in the case of symbiotic products (probiotic + prebiotic) since the hydrolysis of the polysaccharides increases the bifidogenic properties of the prebiotic compounds. Consequently, the probiotic viability is maintained in the functional foods (Alves Filho et al., 2016; Gomes et al., 2017). Some studies indicate that the HIUS is capable of degrading and modifying polymers with high efficiency and low cost. A study realized by Wang et al. (2018) showed that the HIUS changed the structural, functional, and bioactive properties of pectin. This modification improved the bioactivity of the studied compound, improving its prebiotic properties.

Lastly, it is already reported that HIUS can increase the health benefits of prebiotic foods. Although more studies are necessary to prove such benefits, in a study of Guimarães et al. (2019b), in which the effects of the HIUS processing on a soursop whey beverage enriched with inulin were evaluated, they observed an increase in the content of phenolic compounds and an improvement of the antioxidant and antihypertensive activity. However, the inulin chemical structure was not evaluated in the previously mentioned study since it is assumed that the application of HIUS in the nonthermal processing of dairy products does not change the inulin composition. Although USUS used in moderation does not alter the chemical structure of many prebiotics, studies applying this technology to process prebiotic food products observed some changes in the food texture and rheological behavior, which seem to be related to the increased interaction of the prebiotics with the food compounds (Guimarães et al., 2018a).

It is important to highlight that, in prebiotics as well as in probiotics, the effect of the HIUS depends on specific parameters such as the intensity (frequency and power) and processing time; moreover, they must be adjusted properly to achieve the desired effect.

EFFECTS OF HIUS ON PARAPROBIOTIC AND POSTBIOTIC PRODUCTS

In probiotic products, the culture viability must be maintained; then, the HIUS can be used to stimulate or modulate the microbial growth. In paraprobiotic products, culture viability is not necessary. The use of cultures in inactivated form is interesting because there is less interaction with the food matrix and higher safety of use, mainly in cases of strains that could cause inflammatory responses in certain people (Almada et al., 2016).

The HIUS can be used as a technology to inactivate microorganisms to produce paraprobiotics (Barros et al., 2020; Guimarães et al., 2019a). The processing parameters can be adjusted to promote higher sonoporation to the probiotic cell, resulting in impairment, puncture, or even rupture of the cell walls, releasing the intracellular material and leading to cell death (Huang et al., 2017). Although there are not many studies on the application of US technology to produce postbiotics, the HIUS shows the potential to be used since it stimulates the growth and activity of the probiotic cultures increasing the production of metabolites or microbial compounds.

CONCLUSION

In studies on HIUS processing, the subject with greater prominence is microbial inactivation. The approach of a milder method of food processing that ensures food safety and quality draws much attention. However, there are more and more studies exploring the capacity of the HIUS to stimulate microbial growth, mainly with the increasing demand for functional foods in the current days. The effects of the HIUS depend on processing parameters (frequency, power, processing time, and pulse mode), types of probiotic cultures, and food composition; therefore, to achieve the desired objective, it is necessary to quantify and control these factors with precision. As mentioned earlier, the HIUS has shown to be a technology with the potential to be used in the development of functional foods. Its use results in higher probiotic viability, products with lower content of lactose, increased bioactive peptides, speeds up the fermentation of dairy products, and accelerates the proteolysis of cheeses, improving nutritional and sensory properties. The HIUS can also change the microstructure of some prebiotics, facilitating the access of probiotics. Besides that, it may be applied for the inactivation of probiotic cultures to develop paraprobiotic products. After all, more researches are needed for a better understanding of the interactions between the HIUS and the microbial cultures and to observe the effects on them, beyond the better understanding of the US utilization in paraprobiotic and postbiotic foods.

REFERENCES

Abesinghe, A.M.N.L., Islam, N., Vidanarachchi, J.K., Prakash, S., Silva, K.F.S.T., Karim, M.A., 2019. Effects of ultrasound on the fermentation profile of fermented milk products incorporated with lactic acid bacteria. Int. Dairy J. 90, 1–14.

Aguilar-Toalá, J.E., Garcia-Varela, R., Garcia, H.S., Mata-Haro, V., González-Córdova, A.F., Vallejo-Cordoba, B., Hernández-Mendoza, A., 2018. Postbiotics: an evolving term within the functional foods field. Trends Food Sci. Technol. 75, 105–114.

Almada, C.N., Almada, C.N., Martinez, R.C.R., Sant'Ana, A.S., 2016. Paraprobiotics: evidences on their ability to modify biological responses, inactivation methods and perspectives on their application in foods. Trends Food Sci. Technol. 58, 96–114.

Alves Filho, E.G., Cullen, P.J., Frias, J.M., Bourke, P., Tiwari, B.K., Brito, E.S., Rodrigues, S., Fernandes, F.A.N., 2016. Evaluation of plasma, high-pressure and ultrasound processing on the stability of fructooligosaccharides. Int. J. Food Sci. Technol. 51 (9), 2034–2040.

Anaya-Esparza, L.M., Velázquez-Estrada, R.M., Roig, A.X., García-Galindo, H.S., Sayago-Ayerdi, S.G., Montalvo-González, E., 2017. Thermosonication: an alternative processing for fruit and vegetable juices. Trends Food Sci. Technol. 61, 26–37.

Arruda, H.S., Silva, E.K., Pereira, G.A., Meireles, M.A.A., Pastore, G.M., 2020. Inulin thermal stability in prebiotic carbohydrate-enriched araticum whey beverage. LWT 128, 109418.

Arvanitoyannis, I.S., Kotsanopoulos, K.V., Savva, A.G., 2017. Use of ultrasounds in the food industry-methods and effects on quality, safety, and organoleptic characteristics of foods: a review. Crit. Rev. Food Sci. Nutr. 57 (1), 109–128.

Ashokkumar, M., Grieser, F., 1999. Ultrasound assisted chemical processes. Rev. Chem. Eng. 15 (1), 41–83.

Ashokkumar, M., Bhaskaracharya, R., Kentish, S., Lee, J., Palmer, M., Zisu, B., 2009. The ultrasonic processing of dairy products—an overview. Dairy Sci. Technol. 90 (2–3), 147–168.

Avhad, D.N., Rathod, V.K., 2015. Ultrasound assisted production of a fibrinolytic enzyme in a bioreactor. Ultrason. Sonochem. 22, 257–264.

Azam, S.M.R., Ma, H., Xu, B., Devi, S., Siddique, M.A.B., Stanley, S.L., Bhandari, B., Zhu, J., 2020. Efficacy of ultrasound treatment in the removal of pesticide residues from fresh vegetables: a review. Trends Food Sci. Technol. 97, 417–432.

Balthazar, C.F., Silva, H.L.A., Cavalcanti, R.N., Esmerino, E.A., Cappato, L.P., Abud, Y.K.D., Moraes, J., Andrade, M.M., Freitas, M.Q., Sant'Anna, C., Raices, R.S.L., Silva, M.C., Cruz, A.G., 2017a. Prebiotics addition in sheep milk ice cream: a rheological, microstructural and sensory study. J. Funct. Foods 35, 564–573.

Balthazar, C.F., Silva, H.L.A., Vieira, A.H., Neto, R.P.C., Cappato, L.P., Coimbra, P.T., Moraes, J., Andrade, M.M., Calado, V.M.A., Granato, D., Freitas, M.Q., Tavares, M.I.B., Raices, R.S.L., Silva, M.C., Cruz, A.G., 2017b. Assessing the effects of different prebiotic dietary oligosaccharides in sheep milk ice cream. Food Res. Int. 91, 38–46.

Barros, C.P., Guimarães, J.T., Esmerino, E.A., Duarte, M.C.K.H., Silva, M.C., Silva, R., Ferreira, B.M., Sant'Ana, A.S., Freitas, M.Q., Cruz, A.G., 2020. Paraprobiotics and postbiotics: concepts and potential applications in dairy products. Curr. Opin. Food Sci. 32, 1–8.

Barukčić, I., Lisak Jakopović, K., Herceg, Z., Karlović, S., Božanić, R., 2015. Influence of high intensity ultrasound on microbial reduction, physico-chemical characteristics and fermentation of sweet whey. Innovative Food Sci. Emerg. Technol. 27, 94–101.

Beatty, N.F., Walsh, M.K., 2016. Influence of thermosonication on Geobacillus stearothermophilus inactivation in skim milk. Int. Dairy J. 61, 10–17.

Bermúdez-Aguirre, D., 2017. Ultrasound: Advances in Food Processing and Preservation. Academic Press.

Bermúdez-Aguirre, D., Barbosa-Cánovas, G.V., 2008. Study of butter fat content in milk on the inactivation of Listeria innocua ATCC 51742 by thermo-sonication. Innovative Food Sci. Emerg. Technol. 9 (2), 176–185.

Bermúdez-Aguirre, D., Barbosa-Cánovas, G.V., 2010. 'Processing of soft hispanic cheese ("queso fresco") using thermo-sonicated milk: a study of physicochemical characteristics and storage life. J. Food Sci. 75 (9), S548–S558.

Bermúdez-Aguirre, D., Mawson, R., Versteeg, K., Barbosa-Cánovas, G.V., 2009. Composition properties, physicochemical characteristics and shelf life of whole milk after thermal and thermo-sonication treatments. J. Food Qual. 32 (3), 283–302.

Bermúdez-Aguirre, D., Mawson, R., Barbosa-Cánovas, G.V., 2011a. Study of possible mechanisms of inactivation of Listeria innocua in thermo-sonicated milk using scanning electron microscopy and transmission electron microscopy. J. Food Process. Preserv. 35 (6), 767–777.

Bermúdez-Aguirre, D., Mobbs, T., Barbosa-Cánovas, G.V., 2011b. Ultrasound applications in food processing. In: Feng, H., Barbosa-Canovas, G., Weiss, J. (Eds.), Ultrasound Technologies for Food and Bioprocessing. Springer, New York, pp. 65–105.

Bevilacqua, A., Casanova, F.P., Petruzzi, L., Sinigaglia, M., Corbo, M.R., 2016. Using physical approaches for the attenuation of lactic acid bacteria in an organic rice beverage. Food Microbiol. 53, 1–8.

Bezerra, Y.M., Silva, E.K., Strieder, M.M., Pastore, G.M., Meireles, M.A.A., Arruda, H.S., 2018. Processing of Inulin-Enriched Apple Juice Using High-Intensity Ultrasound Technology. Revista dos Trabalhos de Iniciação Científica da UNICAMP, p. 26.

Bleiel, J., 2010. Functional foods from the perspective of the consumer: how to make it a success? Int. Dairy J. 20 (4), 303–306.

Bosiljkov, T., Tripalo, B., Brnčić, M., Ježek, D., Karlović, S., Jagušt, I., 2011. Influence of high intensity ultrasound with different probe diameter on the degree of homogenization (variance) and physical properties of cow milk. Afr. J. Biotechnol. 10 (1), 34–41.

Bubnov, R.V., Babenko, L.P., Lazarenko, L.M., Mokrozub, V.V., Spivak, M.Y., 2018. Specific properties of probiotic strains: relevance and benefits for the host. EPMA J. 9 (2), 205–223.

Capelo-Martínez, J.-L., 2009. Ultrasound in Chemistry: Analytical Applications. John Wiley & Sons.

Champagne, C.P., Gomes da Cruz, A., Daga, M., 2018. Strategies to improve the functionality of probiotics in supplements and foods. Curr. Opin. Food Sci. 22, 160–166.

Chandrapala, J., Leong, T., 2015. Ultrasonic processing for dairy applications: recent advances. Food Eng. Rev. 7 (2), 143–158.

Chandrapala, J., Zisu, B., 2018. Ultrasound Technology in Dairy Processing. SpringerBriefs in Molecular Science, Springer, Cham.

Char, C.D., Mitilinaki, E., Guerrero, S.N., Alzamora, S.M., 2010. Use of high-intensity ultrasound and UV-C light to inactivate some microorganisms in fruit juices. Food Bioprocess Technol. 3 (6), 797–803.

Chen, F., Zhang, M., Yang, C.-h., 2020. Application of ultrasound technology in processing of ready-to-eat fresh food: a review. Ultrason. Sonochem. 63, 104953.

Compare, D., Rocco, A., Coccoli, P., Angrisani, D., Sgamato, C., Iovine, B., Salvatore, U., Nardone, G., 2017. Lactobacillus casei DG and its postbiotic reduce the inflammatory mucosal response: an ex-vivo organ culture model of post-infectious irritable bowel syndrome. BMC Gastroenterol. 17 (1), 53.

Crum, L.A., 1984. Acoustic cavitation series: part five rectified diffusion. Ultrasonics 22 (5), 215–223.

Cruz, A.G., Zacarchenco, P.B., Oliveira, C.A.F., Corassin, C.H., 2017. Processamento de Produtos Lácteos: Queijos, Leites Fermentados, Bebidas Lácteas, Sorvete, Manteiga, Creme de Leite, Doce de Leite, Soro Em pó e Lácteos Funcionais. Elsevier, Brasil.

Dahroud, B.D., Mokarram, R.R., Khiabani, M.S., Hamishehkar, H., Bialvaei, A.Z., Yousefi, M., Kafil, H.S., 2016. Low intensity ultrasound increases the fermentation efficiency of lactobacillus casei subsp. casei ATTC 39392. Int. J. Biol. Macromol. 86, 462–467.

Dai, C., Xiong, F., He, R., Zhang, W., Ma, H., 2017. Effects of low-intensity ultrasound on the growth, cell membrane permeability and ethanol tolerance of *Saccharomyces cerevisiae*. Ultrason. Sonochem. 36, 191–197.

Davani-Davari, D., Negahdaripour, M., Karimzadeh, I., Seifan, M., Mohkam, M., Masoumi, S.J., Berenjian, A., Ghasemi, Y., 2019. Prebiotics: definition, types, sources, mechanisms, and clinical applications. Foods 8 (3), 92.

Espitia, P.J.P., Batista, R.A., Azeredo, H.M.C., Otoni, C.G., 2016. Probiotics and their potential applications in active edible films and coatings. Food Res. Int. 90, 42–52.

Ewe, J.-A., Wan Abdullah, W.-N., Bhat, R., Karim, A.A., Liong, M.-T., 2012. Enhanced growth of lactobacilli and bioconversion of isoflavones in biotin-supplemented soymilk upon ultrasound-treatment. Ultrason. Sonochem. 19 (1), 160–173.

Ferrante, S., Guerrero, S., Alzamora, S.M., 2007. Combined use of ultrasound and natural antimicrobials to inactivate *Listeria monocytogenes* in orange juice. J. Food Prot. 70 (8), 1850–1856.

Gabriel, A.A., 2012. Microbial inactivation in cloudy apple juice by multi-frequency Dynashock power ultrasound. Ultrason. Sonochem. 19 (2), 346–351.

Gao, S., Hemar, Y., Lewis, G.D., Ashokkumar, M., 2014a. Inactivation of *Enterobacter aerogenes* in reconstituted skim milk by high- and low-frequency ultrasound. Ultrason. Sonochem. 21 (6), 2099–2106.

Gao, S., Lewis, G.D., Ashokkumar, M., Hemar, Y., 2014b. Inactivation of microorganisms by low-frequency high-power ultrasound: 1. Effect of growth phase and capsule properties of the bacteria. Ultrason. Sonochem. 21 (1), 446–453.

Gao, S., Lewis, G.D., Ashokkumar, M., Hemar, Y., 2014c. Inactivation of microorganisms by low-frequency high-power ultrasound: 2. A simple model for the inactivation mechanism. Ultrason. Sonochem. 21 (1), 454–460.

Gera, N., Doores, S., 2011. Kinetics and mechanism of bacterial inactivation by ultrasound waves and sonoprotective effect of milk components. J. Food Sci. 76 (2), M111–M119.

Gholamhosseinpour, A., Hashemi, S.M.B., 2019. Ultrasound pretreatment of fermented milk containing probiotic lactobacillus plantarum AF1: carbohydrate metabolism and antioxidant activity. J. Food Process Eng. 42 (1), e12930.

Gibson, G.R., Hutkins, R., Sanders, M.E., Prescott, S.L., Reimer, R.A., Salminen, S.J., Scott, K., Stanton, C., Swanson, K.S., Cani, P.D., Verbeke, K., Reid, G., 2017. Expert consensus document: the International Scientific Association for Probiotics and Prebiotics (ISAPP) consensus statement on the definition and scope of prebiotics. Nat. Rev. Gastroenterol. Hepatol. 14 (8), 491–502.

Gomes, W.F., Tiwari, B.K., Rodriguez, Ó., de Brito, E.S., Fernandes, F.A.N., Rodrigues, S., 2017. Effect of ultrasound followed by high pressure processing on prebiotic cranberry juice. Food Chem. 218, 261–268.

Guimaraes, J.T., Silva, E.K., Alvarenga, V.O., Costa, A.L.R., Cunha, R.L., Sant'Ana, A.S., Freitas, M.Q., Meireles, M.A.A., Cruz, A.G., 2018. Physicochemical changes and microbial inactivation after high-intensity ultrasound processing of prebiotic whey beverage applying different ultrasonic power levels. Ultrason. Sonochem. 44, 251–260.

Guimarães, J.T., Silva, E.K., Alvarenga, V.O., Costa, A.L.R., Cunha, R.L., Sant'Anna, A.S., Freitas, M.Q., Meireles, M.A.A., Cruz, A.G., 2018a. Physicochemical changes and microbial inactivation after high-intensity ultrasound processing of prebiotic whey beverage applying different ultrasonic power levels. Ultrason. Sonochem. 44, 251–260.

Guimarães, J.T., Silva, E.K., Costa, A.L.R., Cunha, R.L., Freitas, M.Q., Meireles, M.A.A., Cruz, A.G., 2018b. Manufacturing a prebiotic whey beverage exploring the influence of degree of inulin polymerization. Food Hydrocoll. 77, 787–795.

Guimarães, J.T., Silva, E.K., Freitas, M.Q., Meireles, M.A.A., Cruz, A.G., 2018c. Non-thermal emerging technologies and their effects on the functional properties of dairy products. Curr. Opin. Food Sci. 22, 62–66.

Guimarães, J.T., Balthazar, C.F., Scudino, H., Pimentel, T.C., Esmerino, E.A., Ashokkumar, M., Freitas, M.Q., Cruz, A.G., 2019a. High-intensity ultrasound: a novel technology for the development of probiotic and prebiotic dairy products. Ultrason. Sonochem. 57, 12–21.

Guimarães, J.T., Silva, E.K., Ranadheera, C.S., Moraes, J., Raices, R.S.L., Silva, M.C., Ferreira, M.S., Freitas, M.Q., Meireles, M.A.A., Cruz, A.G., 2019b. Effect of high-intensity ultrasound on the nutritional profile and volatile compounds of a prebiotic soursop whey beverage. Ultrason. Sonochem. 55, 157–164.

Guimarães, J.T., Balthazar, C.F., Silva, R., Rocha, R.S., Graça, J.S., Esmerino, E.A., Silva, M.C., Sant'Ana, A.S., Duarte, M.C.K.H., Freitas, M.Q., Cruz, A.G., 2020. Impact of probiotics and prebiotics on food texture. Curr. Opin. Food Sci. 33, 38–44.

Halpin, R.M., Cregenzán-Alberti, O., Whyte, P., Lyng, J.G., Noci, F., 2013. Combined treatment with mild heat, manothermosonication and pulsed electric fields reduces microbial growth in milk. Food Control 34 (2), 364–371.

Hill, C., Guarner, F., Reid, G., Gibson, G.R., Merenstein, D.J., Pot, B., Morelli, L., Canani, R.B., Flint, H.J., Salminen, S., Calder, P.C., Sanders, M.E., 2014. The International Scientific Association for Probiotics and Prebiotics consensus statement on the scope and appropriate use of the term probiotic. Nat. Rev. Gastroenterol. Hepatol. 11 (8), 506–514.

Huang, G., Chen, S., Dai, C., Sun, L., Sun, W., Tang, Y., Xiong, F., He, R., Ma, H., 2017. Effects of ultrasound on microbial growth and enzyme activity. Ultrason. Sonochem. 37, 144–149.

Huang, G., Chen, S., Tang, Y., Dai, C., Sun, L., Ma, H., He, R., 2019. Stimulation of low intensity ultrasound on fermentation of skim milk medium for yield of yoghurt peptides by Lactobacillus paracasei. Ultrason. Sonochem. 51, 315–324.

Hughes, D.E., Nyborg, W.L., 1962. Cell disruption by ultrasound. Science 138 (3537), 108–114.

Jambrak, A.R., Mason, T.J., Lelas, V., Paniwnyk, L., Herceg, Z., 2014. Effect of ultrasound treatment on particle size and molecular weight of whey proteins. J. Food Eng. 121, 15–23.

Kentish, S., Feng, H., 2014. Applications of power ultrasound in food processing. Annu. Rev. Food Sci. Technol. 5, 263–284.

Kiang, W.S., Bhat, R., Rosma, A., Cheng, L.H., 2013. Effects of thermosonication on the fate of *Escherichia coli* O157:H7 and Salmonella Enteritidis in mango juice. Lett. Appl. Microbiol. 56 (4), 251–257.

Kobayashi, Y., Sakai, D., Iwashina, T., Iwabuchi, S., Mochida, J., 2009. Low-intensity pulsed ultrasound stimulates cell proliferation, proteoglycan synthesis and expression of growth factor-related genes in human nucleus pulposus cell line. Eur. Cell. Mater. 17 (3), 15–22.

Kolacek, S., Hojsak, I., Berni Canani, R., Guarino, A., Indrio, F., Orel, R., Pot, B., Shamir, R., Szajewska, H., Vandenplas, Y., van Goudoever, J., Weizman, Z., Probiotics, E. W. G. F. and Prebiotics, 2017. Commercial probiotic products: a call for improved quality control. A position paper by the ESPGHAN Working Group for Probiotics and Prebiotics. J. Pediatr. Gastroenterol. Nutr. 65 (1), 117–124.

Lee, B.H., Kermasha, S., Baker, B.E., 1989. Thermal, ultrasonic and ultraviolet inactivation of Salmonella in thin films of aqueous media and chocolate. Food Microbiol. 6 (3), 143–152.

Leighton, T.G., Walton, A.J., Pickworth, M.J.W., 1990. Primary bjerknes forces. Eur. J. Phys. 11 (1), 47.

Leong, T., Ashokkumar, M., Kentish, S., 2011. The fundamentals of power ultrasound—a review. Acoust. Aust. 39, 43–52.

Lollo, P.C.B., Morato, P.N., Moura, C.S., Almada, C.N., Felicio, T.L., Esmerino, E.A., Barros, M.E., Amaya-Farfan, J., Sant'Ana, A.S., Raices, R.R.S., Silva, M.C., Cruz, A.G., 2015. Hypertension parameters are attenuated by the continuous consumption of probiotic Minas cheese. Food Res. Int. 76, 611–617.

Lye, H.-S., Alias, K.A., Rusul, G., Liong, M.-T., 2012. Ultrasound treatment enhances cholesterol removal ability of lactobacilli. Ultrason. Sonochem. 19 (3), 632–641.

Mohammadi, R., Mortazavian, A.M., Khosrokhavar, R., Cruz, A.G., 2011. Probiotic ice cream: viability of probiotic bacteria and sensory properties. Ann. Microbiol. 61 (3), 411–424.

Mohammadi, V., Ghasemi-Varnamkhasti, M., González, L.A., 2017. Analytical measurements of ultrasound propagation in dairy products: a review. Trends Food Sci. Technol. 61, 38–48.

Monteiro, S., Silva, E.K., Alvarenga, V.O., Moraes, J., Freitas, M.Q., Silva, M.C., Raices, R.S.L., Sant'Ana, A.S., Meireles, M.A.A., Cruz, A.G., 2018. Effects of ultrasound energy density on the non-thermal pasteurization of chocolate milk beverage. Ultrason. Sonochem. 42, 1–10.

Mota, M.J., Lopes, R.P., Koubaa, M., Roohinejad, S., Barba, F.J., Delgadillo, I., Saraiva, J.A., 2018. Fermentation at non-conventional conditions in food- and bio-sciences by the application of advanced processing technologies. Crit. Rev. Biotechnol. 38 (1), 122–140.

Moura, C.S., Lollo, P.C.B., Morato, P.N., Esmerino, E.A., Margalho, L.P., Santos-Junior, V.A., Coimbra, P.T., Cappato, L.P., Silva, M.C., Garcia-Gomes, A.S., Granato, D., Bolini, H.M.A., Sant'Ana, A.S., Cruz, A.G., Amaya-Farfan, J., 2016. Assessment of antioxidant activity, lipid profile, general biochemical and immune system responses of Wistar rats fed with dairy dessert containing lactobacillus acidophilus La-5. Food Res. Int. 90, 275–280.

Murata, M., Kondo, J., Iwabuchi, N., Takahashi, S., Yamauchi, K., Abe, F., Miura, K., 2018. Effects of paraprobiotic lactobacillus paracasei MCC1849 supplementation on symptoms of the common cold and mood states in healthy adults. Benefic. Microbes 9 (6), 855–864.

Nguyen, T.M.P., Lee, Y.K., Zhou, W., 2009. Stimulating fermentative activities of bifidobacteria in milk by highintensity ultrasound. Int. Dairy J. 19 (6), 410–416.

Nguyen, T.M.P., Lee, Y.K., Zhou, W., 2012. Effect of high intensity ultrasound on carbohydrate metabolism of bifidobacteria in milk fermentation. Food Chem. 130 (4), 866–874.

Nishida, K., Sawada, D., Kuwano, Y., Tanaka, H., Sugawara, T., Aoki, Y., Fujiwara, S., Rokutan, K., 2017. Daily administration of paraprobiotic Lactobacillus gasseri CP2305 ameliorates chronic stress-associated symptoms in Japanese medical students. J. Funct. Foods 36, 112–121.

Ogueke, C.C., Owuamanam, C.I., Ihediohanma, N.C., Iwouno, J.O., 2010. Probiotics and prebiotics: unfolding prospects for better human health. Pak. J. Nutr. 9 (9), 833–843.

Ojha, K.S., Mason, T.J., O'Donnell, C.P., Kerry, J.P., Tiwari, B.K., 2017. Ultrasound technology for food fermentation applications. Ultrason. Sonochem. 34, 410–417.

Ordoñez, J.A., Sanz, B., Hernandez, P.E., Lopez-Lorenzo, P., 1984. A note on the effect of combined ultrasonic and heat treatments on the survival of thermoduric streptococci. J. Appl. Bacteriol. 56 (1), 175–177.

Pagnossa, J.P., Rocchetti, G., Ribeiro, A.C., Piccoli, R.H., Lucini, L., 2020. Ultrasound: beneficial biotechnological aspects on microorganisms-mediated processes. Curr. Opin. Food Sci. 31, 24–30.

Patist, A., Bates, D., 2008. Ultrasonic innovations in the food industry: from the laboratory to commercial production. Innovative Food Sci. Emerg. Technol. 9 (2), 147–154.

Pereda, J., Ferragut, V., Quevedo, J.M., Guamis, B., Trujillo, A.J., 2009. Heat damage evaluation in ultra-high pressure homogenized milk. Food Hydrocoll. 23 (7), 1974–1979.

Pitt, W.G., Ross, S.A., 2003. Ultrasound increases the rate of bacterial cell growth. Biotechnol. Prog. 19 (3), 1038–1044.

Piyasena, P., Mohareb, E., McKellar, R.C., 2003. Inactivation of microbes using ultrasound: a review. Int. J. Food Microbiol. 87 (3), 207–216.

Racioppo, A., Corbo, M.R., Piccoli, C., Sinigaglia, M., Speranza, B., Bevilacqua, A., 2017. Ultrasound attenuation of lactobacilli and bifidobacteria: effect on some technological and probiotic properties. Int. J. Food Microbiol. 243, 78–83.

Sagong, H.-G., Lee, S.-Y., Chang, P.-S., Heu, S., Ryu, S., Choi, Y.-J., Kang, D.-H., 2011. Combined effect of ultrasound and organic acids to reduce *Escherichia coli* O157:H7, Salmonella Typhimurium, and *Listeria monocytogenes* on organic fresh lettuce. Int. J. Food Microbiol. 145 (1), 287–292.

Salleh-Mack, S.Z., Roberts, J.S., 2007. Ultrasound pasteurization: the effects of temperature, soluble solids, organic acids and pH on the inactivation of *Escherichia coli* ATCC 25922. Ultrason. Sonochem. 14 (3), 323–329.

São José, J.F.B., Dantas Vanetti, M.C., 2012. Effect of ultrasound and commercial sanitizers in removing natural contaminants and *Salmonella enterica* Typhimurium on cherry tomatoes. Food Control 24 (1), 95–99.

Saura, R., Terashi, H., Lee, S., Iwatsubo, A., Ando, Y., Noguchi, M., Hirata, S., Ishikawa, H., 2003. Low intensity pulsed ultrasound exposure increase prostaglandin E2 release in human dermal fibroblasts. Bull. Health Sci. Kobe 19, 121–128.

Scudino, H., Silva, E.K., Gomes, A., Guimarães, J.T., Cunha, R.L., Sant'Ana, A.S., Meireles, M.A.A., Cruz, A.G., 2020. Ultrasound stabilization of raw milk: microbial and enzymatic inactivation, physicochemical properties and kinetic stability. Ultrason. Sonochem. 67, 105185.

Settanni, L., Moschetti, G., 2010. Non-starter lactic acid bacteria used to improve cheese quality and provide health benefits. Food Microbiol. 27 (6), 691–697.

Sfakianakis, P., Topakas, E., Tzia, C., 2014. Comparative study on high-intensity ultrasound and pressure milk homogenization: effect on the kinetics of yogurt fermentation process. Food Bioprocess Technol. 8 (3), 548–557.

Silva, E.K., Arruda, H.S., Pastore, G.M., Meireles, M.A.A., Saldaña, M.D.A., 2020. Xylooligosaccharides chemical stability after high-intensity ultrasound processing of prebiotic orange juice. Ultrason. Sonochem. 63, 104942.

Siró, I., Vén, C., Balla, C., Jónás, G., Zeke, I., Friedrich, L., 2009. Application of an ultrasonic assisted curing technique for improving the diffusion of sodium chloride in porcine meat. J. Food Eng. 91 (2), 353–362.

Soria, A.C., Villamiel, M., 2010. Effect of ultrasound on the technological properties and bioactivity of food: a review. Trends Food Sci. Technol. 21 (7), 323–331.

Sugahara, H., Yao, R., Odamaki, T., Xiao, J., 2017. Differences between live and heat-killed bifidobacteria in the regulation of immune function and the intestinal environment. Benefic. Microbes 8 (3), 463–472.

Suslick, K.S., Didenko, Y., Fang, M.M., Hyeon, T., Kolbeck, K.J., McNamara III, W.B., Mdleleni, M.M., Wong, M., 1999. Acoustic cavitation and its chemical consequences. Philos. Trans. R. Soc. Lond. A Math. Phys. Eng. Sci. 357 (1751), 335–353.

Takeuchi, R., Ryo, A., Komitsu, N., Mikuni-Takagaki, Y., Fukui, A., Takagi, Y., Shiraishi, T., Morishita, S., Yamazaki, Y., Kumagai, K., Aoki, I., Saito, T., 2008. Low-intensity pulsed ultrasound activates the phosphatidylinositol 3 kinase/Akt pathway and stimulates the growth of chondrocytes in three-dimensional cultures: a basic science study. Arthritis Res. Ther. 10 (4), R77.

Tan, H.K., Foo, H.L., Loh, T.C., Alitheen, N.B.M., Rahim, R.A., 2015. Cytotoxic effect of proteinaceous postbiotic metabolites produced by Lactobacillus plantarum I-UL4 cultivated in different media composition on MCF-7 breast cancer cell. Malays. J. Microbiol. 11 (2), 207–214.

Tao, Y., Sun, D.-W., 2015. Enhancement of food processes by ultrasound: a review. Crit. Rev. Food Sci. Nutr. 55 (4), 570–594.

Taverniti, V., Guglielmetti, S., 2011. The immunomodulatory properties of probiotic microorganisms beyond their viability (ghost probiotics: proposal of paraprobiotic concept). Genes Nutr. 6 (3), 261–274.

Tsilingiri, K., Barbosa, T., Penna, G., Caprioli, F., Sonzogni, A., Viale, G., Rescigno, M., 2012. Probiotic and postbiotic activity in health and disease: comparison on a novel polarised ex-vivo organ culture model. Gut 61 (7), 1007.

Villamiel, M., Jong, P., 2000. Influence of high-intensity ultrasound and heat treatment in continuous flow on fat, proteins, and native enzymes of Milk. J. Agric. Food Chem. 48 (2), 472–478.

Villamiel, M., García-Pérez, J.V., Montilla, A., Carcel, J.A., Benedito, J., 2017. Ultrasound in Food Processing: Recent Advances. John Wiley & Sons.

Wang, W., Chen, W., Zou, M., Lv, R., Wang, D., Hou, F., Feng, H., Ma, X., Zhong, J., Ding, T., Ye, X., Liu, D., 2018. Applications of power ultrasound in oriented modification and degradation of pectin: a review. J. Food Eng. 234, 98–107.

Wong, E., Vaillant-Barka, F., Chaves-Olarte, E., 2012. Synergistic effect of sonication and high osmotic pressure enhances membrane damage and viability loss of Salmonella in orange juice. Food Res. Int. 45 (2), 1072–1079.

Wu, H., Hulbert, G.J., Mount, J.R., 2000. Effects of ultrasound on milk homogenization and fermentation with yogurt starter. Innovative Food Sci. Emerg. Technol. 1 (3), 211–218.

Yasmin, A., Butt, M.S., Afzaal, M., van Baak, M., Nadeem, M.T., Shahid, M.Z., 2015. Prebiotics, gut microbiota and metabolic risks: unveiling the relationship. J. Funct. Foods 17, 189–201.

Zhao, L., Zhang, S., Uluko, H., Liu, L., Lu, J., Xue, H., Kong, F., Lv, J., 2014. Effect of ultrasound pretreatment on rennet-induced coagulation properties of goat's milk. Food Chem. 165, 167–174.

Zisu, B., Chandrapala, J., 2015. High power ultrasound processing in milk and dairy products. In: Datta, N., Tomasula, P.M. (Eds.), Emerging Dairy Processing Technologies—Opportunities for the Dairy Industry. John Wiley & Sons, pp. 149–180.

CHAPTER 17

Probiotics and Mycotoxins

CARLOS AUGUSTO FERNANDES DE OLIVEIRA[a] • KHURRAM MUAZ[b] • CLEIDE OLIVEIRA DE ALMEIDA MØLLER[c] • CARLOS HUMBERTO CORASSIN[a] • FERGAL P. RATTRAY[c]
[a]Department of Food Engineering, School of Animal Science and Food Engineering, University of São Paulo, Pirassununga, Brazil, [b]Institute of Food Science & Nutrition, Bahauddin Zakariya University, Multan, Pakistan, [c]Department of Food Science, University of Copenhagen, Frederiksberg, Denmark

INTRODUCTION

Mycotoxins are defined as toxic secondary metabolites that can be released by mycotoxigenic fungi that are able to grow in foods throughout the production chain. Mycotoxins can cause adverse effects on several animal species including humans. The main toxigenic fungi are found in the *Aspergillus*, *Penicillium*, and *Fusarium* genera, and the primary classes of mycotoxin produced by these genera are the aflatoxins (AFs), deoxynivalenol (DON), fumonisins (FBs), ochratoxin A (OTA), and zearalenone (ZEN). Several countries in the world have adopted regulations establishing maximum tolerance levels for AFs, DON, FBs, OTA, and ZEN in food products (Agriopoulou et al., 2020).

AFs are hepatotoxic, carcinogenic mycotoxins classified by the International Agency for Research on Cancer (IARC) as Group 1 (carcinogen to human). AFs are produced by several *Aspergillus* species such as *A. flavus* and *A. parasiticus* on food and feed resulting in huge economic losses (Ismail et al., 2018). The contamination of food and feed with aflatoxins may occur at any stage from preharvest to consumption after fungal invasion of the product. The contamination is promoted by warm and humid conditions. There are more than 20 known metabolites of aflatoxins among which the major ones include AFB_1, AFB_2, AFG_1, AFG_2, and hydroxylated metabolites of AFB_1 and AFB_2 termed as AFM_1 and AFM_2, respectively. Among the known AFs, AFB_1 is considered to be the most toxic (Oliveira et al., 2014).

DON, part of the trichothecenes family of mycotoxins, is mainly produced by *Fusarium graminearum*. DON is also known as vomitoxin due to its emetic effects upon ingestion. It is the most common contaminant of grains and hence becomes part of subsequent products such as bread, snacks, and beer. DON affects animal and human health by causing diarrhea, vomiting, gastrointestinal inflammation, and immunomodulation. FBs are mainly produced by *F. verticillioides* and *F. proliferatum*. Some other *Fusarium* species that produce FBs include *F. ramigenum* and *F. oxysporum*. The FBs are mainly found in corn and corn-based foods. Fumonisin B_1 (FB_1) is most commonly found, while FB_2 and FB_3 may co-occur in concentrations lower than FB_1. FB_2 is known to possess carcinogenic (IARC Group 2B, a probable human carcinogen), hepatotoxic, nephrotoxic, and embryotoxic characteristics. OTA is produced by several fungi including *Aspergillus ochraceus*, *A. carbonarius*, *A. niger*, and *Penicillium verrucosum*. These fungi vary in their optimal temperature and water activity for growth and, hence, contaminate diverse ranges of food and feed commodities including cereals, dried fruits, grapes and wine, coffee, meat, and milk products. The presence of OTA in a wide range of commodities makes it an inevitable threat to humans and animals. OTA is known for its nephrotoxic characteristics causing renal tumors (IARC Group 2B) and various renal diseases in humans. ZEN is a unique mycotoxin possessing estrogenic activity, consequently disrupting the endocrine system in animals and humans. ZEA is mainly produced by *F. graminearum* and *F. culmorum* and is found in all cereal grains worldwide (Coppa et al., 2019).

Mycotoxins are resistant to conventional treatments usually applied in food and feed processing, including pasteurization, sterilization, and other thermal applications. For example, AFs are very stable against high-temperature treatments such as roasting, cooking, and baking with decomposition temperatures ranging from 237°C to 306°C (Ismail et al., 2018). Therefore, good agricultural practices are essential in order to prevent the contamination of grains at pre- and postharvest stages of crop production. Failures in the use of preventive measures may require detoxification methods consequently to remove or degrade mycotoxins from contaminated products by physical, chemical, or biological approaches.

Probiotics and Prebiotics in Foods. https://doi.org/10.1016/B978-0-12-819662-5.00005-7

Biological detoxification using lactic acid bacteria (LAB) has been widely studied due to its safety and their natural presence in various food such as milk. After the discovery of AFs in the 1960s, several LAB species were screened for their in vitro ability to degrade or remove this mycotoxin from media or food matrices (Yousefi and Hosseini, 2017). In recent years, many investigations were carried out to assess the decontamination of DON, FBs, OTA, and ZEN by LAB through various different approaches. However, the application of LAB ultimately depends on the nature of the food substance in addition to the type of mycotoxin and the bacterial strain used. These aspects are discussed in detail later.

LACTIC ACID BACTERIA

Lactic acid bacteria (LAB) were reported to be used for the first time in 1873 by Joseph Lister as a model organism to simulate the cause of infectious disease in humans (Santer, 2010). However, the reason for LAB being widely used is due to their ability to promote a remarkable diversification in flavor and texture in fermented food products. Aware of that, in 1880 the Danish Christian Ditlev Ammentorp Hansen began to produce commercial lactic acid bacteria starters and a year later the French scientist Edmond Fremy produced lactic acid by fermentation, while in the USA the first industrial fermentation of lactic acid occurred (Ghaffar et al., 2014). The selection of LAB for food fermentations has contributed to an increase in the security, safety, and quality of fermented foods worldwide.

LAB are Gram-positive bacteria that can be found in diverse environments ranging from foods to the gastrointestinal tract. This is explained by the fact that many LAB are very robust and can grow in conditions with a wide range of pH values (4.9–8.9) and temperatures (2-53° C), as well as other factors such as salt (as high as 6.0%), which are less tolerated by other microorganisms (Møller et al., 2020). Using advanced tools of detection and identification (Rezac et al., 2018), a large variety of fermented foods contain LAB with the main food matrices are listed in Fig. 1.

The greatest diversity of LAB species is reported to be found in dairy products, and cheese is a category in itself with a large number of species found. Other food matrices such as fermented vegetables and exotic foods, such as Asian and African delicacies, as well as fermented beverages (wine and beer), are also sources of a variety of LAB. Rezac et al. (2018) also pointed out that these food products with live LAB may be considered a good source of probiotic bacteria.

STARTER AND NONSTARTER LAB

In food fermentation, LAB are classified into two main groups the so-called starter LAB (SLAB) and the nonstarter LAB (NSLAB). A good example of a fermented food with a wide range of LAB is cheese, and this food matrix will be used here to illustrate the common classification used for LAB in food processing. The microbial ecology of cheese involves a rich and complex interaction between the SLAB and the NSLAB, which contributes to the final characteristics of the cheese (Czáran et al., 2018). SLAB are intentionally added to the raw materials such as milk and are available commercially from culture suppliers or in the form of

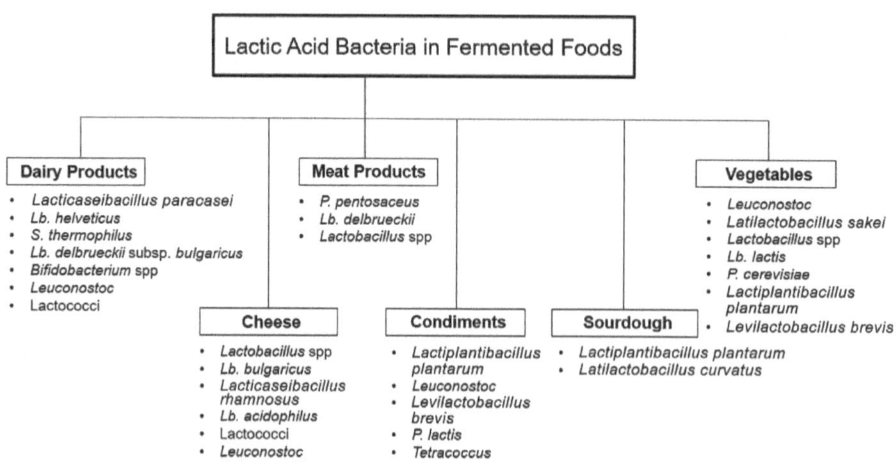

FIG. 1 Main lactic acid bacteria species naturally found in fermented foods.

"inhouse" cultures. For the manufacture of cheese such as Cheddar, the predominant SLAB are lactococci including *Lc. lactis* subsp. *lactis* and *Lc. lactis* subsp. *cremoris*, but *Streptococcus thermophilus* also may be used. Naturally present at very low levels in the raw materials and processing environment, NSLAB are able to grow during cheese ripening period and are predominated by mesophilic lactobacilli such as *Lacticaseibacillus paracasei*, *Lactiplantibacillus plantarum*, *Latilactobacillus curvatus*, and *Lacticaseibacillus rhamnosus*, but *Levilactobacillus brevis*, *Limosilactobacillus fermentum*, *P. pentosaceus*, and *P. acidilactici* also have been reported. Furthermore, pediococci and micrococci constitute the NSLAB flora in certain cheese matrices (Møller et al., 2020).

In addition to this classification based on the food-technological point of view (SLAB and NSLAB), LAB can also be classified according to their morphological, physiological, and metabolic characteristics (Table 1). This classification has been subject to controversy, but from a practical point of view, the main LAB species encompass members of *Aerococcus*, *Carnobacterium*, *Enterococcus*, *Lactobacillus*, *Lactococcus*, *Leuconostoc*, *Oenococcus*, *Pediococcus*, *Streptococcus*, and *Weissella* genera. Since most of the NSLAB belong to the family *Lactobacillaceae*, the polyphasic approach of Zheng et al. (2020) is very relevant when performing studies related to NSLAB. This approach is based on whole-genome sequences and proposes reclassification of the genus *Lactobacillus* into 25 genera.

CELLULAR STRUCTURE OF LAB

LAB cellular structure is composed of extracellular (cell wall) and intracellular structures (mainly including cytoplasmic membrane, cytoplasm, and DNA).

Cell Wall

Peptidoglycan is the main component of LAB cell wall. It consists of glycan chains made of alternating *N*-acetylglucosamine and *N*-acetylmuramic acid that are linked. Peptidic chains are bound covalently through their N-terminus to the lactyl group of *N*-acetylmuramic acid. The peptidoglycan type and sequence in the cell wall differ between LAB species (Table 1). Despite these differences, the structure generated around the cell ensures bacterial integrity. In LAB, the amino acid sequence of the stem peptide is L-Ala-g-D-Glu-X-D-Ala, where the third amino acid (X) is a diamino acid. It is most often L-Lys (in *Lactococcus lactis* and most lactobacilli) but can also be mesodiaminopimelic acid (mDAP) (in *Lactiplantibacillus plantarum*) or L-ornithine (in *Limosilactobacillus fermentum*). These variations in the cell wall glycopolymers confer differences in terms of resistance to enzymes and bacteriocins, activation and inhibition of autolysis as well as regulation of essential metabolic activities (Holzapfel and Wood, 2014).

Cytoplasmic Membrane

The cytoplasmic membrane, composed of phospholipids, is the only membrane present within the cell wall boundary. It forms a diffusion barrier and constitutes a physical barrier between the cytoplasm and the external structure. The barrier property allows the cytoplasmic membrane to serve two important functions: (i) conservation of the integrity of the cytoplasm by preventing metabolites and ions from diffusing uncontrolled out of the cells and of external compounds and ions from diffusing in freely; and (ii) chemiosmotic energy transduction (Konings, 2002).

Cytoplasm

The cytoplasm is the structure where most of the metabolic activities of the LAB cell occur. The most relevant function of the cytoplasm is in pH regulation, which is an important parameter for the enzymatic machinery of the cell (Kashket, 1987).

DNA

The main function of the deoxyribonucleic acid (DNA) is in cell replication. The circular double-stranded DNA molecules of LAB plasmids are capable of replication independently of the chromosome. Plasmids are associated with the production of bacteriocins, phage resistance, citrate utilization, sugar metabolism, proteolysis, and conjugative transfer in *Lactococcus*. With the exception of *Lactobacillus bulgaricus*, plasmids are also a common feature of *Lactobacillus* species (Rixon and Warner, 2003).

METABOLITES FROM LAB

LAB produce many metabolites that can inhibit and inactivate the growth of sensitive microorganisms and therefore are important in the control of pathogenic microorganisms and prevention of off-flavor formation development by certain microorganisms during fermentation. When present in the food matrix, LAB may compete with pathogenic microorganisms for nutrients or produce antimicrobial agents such as organic acids, ethanol, diacetyl, carbon dioxide, hydrogen peroxide, and bacteriocins (Arena et al., 2018).

The combined use of bacteriocin-producing LAB and organic acids improves the antimicrobial effectivity toward pathogenic microorganisms. These antimicrobial peptides may inhibit pathogen growth

TABLE 1
Phenotypic Characteristics of Selected LAB With Potential Mycotoxin Detoxification Properties.

Family	Genus	Morphology	Motility	Co2 From Glucose	Configuration of Lactate	Growth at pH 9.6	Growth in 6.5% NaCl	Arginine Hydrolysis	Growth at 10°C	Growth at 45°C	Peptidoglycan Type in the Cell Wall
Aerococcaceae	*Aerococcus*	Cocci, ovoid, single, pairs, tetrads	–	–	L (+)	ND	+	–/+	+	–	*L*-Lys-*D*-Asp
Carnobacteriaceae	*Carnobacterium*	Rods, single, pairs, chains	+	–	L (+)	+/–	–/+	+	+	–	*m*Dpm
Enterococcaceae	*Enterococcus*	Ovoid cocci in pairs or chains; small groups	–/+	–	L (+)	+	+	+/–	+	+	Lys-*D*-Asp or Lys-Ala_{2-3}
Lactobacillaceae	*Lactobacillus*	Short (coccobacilli) to long rods, straight to irregular, single, pairs, and chains	–/+	–/+	D (–) / DL / L (+)	–	–/+	–/+	–/+	–/+	*L*-Lys-*D*-Asp, Orn-*D*-Asp, Lys-*L*-Ala_2, *m*Dpm
	Pediococcus	Spherical cocci, single, pairs, tetrads	–	–	DL / L (+)	ND	–	–	ND	–	Lys-*D*-Asp
	Leuconostoc	Ovoid cocci in pairs or chains	–	+	D (–)	ND	–	–	+/–	–/+	*L*-Lys-*L*-Ala, *L*-Lys-*L*-Ala_2, *L*-Lys-*L*-Ser-*L*-Ala_2
	Oenococcus	Ellipsoid to spherical cocci in pairs or chains	–	+	D (–)	ND	ND	–/(+)	ND	ND	*L*-Lys-*L*-Ala-*L*-Ser, *L*-Lys-*L*-Ser-*L*-Ser
	Weissella	Short rods or ovoid, pairs, short chains	–	+	D (–) / DL	ND	+/–	+/–	+/–	–/+	Lys-Ala, Lys-Ala_2, Lys-Ala-Ser, Lys-Ser-Ala_2, Lys-*L*-Ala-Gly-*L*-Ala_2
Streptococcaceae	*Lactococcus*	Ovoid cocci in pairs or chains	–	–	L (+)	–	–	–/+	–	–	Lys-*D*-Asp, Lys-Ala-Gly-Ala, Lys-Ser-Ala, Lys-Thr-Ala
	Streptococcus	Ovoid cocci in pairs or chains	–	–	L (+)	–	+/–	–	–	–	Lys-Ala_n, Lys-Ala_{1-3}(Ser), Lys-Ala (Ser), Lys-Thr-Gly, Lys-Thr-Ala, Lys-Ser-Gly

Adapted from Holzapfel, W.H., Wood, B.J.B., 2014. Lactic Acid Bacteria: Biodiversity and Taxonomy. John Wiley & Sons Ltd, Chichester, UK, with modifications based on the classification proposed by Zheng, J., Wittouck, S., Salvetti, E., Franz, C.M.A.P., Harris, H.M.B., Mattarelli, P., O'Toole, P.W., Pot, B., Vandamme, P., Walter, J., Watanabe, K., Wuyts, S., Felis, G.E., Ganzle, M.G., Lebeer, S., 2020. A taxonomic note on the genus Lactobacillus: description of 23 novel genera, emended description of the genus Lactobacillus Beijerinck 1901, and union of Lactobacillaceae and Leuconostocaceae. Int. J. Syst. Evol. Microbiol. 70, 2782–2858.

or interfere in their metabolic activities, thus reducing pathogenicity or inhibiting the synthesis of toxic compounds. Antibiofilm activity of bacteriocins has been reported by Mohapatra and Jeevaratnam (2019) against pathogens such as *Pseudomonas aeruginosa* and *Staphylococcus aureus*. While probiotics are defined as live microorganisms that confer a health benefit on the host, bacteriocin production can also be considered as a probiotic trait.

The potential mechanism of action of bacteriocins is in the interaction of the bacterial cell membrane with the bacteriocin, resulting in a membrane disruption, hydrolysis, permeabilization, or inhibition of cell wall biosynthesis. In addition, it has been reported that bacteriocins may interact with putative vital intracellular targets. A large number of bacteriocins such as nisin, enterocin, streptolysin, glycocin, pediocin, lactococcin, and lacticin may be produced by Gram-positive bacteria. They are classified as antimicrobials according to their chemical structure and mechanism. Nisin is one of the best known and most used bacteriocin in foods. It is produced by strains of *Lactococcus lactis*. It is a small peptide lantibiotic that has antimicrobial action only against Gram-positive bacteria and generally does not inhibit Gram-negative bacteria, yeasts, or molds (Alvarez-Sieiro et al., 2016).

Arena et al. (2018) reported that LAB can have antagonistic activity against filamentous fungi, and although the real mechanisms by which they control the fungal development are still unclear, it seems to be associated with cytoplasmic acidification and disruption of proton motive forces. This is of relevance when considering the issue of mycotoxin production by fungal growth in food matrices. Antifungal metabolites from LAB have been reported to inhibit fungal growth in different kinds of fermented and nonfermented food matrices. *Lactobacillus* species are the most reported bacteria with antifungal activity, especially *Lactiplantibacillus plantarum* (Sadiq et al., 2019).

Some of the metabolites produced by LAB may also be of safety concern, due to their toxicity, such as is the case of biogenic amines. Certain decarboxylase-positive LAB are able to metabolize amino acids to produce biogenic amines. This mechanism is an alternative way to obtain energy and to grow in the absence of carbohydrate sources. Foods related to biogenic amines poisoning included aged cheeses and fermented meats, where histamine and tyramine are the most common products from the decarboxylation of histidine and tyrosine, respectively. The most prevalent histamine-producer in cheese is *Lentilactobacillus parabuchneri* (Møller et al., 2020). Therefore, not all LAB can be considered as probiotic, and before classifying LAB as probiotics, it is important to investigate all the safety aspects of the metabolites produced (Huang et al., 2020).

PROBIOTIC LAB

The Food and Agriculture Organization of the United Nations together with the World Health Organization defines probiotics as live microorganisms able to confer a health benefit on the host, when administered in adequate amounts (Food and Agriculture Organization/World Health Organization, 2006). Although this definition is well accepted by the scientific community, new terminologies have been integrated to probiotics based on pieces of experimental evidence over the years, including "paraprobiotics," which are dead/inactivated (intact or ruptured) cells of probiotics, and "postbiotics," term related to the healthful metabolites of probiotics. In this context, Zendeboodi et al. (2020) proposed three main classes of probiotic including "true probiotic" referring to viable and active probiotic cell, "pseudoprobiotic" referring to viable and inactive cell and "ghost probiotic" referring to dead/nonviable cell, in the forms of intact or ruptured. These definitions are useful to understand the decontamination effects of mycotoxins by probiotic LAB, since not only the viable cells of probiotics but also their dead and ruptured cells can adsorb mycotoxins.

The probiotic property of the well-identified microorganism needs to be documented by applying both in vitro and in vivo models. In the same way, the safety of the microorganism needs to be assessed to avoid the occurrence of transmigration (from the gut to the blood) which may cause bacteremia and sepsis, or transferring of antibiotic resistance, as well as other adverse effects, influencing gastrointestinal physiology or various immunological functions in the host. In addition, it is extremely important that the microorganism is able to survive the path from food processing to exposure at retail and be consumed at the expected concentration to promote the benefits in the host. Finally, the probiotic microorganism must survive passage through the gastrointestinal tract and adhere and colonize in the intestine (Rolim et al., 2020).

LAB strains are the major representatives of probiotics. Many probiotic strains belong to the genera *Lactobacillus* and *Bifidobacterium* and are currently applied in commercial probiotic yogurt and milk-based products throughout the world. Cheeses have been the vehicle for many probiotics too such as *Lactobacillus acidophilus, Bifidobacterium lactis, Lacticaseibacillus rhamnosus, Lacticaseibacillus paracasei, Lacticaseibacillus casei, Levilactobacillus brevis, Lactiplantibacillus plantarum,*

Bifidobacterium longum, and *Lactobacillus crispatus* (Rolim et al., 2020). *Lactiplantibacillus plantarum* and *Levilactobacillus brevis* originating from pickles were described as possessing desirable probiotic properties (Tokatlı et al., 2015). *Lactiplantibacillus plantarum*, *Pediococcus pentosaceus*, *Pediococcus pentosaceus*, *Limosilactobacillus reuteri*, *Lactobacillus johnsonii*, and *Lacticaseibacillus casei* have been reported as probiotic LAB strains with potential health benefits as feed supplements (Huang et al., 2020). Found in different food matrices and many other sources, probiotics have been considered to be very promising for many applications. Probiotics are generally recognized as safe (GRAS status) which explains their use in many parts of the food chain, and in the therapeutic treatment of animals and humans. By extending the use of probiotics it is expected to overcome or at least to reduce issues such as antibiotic resistance, foodborne illness, and the incidence of diseases related to opportunistic microorganisms such as fungi.

The impact of probiotic LAB on the growth of fungi, and consequently on mycotoxin production or even directly on the mycotoxins themselves, has been described (Arena et al., 2018). Mycotoxins are toxic, or in some cases carcinogenic, and are resistant to traditional food processing steps, which make the use of probiotic LAB a promising alternative. The mechanism of the antagonistic effect of probiotic LAB on mycotoxigenic fungi may be related to competition for nutrients and space, interference on metabolites production, or disruption of the fungi by biofilm formation. The mechanisms of mycotoxin detoxification are due to the binding, adsorption, or degradation ability of the probiotic LAB (Afshar et al., 2020; Agriopoulou et al., 2020). Furthermore, temperature, pH, LAB strain, type of food matrix, and type of mycotoxin are effective variables influencing the detoxification ability of LAB (Yousefi and Hosseini, 2017). The most common probiotic LAB effective against the growth of mycotoxigenic fungi are *Limosilactobacillus fermentum*, *Lactiplantibacillus plantarum*, and *Lacticaseibacillus rhamnosus* (Sadiq et al., 2019). Regarding biodetoxification, the most current LAB reported to have a probiotic effect related to (1) binding are *Bifidobacterium lactis*, *Lactobacillus acidophilus*, *Levilactobacillus brevis*, *Lactobacillus bulgaricus*, *Lacticaseibacillus casei*, *Lactobacillus delbrueckii* subsp. bulgaricus, *Limosilactobacillus fermentum*, *Lactobacillus helveticus*, *Lactiplantibacillus pentosus*, *Lactiplantibacillus plantarum*, *Limosilactobacillus reuteri*, *Lacticaseibacillus rhamnosus*, *Lactococcus lactis*, *Pediococcus pentosaceus*, and *Streptococcus thermophilus* (Chlebicz and Śliżewska, 2020; Wacoo et al., 2019; Huang et al., 2017; Sarlak et al., 2017; Sadeghi et al., 2016; Sezer et al., 2013; Kasmani et al., 2012); (2) adsorption are bifidobacteria, *Lacticaseibacillus casei*; *Lentilactobacillus kefiri*, *Lactiplantibacillus plantarum*, and *Lacticaseibacillus rhamnosus* (Huang et al., 2018; Taheur et al., 2017; Meca et al., 2012); and (3) degradation are *Levilactobacillus brevis*, *Lacticaseibacillus paracasei*, *Lactiplantibacillus plantarum*, and *Fructilactobacillus sanfranciscensis* (Złoch et al., 2020; Bartkiene et al., 2018).

The potential of probiotic LAB has not been fully explored yet. There are many sources to be investigated and diverse forms of application to be developed. Emerging probiotic LAB and the broader use of already-known LAB with probiotic potential remains to be fully exploited. Research and development of new trends are very focused on probiotic LAB, which is expected to lead to many improvements and achievements in this area.

DETOXIFICATION OF MYCOTOXINS BY LACTIC ACID BACTERIA

Aflatoxins

Many LAB strains have been analyzed for the purpose of AF decontamination with varying degrees of success reflecting the different decontamination capacities for different strains (Panwar et al., 2019). The introduction of LAB in a media may reduce the production of AFs as a result of their antifungal activities. Furthermore, the decontamination of AFs can also be achieved either through biological degradation of AFs into less toxic substances by LAB metabolism or through binding with various cell wall components (Martínez et al., 2019).

The use of LAB cells in order to degrade AFs into less toxic substances may be applicable in products where fermentative changes are desirable such as various beverages or dairy products. The application of LAB in some of the fermented products can therefore help in the reduction of AFs without the addition of any chemical additives, and consequently making it more appealing to the consumer; nowadays, consumers are more conscious about how products have been treated during processing. The production of yogurt by using a suitable combination of various LAB strains can result in a greater degradation of AFM_1 in the final product. However, the amount of AFM_1 degraded also depends on the storage time. The use of starter cultures for yogurt including *L. acidophilus*, *B. lactis*, *Str. thermophilus*, *Lacticaseibacillus casei*, and *L. bulgaricus* reduced AFM_1 without having a remarkable impact on cell viability (Montaseri et al., 2014) (Table 2).

LAB can also reduce the amounts of AFs in food and feed through binding via their cell wall components. Several lactic acid bacteria species have the ability to

TABLE 2
Aflatoxin Detoxification by Lactic Acid Bacteria (LAB) Strains According to Different Proposed Mechanisms.

Bacterial Strain	Cell Concentration (cfu/mL)	AF	Level (ng/g)	Proposed Mechanism	% Reduction[a]	Reference
Lacticaseibacillus rhamnosus yoba 2012	10^7	AFB_1	1000	Binding in saline solution	69	Byakika et al. (2019)
P. pentosaceus RC006 and *K. marxianus* VM003	NR	AFM_1	50	Binding in milk	61	Martínez et al. (2019)
				Degradation	100	
Lacticaseibacillus rhamnosus LC-4	10^9	AFM_1	25	Binding with acid-treated cells	79	Zhang et al. (2019)
Lacticaseibacillus rhamnosus GG	1.5×10^9	AFM_1	0.1	Binding with biofilm	61	Assaf et al. (2019)
			1		55	
Lactiplantibacillus plantarum CRD7	10^9	AFM_1	10	Binding in an in vitro digestive model	53	Panwar et al. (2019)
Lactiplantibacillus plantarum CECT 749	NR	AFB_1		Inhibition of *A. flavus*	100	Nazareth et al. (2020)
Lacticaseibacillus rhamnosus LOCK 1088	NR	AFB_1	10,000	Binding with LAB cells	80	Chlebicz and Śliżewska (2020)
Limosilactobacillus fermentum PTCC 1744	NR	AFB_1	125	Degradation through LAB extracellular metabolites	91	Ghazvini et al. (2016)
		AFB_2	25		90	
		AFG_1	125		100	
		AFG_2	25		99	
Lacticaseibacillus rhamnosus TISTR 541		AFM_1	50	Binding using heat-treated cells in PBS	94	Elsanhoty et al. (2014)
Lactiplantibacillus plantarum	NR			Degradation/binding in medium	77	
S. thermophilus subsp. *lactis* + *L. bulgaricus* + *Lactiplantibacillus plantarum*				Degradation/binding in yogurt preparation	77	
Lactiplantibacillus plantarum	2×10^6	AFB_1	–	Inhibition of *A. flavus* growth in olive samples	100	Kachouri et al. (2014)

Continued

TABLE 2
Aflatoxin Detoxification by Lactic Acid Bacteria (LAB) Strains According to Different Proposed Mechanisms—cont'd

Bacterial Strain	Cell Concentration (cfu/mL)	AF	Level (ng/g)	Proposed Mechanism	% Reduction[a]	Reference
Lacticaseibacillus rhamnosus	10^{10}	AFB_1	1000	Binding with heat-killed cells in suspension form	46	Bovo et al. (2014b)
				Binding with heat-killed cells in freeze-dried form	367	
				Binding with heat-killed cells in spray-dried form	0	
Lacticaseibacillus rhamnosus	10^{10}	AFB_1	1000	Binding with heat-killed cells	46	Bovo et al. (2014a)
Lacticaseibacillus rhamnosus	10^{10}	AFM_1	150	Binding with heat-killed cells in PBS	46	Bovo et al. (2013)
B. lactis			0.5	Binding with heat-killed cells in skimmed milk	38	
L. delbrueckii subsp. *bulgaricus* + *Lacticaseibacillus rhamnosus* + *B. lactis*	10^{10}	AFM_1	0.5	Binding with heat-killed cells in skimmed Milk	12	Corassin et al. (2013)
Lacticaseibacillus rhamnosus GAF	10^{8}	AFM_1	200	Binding with viable cells in milk	95	Abbès et al. (2013)
Lacticaseibacillus casei	4×10^{6}	AFB_1	50	Removal through fermentation of maize grains	45	Oluwafemi et al. (2010)
Lactiplantibacillus plantarum	NR	AFB_1	80	Binding with heat-killed cells inoculated in maize grains	95	Oluwafemi and Silva (2009)

AF, aflatoxin; NR, not reported.
[a]Maximum value reported.

bind AFs in vitro, but viable LAB cells can be used for AFs binding where the product to be treated is not fermented such as their use in animal feed. The capability of LAB cells to bind AFs can be utilized to reduce the bioaccessibility of the toxins and hence reduce harmful effects. Liu et al. (2018) showed that the addition of LAB cells at concentrations of up to 1.5×10^{10} colony-forming units (cfu)/kg to an AFB_1 contaminated broiler feed resulted in an increased broiler growth performance and reduced AFB_1levels in their serum and organs. The use of LAB in the feed was found to be more beneficial when compared to the use of hydrated sodium calcium aluminosilicate (HSCAS).

Another possibility for AFs decontamination is by biofilms formed by LAB as proposed by Assaf et al. (2019), who used *Lacticaseibacillus rhamnosus* biofilms to achieve up to a 61% reduction in AFM_1 after a contact time of 30 s. *Lacticaseibacillus rhamnosus* was also shown to remove a major portion of AFM_1 from phosphate-buffered saline (PBS) and milk through binding as well as reducing AFM_1-induced effects on total white and red blood cells, and lymphocyte subtypes in the host (Abbès et al., 2013).

AF degradation using LAB cells may not be a suitable option where the LAB activity produces undesirable changes in the food product. Therefore, nonviable LAB cells can be utilized in products that are more vulnerable toward fermentative changes. Milk and milk products are among such food commodities. The cells can be rendered nonviable by heat or acid treatments after which the mycotoxin binding properties still remain. The experiments by Zhang et al. (2019) concluded that the use of heat- and acid-treated cells of *Lacticaseibacillus rhamnosus* LC-4 for decontamination of AFM_1 in milk showed even higher binding efficiencies as compared to untreated cells. The heat-killed cells of *L. delbrueckii* ssp. *bulgaricus* and *Bifidobacterium lactis* also showed AFM_1 reduction to varying degrees when added to milk and PBS (Bovo et al., 2013). Nonviable cells can either be used in liquid or lyophilized form; however, the method used for deactivating or lyophilizing the LAB cells for the subsequent purpose of AF decontamination is of critical importance as any disturbance in cell integrity can alter their binding capacities. Spray-drying can reduce the binding capacities in comparison with freeze-drying due to damage of structural and functional properties of the cell wall (Bovo et al., 2014b).

Deoxynivalenol

DON has been reported to be stable during high-temperature treatment such as baking where temperatures range from 170 °C to 350 °C (Kushiro, 2008). In order to reduce DON levels from contaminated grains, various biological strategies have been tested using LAB (Table 3). Juodeikiene et al. (2018) studied various LAB strains including *Latilactobacillus sakei* KTU05-6; *Pediococcus acidilactici* KTU05-7; and *P. pentosaceus* KTU05-8, KTU05-09, and KTU05-10 for a reducing effect on DON levels in wheat grain. *P. pentosaceus* KTU05-10 showed maximum reduction in the DON content with an average removal of 47% (attributed to degradation). The reduction in DON production by *F. graminearum* of up to 56.8% was achieved through incubation with *Lacticaseibacillus paracasei*, while a complete elimination of DON was achieved by limiting fungal growth with *Lacticaseibacillus rhamnosus* (Gomah and Zohri, 2014). The inhibition of the fungal strain *F. graminearum* IAPAR 2218 to various extents was achieved by using LAB strains that were either isolated from wheat and kefir or obtained commercially. Among these strains, a mixture of *L. rhamnosus* and *L. plantarum* showed maximum inhibition of the fungal growth (Franco et al., 2011).

Although LAB cells possess the capability to reduce DON, their binding efficiencies were reported to be lower toward DON as compared to other mycotoxins. The reduction of DON using various LAB cells was shown to be dependent on incubation time where a maximum of 39% reduction in DON using *Limosilactobacillus reuteri* after 24 h of incubation was achieved compared to 20% and 31% DON reduction after 6 h and 12 h incubation, respectively (Chlebicz and Śliżewska, 2020). Furthermore, it was also shown that cells rendered nonviable through heat treatment had a greater DON-reducing capacity (through binding) compared to viable cells. Heat-treated *Lactiplantibacillus plantarum* cells were found to reduce the DON content up to 71% among various strains studied (Franco et al., 2011). The supplementation of pig feed with LAB resulted in alleviation of the toxic effects of DON with respect to lymphocyte proliferation, liver cytokine synthesis, and intracellular oxidative activity of the granulocytes (Marin et al., 2010).

Fumonisins

FBs are also relatively heat-stable and are not affected by baking and canning techniques; however, processes involving higher temperatures such as frying may result in substantial fumonisin decontamination. Hence, the use of LAB strains for decontamination of FBs has been proposed (Table 4) as a suitable option in various food products that are not highly heat-treated (> 175 °C) during processing. Chlebicz and Śliżewska (2020) showed that LAB strains exhibited the highest reduction efficiencies toward FB_1 and FB_2 among the various mycotoxins studied. The use of *Lactiplantibacillus plantarum* MYS6 resulted in a reduction of 33% and 62% after

TABLE 3
Deoxynivalenol Detoxification by Lactic Acid Bacteria (LAB) Strains According to Different Proposed Mechanisms.

Bacterial Strain	Cell Concentration (cfu/mL)	DON (ng/g)	Proposed Mechanism	% Reduction[a]	Reference
Levilactobacillus brevis ŁOCK 1093	NR	100,000	Binding with LAB cells	94	Chlebicz and Śliżewska (2020)
P. pentosaceus KTU05-10	NR	3370–6930	Binding and degradation using LAB suspension in medium on wheat grains	47	Juodeikiene et al. (2018)
Lacticaseibacillus paracasei	10^7	–	Inhibition of DON production by antifungal activity of LAB	57	Gomah and Zohri (2014)
Lacticaseibacillus rhamnosus				100	
Lactiplantibacillus plantarum	10^{10}	1500	Binding and degradation by LAB cells	53	Franco et al. (2011)
			Binding by pasteurized nonviable cells	67	
			Binding by sterilized nonviable cells	71	

DON, deoxynivalenol; *NR*, not reported.
[a]Maximum value reported.

2h and 4h of incubation, respectively, indicating that the process involving viable cells was time-dependent; however, no degradation residues were found during the analysis (Deepthi et al., 2016). Decontamination through the binding of FBs with cell wall components was confirmed using confocal laser scanning microscopy (CLSM). Moreover, also it has seen that nonviable cells of *Lactiplantibacillus plantarum* (R1096), *Lactiplantibacillus plantarum* (FS2), *P. pentosaceus* (D39), and *L. delbrueckii* (CIP 57.8T) were able to reduce significantly higher levels of FB_1 and similar levels of FB_2 as compared to their viable counterparts (Dawlal et al., 2019). The use of nonviable cells of *L. acidophilus* also showed greater FB_1 binding capacities than viable cells with no significant impact of time on the reduction decontamination capabilities of the strain (Pizzolitto et al., 2012). The addition of certain LAB strains to feed can impart a protective effect against fumonisins by reducing their bioavailability in the gastrointestinal tract. The application of *Lactiplantibacillus plantarum* MYS6 to FB_1 intoxicated broilers was shown to mitigate altered levels of hematological indices and restore the serum parameters (Deepthi et al., 2017).

Nevertheless, Martinez Tuppia et al. (2017) showed losses of up to 33% in FB_1 content by incubation with *Levilactobacillus brevis* for 4 days, among which partial losses were attributed to the degradation of FB_1 into HFB_1. The use of LAB strains including *Lactiplantibacillus plantarum*, *P. pentosaceus*, *Lacticaseibacillus paracasei*, and *Limosilactobacillus fermentum* for the fermentation of porridge samples for 24h resulted in reductions of 14% to 30% FB_1 with a lower pH and backslopping favoring higher decontamination. Lactic acid production by LAB strains during fermentation can be considered as a significant factor in the reduction of FB_1 (Nyamete et al., 2016). The use of *Lacticaseibacillus paracasei* BEJ01 showed protective effects against oxidative stress and immunotoxicity induced by FB_1 in mice partly through binding, and consequently reducing FB_1 bioavailability. However, part of the protective effect can be attributed to the degradation of FB_1 as live bacterial cells tended to reduce FB_1 in higher amounts together with increased time as compared to dead cells in which time had no effect on FB_1 reduction (Abbès et al., 2016).

Variation in the observed modes of FB_1 decontamination by different strains showed that these phenomena are strain-specific. Along with the type of strain used, the extent of decontamination depends also on the time of incubation and the pH of the substrate as well (Pizzolitto et al., 2012).

TABLE 4
Fumonisin Detoxification by Lactic Acid Bacteria (LAB) Strains According to Different Proposed Mechanisms.

Bacterial Strain	Cell Concentration (cfu/mL)	FB	Level (ng/g)	Proposed Mechanism	% Reduction[a]	Reference
Lacticaseibacillus rhamnosus LOCK 1088	NR	FB_1+FB_2	100,000	Binding with viable LAB cells	63	Chlebicz and Śliżewska (2020)
Lactiplantibacillus plantarum FS2	10^{13}	FB_1	10,000	Binding with viable LAB cells	57	Dawlal et al. (2019)
Lactiplantibacillus plantarum R1096		FB_2			93	
Lactiplantibacillus plantarum R1096		FB_1 FB_2		Binding with nonviable LAB cells	77 96	
Levilactobacillus brevis N195	NR	FB_1	500	Binding with LAB cells and degradation into HFB_1	33	Martinez Tuppia et al. (2017)
Lactiplantibacillus plantarum MYS6	10^6	FB_1	10,000	Binding and degradation by LAB cells in PBS	62	Deepthi et al. (2016)
P. pentosaceus	10^7	FB_1	8000	Fermentation of maize porridges by LAB	24	Nyamete et al. (2016)
Lactiplantibacillus plantarum					24	
Lacticaseibacillus paracasei BEJ01	10^8	FB_1	50,000	Binding and degradation by viable LAB cells in PBS	89	Abbès et al. (2016)
				Binding with heat-killed LAB cells	25	
L. acidophilus	4.9×10^8	FB_1	15,000	Binding with viable LAB cells	25	Pizzolitto et al. (2012)
				Binding with heat-killed LAB cells	54	

FB, fumonisin; *NR*, not reported.
[a]Maximum value reported.

Ochratoxin A

OTA has been shown to be stable against temperatures of up to 180°C (Raters and Matissek, 2008). There have been several studies related to the elimination of OTA using biological techniques in order to address the shortcomings of physicochemical methods that may cause deterioration in quality or unwanted interactions between chemicals and nutrients of the food product.

Like other mycotoxins, the reduction of OTA by LAB has also been achieved through adsorption as well as through degradation or by inhibition of fungal growth (Table 5). Ghofrani Tabari et al. (2018) achieved 40% and 60% reduction in OTA using viable and dead cells of *Lacticaseibacillus rhamnosus* GG, respectively. Rapid detoxification of OTA by up to 46% to 60% was achieved through binding with heat-inactivated cells of *Lactiplantibacillus plantarum, Levilactobacillus brevis,* and *Fructilactobacillus sanfranciscensis* after a contact time of 30 min as compared to their viable counterparts that showed 15% to 35% OTA reduction after a contact time of 24 h (Piotrowska, 2014). A kefir culture consisting mainly of LAB strains was able to reduce 94% of OTA in milk through biotransformation and adsorption. Upon isolation of strains from the kefir culture, it was found that *Lentilactobacillus kefiri* reduced the highest amounts of OTA in milk (up to 81%), but only through adsorption. Nonetheless, biotransformation was observed in other strains when experiments were performed in MRS broth (Taheur et al., 2017). The binding

TABLE 5
Ochratoxin A Detoxification by Lactic Acid Bacteria (LAB) Strains According to Different Proposed Mechanisms.

Bacterial Strain	Cell Concentration (cfu/mL)	OTA (ng/g)	Proposed Mechanism	% Reduction[a]	Reference
Lacticaseibacillus rhamnosus ŁOCK 1087	10^9	100,000	Degradation and binding with viable LAB cells	48	Markowiak et al. (2019)
Lacticaseibacillus rhamnosus GG	10^9	2000	Binding with acid-killed LAB cells	62	Ghofrani Tabari et al. (2018)
Lactiplantibacillus plantarum	10^9	25	Degradation by LAB through inoculation in medium with L-malic acid	19	Moncalvo et al. (2018)
Lactiplantibacillus plantarum CECT 749	10^8	600	Degradation through hydrolysis of OTA by CFS	97	Luz et al. (2018)
			Adsorption by LAB cells in medium	2	
Lentilactobacillus kefiri	NR	1000	Binding with LAB cells in milk	81	Taheur et al. (2017)
			Biotransformation by LAB activity in milk	0	
Lactiplantibacillus plantarum	10^8	–	Reduction of OTA production by Antifungal LAB activity in medium	99	Fossi et al. (2016)
P. parvulus	10^9	1000	Biodegradation into OTα by LAB in medium	100	Abrunhosa et al. (2014)
Lactiplantibacillus plantarum LOCK 0862	NR	1000	Binding by viable biomass of LAB	26	Piotrowska (2014)
			Binding by dead biomass of LAB	56	
Levilactobacillus brevis LOCK 0845			Binding by viable biomass of LAB	24	
			Binding by dead biomass of LAB	60	
L. bulgaricus	10^4	NR	Inhibition of OTA production by *A. ochraceous*	79	El-Taher et al. (2012)
L. acidophilus NCC 68	10^8	24.9	Reduction in bioaccessibility in an in vitro digestive model	73	Kabak et al. (2009)

OTA, ochratoxin A; *NR*, not reported.
[a]Maximum value reported.

capacities of OTA vary depending upon the strain and the genetic makeup as proposed by Khattab et al. (2018), who tested several LAB strains among which *L. delbrueckii* NRRL B-1024 was found to be most efficient for OTA reduction. Reduction levels were further increased by introducing genetic variations to the strain.

Another study involving the analysis of various LAB strains for OTA elimination showed that reduction ranges from 30 to 99% were achieved with *Lacticaseibacillus rhamnosus* CECT 278T and *Lactiplantibacillus plantarum* CECT 749; the highest binding capacities of 97% and 99% were obtained at pH 3.5 and 6.5, respectively (Luz et al., 2018). However, the reduction was mainly attributed to the hydrolysis of OTA into ochratoxin alpha (OTα) and phenylalanine. Furthermore, bioavailability of OTA was also reduced by 98% using *L. johnsonii* CECT

289 S. The elimination of OTA by *P. parvulus* (isolated from Douro wines) showed that the rate of OTA reduction was dependent upon the size of inoculum and the incubation temperature together with time. Maximum reductions in OTA of 50% and 90% were obtained after 6 h and 19 h of incubation, respectively. Moreover, dead cells of *P. parvulus* showed OTA adsorption capacities of 1.3%, hence confirming the requirement of viable cells for the biodegradation of OTA into OTα (Abrunhosa et al., 2014). Ochratoxin α (OTα) is considered to be nontoxic and has an elimination half-life that is 10-times less than that of OTA (Li et al., 1997). The growth of OTA-producing fungi was inhibited using various LAB strains among which *Lactiplantibacillus plantarum* (isolated from raffia wine) was found to be the most effective. Significant reduction in OTA content was observed when fungi were mixed with LAB strains (Fossi et al., 2016). The inoculation of *A. ochraceus* with LAB showed a reduction of 80% OTA formation compared to the control (El-Taher et al., 2012).

The use of LAB-based additives to reduce OTA levels in animal feed is considered as an effective strategy as demonstrated by Slizewska and Piotrowska (2014), who inoculated probiotic LAB into chicken feed resulting in a 73% reduction of OTA after fermentation for 6 h. The levels of reduction were similar after 12-h and 24-h incubation. In contrast, Markowiak et al. (2019) showed that the concentration of OTA for some LAB strains continued to decrease with an increase in incubation time to various rates. The different rates of decontamination can be attributed to strain differences as well as the media used for inoculation. Furthermore, the intake of various LAB strains, such as *L. acidophilus* NCC 68, reduced the bioavailability of OTA and consequently its harmful impacts on animal health (Kabak et al., 2009; Markowiak et al., 2019).

Zearalenone

ZEN is a chemically stable compound having the melting point of 164–165 °C. However, it is completely eliminated when exposed at temperatures of up to 225 °C for 30 min in an aqueous buffer (Ryu et al., 2003). Due to their thermal stability or only partial degradation during conventional heat treatments, various biological methods have been tested for minimizing the hazards associated with ZEN.

The antimutagenic properties of LAB strains have encouraged researchers to analyze various LAB in order to address the problems of ZEN in food and feed. Decontamination of ZEN through binding with LAB cells has been studied by a number of researchers (Table 6). ZEN reduction of up to 30% through adsorption was observed with *L. lactis* (Rogowska et al., 2019). Eight different *Lactiplantibacillus pentosus* strains were analyzed for ZEN adsorption with reduction percentages dependent on the initial level of ZEN in the medium. The higher the initial ZEN level, the greater the observed reduction, which may be due to nonsaturation of the binding sites with lower ZEN levels. A strain of *Lactiplantibacillus plantarum* (isolated from plant silage) was shown to detoxify 73% of ZEN after 24 h of incubation with maximum binding with the LAB cells occurring during the first 6 h of incubation (Chlebicz and Śliżewska, 2020). Different LAB strains (isolated from pig rectal swabs) also showed varying adsorption capacities toward ZEN with *Lactiplantibacillus plantarum* exhibiting the highest binding of 82% and 73% by non-viable and viable cells, respectively (Vega et al., 2017). Similarly, experiments by Čvek et al. (2012) showed the elimination of almost 99% ZEN through binding with a *Lactiplantibacillus plantarum* strain (isolated from bovine milk cheese). Binding of ZEN with *L. lactis* also showed a similar pattern of rapid binding of 90% ZEN during the initial 20 min. of contact with the cells, followed by a further of 7% reduction over the next 20 h (Król et al., 2018). Yang et al. (2019) also reported ZEN binding by LAB to be an instant very rapid process; lower reductions were reported for acid-killed cells of *P. pentosaceus* compared to the heat-killed cells, but higher than the viable cells. Zhao et al. (2015) also showed that heat-treated LAB cells were able to bind more ZEN than viable cells.

Partial elimination of ZEN was achieved through the treatment of wheat grains with the suspension of various LAB strains in MRS, with *P. acidilactici* being the most effective. However, the mechanism was not clear and was proposed to be either degradation or binding (Juodeikiene et al., 2018). Similarly, ZEN detoxification capacities by *Lactiplantibacillus plantarum* strains were found to be greater when incubated in MRS medium as compared to when incubated in PBS. However, the amount of ZEN bound by cells did not differ significantly, although the residual ZEN in MRS medium was lower than in PBS, indicating that the reduction was through some other mechanism in the MRS medium (Zhao et al., 2015). The involvement of both binding and degradation mechanisms in the elimination of ZEN by various LAB was confirmed by Chen et al. (2018). Furthermore, Taheur et al. (2017) suggested that the exposure medium was also critical in determining the extent of the adsorption or degradation. This study revealed the elimination of ZEN through both mechanisms in the culture medium, but losses through degradation did not occur when the detoxification was performed in milk. The presence of various agents that impart stress to the cells may also alter the ZEN detoxification capacities as proposed by Rogowska

TABLE 6
Zearalenone Detoxification by Lactic Acid Bacteria (LAB) Strains According to Different Proposed Mechanisms.

Bacterial Strain	Cell Concentration (cfu/mL)	ZEN (ng/g)	Proposed Mechanism	% Reduction[a]	Reference
L. lactis	10^8	800	Binding with LAB cells	30	Rogowska et al. (2019)
Limosilactobacillus reuteri LOCK 1096	NR	100,000	Binding with LAB cells	72	Chlebicz and Śliżewska (2020)
Lacticaseibacillus paracasei	1.8×10^9	2000	Binding and degradation in culture medium added with silver nanoparticles	55	Rogowska et al. (2019)
P. pentosaceus xy46	NR	4000	Binding with viable cells with exposure to ZEA for 15–30 min after 24 h incubation in medium	60	Yang et al. (2019)
			Binding with viable cells with ZEA co-cultivation at the initiation of incubation for 24 h	56	
			Binding with heat-killed cells	94	
Lactiplantibacillus plantarum	10^9	5000	Degradation	45	Chen et al. (2018)
			Binding with LAB cells	7	
Lentilactobacillus kefiri	NR	1000	Binding with LAB cells	100	Taheur et al. (2017)
			Biotransformation by LAB activity in milk	7	
Lactiplantibacillus plantarum	NR	12,300	Binding by LAB cells	64	Vega et al. (2017)
Lactiplantibacillus pentosus JM0812	NR	74,700	Binding with LAB cells	83	Sangsila et al. (2016)
Lactiplantibacillus plantarum Lp22	10^9	100	Binding with LAB cells	47	Zhao et al. (2015)
Lacticaseibacillus paracasei	10^7	NR	Inhibition of ZEA production by LAB	73	Gomah and Zohri (2014)
Lacticaseibacillus rhamnosus				100	

ZEN, zearalenone; *NR*, not reported.
[a]Maximum value reported.

et al. (2019), who showed an increased reduction of ZEN by *Lacticaseibacillus paracasei* and *L. lactis* in the presence of silver nanoparticles; ZEN was metabolized into α- and β-zearalenol (ZOL), with β-ZOL being the predominant form. The α-form is known to be 10 times more toxic than the β-form that has very low toxicity compared to ZEN itself (Oliveira et al., 2014).

Moreover, the reduction of ZEN through the inhibition of fungal growth was achieved through the utilization of two LAB strains among which *Lacticaseibacillus paracasei* was found to reduce the ZEN by 73% of the original amount, while *Lacticaseibacillus rhamnosus* completely prevented the ZEN production by completely suppressing the mold growth (Gomah and

Zohri, 2014). Furthermore, studies have been carried out showing the in vivo effectiveness of LAB strains for the alleviation of the impact of ZEN (Belgacem et al., 2019; Chang et al., 2020).

MECHANISMS INVOLVED IN THE MICROBIOLOGICAL DECONTAMINATION OF MYCOTOXINS

Detoxification of mycotoxins by LAB has been under focus for some time as it is considered to be a safe and effective strategy for the reduction of toxins in food and feed. However, there is still a lot of debate about the mechanisms involved through which LAB decontaminate mycotoxins. Various mechanisms that may contribute toward the reduction of the mycotoxin content include binding of the mycotoxin to the cell wall components of LAB cells, degradation of the mycotoxins into less toxic substances by the LAB or inhibition of mold growth through competition.

Binding With Cell Wall Components

Many studies have been conducted in order to determine the binding abilities of various LAB strains toward mycotoxins. The mycotoxin binding process is dependent upon the initial levels of mycotoxin present in the medium as well as the cell concentration and their state. The possibility for the elimination of mycotoxins through binding was confirmed for heat- and acid-treated nonviable strains that reduced aflatoxin levels even more effectively than compared to viable cells (Yang et al., 2019). As the binding of mycotoxins with LAB can be achieved through viable as well as nonviable cells (Dawlal et al., 2019), it may be possible to use LAB cells in a variety of applications where the deterioration of food through bacterial activity is a concern.

The LAB cell wall consists of the usual Gram-positive structure, which is made up of a thick, multilayered peptidoglycan sacculus in which proteins, teichoic acids, lipoteichoic acids, and polysaccharides are present. Haskard et al. (2001) proposed the binding of aflatoxins to bacterial cell wall through noncovalent interactions. Though the binding mechanism of these strains is not fully understood yet, some authors have proposed that the polysaccharides and the peptidoglycans of the bacterial cell wall may be the major binding sites. This mechanism was proposed based on the lack of binding abilities after treatment of the cells with periodate and pronase E that nonspecifically degrade carbohydrates and proteins, respectively (Lahtinen et al., 2004). The involvement of teichoic acids present in the cell wall of the bacterial strains in the binding of AFB_1 was also proposed by Hernandez-Mendoza et al. (2009). Landersjö et al. (2002) proposed that the high AF-reducing efficiency of *Lacticaseibacillus rhamnosus* was due to the production of exopolysaccharides (EPS) by these strains in the surroundings that also get loosely attached to the bacterial cell and consequently act as a binding site for AF attachment. In contrast, experiments conducted by Lahtinen et al. (2004) revealed that the role of EPS was negligible, with the majority of the binding sites residing on the bacterial cell wall that consists mainly of peptidoglycan. This finding was further confirmed by some shreds of evidence showing that the binding sites of FBs were not surface polysaccharides, lipids, or proteins, but rather the PG or compounds tightly associated with it. The amino acid sequence for peptide linkages of PG varies with bacterial species, hence resulting in different binding capacities for different strains. In addition, the integrity of the cell wall also plays an important factor in the binding of mycotoxins (Hernandez-Mendoza et al., 2009). The presence of metal ions in the medium may also influence the mycotoxin binding efficiencies of LAB strains (Lahtinen et al., 2004). Furthermore, the interaction of mycotoxins with LAB cells is proposed to involve hydrophobic interactions and hence binding depends on the type of mycotoxin and the surface hydrophobicity of the cells (Piotrowska, 2014).

The decontamination of mycotoxins through binding occurs very rapidly and may increase with time in some cases (Piotrowska, 2014; Pizzolitto et al., 2012). However, some studies report the binding to be partially reversible with the release of variable amounts of bound mycotoxin back into the medium occurring depending upon the LAB strain present (Haskard et al., 2001). Reversible binding indicates the presence of weak noncovalent interactions between the mycotoxin and the bacterial cell wall. The presence of cross-linkages between two adjacent cell binding surfaces through mycotoxins has also been proposed as a reason for the release of low amounts of the toxin into the medium (Hernandez-Mendoza et al., 2009).

Biodegradation Into Less Toxic Substances

In addition to binding, the mycotoxins can be degraded into less toxic substances by LAB strains, consequently making the food product safer for consumption. The biodegradation mechanisms are different for different mycotoxins. The reduction in AFM_1 through biodegradation into less toxic substances using metabolites present in cell-free supernatant of LAB strains was shown by Martínez et al. (2019). The degradation of AFM_1 during fermentation is mainly attributed to several by-products such as volatile fatty acids, amino acids, peptides, or aldehydes. Similarly, the degradation of FB_1 into hydrolyzed FB_1 (HFB_1) through LAB activity was found to involve a decarboxylation reaction that cleaves the two tricarballylic acid (TCA) moieties at the C14 and C15 of the main chain. HFB_1 is a less toxic product of FB_1cdegradation (Grenier et al., 2012). The degradation of OTA to nontoxic or less toxic OTα and phenylalanine occurs via the hydrolysis of the amide bond by LAB and is considered the most important OTA biodegradation mechanism (Luz et al., 2018). *P. parvulus* strains have been shown to eliminate 100% of OTA through conversion into OTα after 7 days of incubation in MRS media containing 1 µg of OTA/mL at 30°C (Abrunhosa et al., 2014). The extent of degradation of various mycotoxins by LAB depends on the incubation time and the specific LAB strain used. All LAB strains do not demonstrate degradation capabilities, while for some strains both degradation and binding may occur concurrently as shown by Taheur et al. (2020), who observed various levels of biodegradation and adsorption of AFB_1 in MRS media and black tea by LAB strains (isolated from Kombucha). Another study revealed the production of lactic acid by LAB during incubation to be an important factor for mycotoxin reduction (Nyamete et al., 2016).

Antifungal Activities of LAB

Food and feed can also be made safe by inhibiting the fungal growth and consequently preventing or reducing the production of various mycotoxins. Various LAB strains have been proven to be effective in this regard as they are known to produce antimicrobial agents including organic acids, carbon dioxide, hydrogen peroxides, and bacteriocins. A number of LAB strains have been found to be effective against the growth of various mycotoxigenic fungi including *A. flavus*, *A. ochraceus*, and *A. niger* even after several days of incubation. Gomah and Zohri (2014) showed that *Lacticaseibacillus rhamnosus* was able to completely suppress the mycelium growth of *F. graminearum*, *F. culmorum*, and *F. proliferation*. Deepthi et al. (2016) reported that the incubation of *F. proliferatum* with *Lactiplantibacillus plantarum* MYS6 resulted in a disruption of hyphae structures and the production of less fungal biomass as compared to a control. Various antifungal metabolites including long-chain fatty acids and their esters were suggested to be important for this observation. A significant reduction in OTA was observed due to the inhibition of fungal growth using various LAB strains, among which *Lactiplantibacillus plantarum* exhibited the highest effect. Furthermore, *Lactiplantibacillus plantarum* is known to synthesize phenolic acids, which induce cellular stress with the consequent lysis of the plasma membrane. Other than secondary metabolites of LAB, several factors such as nutritional competition and pH are also considered to play an important role in the inhibition of fungal growth.

Recently, a *Lactiplantibacillus plantarum* strain (CECT 749) was studied by Nazareth et al. (2020) for the inhibition of AFB_1 production in corn by the antifungal activity of its cell-free supernatant (CFS) against *A. niger* resulting in a reduction of 99.7% AFB_1 compared to the control. The antifungal activities of cell-free supernatants of LAB can be attributed to the production of phenolic compounds. A strain of *Lactiplantibacillus plantarum* was also found to be a potential biocontrol agent against the growth of *A. flavus* and AF biosynthesis in various media (Kachouri et al., 2014). Other strains including *Bifidobacterium bifidum*, *Limosilactobacillus fermentum*, and *L. delbrueckii* have also been shown to be effective against AF production by *A. parasiticus* (Ghazvini et al., 2016).

CONCLUDING REMARKS AND FUTURE PERSPECTIVES

Biological methods based on the removal or the degradation of mycotoxins by LAB have a good future potential, with this approach highly desirable in terms of consumer acceptability. At present, a number of LAB cultures (mainly mesophilic *Lactobacillus* species) are specifically marketed for the inhibition of mold (and yeast) growth in fermented milk such as yogurt. As such, these LAB cultures are added with the intention of shelf life extension of the product rather than combating the mycotoxin issue. One of the main barriers in developing commercial LAB products to control mycotoxins lies in the challenge of targeting a broad range of mycotoxins-producing fungi with a small number of LAB strains. Furthermore, the inoculation levels of such LAB cultures are likely to be high, resulting in significant additional costs to the food industry. Nevertheless, further research is necessary to understand the detoxification mechanisms of these microorganisms with a view to optimization and future commercialization. In addition, development work is required in order to determine the more appropriate practical aspects of the use of such LAB cultures in food products and in relation to their possible sensory impact on the food.

ACKNOWLEDGEMENTS

The authors thank Fundação de Amparo à Pesquisa do Estado de São Paulo (FAPESP) for the financial support (Grant # 2019/21603-1).

REFERENCES

Abbès, S., Salah-Abbès, J.B., Sharafi, H., Jebali, R., Noghabi, K.A., Oueslati, R., 2013. Ability of Lactobacillus rhamnosus GAF01 to remove AFM1 in vitro and to counteract AFM1 immunotoxicity in vivo. J. Immunotoxicol. 10, 279–286.

Abbès, S., Salah-Abbes, J.B., Jebali, R., Younes, R.B., Oueslati, R., 2016. Interaction of aflatoxin B1 and fumonisin B1 in mice causes immunotoxicity and oxidative stress: possible protective role using lactic acid bacteria. J. Immunotoxicol. 13, 46–54.

Abrunhosa, L., Inês, A., Rodrigues, A.I., Guimarães, A., Pereira, V.L., Parpot, P., Mendes-Faia, A., Venâncio, A., 2014. Biodegradation of ochratoxin A by Pediococcus parvulus isolated from Douro wines. Int. J. Food Microbiol. 188, 45–52.

Afshar, P., Shokrzadeh, M., Raeisi, S.M., Saraei, G.G.-H., Nasiraii, L.R., 2020. Aflatoxins biodetoxification strategies based on probiotic bacteria. Toxicon 178, 50–58.

Agriopoulou, S., Stamatelopoulou, E., Varzakas, T., 2020. Advances in occurrence, importance, and mycotoxin control strategies: prevention and detoxification in foods. Foods 9, 1–48.

Alvarez-Sieiro, P., Montalbán-López, M., Mu, D., Kuipers, O.P., 2016. Bacteriocins of lactic acid bacteria: extending the family. Appl. Microbiol. Biotechol. 100, 2939–2951.

Arena, M.P., Capozzi, V., Russo, P., Drider, D., Spano, G., Fiocco, D., 2018. Immunobiosis and probiosis: antimicrobial activity of lactic acid bacteria with a focus on their antiviral and antifungal properties. Appl. Microbiol. Biotechnol. 102, 9949–9958.

Assaf, J.C., Khoury, A.E., Chokr, A., Louka, N., Atoui, A., 2019. A novel method for elimination of aflatoxin M1 in milk using Lactobacillus rhamnosus GG biofilm. Int. J. Dairy Technol. 72, 248–256.

Bartkiene, E., Zavistanaviciute, P., Lele, V., Ruzauskas, M., Bartkevics, V., Bernatoniene, J., Gallo, P., Tenore, G.C., Santini, A., 2018. Lactobacillus plantarum LUHS135 and paracasei LUHS244 as functional starter cultures for the food fermentation industry: characterization, mycotoxin-reducing properties, optimization of biomass growth and sustainable encapsulation by using dairy by-products. LWT Food Sci. Technol. 93, 649–658.

Belgacem, H., Salah-Abbès, J.B., Ezzdini, K., Abdel-Wahhab, M.A., Zinedine, A., Abbès, S., 2019. Lactobacillus plantarum MON03 counteracts zearalenone genotoxicity in mice: chromosome aberrations, micronuclei, DNA fragmentation and apoptotic gene expression. Mutat. Res. Genet. Toxicol. Environ. Mutagen. 840, 11–19.

Bovo, F., Corassin, C.H., Rosim, R.E., Oliveira, C.A.F., 2013. Efficiency of lactic acid bacteria strains for decontamination of aflatoxin M1 in phosphate buffer saline solution and in skimmed milk. Food Bioprocess Technol. 6, 2230–2234.

Bovo, F., Franco, L.T., Rosim, R.E., Oliveira, C.A.F., 2014a. Ability of a Lactobacillus rhamnosus strain cultured in milk whey based medium to bind aflatoxin B1. Food Sci. Technol. 34, 566–570.

Bovo, F., Franco, L.T., Rosim, R.E., Trindade, C.S.F., Oliveira, C.A.F., 2014b. The ability of Lactobacillus rhamnosus in solution, spray-dried or lyophilized to bind aflatoxin B1. J. Food Res. 3, 35–42.

Byakika, S., Mukisa, I.M., Wacoo, A.P., Kort, R., Byaruhanga, Y.B., Muyanja, C., 2019. Potential application of lactic acid starters in the reduction of aflatoxin contamination in fermented sorghum-millet beverages. Int. J. Food Contam. 6, 1–8.

Chang, J., Wang, T., Wang, P., Yin, Q., Liu, C., Zhu, Q., Lu, F., Gao, T., 2020. Compound probiotics alleviating aflatoxin B1 and zearalenone toxic effects on broiler production performance and gut microbiota. Ecotoxicol. Environ. Saf. 194, 110420.

Chen, S.W., Hsu, J.T., Chou, Y.A., Wang, H.T., 2018. The application of digestive tract lactic acid bacteria with high esterase activity for zearalenone detoxification. J. Sci. Food Agric. 98, 3870–3879.

Chlebicz, A., Śliżewska, K., 2020. In vitro detoxification of aflatoxin B1, deoxynivalenol, fumonisins, T-2 toxin and zearalenone by probiotic bacteria from genus Lactobacillus and Saccharomyces cerevisiae yeast. Probiotics Antimicrob. Proteins 12, 289–301.

Coppa, C.F.S.C., Mousavi Khaneghah, A., Alvito, P., Assunção, R., Martins, C., Es', I., Gonçalves, B.L., Neeff, D.V., Sant'Ana, A.S., Corassin, C.H., Oliveira, C.A.F., 2019. The occurrence of mycotoxins in breast milk, fruit products and cereal-based infant formula: a review. Trends Food Sci. Technol. 92, 81–93.

Corassin, C.H., Bovo, F., Rosim, R.E., Oliveira, C.A.F., 2013. Efficiency of Saccharomyces cerevisiae and lactic acid bacteria strains to bind aflatoxin M1 in UHT skim milk. Food Control 31, 80–83.

Čvek, D., Markov, K., Frece, J., Friganović, M., Duraković, L., Delaš, F., 2012. Adhesion of zearalenone to the surface of lactic acid bacteria cells. Hrvat. čas za prehrambenu tehnol. biotehnol. nutr. 7, 49–52.

Czáran, T., Rattray, F.P., Møller, C.O.A., Christensen, B.B., 2018. Modelling the influence of metabolite diffusion on nonstarter lactic acid bacteria growth in ripening Cheddar cheese. Int. Dairy J. 80, 35–45.

Dawlal, P., Brabet, C., Thantsha, M.S., Buys, E.M., 2019. Visualisation and quantification of fumonisins bound by lactic acid bacteria isolates from traditional African maize-based fermented cereals, ogi and mahewu. Food Addit. Contam. A 36, 296–307.

Deepthi, B., Rao, K.P., Chennapa, G., Naik, M., Chandrashekara, K., Sreenivasa, M., 2016. Antifungal attributes of Lactobacillus plantarum MYS6 against fumonisin producing Fusarium proliferatum associated with poultry feeds. PLoS One 11, e0155122.

Deepthi, B., Somashekaraiah, R., Poornachandra Rao, K., Deepa, N., Dharanesha, N., Girish, K., Sreenivasa, M., 2017. Lactobacillus plantarum MYS6 ameliorates fumonisin B1-induced hepatorenal damage in broilers. Front. Microbiol. 8, 2317.

Elsanhoty, R.M., Salam, S.A., Ramadan, M.F., Badr, F.H., 2014. Detoxification of aflatoxin M1 in yoghurt using probiotics and lactic acid bacteria. Food Control 43, 129–134.

El-Taher, E.M., El-Ghany, A., Alawlaqi, M., Ashour, M.S., 2012. Biosecurity for reducing ochratoxin A productivity and their impact on germination and ultra-structures of germinated wheat grains. J. Microbiol. Biotechnol. Food Sci. 2, 135.

Food and Agriculture Organization/World Health Organization, 2006. Health and Nutritional Properties and Guidelines for Evaluation. FAO/WHO, Rome. FAO Food and Nutrition Paper 85.

Fossi, B.T., Ngah, B.G., Nchanji, G.T., Wanji, S., Ndjouenkeu, R., 2016. Lactobacilli cultures against ochratoxin A producing moulds isolated from cocoa in the south West Region of Cameroon. Microbiol. Res. J. Int. 15, 1–16.

Franco, T., Garcia, S., Hirooka, E., Ono, Y., Santos, J., 2011. Lactic acid bacteria in the inhibition of Fusarium graminearum and deoxynivalenol detoxification. J. Appl. Microbiol. 111, 739–748.

Ghaffar, T., Irshad, M., Anwar, Z., Aqil, T., Zulifqar, Z., Tarik, A., Kamran, M., Ehsa, N., Mehmood, S., 2014. Recent trends in lactic acid biotechnology: a brief review on production to purification. J. Radiat. Res. Appl. Sci. 7, 222–229.

Ghazvini, R.D., Kouhsari, E., Zibafar, E., Hashemi, S.J., Amini, A., Niknejad, F., 2016. Antifungal activity and aflatoxin degradation of Bifidobacterium bifidum and Lactobacillus fermentum against toxigenic Aspergillus parasiticus. Open Microbiol. J. 10, 197–201.

Ghofrani Tabari, D., Kermanshahi, H., Golian, A., Majidzadeh Heravi, R., 2018. In vitro binding potentials of bentonite, yeast cell wall and lactic acid bacteria for aflatoxin B1 and ochratoxin A. Iran. J. Toxicol. 12, 7–13.

Gomah, N., Zohri, A., 2014. Inhibition of fungal growth and Fusarium toxins by selected cultures of lactic acid bacteria. J. Microb. Biochem. Technol. S7, 1–7.

Grenier, B., Bracarense, A.P.F., Schwartz, H.E., Trumel, C., Cossalter, A.-M., Schatzmayr, G., Kolf-Clauw, M., Moll, W.-D., Oswald, I.P., 2012. The low intestinal and hepatic toxicity of hydrolyzed fumonisin B1 correlates with its inability to alter the metabolism of sphingolipids. Biochem. Pharmacol. 83, 1465–1473.

Haskard, C.A., El-Nezami, H.S., Kankaanpää, P.E., Salminen, S., Ahokas, J.T., 2001. Surface binding of aflatoxin B1 by lactic acid bacteria. Appl. Environ. Microbiol. 67, 3086–3091.

Hernandez-Mendoza, A., Guzman-de-Peña, D., Garcia, H., 2009. Key role of teichoic acids on aflatoxin B1 binding by probiotic bacteria. J. Appl. Microbiol. 107, 395–403.

Holzapfel, W.H., Wood, B.J.B., 2014. Lactic Acid Bacteria: Biodiversity and Taxonomy. John Wiley & Sons Ltd, Chichester, UK.

Huang, L., Duan, C., Zhao, Y., Gao, L., Niu, C., Xu, J., Li, S., 2017. Reduction of aflatoxin B1 toxicity by Lactobacillus plantarum C88: a potential probiotic strain isolated from Chinese traditional fermented food "tofu". PLoS One 12, e0170109.

Huang, W., Chang, J., Wang, P., Liu, C., Yin, Q., Zhu, Q., Lu, F., Gao, T., 2018. Effect of the combined compound probiotics with mycotoxin-degradation enzyme on detoxifying aflatoxin B1 and zearalenone. J. Toxicol. Sci. 43, 377–385.

Huang, J., Zhang, W., Hu, Z., Liu, Z., Du, T., Dai, Y., Xiong, T., 2020. Isolation, characterization and selection of potential probiotic lactic acid bacteria from feces of wild boar, native pig and commercial pig. Livest. Sci. 237, 1–10.

Ismail, A., Gonçalves, B.L., Neeff, D.V., Ponzilacqua, B., Coppa, C.F.S.C., Hintzsche, H., Sajid, M., Cruz, A.G., Corassin, C.H., Oliveira, C.A.F., 2018. Aflatoxin in foodstuffs: occurrence and recent advances in decontamination. Food Res. Int. 113, 74–85.

Juodeikiene, G., Bartkiene, E., Cernauskas, D., Cizeikiene, D., Zadeike, D., Lele, V., Bartkevics, V., 2018. Antifungal activity of lactic acid bacteria and their application for Fusarium mycotoxin reduction in malting wheat grains. LWT Food Sci. Technol. 89, 307–314.

Kabak, B., Brandon, E.F., Var, I., Blokland, M., Sips, A.J., 2009. Effects of probiotic bacteria on the bioaccessibility of aflatoxin B1 and ochratoxin A using an in vitro digestion model under fed conditions. J. Environ. Sci. Health B 44, 472–480.

Kachouri, F., Ksontini, H., Hamdi, M., 2014. Removal of aflatoxin B1 and inhibition of Aspergillus flavus growth by the use of Lactobacillus plantarum on olives. J. Food Prot. 77, 1760–1767.

Kashket, E.R., 1987. Bioenergetics of lactic acid bacteria: cytoplasmic pH and osmotolerance. FEMS Microbiol. Rev. 46, 233–244.

Kasmani, F., Karimi Torshizi, M.A., Allameh, A.A., Shariatmadari, F., 2012. Aflatoxin detoxification potential of lactic acid bacteria isolated from Iranian poultry. Iran. J. Vet. Res. 13, 152–155.

Khattab, A., Ibrahim, M., El-Kady, A., 2018. Ochratoxin A biosorption onto genetically improved of Lactobacillus delbrueckii mutants. Int. Food Res. J. 25, 515–522.

Konings, W.N., 2002. The cell membrane and the struggle for life of lactic acid bacteria. Antonie Van Leeuwenhoek 82, 3–27.

Król, A., Pomastowski, P., Rafińska, K., Railean-Plugaru, V., Walczak, J., Buszewski, B., 2018. Microbiology neutralization of zearalenone using Lactococcus lactis and Bifidobacterium sp. Anal. Bioanal. Chem. 410, 943–952.

Kushiro, M., 2008. Effects of milling and cooking processes on the deoxynivalenol content in wheat. Int. J. Mol. Sci. 9, 2127–2145.

Lahtinen, S., Haskard, C., Ouwehand, A., Salminen, S., Ahokas, J., 2004. Binding of aflatoxin B1 to cell wall components of Lactobacillus rhamnosus strain GG. Food Addit. Contam. 21, 158–164.

Landersjö, C., Yang, Z., Huttunen, E., Widmalm, G., 2002. Structural studies of the exopolysaccharide produced by Lactobacillus rhamnosus strain GG (ATCC 53103). Biomacromolecules 3, 880–884.

Li, S., Marquardt, R., Frohlich, A., Vitti, T., Crow, G., 1997. Pharmacokinetics of ochratoxin A and its metabolites in rats. Toxicol. Appl. Pharmacol. 145, 82–90.

Liu, N., Wang, J., Deng, Q., Gu, K., Wang, J., 2018. Detoxification of aflatoxin B1 by lactic acid bacteria and hydrated sodium calcium aluminosilicate in broiler chickens. Livest. Sci. 208, 28–32.

Luz, C., Ferrer, J., Mañes, J., Meca, G., 2018. Toxicity reduction of ochratoxin A by lactic acid bacteria. Food Chem. Toxicol. 112, 60–66.

Marin, D.E., Taranu, I., Motiu, M., Manda, G., 2010. Effect of Lactobacillus feed supplement in deoxynivalenol intoxicated piglets. Arch. Zootech. 13, 12–22.

Markowiak, P., Śliżewska, K., Nowak, A., Chlebicz, A., Żbikowski, A., Pawłowski, K., Szeleszczuk, P., 2019. Probiotic microorganisms detoxify ochratoxin A in both a chicken liver cell line and chickens. J. Sci. Food Agric. 99, 4309–4318.

Martinez Tuppia, C., Atanasova-Penichon, V., Chereau, S., Ferrer, N., Marchegay, G., Savoie, J.M., Richard-Forget, F., 2017. Yeast and bacteria from ensiled high moisture maize grains as potential mitigation agents of fumonisin B1. J. Sci. Food Agric. 97, 2443–2452.

Martínez, M., Magnoli, A., Pereyra, M.G., Cavaglieri, L., 2019. Probiotic bacteria and yeasts adsorb aflatoxin M1 in milk and degrade it to less toxic AFM1-metabolites. Toxicon 172, 1–7.

Meca, G., Ritieni, A., Mañes, J., 2012. Reduction in vitro of the minor Fusarium mycotoxin beauvericin employing different strains of probiotic bacteria. Food Control 28, 435–440.

Mohapatra, A.R., Jeevaratnam, K., 2019. Inhibiting bacterial colonization on catheters: antibacterial and antibiofilm activities of bacteriocins from Lactobacillus plantarum SJ33. J. Global Antimicrob. Resist. 19, 85–92.

Møller, C.O.A., Ücok, E.F., Rattray, F.P., 2020. Histamine forming behavior of bacterial isolates from aged cheese. Food Res. Int. 128, 108719.

Moncalvo, A., Dordoni, R., Silva, A., Fumi, M., Di Piazza, S., Spigno, G., 2018. Ochratoxin A removal by Lactobacillus plantarum V22 in synthetic substrates. Open Biotechnol. J. 12, 282–287.

Montaseri, H., Arjmandtalab, S., Dehghanzadeh, G., Karami, S., Razmjoo, M., Sayadi, M., Oryan, A., 2014. Effect of production and storage of probiotic yogurt on aflatoxin M1 residue. J. Food Qual. Hazards Control 1, 7–14.

Nazareth, T.D.M., Luz, C., Torrijos, R., Quiles, J.M., Luciano, F.B., Mañes, J., Meca, G., 2020. Potential application of lactic acid bacteria to reduce aflatoxin B1 and fumonisin B1 occurrence on corn kernels and corn ears. Toxins 12, 21.

Nyamete, F., Mourice, B., Mugula, J., 2016. Fumonisin B 1 reduction in lactic acid bacteria fermentation of maize porridges. Tanzania J. Agric. Sci. 15, 13–20.

Oliveira, C.A.F., Corassin, C.H., Corrêa, B., Oswald, I.P., 2014. Animal health: mycotoxins. In: Van Alfen, N. (Ed.), Encyclopedia of Agriculture and Food Systems, second ed. Elsevier, Oxford, UK, pp. 358–377.

Oluwafemi, F., Silva, F., 2009. Removal of aflatoxins by viable and heat-killed Lactobacillus species isolated from fermented maize. J. Appl. Biosci. 16, 871–876.

Oluwafemi, F., Kumar, M., Bandyopadhyay, R., Ogunbanwo, T., Ayanwande, K.B., 2010. Bio-detoxification of aflatoxin B1 in artificially contaminated maize grains using lactic acid bacteria. Toxin Rev. 29, 115–122.

Panwar, R., Kumar, N., Kashyap, V., Ram, C., Kapila, R., 2019. Aflatoxin M1 detoxification ability of probiotic lactobacilli of Indian origin in in vitro digestion model. Probiotics Antimicrob. Proteins 11, 460–469.

Piotrowska, M., 2014. The adsorption of ochratoxin A by Lactobacillus species. Toxins 6, 2826–2839.

Pizzolitto, R.P., Salvano, M.A., Dalcero, A.M., 2012. Analysis of fumonisin B1 removal by microorganisms in co-occurrence with aflatoxin B1 and the nature of the binding process. Int. J. Food Microbiol. 156, 214–221.

Raters, M., Matissek, R., 2008. Thermal stability of aflatoxin B1 and ochratoxin A. Mycotoxin Res. 24, 130–134.

Rezac, S., Kok, C.R., Heermann, M., Hutkins, R., 2018. Fermented foods as a dietary source of live organisms. Front. Microbiol. 9, 1785.

Rixon, J.E., Warner, P.J., 2003. Introduction: background, relevant genetic techniques, and terms. In: Wood, B.J.B., Warner, P.J. (Eds.), Genetics of Lactic Acid Bacteria. Springer, New York, pp. 1–24.

Rogowska, A., Pomastowski, P., Walczak, J., Railean-Plugaru, V., Rudnicka, J., Buszewski, B., 2019. Investigation of zearalenone adsorption and biotransformation by microorganisms cultured under cellular stress conditions. Toxins 11, 463.

Rolim, F.R.L., Freitas Neto, O.C., Oliveira, M.E.G., Oliveira, C.J.B., Queiroga, R.C.R.E., 2020. Cheeses as food matrices for probiotics: in vitro and in vivo tests. Trends Food Sci. Technol. 100, 138–154.

Ryu, D., Hanna, M.A., Eskridge, K.M., Bullerman, L.B., 2003. Heat stability of zearalenone in an aqueous buffered model system. J. Agric. Food Chem. 51, 1746–1748.

Sadeghi, A.R., Ebrahimi, M., Sadeghi, B., 2016. Effect of isolated Lactobacillus acidophilus and Lactobacillus brevis on growth of Aspergillus flavus and reduction of aflatoxin B1. J. Rafsanjan Univ. Med. Sci. 15, 3–16.

Sadiq, F.A., Yan, B., Tian, F., Zhao, J., Zhang, H., Chen, W., 2019. Lactic acid bacteria as antifungal and anti-mycotoxigenic agents: a comprehensive review. Compr. Rev. Food Sci. Food Saf. 18, 1403–1436.

Sangsila, A., Faucet-Marquis, V., Pfohl-Leszkowicz, A., Itsaranuwat, P., 2016. Detoxification of zearalenone by Lactobacillus pentosaceus strains. Food Control 62, 187–192.

Santer, M., 2010. Joseph Lister: first use of a bacterium as a 'model organism' to illustrate the cause of infectious disease of humans. Notes Rec. R. Soc. Lond. 64, 59–65.

Sarlak, Z., Rouhi, M., Mohammadi, R., Khaksar, R., Mortazavian, A.M., Sohrabvandi, S., Garavand, F., 2017. Probiotic biological strategies to decontaminate aflatoxin M1 in a traditional Iranian fermented milk drink (Doogh). Food Control 71, 152–159.

Sezer, Ç., Güven, A., Bilge Oral, N.B., Vatansever, L., 2013. Detoxification of aflatoxin B1 by bacteriocins and bacteriocinogenic lactic acid bacteria. Turk. J. Vet. Anim. Sci. 37, 594–601.

Slizewska, K., Piotrowska, M., 2014. Reduction of ochratoxin A in chicken feed using probiotic. Ann. Agric. Environ. Med. 21, 676–680.

Taheur, F.B., Fedhila, K., Chaieb, K., Kouidhi, B., Bakhrouf, A., Abrunhosa, L., 2017. Adsorption of aflatoxin B1, zearalenone and ochratoxin A by microorganisms isolated from kefir grains. Int. J. Food Microbiol. 251, 1–7.

Taheur, F.B., Mansour, C., Jeddou, K.B., Machreki, Y., Kouidhi, B., Abdulhakim, J.A., Chaieb, K., 2020. Aflatoxin B1 degradation by microorganisms isolated from Kombucha culture. Toxicon 179, 76–83.

Tokatlı, M., Gülgör, G., Elmacı, S.B., İşleyen, N.A., Özçelik, F., 2015. In vitro properties of potential probiotic indigenous lactic acid bacteria originating from traditional pickles. Biomed. Res. Int. 315819, 1–8.

Vega, M.F., Dieguez, S.N., Riccio, B., Aranguren, S., Giordano, A., Denzoin, L., Soraci, A.L., Tapia, M.O., Ross, R., Apás, A., 2017. Zearalenone adsorption capacity of lactic acid bacteria isolated from pigs. Braz. J. Microbiol. 48, 715–723.

Wacoo, A.P., Mukisa, I.M., Meeme, R., Byakika, S., Wendiro, D., Sybesma, W., Kort, R., 2019. Probiotic enrichment and reduction of aflatoxins in a traditional African maize-based fermented food. Nutrients 11, 265.

Yang, S., Gong, P., Pan, J., Wang, N., Tong, J., Wang, M., Long, M., Li, P., He, J., 2019. Pediococcus pentosaceus xy46 can absorb zearalenone and alleviate its toxicity to the reproductive systems of male mice. Microorganisms 7, 266.

Yousefi, M., Hosseini, H., 2017. Detoxification of AFM1 in Milk and Milk products. In: Mortazavian, A.M., Khorshidian, N. (Eds.), Aflatoxin M1 in Milk and Milk Products. Lambert Academic Publishing, Saarbrücken, Germany, pp. 72–103.

Zendeboodi, F., Khorshidian, N., Mortazavian, A.M., Cruz, A.G., 2020. Probiotic: conceptualization from a new approach. Curr. Opin. Food Sci. 32, 103–123.

Zhang, G., Li, J., Lv, J., Liu, L., Li, C., Liu, L., 2019. Decontamination of aflatoxin M1 in yogurt using Lactobacillus rhamnosus LC-4. J. Food Saf. 39, e12673.

Zhao, L., Jin, H., Lan, J., Zhang, R., Ren, H., Zhang, X., Yu, G., 2015. Detoxification of zearalenone by three strains of Lactobacillus plantarum from fermented food in vitro. Food Control 54, 158–164.

Zheng, J., Wittouck, S., Salvetti, E., Franz, C.M.A.P., Harris, H.M.B., Mattarelli, P., O'Toole, P.W., Pot, B., Vandamme, P., Walter, J., Watanabe, K., Wuyts, S., Felis, G.E., Ganzle, M.G., Lebeer, S., 2020. A taxonomic note on the genus Lactobacillus: description of 23 novel genera, emended description of the genus Lactobacillus Beijerinck 1901, and union of Lactobacillaceae and Leuconostocaceae. Int. J. Syst. Evol. Microbiol. 70, 2782–2858.

Złoch, M., Rogowska, A., Pomastowski, P., Railean-Plugaru, V., Walczak-Skierska, J., Rudnicka, J., Buszewski, B., 2020. Use of Lactobacillus paracasei strain for zearalenone finding and metabolization. Toxicon 181, 9–18.

Index

Note: Page numbers followed by *f* indicate figures, *t* indicate tables, and *b* indicate boxes.

MIX
Papier | Fördert
gute Waldnutzung
FSC® C083411

Kolibri 360 GmbH
Ferdinand-Jühlke-Straße 7
99095 Erfurt, Deutschland
produktsicherheit@kolibri360.de